# Probability and Its Applications

*Published in association with the Applied Probability Trust*

*Editors:* J. Gani, C.C. Heyde, P. Jagers, T.G. Kurtz

# Probability and Its Applications

David Nualart

# The Malliavin Calculus and Related Topics

 Springer

David Nualart

Department of Mathematics, University of Kansas, 405 Snow Hall, 1460 Jayhawk Blvd, Lawrence, Kansas 66045-7523, USA

*Series Editors*

J. Gani

Stochastic Analysis Group, CMA
Australian National University
Canberra ACT 0200
Australia

C.C. Heyde

Stochastic Analysis Group, CMA
Australian National University
Canberra ACT 0200
Australia

P. Jagers

Mathematical Statistics
Chalmers University of Technology
SE-412 96 Göteborg
Sweden

T.G. Kurtz

Department of Mathematics
University of Wisconsim
480 Lincoln Drive
Madison, WI 53706
USA

ISBN 13 978-3-642-06651-1          e-ISBN 13 978-3-540-28329-4

Mathematics Subject Classification (2000): 60H07, 60H10, 60H15, 60-02

Springer is a part of Springer Science+Business Media
springer.com
© Springer-Verlag Berlin Heidelberg 2010
Printed in The Netherlands

Cover design: *Erich Kirchner*, Heidelberg

To my wife Maria Pilar

# Preface to the second edition

There have been ten years since the publication of the first edition of this book. Since then, new applications and developments of the Malliavin calculus have appeared. In preparing this second edition we have taken into account some of these new applications, and in this spirit, the book has two additional chapters that deal with the following two topics: Fractional Brownian motion and Mathematical Finance.

The presentation of the Malliavin calculus has been slightly modified at some points, where we have taken advantage of the material from the lectures given in Saint Flour in 1995 (see reference [248]). The main changes and additional material are the following:

In Chapter 1, the derivative and divergence operators are introduced in the framework of an isonormal Gaussian process associated with a general Hilbert space $H$. The case where $H$ is an $L^2$-space is trated in detail afterwards (white noise case). The Sobolev spaces $\mathbb{D}^{s,p}$, with $s$ is an arbitrary real number, are introduced following Watanabe's work.

Chapter 2 includes a general estimate for the density of a one-dimensional random variable, with application to stochastic integrals. Also, the composition of tempered distributions with nondegenerate random vectors is discussed following Watanabe's ideas. This provides an alternative proof of the smoothness of densities for nondegenerate random vectors. Some properties of the support of the law are also presented.

In Chapter 3, following the work by Alòs and Nualart [10], we have included some recent developments on the Skorohod integral and the associated change-of-variables formula for processes with are differentiable in future times. Also, the section on substitution formulas has been rewritten and an Itô-Ventzell formula has been added, following [248]. This formula allows us to solve anticipating stochastic differential equations in Stratonovich sense with random initial condition.

There have been only minor changes in Chapter 4, and two additional chapters have been included. Chapter 5 deals with the stochastic calculus with respect to the fractional Brownian motion. The fractional Brownian motion is a self-similar Gaussian process with stationary increments and variance $t^{2H}$. The parameter $H \in (0,1)$ is called the Hurst parameter. The main purpose of this chapter is to use the the Malliavin Calculus techniques to develop a stochastic calculus with respect to the fractional Brownian motion.

Finally, Chapter 6 contains some applications of Malliavin Calculus in Mathematical Finance. The integration-by-parts formula is used to compute "greeks", sensitivity parameters of the option price with respect to the underlying parameters of the model. We also discuss the application of the Clark-Ocone formula in hedging derivatives and the additional expected logarithmic utility for insider traders.

August 20, 2005                                                    *David Nualart*

# Preface

The origin of this book lies in an invitation to give a series of lectures on Malliavin calculus at the Probability Seminar of Venezuela, in April 1985. The contents of these lectures were published in Spanish in [245]. Later these notes were completed and improved in two courses on Malliavin calculus given at the University of California at Irvine in 1986 and at École Polytechnique Fédérale de Lausanne in 1989. The contents of these courses correspond to the material presented in Chapters 1 and 2 of this book. Chapter 3 deals with the anticipating stochastic calculus and it was developed from our collaboration with Moshe Zakai and Etienne Pardoux. The series of lectures given at the Eighth Chilean Winter School in Probability and Statistics, at Santiago de Chile, in July 1989, allowed us to write a pedagogical approach to the anticipating calculus which is the basis of Chapter 3. Chapter 4 deals with the nonlinear transformations of the Wiener measure and their applications to the study of the Markov property for solutions to stochastic differential equations with boundary conditions. The presentation of this chapter was inspired by the lectures given at the Fourth Workshop on Stochastic Analysis in Oslo, in July 1992. I take the opportunity to thank these institutions for their hospitality, and in particular I would like to thank Enrique Cabaña, Mario Wschebor, Joaquín Ortega, Süleyman Üstünel, Bernt Øksendal, Renzo Cairoli, René Carmona, and Rolando Rebolledo for their invitations to lecture on these topics.

We assume that the reader has some familiarity with the Itô stochastic calculus and martingale theory. In Section 1.1.3 an introduction to the Itô calculus is provided, but we suggest the reader complete this outline of the classical Itô calculus with a review of any of the excellent presentations of this theory that are available (for instance, the books by Revuz and Yor [292] and Karatzas and Shreve [164]).

In the presentation of the stochastic calculus of variations (usually called the Malliavin calculus) we have chosen the framework of an arbitrary centered Gaussian family, and have tried to focus our attention on the notions and results that depend only on the covariance operator (or the associated Hilbert space). We have followed some of the ideas and notations developed by Watanabe in [343] for the case of an abstract Wiener space. In addition to Watanabe's book and the survey on the stochastic calculus of variations written by Ikeda and Watanabe in [144] we would like to mention the book by Denis Bell [22] (which contains a survey of the different approaches to the Malliavin calculus), and the lecture notes by Dan Ocone in [270]. Readers interested in the Malliavin calculus for jump processes can consult the book by Bichteler, Gravereaux, and Jacod [35].

The objective of this book is to introduce the reader to the Sobolev differential calculus for functionals of a Gaussian process. This is called the analysis on the Wiener space, and is developed in Chapter 1. The other chapters are devoted to different applications of this theory to problems such as the smoothness of probability laws (Chapter 2), the anticipating stochastic calculus (Chapter 3), and the shifts of the underlying Gaussian process (Chapter 4). Chapter 1, together with selected parts of the subsequent chapters, might constitute the basis for a graduate course on this subject.

I would like to express my gratitude to the people who have read the several versions of the manuscript, and who have encouraged me to complete the work, particularly I would like to thank John Walsh, Giuseppe Da Prato, Moshe Zakai, and Peter Imkeller. My special thanks go to Michael Röckner for his careful reading of the first two chapters of the manuscript.

March 17, 1995                                                          *David Nualart*

# Contents

# Introduction

The Malliavin calculus (also known as the stochastic calculus of variations) is an infinite-dimensional differential calculus on the Wiener space. It is tailored to investigate regularity properties of the law of Wiener functionals such as solutions of stochastic differential equations. This theory was initiated by Malliavin and further developed by Stroock, Bismut, Watanabe, and others. The original motivation, and the most important application of this theory, has been to provide a probabilistic proof of Hörmander's "sum of squares" theorem.

One can distinguish two parts in the Malliavin calculus. First is the theory of the differential operators defined on suitable Sobolev spaces of Wiener functionals. A crucial fact in this theory is the integration-by-parts formula, which relates the derivative operator on the Wiener space and the Skorohod extended stochastic integral. A second part of this theory deals with establishing general criteria in terms of the "Malliavin covariance matrix" for a given random vector to possess a density or, even more precisely, a smooth density. In the applications of Malliavin calculus to specific examples, one usually tries to find sufficient conditions for these general criteria to be fulfilled.

In addition to the study of the regularity of probability laws, other applications of the stochastic calculus of variations have recently emerged. For instance, the fact that the adjoint of the derivative operator coincides with a noncausal extension of the Itô stochastic integral introduced by Skorohod is the starting point in developing a stochastic calculus for nonadapted processes, which is similar in some aspects to the Itô calculus. This anticipating stochastic calculus has allowed mathematicians to formulate and discuss stochastic differential equations where the solution is not adapted to the Brownian filtration.

The purposes of this monograph are to present the main features of the Malliavin calculus, including its application to the proof of Hörmander's theorem, and to discuss in detail its connection with the anticipating stochastic calculus. The material is organized in the following manner:

In Chapter 1 we develop the analysis on the Wiener space (Malliavin calculus). The first section presents the Wiener chaos decomposition. In Sections 2,3, and 4 we study the basic operators $D$, $\delta$, and $L$, respectively. The operator $D$ is the derivative operator, $\delta$ is the adjoint of $D$, and $L$ is the generator of the Ornstein-Uhlenbeck semigroup. The last section of this chapter is devoted to proving Meyer's equivalence of norms, following a simple approach due to Pisier. We have chosen the general framework of an isonormal Gaussian process $\{W(h), h \in H\}$ associated with a Hilbert space $H$. The particular case where $H$ is an $L^2$ space over a measure space $(T, \mathcal{B}, \mu)$ (white noise case) is discussed in detail.

Chapter 2 deals with the regularity of probability laws by means of the Malliavin calculus. In Section 3 we prove Hörmander's theorem, using the general criteria established in the first sections. Finally, in the last section we discuss the regularity of the probability law of the solutions to hyperbolic and parabolic stochastic partial differential equations driven by a space-time white noise.

In Chapter 3 we present the basic elements of the stochastic calculus for anticipating processes, and its application to the solution of anticipating stochastic differential equations. Chapter 4 examines different extensions of the Girsanov theorem for nonlinear and anticipating transformations of the Wiener measure, and their application to the study of the Markov property of solution to stochastic differential equations with boundary conditions.

Chapter 5 deals with some recent applications of the Malliavin Calculus to develop a stochastic calculus with respect to the fractional Brownian motion. Finally, Chapter 6 presents some applications of the Malliavin Calculus in Mathematical Finance.

The appendix contains some basic results such as martingale inequalities and continuity criteria for stochastic processes that are used along the book.

# 1
# Analysis on the Wiener space

In this chapter we study the differential calculus on a Gaussian space. That is, we introduce the derivative operator and the associated Sobolev spaces of weakly differentiable random variables. Then we prove the equivalence of norms established by Meyer and discuss the relationship between the basic differential operators: the derivative operator, its adjoint (which is usually called the Skorohod integral), and the Ornstein-Uhlenbeck operator.

## 1.1 Wiener chaos and stochastic integrals

This section describes the basic framework that will be used in this monograph. The general context consists of a probability space $(\Omega, \mathcal{F}, P)$ and a Gaussian subspace $\mathcal{H}_1$ of $L^2(\Omega, \mathcal{F}, P)$. That is, $\mathcal{H}_1$ is a closed subspace whose elements are zero-mean Gaussian random variables. Often it will be convenient to assume that $\mathcal{H}_1$ is isometric to an $L^2$ space of the form $L^2(T, \mathcal{B}, \mu)$, where $\mu$ is a $\sigma$-finite measure without atoms. In this way the elements of $\mathcal{H}_1$ can be interpreted as stochastic integrals of functions in $L^2(T, \mathcal{B}, \mu)$ with respect to a random Gaussian measure on the parameter space $T$ (Gaussian white noise).

In the first part of this section we obtain the orthogonal decomposition into the Wiener chaos for square integrable functionals of our Gaussian process. The second part is devoted to the construction and main properties of multiple stochastic integrals with respect to a Gaussian white noise. Finally, in the third part we recall some basic facts about the Itô integral.

### 1.1.1   The Wiener chaos decomposition

Suppose that $H$ is a real separable Hilbert space with scalar product denoted by $\langle \cdot, \cdot \rangle_H$. The norm of an element $h \in H$ will be denoted by $\|h\|_H$.

**Definition 1.1.1** *We say that a stochastic process $W = \{W(h), h \in H\}$ defined in a complete probability space $(\Omega, \mathcal{F}, P)$ is an isonormal Gaussian process (or a Gaussian process on $H$) if $W$ is a centered Gaussian family of random variables such that $E(W(h)W(g)) = \langle h, g \rangle_H$ for all $h, g \in H$.*

**Remarks:**

**1.** Under the above conditions, the mapping $h \rightarrow W(h)$ is linear. Indeed, for any $\lambda$, $\mu \in \mathbb{R}$, and $h, g \in H$, we have

$$E\left((W(\lambda h + \mu g) - \lambda W(h) - \mu W(g))^2\right) = \|\lambda h + \mu g\|_H^2$$
$$+ \lambda^2 \|h\|_H^2 + \mu^2 \|g\|_H^2 - 2\lambda \langle \lambda h + \mu g, h \rangle_H$$
$$- 2\mu \langle \lambda h + \mu g, g \rangle_H + 2\lambda \mu \langle h, g \rangle_H = 0.$$

The mapping $h \rightarrow W(h)$ provides a linear isometry of $H$ onto a closed subspace of $L^2(\Omega, \mathcal{F}, P)$ that we will denote by $\mathcal{H}_1$. The elements of $\mathcal{H}_1$ are zero-mean Gaussian random variables.

**2.** In Definition 1.1.1 it is enough to assume that each random variable $W(h)$ is Gaussian and centered, since by Remark 1 the mapping $h \rightarrow W(h)$ is linear, which implies that $\{W(h)\}$ is a Gaussian family.

**3.** By Kolmogorov's theorem, given the Hilbert space $H$ we can always construct a probability space and a Gaussian process $\{W(h)\}$ verifying the above conditions.

Let $H_n(x)$ denote the $n$th *Hermite polynomial*, which is defined by

$$H_n(x) = \frac{(-1)^n}{n!} e^{\frac{x^2}{2}} \frac{d^n}{dx^n} (e^{-\frac{x^2}{2}}), \quad n \geq 1,$$

and $H_0(x) = 1$. These polynomials are the coefficients of the expansion in powers of $t$ of the function $F(x, t) = \exp(tx - \frac{t^2}{2})$. In fact, we have

$$
\begin{aligned}
F(x, t) &= \exp[\frac{x^2}{2} - \frac{1}{2}(x - t)^2] \\
&= e^{\frac{x^2}{2}} \sum_{n=0}^{\infty} \frac{t^n}{n!} (\frac{d^n}{dt^n} e^{-\frac{(x-t)^2}{2}})|_{t=0} \qquad (1.1) \\
&= \sum_{n=0}^{\infty} t^n H_n(x).
\end{aligned}
$$

Using this development, one can easily show the following properties:

$$H'_n(x) = H_{n-1}(x), \qquad n \geq 1, \tag{1.2}$$

$$(n+1)H_{n+1}(x) = xH_n(x) - H_{n-1}(x), \qquad n \geq 1, \tag{1.3}$$

$$H_n(-x) = (-1)^n H_n(x), \qquad n \geq 1. \tag{1.4}$$

Indeed, (1.2) and (1.3) follow from $\frac{\partial F}{\partial x} = tF$, respectively, and $\frac{\partial F}{\partial t} = (x - t)F$, and (1.4) is a consequence of $F(-x,t) = F(x,-t)$.

The first Hermite polynomials are $H_1(x) = x$ and $H_2(x) = \frac{1}{2}(x^2 - 1)$. From (1.3) it follows that the highest-order term of $H_n(x)$ is $\frac{x^n}{n!}$. Also, from the expansion of $F(0,t) = \exp(-\frac{t^2}{2})$ in powers of $t$, we get $H_n(0) = 0$ if $n$ is odd and $H_{2k}(0) = \frac{(-1)^k}{2^k k!}$ for all $k \geq 1$. The relationship between Hermite polynomials and Gaussian random variables is explained by the following result.

**Lemma 1.1.1** *Let $X, Y$ be two random variables with joint Gaussian distribution such that $E(X) = E(Y) = 0$ and $E(X^2) = E(Y^2) = 1$. Then for all $n, m \geq 0$ we have*

$$E(H_n(X)H_m(Y)) = \begin{cases} 0 & \text{if } n \neq m. \\ \frac{1}{n!}(E(XY))^n & \text{if } n = m. \end{cases}$$

*Proof:*    For all $s, t \in \mathbb{R}$ we have

$$E\left(\exp(sX - \frac{s^2}{2})\exp(tY - \frac{t^2}{2})\right) = \exp(stE(XY)).$$

Taking the $(n+m)$th partial derivative $\frac{\partial^{n+m}}{\partial s^n \partial t^m}$ at $s = t = 0$ in both sides of the above equality yields

$$E(n!m!H_n(X)H_m(Y)) = \begin{cases} 0 & \text{if } n \neq m. \\ n!(E(XY))^n & \text{if } n = m. \end{cases}$$

$\square$

We will denote by $\mathcal{G}$ the $\sigma$-field generated by the random variables $\{W(h), h \in H\}$.

**Lemma 1.1.2** *The random variables $\{e^{W(h)}, h \in H\}$ form a total subset of $L^2(\Omega, \mathcal{G}, P)$.*

*Proof:*    Let $X \in L^2(\Omega, \mathcal{G}, P)$ be such that $E(Xe^{W(h)}) = 0$ for all $h \in H$. The linearity of the mapping $h \to W(h)$ implies

$$E\left(X \exp \sum_{i=1}^{m} t_i W(h_i)\right) = 0 \tag{1.5}$$

for any $t_1, \ldots, t_m \in \mathbb{R}$, $h_1, \ldots, h_m \in H$, $m \geq 1$. Suppose that $m \geq 1$ and $h_1, \ldots, h_m \in H$ are fixed. Then Eq. (1.5) says that the Laplace transform of the signed measure

$$\nu(B) = E\left(X\mathbf{1}_B\left(W(h_1), \ldots, W(h_m)\right)\right),$$

where $B$ is a Borel subset of $\mathbb{R}^m$, is identically zero on $\mathbb{R}^m$. Consequently, this measure is zero, which implies $E(X\mathbf{1}_G) = 0$ for any $G \in \mathcal{G}$. So $X = 0$, completing the proof of the lemma.    □

For each $n \geq 1$ we will denote by $\mathcal{H}_n$ the closed linear subspace of $L^2(\Omega, \mathcal{F}, P)$ generated by the random variables $\{H_n(W(h)), h \in H, \|h\|_H = 1\}$. $\mathcal{H}_0$ will be the set of constants. For $n = 1$, $\mathcal{H}_1$ coincides with the set of random variables $\{W(h), h \in H\}$. From Lemma 1.1.1 we deduce that the subspaces $\mathcal{H}_n$ and $\mathcal{H}_m$ are orthogonal whenever $n \neq m$. The space $\mathcal{H}_n$ is called the Wiener chaos of order $n$, and we have the following orthogonal decomposition.

**Theorem 1.1.1** *Then the space $L^2(\Omega, \mathcal{G}, P)$ can be decomposed into the infinite orthogonal sum of the subspaces $\mathcal{H}_n$:*

$$L^2(\Omega, \mathcal{G}, P) = \oplus_{n=0}^{\infty}\mathcal{H}_n.$$

*Proof:*    Let $X \in L^2(\Omega, \mathcal{G}, P)$ such that $X$ is orthogonal to $\mathcal{H}_n$ for all $n \geq 0$. We want to show that $X = 0$. We have $E(XH_n(W(h))) = 0$ for all $h \in H$ with $\|h\|_H = 1$. Using the fact that $x^n$ can be expressed as a linear combination of the Hermite polynomials $H_r(x)$, $0 \leq r \leq n$, we get $E(XW(h)^n) = 0$ for all $n \geq 0$, and therefore $E(X \exp(tW(h))) = 0$ for all $t \in \mathbb{R}$, and for all $h \in H$ of norm one. By Lemma 1.1.2 we deduce $X = 0$, which completes the proof of the theorem.    □

For any $n \geq 1$ we can consider the space $\mathcal{P}_n^0$ formed by the random variables $p(W(h_1), \ldots, W(h_k))$, where $k \geq 1$, $h_1, \ldots, h_k \in H$, and $p$ is a real polynomial in $k$ variables of degree less than or equal to $n$. Let $\mathcal{P}_n$ be the closure of $\mathcal{P}_n^0$ in $L^2$. Then it holds that $\mathcal{H}_0 \oplus \mathcal{H}_1 \oplus \cdots \oplus \mathcal{H}_n = \mathcal{P}_n$. In fact, the inclusion $\oplus_{i=0}^{n}\mathcal{H}_i \subset \mathcal{P}_n$ is immediate. To prove the converse inclusion, it suffices to check that $\mathcal{P}_n$ is orthogonal to $\mathcal{H}_m$ for all $m > n$. We want to show that $E(p(W(h_1), \ldots, W(h_k))H_m(W(h))) = 0$, where $\|h\|_H = 1$, $p$ is a polynomial of degree less than or equal to $n$, and $m > n$. We can replace $p(W(h_1), \ldots, W(h_k))$ by $q(W(e_1), \ldots, W(e_j), W(h))$, where $\{e_1, \ldots, e_j, h\}$ is an orthonormal family and the degree of $q$ is less than or equal to $n$. Then it remains to show only that $E(W(h)^r H_m(W(h))) = 0$ for all $r \leq n < m$; this is immediate because $x^r$ can be expressed as a linear combination of the Hermite polynomials $H_q(x)$, $0 \leq q \leq r$.

We denote by $J_n$ the projection on the $n$th Wiener chaos $\mathcal{H}_n$.

**Example 1.1.1** *Consider the following simple example, which corresponds to the case where the Hilbert space $H$ is one-dimensional. Let $(\Omega, \mathcal{F}, P) =$*

$(\mathbb{R}, \mathcal{B}(\mathbb{R}), \nu)$, *where $\nu$ is the standard normal law $N(0,1)$. Take $H = \mathbb{R}$, and for any $h \in \mathbb{R}$ set $W(h)(x) = hx$. There are only two elements in $H$ of norm one: 1 and $-1$. We associate with them the random variables $x$ and $-x$, respectively. From (1.4) it follows that $\mathcal{H}_n$ has dimension one and is generated by $H_n(x)$. In this context, Theorem 1.1.1 means that the Hermite polynomials form a complete orthonormal system in $L^2(\mathbb{R}, \nu)$.*

Suppose now that $H$ is infinite-dimensional (the finite-dimensional case would be similar and easier), and let $\{e_i, i \geq 1\}$ be an orthonormal basis of $H$. We will denote by $\Lambda$ the set of all sequences $a = (a_1, a_2, \dots)$, $a_i \in \mathbb{N}$, such that all the terms, except a finite number of them, vanish. For $a \in \Lambda$ we set $a! = \prod_{i=1}^{\infty} a_i!$ and $|a| = \sum_{i=1}^{\infty} a_i$. For any multiindex $a \in \Lambda$ we define the generalized Hermite polynomial $H_a(x)$, $x \in \mathbb{R}^{\mathbb{N}}$, by

$$H_a(x) = \prod_{i=1}^{\infty} H_{a_i}(x_i).$$

The above product is well defined because $H_0(x) = 1$ and $a_i \neq 0$ only for a finite number of indices.

For any $a \in \Lambda$ we define

$$\Phi_a = \sqrt{a!} \prod_{i=1}^{\infty} H_{a_i}(W(e_i)). \tag{1.6}$$

The family of random variables $\{\Phi_a, a \in \Lambda\}$ is an orthonormal system. Indeed, for any $a, b \in \Lambda$ we have

$$E\left(\prod_{i=1}^{\infty} H_{a_i}(W(e_i)) H_{b_i}(W(e_i))\right) = \prod_{i=1}^{\infty} E(H_{a_i}(W(e_i)) H_{b_i}(W(e_i)))$$

$$= \begin{cases} \frac{1}{a!} & \text{if } a = b. \\ 0 & \text{if } a \neq b. \end{cases} \tag{1.7}$$

**Proposition 1.1.1** *For any $n \geq 1$ the random variables*

$$\{\Phi_a, a \in \Lambda, |a| = n\} \tag{1.8}$$

*form a complete orthonormal system in $\mathcal{H}_n$.*

*Proof:* Observe that when $n$ varies, the families (1.8) are mutually orthogonal in view of (1.7). On the other hand, the random variables of the family (1.8) belong to $\mathcal{P}_n$. Then it is enough to show that every polynomial random variable $p(W(h_1), \dots, W(h_k))$ can be approximated by polynomials in $W(e_i)$, which is clear because $\{e_i, i \geq 1\}$ is a basis of $H$. □

As a consequence of Proposition 1.1.1 the family $\{\Phi_a, a \in \Lambda\}$ is a complete orthonormal system in $L^2(\Omega, \mathcal{G}, P)$.

Let $a \in \Lambda$ be a multiindex such that $|a| = n$. The mapping

$$I_n \left( \text{symm} \left( \otimes_{i=1}^{\infty} e_i^{\otimes a_i} \right) \right) = \sqrt{a!} \Phi_a \qquad (1.9)$$

provides an isometry between the symmetric tensor product $H^{\widehat{\otimes}n}$, equipped with the norm $\sqrt{n!} \|\cdot\|_{H^{\otimes n}}$, and the $n$th Wiener chaos $\mathcal{H}_n$. In fact,

$$\left\| \text{symm} \left( \otimes_{i=1}^{\infty} e_i^{\otimes a_i} \right) \right\|_{H^{\otimes n}}^2 = \left( \frac{a!}{n!} \right)^2 \frac{n!}{a!} \left\| \otimes_{i=1}^{\infty} e_i^{\otimes a_i} \right\|_{H^{\otimes n}}^2 = \frac{a!}{n!}$$

and

$$\left\| \sqrt{a!} \Phi_a \right\|_2^2 = a!.$$

As a consequence, the space $L^2(\Omega, \mathcal{G}, P)$ is isometric to the Fock space, defined as the orthogonal sum $\bigoplus_{n=0}^{\infty} \sqrt{n!} H^{\widehat{\otimes}n}$. In the next section we will see that if $H$ is an $L^2$ space of the form $L^2(T, \mathcal{B}, \mu)$, then $I_n$ coincides with a multiple stochastic integral.

## 1.1.2  The white noise case: Multiple Wiener-Itô integrals

Assume that the underlying separable Hilbert space $H$ is an $L^2$ space of the form $L^2(T, \mathcal{B}, \mu)$, where $(T, \mathcal{B})$ is a measurable space and $\mu$ is a $\sigma$-finite measure without atoms. In that case the Gaussian process $W$ is characterized by the family of random variables $\{W(A), A \in \mathcal{B}, \mu(A) < \infty\}$, where $W(A) = W(\mathbf{1}_A)$. We can consider $W(A)$ as an $L^2(\Omega, \mathcal{F}, P)$-valued measure on the parameter space $(T, \mathcal{B})$, which takes independent values on any family of disjoint subsets of $T$, and such that any random variable $W(A)$ has the distribution $N(0, \mu(A))$ if $\mu(A) < \infty$. We will say that $W$ is an $L^2(\Omega)$-valued Gaussian measure (or a Brownian measure) on $(T, \mathcal{B})$. This measure will be also called the *white noise* based on $\mu$. In that sense, $W(h)$ can be regarded as the stochastic integral (Wiener integral) of the function $h \in L^2(T)$ with respect to $W$. We will write $W(h) = \int_T h \, dW$, and observe that this stochastic integral cannot be defined pathwise, because the paths of $\{W(A)\}$ are not $\sigma$-additive measures on $T$. More generally, we will see in this section that the elements of the $n$th Wiener chaos $\mathcal{H}_n$ can be expressed as multiple stochastic integrals with respect to $W$. We start with the construction of multiple stochastic integrals.

Fix $m \geq 1$. Set $\mathcal{B}_0 = \{A \in \mathcal{B} : \mu(A) < \infty\}$. We want to define the multiple stochastic integral $I_m(f)$ of a function $f \in L^2(T^m, \mathcal{B}^m, \mu^m)$. We denote by $\mathcal{E}_m$ the set of elementary functions of the form

$$f(t_1, \ldots, t_m) = \sum_{i_1, \ldots, i_m=1}^{n} a_{i_1 \cdots i_m} \mathbf{1}_{A_{i_1} \times \cdots \times A_{i_m}} (t_1, \ldots, t_m), \qquad (1.10)$$

where $A_1, A_2, \ldots, A_n$ are pairwise-disjoint sets belonging to $\mathcal{B}_0$, and the coefficients $a_{i_1 \cdots i_m}$ are zero if any two of the indices $i_1, \ldots, i_m$ are equal.

The fact that $f$ vanishes on the rectangles that intersect any diagonal subspace $\{t_i = t_j, i \neq j\}$ plays a basic role in the construction of the multiple stochastic integral.

For a function of the form (1.10) we define

$$I_m(f) = \sum_{i_1,\ldots,i_m=1}^{n} a_{i_1\cdots i_m} W(A_{i_1}) \cdots W(A_{i_m}).$$

This definition does not depend on the particular representation of $f$, and the following properties hold:

(i) $I_m$ is linear,

(ii) $I_m(f) = I_m(\widetilde{f})$, where $\widetilde{f}$ denotes the symmetrization of $f$, which means

$$\widetilde{f}(t_1,\ldots,t_m) = \frac{1}{m!} \sum_{\sigma} f(t_{\sigma(1)},\ldots,t_{\sigma(m)}),$$

$\sigma$ running over all permutations of $\{1,\ldots,m\}$,

(iii) $E(I_m(f)I_q(g)) = \begin{cases} 0 & \text{if } m \neq q, \\ m!\langle \widetilde{f},\widetilde{g}\rangle_{L^2(T^m)} & \text{if } m = q. \end{cases}$

*Proof of these properties:*

Property (i) is clear. In order to show (ii), by linearity we may assume that $f(t_1,\ldots,t_m) = \mathbf{1}_{A_{i_1} \times \cdots \times A_{i_m}}(t_1,\ldots,t_m)$, and in this case the property is immediate. In order to show property (iii), consider two symmetric functions $f \in \mathcal{E}_m$ and $g \in \mathcal{E}_q$. We can always assume that they are associated with the same partition $A_1,\ldots,A_n$. The case $m \neq q$ is easy. Finally, let $m = q$ and suppose that the functions $f$ and $g$ are given by (1.10) and by

$$g(t_1,\ldots,t_m) = \sum_{i_1,\ldots,i_m=1}^{n} b_{i_1\cdots i_m} \mathbf{1}_{A_{i_1} \times \cdots \times A_{i_m}}(t_1,\ldots,t_m),$$

respectively. Then we have

$$
\begin{aligned}
E(I_m(f)I_m(g)) &= E\Big(\big(\sum_{i_1<\cdots<i_m} m!\, a_{i_1\cdots i_m} W(A_{i_1}) \cdots W(A_{i_m})\big) \\
&\quad \times \big(\sum_{i_1<\cdots<i_m} m!\, b_{i_1\cdots i_m} W(A_{i_1}) \cdots W(A_{i_m})\big)\Big) \\
&= \sum_{i_1<\cdots<i_m} (m!)^2 a_{i_1\cdots i_m} b_{i_1\cdots i_m} \mu(A_{i_1}) \cdots \mu(A_{i_m}) \\
&= m!\langle f,g\rangle_{L^2(T^m)}.
\end{aligned}
$$

In order to extend the multiple stochastic integral to the space $L^2(T^m)$, we have to prove that the space $\mathcal{E}_m$ of elementary functions is dense in

$L^2(T^m)$. To do this it suffices to show that the characteristic function of any set $A = A_1 \times A_2 \times \cdots \times A_m$, $A_i \in \mathcal{B}_0$, $1 \le i \le m$, can be approximated by elementary functions in $\mathcal{E}_m$. Using the nonexistence of atoms for the measure $\mu$, for any $\epsilon > 0$ we can determine a system of pairwise-disjoint sets $\{B_1, \ldots, B_n\} \subset \mathcal{B}_0$, such that $\mu(B_i) < \epsilon$ for any $i = 1, \ldots, n$, and each $A_i$ can be expressed as the disjoint union of some of the $B_j$. This is possible because for any set $A \in \mathcal{B}_0$ of measure different from zero and any $0 < \gamma < \mu(A)$ we can find a measurable set $B \subset A$ of measure $\gamma$. Set $\mu(\cup_{i=1}^m A_i) = \alpha$. We have

$$\mathbf{1}_A = \sum_{i_1, \ldots, i_m = 1}^n \epsilon_{i_1 \cdots i_m} \mathbf{1}_{B_{i_1} \times \cdots \times B_{i_m}},$$

where $\epsilon_{i_1 \cdots i_m}$ is 0 or 1. We divide this sum into two parts. Let $I$ be the set of $m$ples $(i_1, \ldots, i_m)$, where all the indices are different, and let $J$ be the set of the remaining $m$ples. We set

$$\mathbf{1}_B = \sum_{(i_1, \ldots, i_m) \in I} \epsilon_{i_1 \cdots i_m} \mathbf{1}_{B_{i_1} \times \cdots \times B_{i_m}}.$$

Then $\mathbf{1}_B$ belongs to the space $\mathcal{E}_m$, $B \subset A$, and we have

$$\begin{aligned} \|\mathbf{1}_A - \mathbf{1}_B\|_{L^2(T^m)}^2 &= \sum_{(i_1, \ldots, i_m) \in J} \epsilon_{i_1 \cdots i_m} \mu(B_{i_1}) \cdots \mu(B_{i_m}) \\ &\le \binom{m}{2} \sum_{i=1}^n \mu(B_i)^2 \left( \sum_{i=1}^n \mu(B_i) \right)^{m-2} \\ &\le \binom{m}{2} \epsilon \alpha^{m-1}, \end{aligned}$$

which shows the desired approximation.

Letting $f = g$ in property (iii) obtains

$$E(I_m(f)^2) = m! \|\tilde{f}\|_{L^2(T^m)}^2 \le m! \|f\|_{L^2(T^m)}^2.$$

Therefore, the operator $I_m$ can be extended to a linear and continuous operator from $L^2(T^m)$ to $L^2(\Omega, \mathcal{F}, P)$, which satisfies properties (i), (ii), and (iii). We will also write $\int_{T^m} f(t_1, \ldots, t_m) W(dt_1) \cdots W(dt_m)$ for $I_m(f)$.

If $f \in L^2(T^p)$ and $g \in L^2(T^q)$ are symmetric functions, for any $1 \le r \le \min(p, q)$ the contraction of $r$ indices of $f$ and $g$ is denoted by $f \otimes_r g$ and is defined by

$$\begin{aligned} &(f \otimes_r g)(t_1, \ldots, t_{p+q-2r}) \\ &= \int_{T^r} f(t_1, \ldots, t_{p-r}, s) g(t_{p+1}, \ldots, t_{p+q-r}, s) \mu^r(ds). \end{aligned}$$

Notice that $f \otimes_r g \in L^2(T^{p+q-2r})$.

The tensor product $f \otimes g$ and the contractions $f \otimes_r g$, $1 \leq r \leq \min(p,q)$, are not necessarily symmetric even though $f$ and $g$ are symmetric. We will denote their symmetrizations by $f \widetilde{\otimes} g$ and $f \widetilde{\otimes}_r g$, respectively.

The next formula for the multiplication of multiple integrals will play a basic role in the sequel.

**Proposition 1.1.2** *Let $f \in L^2(T^p)$ be a symmetric function and let $g \in L^2(T)$. Then,*

$$I_p(f)I_1(g) = I_{p+1}(f \otimes g) + pI_{p-1}(f \otimes_1 g). \tag{1.11}$$

*Proof:* By the density of elementary functions if $L^2(T^p)$ and by linearity we can assume that $f$ is the symmetrization of the characteristic function of $A_1 \times \cdots \times A_p$, where the $A_i$ are pairwise-disjoint sets of $\mathcal{B}_0$, and $g = \mathbf{1}_{A_1}$ or $\mathbf{1}_{A_0}$, where $A_0$ is disjoint with $A_1, \ldots, A_p$. The case $g = \mathbf{1}_{A_0}$ is immediate because the tensor product $f \otimes g$ belongs to $\mathcal{E}_{p+1}$, and $f \otimes_1 g = 0$. So, we assume $g = \mathbf{1}_{A_1}$. Set $\beta = \mu(A_1) \cdots \mu(A_p)$. Given $\epsilon > 0$, we can consider a measurable partition $A_1 = B_1 \cup \cdots \cup B_n$ such that $\mu(B_i) < \epsilon$. Now we define the elementary function

$$h_\epsilon = \sum_{i \neq j} \mathbf{1}_{B_i \times B_j \times A_2 \times \cdots \times A_p}.$$

Then we have

$$
\begin{aligned}
I_p(f)I_1(g) &= W(A_1)^2 W(A_2) \cdots W(A_p) \\
&= \sum_{i \neq j} W(B_i) W(B_j) W(A_2) \cdots W(A_p) \\
&\quad + \sum_{i=1}^n (W(B_i)^2 - \mu(B_i)) W(A_2) \cdots W(A_p) \quad (1.12) \\
&\quad + \mu(A_1) W(A_2) \cdots W(A_p) \\
&= I_{p+1}(h_\epsilon) + R_\epsilon + pI_{p-1}(f \otimes_1 g).
\end{aligned}
$$

Indeed,

$$f \otimes_1 g = \frac{1}{p} \widetilde{\mathbf{1}}_{A_2 \times \cdots \times A_p} \mu(A_1).$$

We have

$$
\begin{aligned}
\|\widetilde{h}_\epsilon - f \widetilde{\otimes} g\|^2_{L^2(T^{p+1})} &= \|\widetilde{h}_\epsilon - \widetilde{\mathbf{1}}_{A_1 \times A_1 \times A_2 \times \cdots \times A_p}\|^2_{L^2(T^{p+1})} \\
&\leq \|h_\epsilon - \mathbf{1}_{A_1 \times A_1 \times A_2 \times \cdots \times A_p}\|^2_{L^2(T^{p+1})} \\
&= \sum_{i=1}^n \mu(B_i)^2 \mu(A_2) \cdots \mu(A_p) \leq \epsilon\beta
\end{aligned}
$$

and

$$E(R_\epsilon^2) = 2 \sum_{i=1}^{n} \mu(B_i)^2 \mu(A_2) \cdots \mu(A_p) \le 2\epsilon\beta,$$

and letting $\epsilon$ tend to zero in (1.12) we obtain the desired result.    □

Formula (1.11) can be generalized as follows.

**Proposition 1.1.3** *Let $f \in L^2(T^p)$ and $g \in L^2(T^q)$ be two symmetric functions. Then*

$$I_p(f)I_q(g) = \sum_{r=0}^{p \wedge q} r! \binom{p}{r} \binom{q}{r} I_{p+q-2r}(f \otimes_r g). \tag{1.13}$$

*Proof:*    The proof can be done by induction with respect to the index $q$. We will assume that $p \ge q$. For $q = 1$ it reduces to (1.11). Suppose it holds for $q - 1$. By a density argument we can assume that the function $g$ is of the form $g = g_1 \tilde{\otimes} g_2$, where $g_1$ and $g_2$ are symmetric functions of $q - 1$ and one variable, respectively, such that $g_1 \otimes_1 g_2 = 0$. By (1.11) we have

$$I_q(g_1 \tilde{\otimes} g_2) = I_{q-1}(g_1)I_1(g_2).$$

Thus by the induction hypothesis, and using (1.11), we obtain

$$
\begin{aligned}
I_p(f)I_q(g) &= I_p(f)I_{q-1}(g_1)I_1(g_2) \\
&= \sum_{r=0}^{q-1} r! \binom{p}{r} \binom{q-1}{r} I_{p+q-1-2r}(f \otimes_r g_1)I_1(g_2) \\
&= \sum_{r=0}^{q-1} r! \binom{p}{r} \binom{q-1}{r} \Big[ I_{p+q-2r}((f \tilde{\otimes}_r g_1) \otimes g_2) \\
&\qquad + (p+q-1-2r)I_{p+q-2r-2}((f \tilde{\otimes}_r g_1) \otimes_1 g_2) \Big] \\
&= \sum_{r=0}^{q-1} r! \binom{p}{r} \binom{q-1}{r} I_{p+q-2r}((f \tilde{\otimes}_r g_1) \otimes g_2) \\
&\qquad + \sum_{r=1}^{q} (r-1)! \binom{p}{r-1} \binom{q-1}{r-1} \\
&\qquad \times (p+q-2r+1)I_{p+q-2r}((f \tilde{\otimes}_{r-1} g_1) \otimes_1 g_2).
\end{aligned}
$$

For any $1 \le r \le q$, one can show the following equality:

$$q(f \tilde{\otimes}_r g) = \frac{r(p+q-2r+1)}{p-r+1}(f \tilde{\otimes}_{r-1} g_1) \otimes_1 g_2 + (q-r)((f \tilde{\otimes}_r g_1) \tilde{\otimes} g_2). \tag{1.14}$$

Substituting (1.14) into the above summations yields (1.13).    □

The next result gives the relationship between Hermite polynomials and multiple stochastic integrals.

**Proposition 1.1.4** *Let $H_m(x)$ be the mth Hermite polynomial, and let $h \in H = L^2(T)$ be an element of norm one. Then it holds that*

$$m! \, H_m(W(h)) = \int_{T^m} h(t_1) \cdots h(t_m) W(dt_1) \cdots W(dt_m). \qquad (1.15)$$

*As a consequence, the multiple integral $I_m$ maps $L^2(T^m)$ onto the Wiener chaos $\mathcal{H}_m$.*

*Proof:*     Eq. (1.15) will be proved by induction on $m$. For $m = 1$ it is immediate. Assume it holds for $1, 2, \ldots, m$. Using the recursive relation for the Hermite polynomials (1.3) and the product formula (1.11), we have

$$
\begin{aligned}
I_{m+1}(h^{\otimes(m+1)}) &= I_m(h^{\otimes m})I_1(h) - mI_{m-1}\left(h^{\otimes(m-1)}\int_T h(t)^2 \mu(dt)\right) \\
&= m! \, H_m(W(h))W(h) - m(m-1)! \, H_{m-1}(W(h)) \\
&= m!(m+1)H_{m+1}(W(h)) = (m+1)! \, H_{m+1}(W(h)),
\end{aligned}
$$

where $h^{\otimes m}$ denotes the function of $m$ variables defined by

$$h^{\otimes m}(t_1, \ldots, t_m) = h(t_1) \cdots h(t_m).$$

Denote by $L_S^2(T^m)$ the closed subspace of $L^2(T^m)$ formed by symmetric functions. The multiple integral $I_m$ verifies $E(I_m(f)^2) = m! \, \|f\|^2_{L^2(T^m)}$ on $L_S^2(T^m)$. So the image $I_m(L_S^2(T^m))$ is closed, and by (1.15) it contains the random variables $H_m(W(h))$, $h \in H$, and $\|h\|_H = 1$. Consequently, $\mathcal{H}_m \subset I_m(L_S^2(T^m))$. Due to the orthogonality between multiple integrals of different order, we have that $I_m(L_S^2(T^m))$ is orthogonal to $\mathcal{H}_n$, $n \neq m$. So, $I_m(L_S^2(T^m)) = \mathcal{H}_m$, which completes the proof of the proposition. $\quad\square$

As a consequence we deduce the following version of the Wiener chaos expansion.

**Theorem 1.1.2** *Any square integrable random variable $F \in L^2(\Omega, \mathcal{G}, P)$ (recall that $\mathcal{G}$ denotes the $\sigma$-field generated by $W$) can be expanded into a series of multiple stochastic integrals:*

$$F = \sum_{n=0}^{\infty} I_n(f_n).$$

*Here $f_0 = E(F)$, and $I_0$ is the identity mapping on the constants. Furthermore, we can assume that the functions $f_n \in L^2(T^n)$ are symmetric and, in this case, uniquely determined by $F$.*

Let $\{e_i, i \geq 1\}$ be an orthonormal basis of $H$, and fix a miltiindex $a = (a_1, \ldots, a_M, 0, \ldots)$ such that $|a| = a_1 + \cdots + a_M = n$. From (1.15) and

(1.13) it follows that

$$a! \prod_{i=1}^{M} H_{a_i}(W(e_i)) = \prod_{i=1}^{M} I_{a_i}(e_i^{\otimes a_i})$$

$$= I_n \left( e_1^{\otimes a_1} \otimes e_2^{\otimes a_2} \otimes \cdots \otimes e_M^{\otimes a_M} \right).$$

Hence, the multiple stochastic integral $I_n$ coincides with the isometry between the symmetric tensor product $H^{\otimes n}$ (equipped with the modified norm $\sqrt{n!}\,\|\cdot\|_{H^{\otimes n}}$) and the $n$th Wiener chaos $\mathcal{H}_n$ introduced in (1.9). Notice that $H^{\hat{\otimes} n}$ is isometric to $L_S^2(T^n)$.

**Example 1.1.2** *Suppose that the parameter space is $T = \mathbb{R}_+ \times \{1, \ldots, d\}$ and that the measure $\mu$ is the product of the Lebesgue measure times the uniform measure, which gives mass one to each point $1, 2, \ldots, d$. Then we have $H = L^2(\mathbb{R}_+ \times \{1, \ldots, d\}, \mu) \cong L^2(\mathbb{R}_+; \mathbb{R}^d)$. In this situation we have that $W^i(t) = W([0, t] \times \{i\})$, $0 \leq t \leq 1$, $1 \leq i \leq d$, is a standard $d$-dimensional Brownian motion. That is, $\{W^i(t), t \in \mathbb{R}_+\}$, $i = 1, \ldots, d$, are independent zero-mean Gaussian processes with covariance function $E(W^i(s)W^i(t)) = s \wedge t$. Furthermore, for any $h \in H$, the random variable $W(h)$ can be obtained as the stochastic integral $\sum_{i=1}^{d} \int_0^\infty h_t^i dW_t^i$.*

*The Brownian motion verifies*

$$E(|W^i(t) - W^i(s)|^2) = |t - s|$$

*for any $s, t \geq 0$, $i = 1, \ldots, d$. This implies that*

$$E(|W^i(t) - W^i(s)|^{2k}) = \frac{(2k)!}{2^k k!} |t - s|^k$$

*for any integer $k \geq 2$. From Kolmogorov's continuity criterion (see the appendix, Section A.3) it follows that $W$ possesses a continuous version. Consequently, we can define the $d$-dimensional Brownian motion on the canonical space $\Omega = C_0(\mathbb{R}_+; \mathbb{R}^d)$. The law of the process $W$ is called the Wiener measure.*

In this example multiple stochastic integrals can be considered as iterated Itô stochastic integrals with respect to the Brownian motion, as we shall see in the next section.

**Example 1.1.3** *Take $T = \mathbb{R}_+^2$ and $\mu$ equal to the Lebesgue measure. Let $W$ be a white noise on $T$. Then $W(s, t) = W([0, s] \times [0, t])$, $s, t \in \mathbb{R}_+$, defines a two-parameter, zero-mean Gaussian process with covariance given by*

$$E(W(s, t)W(s', t')) = (s \wedge s')(t \wedge t'),$$

*which is called the Wiener sheet or the two-parameter Wiener process. The process $W$ has a version with continuous paths. This follows easily from Kolmogorov's continuity theorem, taking into account that*

$$E(|W(s, t) - W(s', t')|^2) \leq \max(s, s', t, t')(|s - s'| + |t - t'|).$$

### 1.1.3 Itô stochastic calculus

In this section we survey some of the basic properties of the stochastic integral of adapted processes with respect to the Brownian motion, introduced by Itô.

Suppose that $W = \{W(t), t \geq 0\}$ is a standard Brownian motion defined on the canonical probability space $(\Omega, \mathcal{F}, P)$. That is, $\Omega = C_0(\mathbb{R}_+)$ and $P$ is a probability measure on the Borel $\sigma$-field $\mathcal{B}(\Omega)$ such that the canonical process $W_t(\omega) = \omega(t)$ is a zero-mean Gaussian process with covariance $E(W_s W_t) = s \wedge t$. The $\sigma$-field $\mathcal{F}$ will be the completion of $\mathcal{B}(\Omega)$ with respect to $P$. We know that the sequence

$$S_n(t) = \sum_{1 \leq k \leq 2^n} [W(tk2^{-n}) - W(t(k-1)2^{-n})]^2$$

converges almost surely and in $L^2(\Omega)$ to the constant $t$, as $n$ tends to infinity. In other words, the paths of the Brownian motion have a quadratic variation equal to $t$. This property, together with the continuity of the paths, implies that the paths of $W$ have infinite total variation on any bounded interval. Consequently, we cannot define path-wise a stochastic integral of the form

$$\int_0^t u(s) W(ds),$$

where $u = \{u(t), t \geq 0\}$ is a given stochastic process. If the paths of the process $u$ have finite total variation on bounded intervals, we can overcome this difficulty by letting

$$\int_0^t u(s) W(ds) = u(t) W(t) - \int_0^t W(s) u(ds).$$

However, most of the processes that we will find (like $W$ itself) do not have paths with finite total variation on bounded intervals.

For each $t \geq 0$ we will denote by $\mathcal{F}_t$ the $\sigma$-field generated by the random variables $\{W(s), 0 \leq s \leq t\}$ and the null sets of $\mathcal{F}$. Then a stochastic process $u = \{u(t), t \geq 0\}$ will be called adapted or nonanticipative if $u(t)$ is $\mathcal{F}_t$-measurable for any $t \geq 0$.

We will fix a time interval, denoted by $T$, which can be $[0, t_0]$ or $\mathbb{R}_+$. We will denote by $L^2(T \times \Omega) = L^2(T \times \Omega, \mathcal{B}(T) \otimes \mathcal{F}, \lambda^1 \times P)$ (where $\lambda^1$ denotes the Lebesgue measure) the set of square integrable processes, and $L_a^2(T \times \Omega)$ will represent the subspace of adapted processes.

Let $\mathcal{E}$ be the class of elementary adapted processes. That is, a process $u$ belongs to $\mathcal{E}$ if it can be written as

$$u(t) = \sum_{i=1}^n F_i \mathbf{1}_{(t_i, t_{i+1}]}(t), \qquad (1.16)$$

where $0 \leq t_1 < \cdots < t_{n+1}$ are points of $T$, and every $F_i$ is an $\mathcal{F}_{t_i}$-measurable and square integrable random variable. Then we have the following result.

**Lemma 1.1.3** *The class $\mathcal{E}$ is dense in $L_a^2(T \times \Omega)$.*

*Proof:*    Suppose $T = [0, 1]$. Let $u$ be a process in $L_a^2(T \times \Omega)$, and consider the sequence of processes defined by

$$\widetilde{u}^n(t) = \sum_{i=1}^{2^n-1} 2^n \left( \int_{(i-1)2^{-n}}^{i2^{-n}} u(s)ds \right) \mathbf{1}_{(i2^{-n},(i+1)2^{-n}]}(t). \qquad (1.17)$$

We claim that the sequence $\widetilde{u}^n$ converges to $u$ in $L^2(T \times \Omega)$. In fact, define $P_n(u) = \widetilde{u}^n$. Then $P_n$ is a linear operator in $L^2(T \times \Omega)$ with norm bounded by one, such that $P_n(u) \to u$ as $n$ tends to infinity whenever the process $u$ is continuous in $L^2(\Omega)$. The proof now follows easily.    □

**Remark:**    A measurable process $u : T \times \Omega \to \mathbb{R}$ is called *progressively measurable* if the restriction of $u$ to the product $[0, t] \times \Omega$ is $\mathcal{B}([0, t]) \otimes \mathcal{F}_t$-measurable for all $t \in T$. One can show (see [225, Theorem 4.6]) that any adapted process has a progressively measurable version, and we will always assume that we are dealing with this kind of version. This is necessary, for instance, to ensure that the approximating processes $\widetilde{u}^n$ introduced in Lemma 1.1.3 are adapted.

For a nonanticipating process of the form (1.16), the random variable

$$\int_T u(t)dW_t = \sum_{i=1}^{n} F_i(W(t_{i+1}) - W(t_i)) \qquad (1.18)$$

will be called the *stochastic integral* (or the Itô integral) of $u$ with respect to the Brownian motion $W$.

The Itô integral of elementary processes is a linear functional that takes values on $L^2(\Omega)$ and has the following basic properties:

$$E\left( \int_T u(t)dW_t \right) = 0, \qquad (1.19)$$

$$E\left( \left| \int_T u(t)dW_t \right|^2 \right) = E\left( \int_T u(t)^2 dt \right). \qquad (1.20)$$

Property (1.19) is immediate from both (1.18) and the fact that for each $i = 1, \ldots, n$ the random variables $F_i$ and $W(t_{i+1}) - W(t_i)$ are independent.

*Proof of (1.20):* We have

$$E(|\int_T u(t)dW_t|^2) = \sum_{i=1}^{n} E(|F_i(W(t_{i+1}) - W(t_i))|^2)$$

$$+ 2\sum_{i<j} E(F_i F_j(W(t_{j+1}) - W(t_i))$$

$$\times (W(t_{j+1}) - W(t_j)))$$

$$= \sum_{i=1}^{n} E(F_i^2)(t_{i+1} - t_i) = E(\int_T u(t)^2 dt),$$

because whenever $i < j$, $W(t_{j+1}) - W(t_j)$ is independent of $F_i F_j(W(t_{i+1}) - W(t_i))$. $\qquad\square$

The isometry property (1.20) allows us to extend the Itô integral to the class $L_a^2(T \times \Omega)$ of adapted square integrable processes, and the above properties still hold in this class.

The Itô integral verifies the following *local property:*

$$\int_T u(t)dW_t = 0,$$

almost surely (a.s.) on the set $G = \{\int_T u(t)^2 dt = 0\}$. In fact, on the set $G$ the processes $\{\tilde{u}^n\}$ introduced in (1.17) vanish, and therefore $\int_T \tilde{u}^n(t)dW_t = 0$ on $G$. Then the result follows from the convergence of $\int_T \tilde{u}^n(t)dW_t$ to $\int_T u(t)dW_t$ in $L^2(\Omega)$.

We also have for any $u \in L_a^2(T \times \Omega)$, $\epsilon > 0$, and $K > 0$,

$$P\left\{\left|\int_T u(t)dW_t\right| > \epsilon\right\} \leq P\left\{\int_T u(t)^2 dt > K\right\} + \frac{K}{\epsilon^2}. \qquad (1.21)$$

*Proof of (1.21):* Define

$$\tilde{u}(t) = u(t)\mathbf{1}_{\{\int_0^t u(s)^2 ds \leq K\}}.$$

The process $\tilde{u}$ belongs to $L_a^2(T \times \Omega)$, and using the local property of the Itô integral we obtain

$$P\left\{\left|\int_T u(t)dW_t\right| > \epsilon, \int_T u(t)^2 dt \leq K\right\}$$

$$= P\left\{\left|\int_T \tilde{u}(t)dW_t\right| > \epsilon, \int_T u(t)^2 dt \leq K\right\}$$

$$\leq P\left\{\left|\int_T \tilde{u}(t)dW_t\right| > \epsilon\right\} \leq \frac{1}{\epsilon^2} E\left(\int_T \tilde{u}(t)^2 dt\right) \leq \frac{K}{\epsilon^2}.$$

$$\square$$

Using property (1.21), one can extend the Itô integral to the class of measurable and adapted processes such that

$$\int_T u(t)^2 dt < \infty \qquad \text{a.s.,}$$

and the local property still holds for these processes.

Suppose that $u$ belongs to $L_a^2(T \times \Omega)$. Then the indefinite integral

$$\int_0^t u(s) dW_s = \int_T u(s) \mathbf{1}_{[0,t]}(s) dW_s, \qquad t \in T,$$

is a martingale with respect to the increasing family of $\sigma$-fields $\{\mathcal{F}_t, t \geq 0\}$. Indeed, the martingale property is easy to check for elementary processes and is transferred to general adapted processes by $L^2$ convergence.

If $u$ is an elementary process of the form (1.16), the martingale

$$\int_0^t u(s) dW_s = \sum_{i=1}^n F_i(W(t_{i+1} \wedge t) - W(t_i \wedge t))$$

clearly possesses a continuous version. The existence of a continuous version for $\{\int_0^t u(s) dW_s\}$ in the general case $u \in L_a^2(T \times \Omega)$ follows from Doob's maximal inequality for martingales (see (A.2)) and from the Borel-Cantelli lemma.

If $u$ is an adapted and measurable process such that $\int_T u(t)^2 dt < \infty$, then the indefinite integral is a continuous local martingale. That is, if we define the random times

$$T_n = \inf\{t \geq 0 : \int_0^t u(s)^2 ds \geq n\}, \qquad n \geq 1,$$

then:

(i) For each $n \geq 1$, $T_n$ is a stopping time (i.e., $\{T_n \leq t\} \in \mathcal{F}_t$ for any $t \geq 0$).

(ii) $T_n \uparrow \infty$ as $n$ tends to infinity.

(iii) The processes

$$M_n(t) = \int_0^t u(s) \mathbf{1}_{\{s \leq T_n\}} dW_s$$

are continuous square integrable martingales such that

$$M_n(t) = \int_0^t u(s) dW_s$$

whenever $t \leq T_n$. In fact, $u \mathbf{1}_{[0,T_n]} \in L_a^2(T \times \Omega)$ for each $n$.

Let $u$ be an adapted and measurable process such that $\int_T u(t)^2 dt < \infty$, and consider the continuous local martingale $M(t) = \int_0^t u(s)dW_s$. Define

$$\langle M \rangle_t = \int_0^t u(s)^2 ds.$$

Then $M_t^2 - \langle M \rangle_t$ is a martingale when $u \in L_a^2(T \times \Omega)$. This is clear if $u$ is an elementary process of the form (1.16), and in the general case it holds by approximation.

The increasing process $\langle M \rangle_t$ is called the *quadratic variation* of the local martingale $M$. That is, the family $\sum_{i=0}^{n-1}(M_{t_{i+1}} - M_{t_i})^2$, when $\pi = \{0 = t_0 < t_1 < \cdots < t_n = t\}$ runs over all the partitions of $[0, t]$, converges in probability to $\int_0^t u(s)^2 ds$ as $|\pi| = \max_i(t_{i+1} - t_i)$ tends to zero. Indeed, by a localization argument, it suffices to prove the convergence when $\int_0^t u(s)^2 ds \leq K$ for some constant $K > 0$, and in this case it holds in $L^2(\Omega)$ due to Burkholder's inequality (A.3) and the fact that $M_t^2 - \langle M \rangle_t$ is a square integrable martingale. In fact, we have

$$E\left(\left|\sum_{i=0}^{n-1}\int_{t_i}^{t_{i+1}} u^2(s)ds - (M_{t_{i+1}} - M_{t_i})^2\right|^2\right)$$

$$= \sum_{i=0}^{n-1} E\left(\left|\int_{t_i}^{t_{i+1}} u^2(s)ds - (M_{t_{i+1}} - M_{t_i})^2\right|^2\right)$$

$$\leq c\sum_{i=0}^{n-1} E\left(\left|\int_{t_i}^{t_{i+1}} u^2(s)ds\right|^2\right) \leq cKE\left(\sup_{|s-r|\leq|\pi|}\int_r^s u^2(\theta)d\theta\right)$$

for some constant $c > 0$, and this converges to zero as $|\pi|$ tends to zero.

One of the most important tools in the stochastic calculus is the change-of-variable formula, or *Itô's formula*.

**Proposition 1.1.5** *Let $F : \mathbb{R} \to \mathbb{R}$ be a twice continuously differentiable function. Suppose that $u$ and $v$ are measurable and adapted processes verifying $\int_0^\tau u(t)^2 dt < \infty$ a.s. and $\int_0^\tau |v(t)|dt < \infty$ a.s. for every $\tau \in T$. Set $X(t) = X(0) + \int_0^t u(s)dW_s + \int_0^t v(s)ds$. Then we have*

$$F(X_t) - F(X_0) = \int_0^t F'(X_s)u_s dW_s + \int_0^t F'(X_s)v_s ds$$

$$+ \frac{1}{2}\int_0^t F''(X_s)u_s^2 ds. \tag{1.22}$$

The proof of (1.22) comes from the fact that the quadratic variation of the process $X(t)$ is equal to $\int_0^t u_s^2 ds$; consequently, when we develop by Taylor's expansion the function $F(X(t))$, there is a contribution from the second-order term, which produces the additional summand in Itô's formula.

*Proof:*  By a localization procedure we can assume $F \in C_b^2(\mathbb{R})$ and

$$\sup\{\int_T u(t)^2 dt, \int_T |v(t)| dt\} \leq K$$

for some constant $K > 0$. Fix $t > 0$. For any partition $\pi = \{0 = t_0 < t_1 < \cdots < t_n = t\}$ we can write, using Taylor's formula,

$$
\begin{aligned}
F(X_t) - F(X_0) &= \sum_{i=0}^{n-1} (F(X_{t_{i+1}}) - F(X_{t_i})) \\
&= \sum_{i=0}^{n-1} F'(X_{t_i})(X_{t_{i+1}} - X_{t_i}) \\
&\quad + \frac{1}{2} \sum_{i=0}^{n-1} F''(\overline{X}_i)(X_{t_{i+1}} - X_{t_i})^2,
\end{aligned}
$$

where $\overline{X}_i$ is a random point between $X_{t_i}$ and $X_{t_{i+1}}$. The first summand in the above expression converges to $\int_0^t F'(X_s)u_s dW_s + \int_0^t F'(X_s)v_s ds$ in $L^2(\Omega)$ as $|\pi| = \max_i(t_{i+1} - t_i)$ tends to zero. For the second summand we use the decomposition

$$
\begin{aligned}
(X_{t_{i+1}} - X_{t_i})^2 &= \left( \int_{t_i}^{t_{i+1}} u_s dW_s \right)^2 + 2 \left( \int_{t_i}^{t_{i+1}} u_s dW_s \right) \left( \int_{t_i}^{t_{i+1}} v_s ds \right) \\
&\quad + \left( \int_{t_i}^{t_{i+1}} v_s ds \right)^2.
\end{aligned}
$$

Only the first term produces a nonzero contribution. Then we can write

$$
\begin{aligned}
\int_0^t F''(X_s)u_s^2 ds &- \sum_{i=0}^{n-1} F''(\overline{X}_i) \left( \int_{t_i}^{t_{i+1}} u_s dW_s \right)^2 \\
&= \sum_{i=0}^{n-1} \int_{t_i}^{t_{i+1}} [F''(X_s) - F''(X_{t_i})] u_s^2 ds \\
&\quad + \sum_{i=0}^{n-1} F''(X_{t_i}) \left( \int_{t_i}^{t_{i+1}} u_s^2 ds - \left( \int_{t_i}^{t_{i+1}} u_s dW_s \right)^2 \right) \\
&\quad + \sum_{i=0}^{n-1} [F''(X_{t_i}) - F''(\overline{X}_i)] \left( \int_{t_i}^{t_{i+1}} u_s dW_s \right)^2 \\
&= a_1 + a_2 + a_3.
\end{aligned}
$$

We have

$$|a_1| \leq K \sup_{|s-r| \leq |\pi|} |F''(X_s) - F''(X_r)|,$$

$$|a_3| \leq \sup_{|s-r| \leq |\pi|} |F''(X_s) - F''(X_r)| \left( \sum_{i=0}^{n-1} \left( \int_{t_i}^{t_{i+1}} u_s dW_s \right)^2 \right).$$

These expressions converge to zero in probability as $|\pi|$ tends to zero. Finally, applying Burkholder's inequality (A.3) and the martingale property of $(\int_0^t u_s dW_s)^2 - \int_0^t u_s^2 ds$, we can get a constant $c > 0$ such that

$$
\begin{aligned}
E(|a_2|^2) &= E \left( \sum_{i=0}^{n-1} F''(X_{t_i})^2 \left( \int_{t_i}^{t_{i+1}} u_s^2 ds - \left( \int_{t_i}^{t_{i+1}} u_s dW_s \right)^2 \right)^2 \right) \\
&\leq c\|F''\|_\infty \sum_{i=0}^{n-1} E \left( \left( \int_{t_i}^{t_{i+1}} u_s^2 ds \right)^2 \right) \\
&\leq Kc\|F''\|_\infty E \left( \sup_{|s-r| \leq |\pi|} \int_r^s u_\theta^2 d\theta \right),
\end{aligned}
$$

and this converges to zero as $|\pi|$ tends to zero. □

Consider two adapted processes $\{u_t, t \in T\}$ and $\{v_t, t \in T\}$ such that $\int_0^t u(s)^2 ds < \infty$ a.s. and $\int_0^t |v(s)| ds < \infty$ a.s. for all $t \in T$. Let $X_0 \in \mathbb{R}$. The process

$$X_t = X_0 + \int_0^t u_s dW_s + \int_0^t v_s ds \qquad (1.23)$$

is called a continuous semimartingale, and $M_t = \int_0^t u_s dW_s$ and $V_t = \int_0^t v_s ds$ are the local martingale part and bounded variation part of $X$, respectively. Itô's formula tells us that this class of processes is stable by the composition with twice continuously differentiable functions.

Let $\pi = \{0 = t_0 < t_1 < \cdots < t_n = t\}$ be a partition of the interval $[0, t]$. The sums

$$\sum_{i=0}^{n-1} \frac{1}{2}(X_{t_i} + X_{t_{i+1}})(W_{t_{i+1}} - W_{t_i}) \qquad (1.24)$$

converge in probability as $|\pi|$ tends to zero to

$$\int_0^t X_s dW_s + \frac{1}{2} \int_0^t u_s ds.$$

This expression is called the Stratonovich integral of $X$ with respect to $W$ and is denoted by $\int_0^t X_s \circ dW_s$.

The convergence of the sums in (1.24) follows easily from the decomposition

$$\frac{1}{2}(X_{t_i} + X_{t_{i+1}})(W_{t_{i+1}} - W_{t_i}) = X_{t_i}(W_{t_{i+1}} - W_{t_i})$$

$$+ \frac{1}{2}(X_{t_{i+1}} - X_{t_i})(W_{t_{i+1}} - W_{t_i}),$$

and the fact that the joint quadratic variation of the processes $X$ and $W$ (denoted by $\langle X, W \rangle_t$) is equal to $\frac{1}{2}\int_0^t u_s ds$.

Let $u \in L_a^2(T \times \Omega)$. Set $M_u(t) = \exp(\int_0^t u_s dW_s - \frac{1}{2}\int_0^t u_s^2 ds)$. As an application of Itô's formula we deduce

$$M_u(t) = 1 + \int_0^t M_u(s)u(s)dW_s. \tag{1.25}$$

That means $M_u$ is a local martingale. In particular, if $u = h$ is a deterministic square integrable function of the space $H = L^2(T)$, then $M_h$ is a square integrable martingale. Formula (1.25) shows that $\exp(W_t - \frac{t}{2})$ plays the role of the customary exponentials in the stochastic calculus.

The following result provides an integral representation of any square functional of the Brownian motion. Set $\mathcal{F}_T = \sigma\{W(s), s \in T\}$.

**Theorem 1.1.3** *Let $F$ be a square integrable random variable. Then there exists a unique process $u \in L_a^2(T \times \Omega)$ such that*

$$F = E(F) + \int_T u_t dW_t. \tag{1.26}$$

*Proof:*    To prove the theorem it suffices to show that any zero-mean square integrable random variable $G$ that is orthogonal to all the stochastic integrals $\int_T u_t dW_t$, $u \in L_a^2(T \times \Omega)$ must be zero. In view of formula (1.25), such a random variable $G$ is orthogonal to the exponentials

$$\mathcal{E}(h) = \exp(\int_T h_s dW_s - \frac{1}{2}\int_T h_s^2 ds),$$

$h \in L^2(T)$. Finally, because these exponentials form a total subset of $L^2(\Omega, \mathcal{F}_T, P)$ by Lemma 1.1.2, we can conclude this proof.    □

As a consequence of this theorem, any square integrable martingale on the time interval $T$ can be represented as an indefinite Itô integral. In fact, such a martingale has the form $M_t = E(F|\mathcal{F}_t)$ for some random variable $F \in L^2(\Omega, \mathcal{F}_T, P)$. Then, taking conditional expectations with respect to the $\sigma$-field $\mathcal{F}_t$ in Eq. (1.26), we obtain

$$E(F|\mathcal{F}_t) = E(F) + \int_0^t u_s dW_s.$$

Let $f_n : T^n \to \mathbb{R}$ be a symmetric and square integrable function. For these functions the multiple stochastic integral $I_n(f_n)$ with respect to the Gaussian process $\{W(h) = \int_T h_s dW_s, h \in L^2(T)\}$ introduced in Section 1.1.2 coincides with an iterated Itô integral. That is, assuming $T = \mathbb{R}_+$, we have

$$I_n(f_n) = n! \int_0^\infty \int_0^{t_n} \cdots \int_0^{t_2} f_n(t_1, \ldots, t_n) dW_{t_1} \cdots dW_{t_n}. \qquad (1.27)$$

Indeed, this equality is clear if $f_n$ is an elementary function of the form (1.10), and in the general case the equality will follow by a density argument, taking into account that the iterated stochastic Itô integral verifies the same isometry property as the multiple stochastic integral.

Let $\{W(t), t \geq 0\}$ be a $d$-dimensional Brownian motion. In this case the multiple stochastic integral $I_n(f_n)$ is defined for square integrable kernels $f_n((t_1, i_1), \ldots, (t_n, i_n))$, which are symmetric in the variables $(t_j, i_j) \in \mathbb{R}_+ \times \{1, \ldots, d\}$, and it can be expressed as a sum of iterated Itô integrals:

$$
\begin{aligned}
I_n(f_n) = {} & n! \sum_{i_1, \ldots, i_n = 1}^d \int_0^\infty \int_0^{t_n} \cdots \int_0^{t_2} f_n((t_1, i_1), \ldots, (t_n, i_n)) \\
& \times dW_{t_1}^{i_1} \cdots dW_{t_n}^{i_n}.
\end{aligned}
$$

## Exercises

**1.1.1** For every $n$ let us define the Hermite polynomial $H_n(\lambda, x)$ by

$$H_n(\lambda, x) = \lambda^{\frac{n}{2}} H_n(\frac{x}{\sqrt{\lambda}}), \text{where } x \in \mathbb{R} \text{ and } \lambda > 0.$$

Check that

$$\exp(tx - \frac{t^2 \lambda}{2}) = \sum_{n=0}^\infty t^n H_n(\lambda, x).$$

Let $W$ be a white noise on a measure space $(T, \mathcal{B}, \mu)$. Show that

$$H_m(\|h\|_H^2, W(h)) = \frac{1}{m!} I_m(h^{\otimes m})$$

for any $h \in L^2(T, \mathcal{B}, \mu)$.

**1.1.2** Using the recursive formula (1.2), deduce the following explicit expression for the Hermite polynomials

$$H_n(x) = \sum_{k=0}^{[n/2]} \frac{(-1)^k \, x^{n-2k}}{k! \, (n - 2k)! \, 2^k}.$$

As an application show that if $Y$ is a random variable with distribution $N(0, \sigma^2)$, then

$$E(H_{2m}(Y)) = \frac{(\sigma^2 - 1)^m}{2^m \, m!},$$

and $E(H_n(Y)) = 0$ if $n$ is odd.

**1.1.3** Let $\{W_t, t \geq 0\}$ be a one-dimensional Brownian motion. Show that the process $\{H_n(t, W_t), t \geq 0\}$ (where $H_n(t, x)$ is the Hermite polynomial introduced in Exercise 1.1.1) is a martingale.

**1.1.4** Let $W = \{W(h), h \in H\}$ be an isonormal Gaussian process defined on the probability space $(\Omega, \mathcal{F}, P)$, where $\mathcal{F}$ is generated by $W$. Let $V$ be a real separable Hilbert space. Show the Wiener chaos expansion

$$L^2(\Omega; V) = \bigoplus_{n=0}^{\infty} \mathcal{H}_n(V),$$

where $\mathcal{H}_n(V)$ is the closed subspace of $L^2(\Omega; V)$ generated by the $V$-valued random variables of the form $\sum_{j=1}^{m} F_j v_j$, $F_j \in \mathcal{H}_n$ and $v_j \in V$. Construct an isometry between $H^{\hat{\otimes} n} \otimes V$ and $\mathcal{H}_n(V)$ as in (1.9).

**1.1.5** By iteration of the representation formula (1.26) and using expression (1.27) show that any random variable $F \in L^2(\Omega, \mathcal{F}, P)$ (where $\mathcal{F}$ is generated by $W$) can be expressed as an infinite sum of orthogonal multiple stochastic integrals. This provides an alternative proof of the Wiener chaos expansion for Brownian functionals.

**1.1.6** Prove Eq. (1.14).

**1.1.7** Let us denote by $\mathcal{P}$ the family of random variables of the form $p(W(h_1), \ldots, W(h_n))$, where $h_i \in H$ and $p$ is a polynomial. Show that $\mathcal{P}$ is dense in $L^r(\Omega)$ for all $r \geq 1$.

*Hint:* Assume that $r > 1$ and let $q$ be the conjugate of $r$. As in the proof of Theorem 1.1.1 show that if $Z \in L^q(\Omega)$ verifies $E(ZY) = 0$ for all $Y \in \mathcal{P}$, then $Z = 0$.

## 1.2  The derivative operator

This section will be devoted to the properties of the derivative operator. Let $W = \{W(h), h \in H\}$ denote an isonormal Gaussian process associated with the Hilbert space $H$. We assume that $W$ is defined on a complete probability space $(\Omega, \mathcal{F}, P)$, and that $\mathcal{F}$ is generated by $W$.

We want to introduce the derivative $DF$ of a square integrable random variable $F : \Omega \to \mathbb{R}$. This means that we want to differentiate $F$ with respect to the chance parameter $\omega \in \Omega$. In the usual applications of this theory, the space $\Omega$ will be a topological space. For instance, in the example of the $d$-dimensional Brownian motion, $\Omega$ is the Fréchet space $C_0(\mathbb{R}_+; \mathbb{R}^d)$. However, we will be interested in random variables $F$ that are defined $P$ a.s. and that do not possess a continuous version (see Exercise 1.2.1). For this reason we will introduce a notion of derivative defined in a weak sense, and without assuming any topological structure on the space $\Omega$.

We denote by $C_p^\infty(\mathbb{R}^n)$ the set of all infinitely continuously differentiable functions $f : \mathbb{R}^n \to \mathbb{R}$ such that $f$ and all of its partial derivatives have polynomial growth.

Let $S$ denote the class of smooth random variables such that a random variable $F \in S$ has the form

$$F = f(W(h_1), \ldots, W(h_n)), \qquad (1.28)$$

where $f$ belongs to $C_p^\infty(\mathbb{R}^n)$, $h_1, \ldots, h_n$ are in $H$, and $n \geq 1$.

We will make use of the notation $\partial_i f = \frac{\partial f}{\partial x_i}$ and $\nabla f = (\partial_1 f, \ldots, \partial_n f)$, whenever $f \in C^1(\mathbb{R}^n)$.

We will denote by $S_b$ and $S_0$ the classes of smooth random variables of the form (1.28) such that the function $f$ belongs to $C_b^\infty(\mathbb{R}^n)$ ($f$ and all of its partial derivatives are bounded) and to $C_0^\infty(\mathbb{R}^n)$ ($f$ has compact support), respectively. Moreover, we will denote by $\mathcal{P}$ the class of random variables of the form (1.28) such that $f$ is a polynomial. Note that $\mathcal{P} \subset S$, $S_0 \subset S_b \subset S$, and that $\mathcal{P}$ and $S_0$ are dense in $L^2(\Omega)$.

**Definition 1.2.1** *The derivative of a smooth random variable $F$ of the form (1.28) is the $H$-valued random variable given by*

$$DF = \sum_{i=1}^{n} \partial_i f(W(h_1), \ldots, W(h_n)) h_i. \qquad (1.29)$$

For example, $DW(h) = h$. In order to interpret $DF$ as a directional derivative, note that for any element $h \in H$ we have

$$\langle DF, h \rangle_H = \lim_{\epsilon \to 0} \frac{1}{\epsilon} [f(W(h_1) + \epsilon \langle h_1, h \rangle_H, \ldots, W(h_n) + \epsilon \langle h_n, h \rangle_H)$$
$$- f(W(h_1), \ldots, W(h_n))].$$

Roughly speaking, the scalar product $\langle DF, h \rangle_H$ is the derivative at $\epsilon = 0$ of the random variable $F$ composed with shifted process $\{W(g) + \epsilon \langle g, h \rangle_H, g \in H\}$.

The following result is an integration-by-parts formula that will play a fundamental role along this chapter.

**Lemma 1.2.1** *Suppose that $F$ is a smooth random variable and $h \in H$. Then*

$$E(\langle DF, h \rangle_H) = E(FW(h)). \qquad (1.30)$$

*Proof:*    First notice that we can normalize Eq. (1.30) and assume that the norm of $h$ is one. There exist orthonormal elements of $H$, $e_1, \ldots, e_n$, such that $h = e_1$ and $F$ is a smooth random variable of the form

$$F = f(W(e_1), \ldots, W(e_n)),$$

where $f$ is in $C_p^\infty(\mathbb{R}^n)$. Let $\phi(x)$ denote the density of the standard normal distribution on $\mathbb{R}^n$, that is,

$$\phi(x) = (2\pi)^{-\frac{n}{2}} \exp(-\frac{1}{2} \sum_{i=1}^{n} x_i^2).$$

Then we have

$$
\begin{aligned}
E(\langle DF, h \rangle_H) &= \int_{\mathbb{R}^n} \partial_1 f(x)\phi(x)dx \\
&= \int_{\mathbb{R}^n} f(x)\phi(x)x_1 dx \\
&= E(FW(e_1)) = E(FW(h)),
\end{aligned}
$$

which completes the proof of the lemma.    □

Applying the previous result to a product $FG$, we obtain the following consequence.

**Lemma 1.2.2** *Suppose that $F$ and $G$ are smooth random variables, and let $h \in H$. Then we have*

$$E(G\langle DF, h \rangle_H) = E(-F\langle DG, h \rangle_H + FGW(h)). \qquad (1.31)$$

As a consequence of the above lemma we obtain the following result.

**Proposition 1.2.1** *The operator $D$ is closable from $L^p(\Omega)$ to $L^p(\Omega; H)$ for any $p \geq 1$.*

*Proof:*    Let $\{F_N, N \geq 1\}$ be a sequence of smooth random variables such that $F_N$ converges to zero in $L^p(\Omega)$ and the sequence of derivatives $DF_N$ converges to $\eta$ in $L^p(\Omega; H)$. Then, from Lemma 1.2.2 it follows that $\eta$ is equal to zero. Indeed, for any $h \in H$ and for any smooth random variable $F \in S_b$ such that $FW(h)$ is bounded (for intance, $F = Ge^{-\varepsilon W(h)^2}$ where $G \in S_b$ and $\varepsilon > 0$), we have

$$
\begin{aligned}
E(\langle \eta, h \rangle_H F) &= \lim_{N \to \infty} E(\langle DF_N, h \rangle_H F) \\
&= \lim_{N \to \infty} E(-F_N \langle DF, h \rangle_H + F_N FW(h)) = 0,
\end{aligned}
$$

because $F_N$ converges to zero in $L^p(\Omega)$ as $N$ tends to infinity, and the random variables $\langle DF, h \rangle_H$ and $FW(h)$ are bounded. This implies $\eta = 0$.    □

For any $p \geq 1$ we will denote the domain of $D$ in $L^p(\Omega)$ by $\mathbb{D}^{1,p}$, meaning that $\mathbb{D}^{1,p}$ is the closure of the class of smooth random variables $\mathcal{S}$ with respect to the norm

$$\|F\|_{1,p} = [E(|F|^p) + E(\|DF\|_H^p)]^{\frac{1}{p}}.$$

For $p = 2$, the space $\mathbb{D}^{1,2}$ is a Hilbert space with the scalar product

$$\langle F, G \rangle_{1,2} = E(FG) + E(\langle DF, DG \rangle_H).$$

We can define the iteration of the operator $D$ in such a way that for a smooth random variable $F$, the iterated derivative $D^k F$ is a random variable with values in $H^{\otimes k}$. Then for every $p \geq 1$ and any natural number $k \geq 1$ we introduce the seminorm on $\mathcal{S}$ defined by

$$\|F\|_{k,p} = \left[ E(|F|^p) + \sum_{j=1}^{k} E(\|D^j F\|_{H^{\otimes j}}^p) \right]^{\frac{1}{p}}. \tag{1.32}$$

This family of seminorms verifies the following properties:

(i) *Monotonicity:*  $\|F\|_{k,p} \leq \|F\|_{j,q}$, for any $F \in \mathcal{S}$, if $p \leq q$ and $k \leq j$.

(ii) *Closability:*  The operator $D^k$ is closable from $\mathcal{S}$ into $L^p(\Omega; H^{\otimes k})$, for all $p \geq 1$.

   *Proof:*  The proof is analogous to the case where $k = 1$ (see Exercise 1.2.3). $\qquad\square$

(iii) *Compatibility:*  Let $p, q \geq 1$ be real numbers and $k, j$ be natural numbers. Suppose that $F_n$ is a sequence of smooth random variables such that $\|F_n\|_{k,p}$ converges to zero as $n$ tends to infinity, and $\|F_n - F_m\|_{j,q}$ converges to zero as $n, m$ tend to infinity. Then $\|F_n\|_{j,q}$ tends to zero as $n$ tends to infinity.

   *Proof:*  This is an immediate consequence of the closability of the operators $D^i$, $i \geq 1$, on $\mathcal{S}$. $\qquad\square$

We will denote by $\mathbb{D}^{k,p}$ the completion of the family of smooth random variables $\mathcal{S}$ with respect to the norm $\| \cdot \|_{k,p}$. From property (i) it follows that $\mathbb{D}^{k+1,p} \subset \mathbb{D}^{k,q}$ if $k \geq 0$ and $p > q$. For $k = 0$ we put $\| \cdot \|_{0,p} = \| \cdot \|_p$ and $\mathbb{D}^{0,p} = L^p(\Omega)$.

Fix an element $h \in H$. We can define the operator $D^h$ on the set $\mathcal{S}$ of smooth random variables by

$$D^h F = \langle DF, h \rangle_H. \tag{1.33}$$

By Lemma 1.2.2 this operator is closable from $L^p(\Omega)$ into $L^p(\Omega)$, for any $p \geq 1$, and it has a domain that we will denote by $\mathbb{D}^{h,p}$.

The following result characterizes the domain of the derivative operator $\mathbb{D}^{1,2}$ in terms of the Wiener chaos expansion.

**Proposition 1.2.2** *Let $F$ be a square integrable random variable with the Wiener chaos expansion $F = \sum_{n=0}^{\infty} J_n F$. Then $F \in \mathbb{D}^{1,2}$ if and only if*

$$E(\|DF\|_H^2) = \sum_{n=1}^{\infty} n \|J_n F\|_2^2 < \infty. \qquad (1.34)$$

*Moreover, if (1.34) holds, then for all $n \geq 1$ we have $D(J_n F) = J_{n-1}(DF)$.*

*Proof:*    The derivative of a random variable of the form $\Phi_a$, defined in (1.6), can be computed using (1.2):

$$D(\Phi_a) = \sqrt{a!} \sum_{j=1}^{\infty} \prod_{i=1, i \neq j}^{\infty} H_{a_i}(W(e_i)) H_{a_j-1}(W(e_j)) e_j.$$

Then, $D(\Phi_a) \in \mathcal{H}_{n-1}(H)$ (see Execise 1.1.4) if $|a| = n$, and

$$E\left(\|D(\Phi_a)\|_H^2\right) = \sum_{j=1}^{\infty} \frac{a!}{\prod_{i=1, i \neq j}^{\infty} a_i!(a_j - 1)!} = |a|.$$

The proposition follows easily from Proposition 1.1.1.    □

By iteration we obtain $D^k(J_n F) = J_{n-k}(D^k F)$ for all $k \geq 2$ and $n \geq k$. Hence,

$$E(\|D^k F\|_{H^{\otimes k}}^2) = \sum_{n=k}^{\infty} n(n-1) \cdots (n-k+1) \|J_n F\|_2^2,$$

and $F \in \mathbb{D}^{k,2}$ if and only if $\sum_{n=1}^{\infty} n^k \|J_n F\|_2^2 < \infty$.

The following result is the chain rule, which can be easily proved by approximating the random variable $F$ by smooth random variables and the function $\varphi$ by $\varphi * \psi_\epsilon$, where $\{\psi_\epsilon\}$ is an approximation of the identity.

**Proposition 1.2.3** *Let $\varphi : \mathbb{R}^m \to \mathbb{R}$ be a continuously differentiable function with bounded partial derivatives, and fix $p \geq 1$. Suppose that $F = (F^1, \ldots, F^m)$ is a random vector whose components belong to the space $\mathbb{D}^{1,p}$. Then $\varphi(F) \in \mathbb{D}^{1,p}$, and*

$$D(\varphi(F)) = \sum_{i=1}^{m} \partial_i \varphi(F) DF^i.$$

Let us prove the following technical result.

**Lemma 1.2.3** *Let $\{F_n, n \geq 1\}$ be a sequence of random variables in $\mathbb{D}^{1,2}$ that converges to $F$ in $L^2(\Omega)$ and such that*

$$\sup_n E\left(\|DF_n\|_H^2\right) < \infty.$$

*Then $F$ belongs to $\mathbb{D}^{1,2}$, and the sequence of derivatives $\{DF_n, n \geq 1\}$ converges to $DF$ in the weak topology of $L^2(\Omega; H)$.*

*Proof:*    There exists a subsequence $\{F_{n(k)}, k \geq 1\}$ such that the sequence of derivatives $DF_{n(k)}$ converges in the weak topology of $L^2(\Omega; H)$ to some element $\alpha \in L^2(\Omega; H)$. By Proposition 1.2.2, the projections of $DF_{n(k)}$ on any Wiener chaos converge in the weak topology of $L^2(\Omega)$, as $k$ tends to infinity, to those of $\alpha$. Consequently, Proposition 1.2.2 implies $F \in \mathbb{D}^{1,2}$ and $\alpha = DF$. Moreover, for any weakly convergent subsequence the limit must be equal to $\alpha$ by the preceding argument, and this implies the weak convergence of the whole sequence.    □

The chain rule can be extended to the case of a Lipschitz function:

**Proposition 1.2.4** *Let $\varphi : \mathbb{R}^m \to \mathbb{R}$ be a function such that*

$$|\varphi(x) - \varphi(y)| \leq K|x - y|$$

*for any $x, y \in \mathbb{R}^m$. Suppose that $F = (F^1, \ldots, F^m)$ is a random vector whose components belong to the space $\mathbb{D}^{1,2}$. Then $\varphi(F) \in \mathbb{D}^{1,2}$, and there exists a random vector $G = (G_1, \ldots, G_m)$ bounded by $K$ such that*

$$D(\varphi(F)) = \sum_{i=1}^{m} G_i DF^i. \tag{1.35}$$

*Proof:*    If the function $\varphi$ is continuously differentiable, then the result reduces to that of Proposition 1.2.3 with $G_i = \partial_i \varphi(F)$. Let $\alpha_n(x)$ be a sequence of regularization kernels of the form $\alpha_n(x) = n^m \alpha(nx)$, where $\alpha$ is a nonnegative function belonging to $C_0^\infty(\mathbb{R}^m)$ whose support is the unit ball and such that $\int_{\mathbb{R}^m} \alpha(x)dx = 1$. Set $\varphi_n = \varphi * \alpha_n$. It is easy to check that $\lim_n \varphi_n(x) = \varphi(x)$ uniformly with respect to $x$, and the functions $\varphi_n$ are $C^\infty$ with $|\nabla \varphi_n| \leq K$. For each $n$ we have

$$D(\varphi_n(F)) = \sum_{i=1}^{m} \partial_i \varphi_n(F) DF^i. \tag{1.36}$$

The sequence $\varphi_n(F)$ converges to $\varphi(F)$ in $L^2(\Omega)$ as $n$ tends to infinity. On the other hand, the sequence $\{D(\varphi_n(F)), n \geq 1\}$ is bounded in $L^2(\Omega; H)$. Hence, by Lemma 1.2.3 $\varphi(F) \in \mathbb{D}^{1,2}$ and $\{D(\varphi_n(F)), n \geq 1\}$ converges in the weak topology of $L^2(\Omega; H)$ to $D(\varphi(F))$. On the other hand, the sequence $\{\nabla \varphi_n(F), n \geq 1\}$ is bounded by $K$. Hence, there exists a subsequence $\{\nabla \varphi_{n(k)}(F), k \geq 1\}$ that converges to some random vector $G = (G_1, \ldots, G_m)$ in the weak topology $\sigma(L^2(\Omega; \mathbb{R}^m))$. Moreover, $G$ is bounded by $K$. Then, taking the limit in (1.36), we obtain Eq. (1.35). The proof of the lemma is now complete.    □

If the law of the random vector $F$ is absolutely continuous with respect to the Lebesgue measure on $\mathbb{R}^m$, then $G^i = \partial_i \varphi(F)$ in (1.35). Proposition 1.2.4 and Lemma 1.2.3 still hold if we replace $\mathbb{D}^{1,2}$ by $\mathbb{D}^{1,p}$ for any $p > 1$. In fact, this follows from Lemma 1.5.3 and the duality relationship between $D$ and $\delta$.

We will make use of the following technical result.

**Lemma 1.2.4** *The family of random variables* $\{1, W(h)G - D^hG, G \in \mathcal{S}_b, h \in H\}$ *is total in* $L^2(\Omega)$.

*Proof:*    Fix $h \in H$, $n, N \geq 1$, and set $G_N = W(h)^n \psi_N(W(h))$, where $\psi_N$ is an infinitely differentiable function such that $0 \leq \psi_N \leq 1$, $\psi_N(x) = 0$ if $|x| \geq N + 1$, $\psi_N(x) = 1$ if $|x| \leq N$, and $\sup_{x,N} |\psi'_N(x)| < \infty$. Then, $W(h)G_N - D^hG_N$ converges in $L^2(\Omega)$ to $W(h)^{n+1} - n \|h\|_H^2 W(h)^{n-1}$ as $N$ tends to infinity. Hence the closed linear span of the family contains all powers $W(h)^n$, $n \geq 1$, $h \in H$, which implies the result.    $\square$

**Proposition 1.2.5** *Let $F$ be a random variable of the space* $\mathbb{D}^{1,1}$ *such that $DF = 0$. Then $F = E(F)$.*

*Proof:*    If $F \in \mathbb{D}^{1,2}$, then the result follows directly from Proposition 1.2.2. In the general case, let $\psi_N$ be a function in $C_b^\infty(\mathbb{R})$ such that $\psi_N(x) = 0$ if $|x| \geq N+1$, $\psi_N(x) = x$ if $|x| \leq N$. Let $F_n$ be a sequence of smooth random variables converging in $L^1(\Omega)$ to $F$ and such that $E(\|DF_n\|_H)$ tends to zero as $n$ tends to infinity. Then using Lemma 1.2.1 we obtain for any $G \in \mathcal{S}_b$ and any $h \in H$

$$
\begin{aligned}
E\left[\psi_N(F_n)\left(W(h)G - D^hG\right)\right] &= E\left[\psi_N(F_n)W(h)G - D^h\left(G\psi_N(F_n)\right)\right] \\
&\quad + E\left[GD^h\left(\psi_N(F_n)\right)\right] \\
&= E\left[GD^h\left(\psi_N(F_n)\right)\right].
\end{aligned}
$$

Taking the limit as $n$ tends to infinity yields

$$
E\left[\psi_N(F)\left(W(h)G - D^hG\right)\right] = 0.
$$

As a consequence, by Lemma 1.2.4 $E\left[\psi_N(F)\right] = \psi_N(F)$ for each $N$. Hence, $F = E(F)$.    $\square$

**Proposition 1.2.6** *Let $A \in \mathcal{F}$. Then the indicator function of $A$ belongs to* $\mathbb{D}^{1,1}$ *if and only if $P(A)$ is equal to zero or one.*

*Proof:*    By the chain rule (Proposition 1.2.3) applied to to a function $\varphi \in C_0^\infty(\mathbb{R})$, which is equal to $x^2$ on $[0, 1]$, we have

$$
D\mathbf{1}_A = D(\mathbf{1}_A)^2 = 2\mathbf{1}_A D\mathbf{1}_A
$$

and, therefore, $D\mathbf{1}_A = 0$ because from the above equality we get that this derivative is zero on $A^c$ and equal to twice its value on $A$. So, by Proposition 1.2.5 we obtain $\mathbf{1}_A = P(A)$.

**Remarks:**

**1.** If the underlying Hilbert space $H$ is finite-dimensional, then the spaces $\mathbb{D}^{k,p}$ can be identified as ordinary Sobolev spaces of functions on $\mathbb{R}^n$ that together with their $k$ first partial derivatives have moments of order $p$ with respect to the standard normal law. We refer to Ocone [270] for a detailed discussion of this fact. See also Exercise 1.2.8.

**2.** The above definitions can be exended to Hilbert-valued random variables. Let $V$ be a real separable Hilbert space. Consider the family $\mathcal{S}_V$ of $V$-valued smooth random variables of the form

$$F = \sum_{j=1}^{n} F_j v_j, \qquad v_j \in V, \qquad F_j \in \mathcal{S}.$$

Define $D^k F = \sum_{j=1}^{n} D^k F_j \otimes v_j$, $k \geq 1$. Then $D^k$ is a closable operator from $\mathcal{S}_V \subset L^p(\Omega; V)$ into $L^p(\Omega; H^{\otimes k} \otimes V)$ for any $p \geq 1$. For any integer $k \geq 1$ and any real number $p \geq 1$ we can define the seminorm on $\mathcal{S}_V$

$$\|F\|_{k,p,V} = \left[ E(\|F\|_V^p) + \sum_{j=1}^{k} E(\|D^j F\|_{H^{\otimes j} \otimes V}^p) \right]^{\frac{1}{p}}. \qquad (1.37)$$

The operator $D^k$ and the seminorms $\|\cdot\|_{k,p,V}$ verify properties (i), (ii), and (iii) . We define the space $\mathbb{D}^{k,p}(V)$ as the completion of $\mathcal{S}_V$ with respect to the norm $\|\cdot\|_{k,p,V}$. For $k = 0$ we put $\|F\|_{0,p,V} = [E(\|F\|_V^p)]^{\frac{1}{p}}$, and $\mathbb{D}^{0,p}(V) = L^p(\Omega; V)$.

## 1.2.1   The derivative operator in the white noise case

We will suppose in this subsection that the separable Hilbert space $H$ is an $L^2$ space of the form $H = L^2(T, \mathcal{B}, \mu)$, where $\mu$ is a $\sigma$-finite atomless measure on a measurable space $(T, \mathcal{B})$.

The derivative of a random variable $F \in \mathbb{D}^{1,2}$ will be a stochastic process denoted by $\{D_t F, t \in T\}$ due to the identification between the Hilbert spaces $L^2(\Omega; H)$ and $L^2(T \times \Omega)$. Notice that $D_t F$ is defined almost everywhere (a.e.) with respect to the measure $\mu \times P$. More generally, if $k \geq 2$ and $F \in \mathbb{D}^{k,2}$, the derivative

$$D^k F = \{D^k_{t_1, \ldots, t_k} F, t_i \in T\},$$

is a measurable function on the product space $T^k \times \Omega$, which is defined a.e. with respect to the measure $\mu^k \times P$.

**Example 1.2.1** *Consider the example of a d-dimensional Brownian motion on the interval $[0, 1]$, defined on the canonical space $\Omega = C_0([0, 1]; \mathbb{R}^d)$.*

*In this case $\langle DF, h \rangle_H$ can be interpreted as a directional Fréchet derivative. In fact, let us introduce the subspace $H^1$ of $\Omega$ which consists of all absolutely continuous functions $x : [0,1] \to \mathbb{R}^d$ with a square integrable derivative, i.e., $x(t) = \int_0^t \dot{x}(s)ds$, $\dot{x} \in H = L^2([0,1]; \mathbb{R}^d)$. The space $H^1$ is usually called the Cameron-Martin space. We can transport the Hilbert space structure of $H$ to the space $H^1$ by putting*

$$\langle x, y \rangle_{H^1} = \langle \dot{x}, \dot{y} \rangle_H = \sum_{i=1}^d \int_0^1 \dot{x}^i(s)\dot{y}^i(s)ds.$$

*In this way $H^1$ becomes a Hilbert space isomorphic to $H$. The injection of $H^1$ into $\Omega$ is continuous because we have*

$$\sup_{0 \le t \le 1} |x(t)| \le \int_0^1 |\dot{x}(s)|ds \le \|\dot{x}\|_H = \|x\|_{H^1}.$$

*Assume $d = 1$ and consider a smooth functional of the particular form $F = f(W(t_1), \ldots, W(t_n))$, $f \in C_p^\infty(\mathbb{R}^n)$, $0 \le t_1 < \cdots < t_n \le 1$, where $W(t_i) = \int_0^{t_i} dW_t = W(\mathbf{1}_{[0,t_i]})$. Notice that such a functional is continuous in $\Omega$. Then, for any function $h$ in $H$, the scalar product $\langle DF, h \rangle_H$ coincides with the directional derivative of $F$ in the direction of the element $\int_0^\cdot h(s)ds$, which belongs to $H^1$. In fact,*

$$\begin{aligned}
\langle DF, h \rangle_H &= \sum_{i=1}^n \partial_i f(W(t_1), \ldots, W(t_n)) \langle \mathbf{1}_{[0,t_i]}, h \rangle_H \\
&= \sum_{i=1}^n \partial_i f(W(t_1), \ldots, W(t_n)) \int_0^{t_i} h(s)ds \\
&= \frac{d}{d\epsilon} F(\omega + \epsilon \int_0^\cdot h(s)ds)|_{\epsilon=0}.
\end{aligned}$$

*On the other hand, if $F$ is Fréchet differentiable and $\lambda^F$ denotes the signed measure associated with the Fréchet derivative of $F$, then $D_t F = \lambda^F((t,1])$. In fact, for any $h \in H$ we have*

$$\langle DF, h \rangle_H = \int_0^1 \lambda^F(dt)(\int_0^t h(s)ds)dt = \int_0^1 \lambda^F((t,1])h(t)dt.$$

Suppose that $F$ is a square integrable random variable having an orthogonal Wiener series of the form

$$F = \sum_{n=0}^\infty I_n(f_n), \tag{1.38}$$

where the kernels $f_n$ are symmetric functions of $L^2(T^n)$. The derivative $D_t F$ can be easily computed using this expression.

**Proposition 1.2.7** *Let $F \in \mathbb{D}^{1,2}$ be a square integrable random variable with a development of the form (1.38). Then we have*

$$D_t F = \sum_{n=1}^{\infty} n I_{n-1}(f_n(\cdot, t)). \tag{1.39}$$

*Proof:* Suppose first that $F = I_m(f_m)$, where $f_m$ is a symmetric and elementary function of the form (1.10). Then

$$D_t F = \sum_{j=1}^{m} \sum_{i_1,\ldots,i_m=1}^{m} a_{i_1 \cdots i_m} W(A_{i_1}) \cdots \mathbf{1}_{A_{i_j}}(t) \cdots W(A_{i_m}) = m I_{m-1}(f_m(\cdot, t)).$$

Then the result follows easily.  □

The heuristic meaning of the preceding proposition is clear. Suppose that $F$ is a multiple stochastic integral of the form $I_n(f_n)$, which has also been denoted by

$$F = \int_T \cdots \int_T f_n(t_1, \ldots, t_n) W(dt_1) \cdots W(dt_n).$$

Then, $F$ belongs to the domain of the derivation operator and $D_t F$ is obtained simply by removing one of the stochastic integrals, letting the variable $t$ be free, and multiplying by the factor $n$.

Now we will compute the derivative of a conditional expectation with respect to a $\sigma$-field generated by Gaussian stochastic integrals. Let $A \in \mathcal{B}$. We will denote by $\mathcal{F}_A$ the $\sigma$-field (completed with respect to the probability $P$) generated by the random variables $\{W(B), B \subset A, B \in \mathcal{B}_0\}$. We need the following technical result:

**Lemma 1.2.5** *Suppose that $F$ is a square integrable random variable with the representation (1.38). Let $A \in \mathcal{B}$. Then*

$$E(F|\mathcal{F}_A) = \sum_{n=0}^{\infty} I_n(f_n \mathbf{1}_A^{\otimes n}). \tag{1.40}$$

*Proof:* It suffices to assume that $F = I_n(f_n)$, where $f_n$ is a function in $\mathcal{E}_n$. Also, by linearity we can assume that the kernel $f_n$ is of the form $\mathbf{1}_{B_1 \times \cdots \times B_n}$, where $B_1, \ldots, B_n$ are mutually disjoint sets of finite measure. In this case we have

$$E(F|\mathcal{F}_A) = E(W(B_1) \cdots W(B_m)|\mathcal{F}_A)$$
$$= E\left( \prod_{i=1}^{n} (W(B_i \cap A) + W(B_i \cap A^c)) \,|\, \mathcal{F}_A \right)$$
$$= I_n(\mathbf{1}_{(B_1 \cap A) \times \cdots \times (B_n \cap A)}).$$

□

**Proposition 1.2.8** *Suppose that $F$ belongs to $\mathbb{D}^{1,2}$, and let $A \in \mathcal{B}$. Then the conditional expectation $E(F|\mathcal{F}_A)$ also belongs to the space $\mathbb{D}^{1,2}$, and we have:*

$$D_t(E(F|\mathcal{F}_A)) = E(D_t F|\mathcal{F}_A)\mathbf{1}_A(t)$$

*a.e. in $T \times \Omega$.*

*Proof:*    By Lemma 1.2.5 and Proposition 1.2.7 we obtain

$$D_t(E(F|\mathcal{F}_A)) = \sum_{n=1}^{\infty} n I_{n-1}(f_n(\cdot,t)\mathbf{1}_A^{\otimes(n-1)})\mathbf{1}_A(t) = E(D_t F|\mathcal{F}_A)\mathbf{1}_A(t).$$

$\square$

**Corollary 1.2.1** *Let $A \in \mathcal{B}$ and suppose that $F \in \mathbb{D}^{1,2}$ is $\mathcal{F}_A$-measurable. Then $D_t F$ is zero almost everywhere in $A^c \times \Omega$.*

Given a measurable set $A \in \mathcal{B}$, we can introduce the space $\mathbb{D}^{A,2}$ of random variables which are differentiable on $A$ as the closure of $\mathcal{S}$ with respect to the seminorm

$$\|F\|_{A,2}^2 = E(F^2) + E\left(\int_A (D_t F)^2 \, \mu(dt)\right).$$

## *Exercises*

**1.2.1** Let $W = \{W(t), 0 \le t \le 1\}$ be a one-dimensional Brownian motion. Let $h \in L^2([0,1])$, and consider the stochastic integral $F = \int_0^1 h_t dW_t$. Show that $F$ has a continuous modification on $C_0([0,1])$ if and only if there exists a signed measure $\mu$ on $(0,1]$ such that $h(t) = \mu((t,1])$, for all $t \in [0,1]$, almost everywhere with respect to the Lebesgue measure.

*Hint:* If $h$ is given by a signed measure, the result is achieved through integrating by parts. For the converse implication, show first that the continuous modification of $F$ must be linear, and then use the Riesz representation theorem of linear continuous functionals on $C([0,1])$. For a more general treatment of this problem, refer to Nualart and Zakai [268].

**1.2.2** Show that the expression of the derivative given in Definition 1.2.1 does not depend on the particular representation of $F$ as a smooth functional.

**1.2.3** Show that the operator $D^k$ is closable from $\mathcal{S}$ into $L^p(\Omega; H^{\otimes k})$.

*Hint:* Let $\{F_N, N \ge 1\}$ be a sequence of smooth functionals that converges to zero in $L^p$ and such that $D^k F_N$ converges to some $\eta$ in $L^p(\Omega; H^{\otimes k})$. Iterating the integration-by-parts formula (1.31), show that $E(\langle \eta, h_1 \otimes \cdots \otimes h_k \rangle F) = 0$ for all $h_1, \dots, h_k \in H$, $F \in \mathcal{S}_b$, and

$$\xi = \exp\left(-\epsilon \sum_{i=1}^{k} W(h_i)^2\right).$$

**1.2.4** Let $f_n$ be a symmetric function in $L^2([0,1]^n)$. Deduce the following expression for the derivative of $F = I_n(f_n)$:

$$
\begin{aligned}
D_t F \;=\; & n! \sum_{i=1}^{n} \int_{\{t_1 < \cdots < t_{i-1} < t < t_i \cdots < t_{n-1}\}} f_n(t_1, \ldots, t_{n-1}, t) \\
& \times\, dW_{t_1} \cdots dW_{t_{n-1}},
\end{aligned}
$$

with the convention $t_n = 1$.

**1.2.5** Let $F \in \mathbb{D}^{k,2}$ be given by the expansion $F = \sum_{n=0}^{\infty} I_n(f_n)$. Show that

$$
D_{t_1, \ldots, t_k}^{k} F = \sum_{n=k}^{\infty} n(n-1) \cdots (n-k+1) I_{n-k}(f_n(\cdot, t_1, \ldots, t_k)),
$$

and

$$
E(\|D^k F\|_{L^2(T^k)}^2) = \sum_{n=k}^{\infty} \frac{n!^2}{(n-k)!} \|f_n\|_{L^2(T^n)}^2.
$$

**1.2.6** Suppose that $F = \sum_{n=0}^{\infty} I_n(f_n)$ is a random variable belonging to the space $\mathbb{D}^{\infty,2} = \cap_k \mathbb{D}^{k,2}$. Show that $f_n = \frac{1}{n!} E(D^n F)$ for every $n \ge 0$ (cf. Stroock [321]).

**1.2.7** Let $F = \exp(W(h) - \frac{1}{2} \int_T h_s^2 \mu(ds))$, $h \in L^2(T)$. Compute the iterated derivatives of $F$ and the kernels of its expansion into the Wiener chaos.

**1.2.8** Let $e_1, \ldots, e_n$ be orthonormal elements in the Hilbert space $H$. Denote by $\mathcal{F}_n$ the $\sigma$-field generated by the random variables $W(e_1), \ldots, W(e_n)$. Show that an $\mathcal{F}_n$-measurable random variable $F$ belongs to $\mathbb{D}^{1,2}$ if and only if there exists a function $f$ in the weighted Sobolev space $\mathbb{W}^{1,2}(\mathbb{R}^n, N(0, I_n))$ such that

$$
F = f(W(e_1), \ldots, W(e_n)).
$$

Moreover, it holds that $DF = \sum_{i=1}^{n} \partial_i f(W(e_1), \ldots, W(e_n)) e_i$.

**1.2.9** Let $(\Omega, \mathcal{F}, P)$ be the canonical probability space of the standard Brownian motion on the time interval $[0,1]$. Let $F$ be a random variable that satisfies the following Lipschitz property:

$$
\left| F\left(\omega + \int_0^{\cdot} h_s ds\right) - F(\omega) \right| \le c \|h\|_H \quad \text{a.s.}, \quad h \in H = L^2([0,1]).
$$

Show that $F \in \mathbb{D}^{1,2}$ and $\|DF\|_H \le c$ a.s. In [92] Enchev and Stroock proved the reciprocal implication.

*Hint:* Suppose that $F \in L^2(\Omega)$ (the general case is treated by a truncation argument). Consider a complete orthonormal system $\{e_i, i \ge 1\}$ in $H$. Define $F_n = E(F|\mathcal{F}_n)$, where $\mathcal{F}_n$ is the $\sigma$-field generated by the random variables $W(e_1), \ldots, W(e_n)$. Show that $F_n = f_n(W(e_1), \ldots, W(e_n))$, where

$f_n$ is a Lipschitz function with a Lipschitz constant bounded by $c$. Use Exercise 1.2.8 to prove that $F_n$ belongs to $\mathbb{D}^{1,2}$ and $\|DF_n\|_H \leq c$, a.s. Conclude using Lemma 1.2.3.

**1.2.10** Show that the operator defined in (1.33) is closable in $L^p(\Omega)$, for all $p \geq 1$.

**1.2.11** Suppose that $W = \{W(t), 0 \leq t \leq 1\}$ is a standard one-dimensional Brownian motion. Show that the random variable $M = \sup_{0 \leq t \leq 1} W(t)$ belongs to the space $\mathbb{D}^{1,2}$, and $D_t M = \mathbf{1}_{[0,T]}(t)$, where $T$ is the a.s. unique point where $W$ attains its maximum.

*Hint:* Approximate the supremum of $W$ by the maximum on a finite set (see Section 2.1.4).

**1.2.12** Let $F_1$ and $F_2$ be two elements of $\mathbb{D}^{1,2}$ such that $F_1$ and $\|DF_1\|_H$ are bounded. Show that $F_1 F_2 \in \mathbb{D}^{1,2}$ and $D(F_1 F_2) = F_1 DF_2 + F_2 DF_1$.

**1.2.13** Show the following Leibnitz rule for the operator $D^k$:

$$D^k_{t_1,\ldots,t_k}(FG) = \sum_{I \subset \{t_1,\ldots,t_k\}} D^{|I|}_I(F) D^{k-|I|}_{I^c}(G), \quad F, G \in \mathcal{S},$$

where for any subset $I$ of $\{t_1,\ldots,t_k\}$, $|I|$ denotes the cardinality of $I$.

**1.2.14** Show that the set $\mathcal{S}_0$ is dense in $\mathbb{D}^{k,p}$ for any $k \geq 1$, $p \geq 1$.

## 1.3   The divergence operator

In this section we consider the divergence operator, defined as the adjoint of the derivative operator. If the underlying Hilbert space $H$ is an $L^2$ space of the form $L^2(T, \mathcal{B}, \mu)$, where $\mu$ is a $\sigma$-finite atomless measure, we will interpret the divergence operator as a stochastic integral and we will call it the Skorohod integral because in the Brownian motion case it coincides with the generalization of the Itô stochastic integral to anticipating integrands introduced by Skorohod [315]. We will deduce the expression of the Skorohod integral in terms of the Wiener chaos expansion as well as prove some of its basic properties.

We will first introduce the divergence operator in the framework of a Gaussian isonormal process $W = \{W(h), h \in H\}$ associated with the Hilbert space $H$. We assume that $W$ is defined on a complete probability space $(\Omega, \mathcal{F}, P)$, and that $\mathcal{F}$ is generated by $W$.

We recall that the derivative operator $D$ is a closed and unbounded operator with values in $L^2(\Omega; H)$ defined on the dense subset $\mathbb{D}^{1,2}$ of $L^2(\Omega)$.

**Definition 1.3.1** *We denote by $\delta$ the adjoint of the operator $D$. That is, $\delta$ is an unbounded operator on $L^2(\Omega; H)$ with values in $L^2(\Omega)$ such that:*

(i) *The domain of $\delta$, denoted by* $\mathrm{Dom}\,\delta$, *is the set of $H$-valued square integrable random variables* $u \in L^2(\Omega; H)$ *such that*

$$|E(\langle DF, u \rangle_H)| \le c\|F\|_2, \tag{1.41}$$

*for all $F \in \mathbb{D}^{1,2}$, where $c$ is some constant depending on $u$.*

(ii) *If $u$ belongs to* $\mathrm{Dom}\,\delta$, *then $\delta(u)$ is the element of $L^2(\Omega)$ characterized by*

$$E(F\delta(u)) = E(\langle DF, u \rangle_H) \tag{1.42}$$

*for any $F \in \mathbb{D}^{1,2}$.*

The operator $\delta$ is called the *divergence operator* and is closed as the adjoint of an unbounded and densely defined operator. Let us study some basic properties of this operator.

## 1.3.1 Properties of the divergence operator

Taking $F = 1$ in (1.42) we obtain $E(\delta(u)) = 0$ if $u \in \mathrm{Dom}\,\delta$. Also, $\delta$ is a linear operator in $\mathrm{Dom}\,\delta$. We denote by $\mathcal{S}_H$ the class of smooth elementary elements of the form

$$u = \sum_{j=1}^{n} F_j h_j, \tag{1.43}$$

where the $F_j$ are smooth random variables, and the $h_j$ are elements of $H$. From the integration-by-parts formula established in Lemma 1.2.2 we deduce that an element $u$ of this type belongs to the domain of $\delta$ and moreover that

$$\delta(u) = \sum_{j=1}^{n} F_j W(h_j) - \sum_{j=1}^{n} \langle DF_j, h_j \rangle_H. \tag{1.44}$$

The following proposition provides a large class of $H$-valued random variables in the domain of the divergence. Note that if $u \in \mathbb{D}^{1,2}(H)$ then the derivative $Du$ is a square integrable random variable with values in the Hilbert space $H \otimes H$, which can be indentified with the space of Hilbert-Schmidt operators from $H$ to $H$.

**Proposition 1.3.1** *The space $\mathbb{D}^{1,2}(H)$ is included in the domain of $\delta$. If $u, v \in \mathbb{D}^{1,2}(H)$, then*

$$E(\delta(u)\delta(v))) = E(\langle u, v \rangle_H) + E(\mathrm{Tr}(Du \circ Dv)). \tag{1.45}$$

In order to prove Proposition 1.3.1 we need the following commutativity relationship between the derivative and divergence operators. Let $u \in \mathcal{S}_H$, $F \in \mathcal{S}$ and $h \in H$. Then

$$D^h(\delta(u)) = \langle u, h \rangle_H + \delta(D^h u). \tag{1.46}$$

*Proof of (1.46):* Suppose that $u$ has the form (1.43). From (1.44) we deduce

$$
\begin{aligned}
D^h(\delta(u)) &= \sum_{j=1}^{n} F_j \langle h, h_j \rangle_H + \sum_{j=1}^{n} \left( D^h F_j W(h_j) - \langle D\left( D^h F_j \right), h_j \rangle_H \right) \\
&= \langle u, h \rangle_H + \delta(D^h u).
\end{aligned}
$$

$\square$

Notice that (1.46) is just a "Heisenberg commutativity relationship" that can be written, using commutator brackets, as $[D^h, \delta]u = \langle u, h \rangle_H$.

*Proof of Proposition (1.3.1):* Suppose first that $u, v \in \mathcal{S}_H$. Let $\{e_i, i \geq 1\}$ be a complete orthonormal system on $H$. Using the duality relationship (1.42) and property (1.46) we obtain

$$
\begin{aligned}
E\left( \delta(u)\delta(v) \right) &= E\left( \langle v, D(\delta(u)) \rangle_H \right) = E\left( \sum_{i=1}^{\infty} \langle v, e_i \rangle_H D^{e_i}(\delta(u)) \right) \\
&= E\left( \sum_{i=1}^{\infty} \langle v, e_i \rangle_H \left( \langle u, e_i \rangle_H + \delta(D^{e_i} u) \right) \right) \\
&= E\left( \langle u, v \rangle_H \right) + E\left( \sum_{i,j=1}^{\infty} D^{e_i} \langle u, e_j \rangle_H \, D^{e_j} \langle v, e_i \rangle_H \right) \\
&= E\left( \langle u, v \rangle_H \right) + E\left( \mathrm{Tr}\left( Du \circ Dv \right) \right).
\end{aligned}
$$

As a consequence, we obtain the estimate

$$
E\left( \delta(u)^2 \right) \leq E\left( \|u\|_H^2 \right) + E\left( \|Du\|_{H \otimes H}^2 \right) = \|u\|_{1,2,H}^2. \tag{1.47}
$$

This implies that the space $\mathbb{D}^{1,2}(H)$ is included in the domain of $\delta$. In fact, if $u \in \mathbb{D}^{1,2}(H)$, there exists a sequence $u^n \in \mathcal{S}_H$ such that $u^n$ converges to $u$ in $L^2(\Omega)$ and $Du^n$ converges to $Du$ in $L^2(\Omega; H \otimes H)$. Therefore, $\delta(u^n)$ converges in $L^2(\Omega)$ and its limit is $\delta(u)$. Moreover, (1.45) holds for any $u, v \in \mathbb{D}^{1,2}(H)$. $\square$

In order to extend the equality (1.46) to a more general class of random variables we need the following technical lemma.

**Lemma 1.3.1** *Let $G$ be a square integrable random variable. Suppose there exists $Y \in L^2(\Omega)$ such that*

$$
E(G\delta(hF)) = E(YF),
$$

*for all $F \in \mathbb{D}^{1,2}$. Then $G \in \mathbb{D}^{h,2}$ and $D^h G = Y$.*

*Proof:*    We have

$$E\left(YF\right) = E(G\delta(hF)) = \sum_{n=1}^{\infty} E\left((J_nG)\,\delta(hF)\right) = \sum_{n=1}^{\infty} E(FD^h\left(J_nG\right)),$$

hence, $J_{n-1}Y = D^h(J_nG)$ for each $n \geq 1$ and this implies the result.    $\square$

**Proposition 1.3.2** *Suppose that* $u \in \mathbb{D}^{1,2}(H)$, *and* $D^hu$ *belongs to the domain of the divergence. Then* $\delta(u) \in \mathbb{D}^{h,2}$, *and the commutation relation (1.46) holds.*

*Proof:*    For all $F \in \mathbb{D}^{1,2}$ we have using (1.45) and the duality relationship (1.42)

$$\begin{aligned} E(\delta(u)\delta(hF)) &= E\left(\langle u, h\rangle_H\, F + \langle D^hu, DF\rangle_H\right) \\ &= E\left((\langle u, h\rangle_H + \delta(D^hu))\, F\right), \end{aligned}$$

which implies the result, taking into account Lemma 1.3.1    $\square$

The following proposition allows us to factor out a scalar random variable in a divergence.

**Proposition 1.3.3** *Let* $F \in \mathbb{D}^{1,2}$ *and* $u$ *be in the domain of* $\delta$ *such that* $Fu \in L^2(\Omega; H)$. *Then* $Fu$ *belongs to the domain of* $\delta$ *and the following equality is true*

$$\delta(Fu) = F\delta(u) - \langle DF, u\rangle_H, \tag{1.48}$$

*provided the right-hand side of (1.48) is square integrable.*

*Proof:*    For any smooth random variable $G \in \mathcal{S}_0$ we have

$$\begin{aligned} E\left(\langle DG, Fu\rangle_H\right) &= E\left(\langle u, D(FG) - GDF\rangle_H\right) \\ &= E\left((\delta(u)F - \langle u, DF\rangle_H)\, G\right), \end{aligned}$$

which implies the desired result.    $\square$

The next proposition is a version of Proposition 1.3.3, where $u$ is replaced by a deterministic element $h \in H$. In this case it suffices to  impose that $F$ is differentiable in the direction of $h$ (see Lemma 1.3.2 for a related result).

**Proposition 1.3.4** *Let* $h \in H$ *and* $F \in \mathbb{D}^{h,2}$. *Then* $Fh$ *belongs to the domain of* $\delta$ *and the following equality is true*

$$\delta(Fh) = FW(h) - D^hF.$$

*Proof:*    Suppose first that $F \in \mathcal{S}$. Then, the result is clearly true and using (1.45) yields

$$E\left[(\delta(Fh))^2\right] = E\left(F^2\,\|h\|_H^2\right) + E\left[(D^hF)^2\right]. \tag{1.49}$$

Finally, if $F_n \in S$ is a sequence of smooth random variables converging to $F$ in $L^2(\Omega)$ and such that $D^h F_n$ converges to $D^h F$ in $L^2(\Omega)$, then by (1.49) the sequence $\delta(F_n h)$ is convergent in $L^2(\Omega)$.  □

The following extension of Proposition 1.3.3 will be useful.

**Proposition 1.3.5** *Suppose that $H = L^2(T, \mathcal{B}, \mu)$. Let $A \in \mathcal{B}$, and consider a random variable $F \in \mathbb{D}^{A,2}$. Let $u$ be an element of $L^2(\Omega; H)$ such that $u\mathbf{1}_A$ belongs to the domain of $\delta$ and such that $Fu\mathbf{1}_A \in L^2(\Omega; H)$. Then $Fu\mathbf{1}_A$ belongs to the domain of $\delta$ and the following equality is true*

$$\delta(Fu\mathbf{1}_A) = F\delta(u\mathbf{1}_A) - \int_A D_t F u_t \mu(dt), \qquad (1.50)$$

*provided the right-hand side of (1.48) is square integrable.*

The next proposition provides a useful criterion to for the existence of the divergence.

**Proposition 1.3.6** *Consider an element $u \in L^2(\Omega; H)$ such that there exists a sequence $u^n \in \mathrm{Dom}\,\delta$ which converges to $u$ in $L^2(\Omega; H)$. Suppose that there exists $G \in L^2(\Omega)$ such that $\lim_{n\to\infty} E(\delta(u^n)F) = E(GF)$ for all $F \in S$. Then, $u$ belongs to $\mathrm{Dom}\,\delta$ and $\delta(u) = G$.*

## 1.3.2   The Skorohod integral

We will suppose in this subsection that the separable Hilbert space $H$ is an $L^2$ space of the form $H = L^2(T, \mathcal{B}, \mu)$, where $\mu$ is a $\sigma$-finite atomless measure on a measurable space $(T, \mathcal{B})$.

In this case the elements of $\mathrm{Dom}\,\delta \subset L^2(T \times \Omega)$ are square integrable processes, and the divergence $\delta(u)$ is called the *Skorohod stochastic integral* of the process $u$. We will use the following notation:

$$\delta(u) = \int_T u_t dW_t.$$

Any element $u \in L^2(T \times \Omega)$ has a Wiener chaos expansion of the form

$$u(t) = \sum_{n=0}^{\infty} I_n(f_n(\cdot, t)), \qquad (1.51)$$

where for each $n \geq 1$, $f_n \in L^2(T^{n+1})$ is a symmetric function in the first $n$ variables. Furthermore

$$E\left(\int_T u(t)^2 \mu(dt)\right) = \sum_{n=0}^{\infty} n! \|f_n\|_{L^2(T^{n+1})}^2.$$

The following result expresses the operator $\delta$ in terms of the Wiener chaos decomposition.

**Proposition 1.3.7** *Let $u \in L^2(T \times \Omega)$ with the expansion (1.51). Then $u$ belongs to $\text{Dom}\,\delta$ if and only if the series*

$$\delta(u) = \sum_{n=0}^{\infty} I_{n+1}(\widetilde{f_n}) \tag{1.52}$$

*converges in $L^2(\Omega)$.*

Observe that the $(n+1)$-dimensional kernels $f_n$ appearing in formula (1.51) are not symmetric functions of all its variables (only on the first $n$ variables). For this reason, the symmetrization of $f_n$ in all its variables will be given by

$$
\begin{aligned}
\widetilde{f_n}(t_1, \ldots, t_n, t) \;=\; & \frac{1}{n+1} [f_n(t_1, \ldots, t_n, t) \\
& + \sum_{i=1}^{n} f_n(t_1, \ldots, t_{i-1}, t, t_{i+1}, \ldots, t_n, t_i)].
\end{aligned}
$$

Equation (1.52) can also be written without symmetrization, because for each $n$, $I_{n+1}(f_n) = I_{n+1}(\widetilde{f_n})$. However, the symmetrization is needed in order to compute the $L^2$ norm of the stochastic integrals (see formula (1.53) ahead).

*Proof:* Suppose that $G = I_n(g)$ is a multiple stochastic integral of order $n \geq 1$ where $g$ is symmetric. Then we have the following equalities:

$$
\begin{aligned}
E\left(\int_T u_t D_t G \mu(dt)\right) &= \sum_{m=0}^{\infty} \int_T E\left(I_m(f_m(\cdot, t)) n I_{n-1}(g(\cdot, t))\right) \mu(dt) \\
&= \int_T E\left(I_{n-1}(f_{n-1}(\cdot, t)) n I_{n-1}(g(\cdot, t))\right) \mu(dt) \\
&= n(n-1)! \int_T \langle f_{n-1}(\cdot, t), g(\cdot, t) \rangle_{L^2(T^{n-1})} \mu(dt) \\
&= n! \langle f_{n-1}, g \rangle_{L^2(T^n)} = n! \langle \widetilde{f}_{n-1}, g \rangle_{L^2(T^n)} \\
&= E\left(I_n(\widetilde{f}_{n-1}) I_n(g)\right) = E\left(I_n(\widetilde{f}_{n-1}) G\right).
\end{aligned}
$$

Suppose first that $u \in \text{Dom}\,\delta$. Then from the above computations and from formula (1.42) we deduce that

$$E(\delta(u)G) = E(I_n(\widetilde{f}_{n-1})G)$$

for every multiple stochastic integral $G = I_n(g)$. This implies that $I_n(\widetilde{f}_{n-1})$ coincides with the projection of $\delta(u)$ on the $n$th Wiener chaos. Consequently, the series in (1.52) converges in $L^2(\Omega)$ and its sum is equal to $\delta(u)$.

Conversely, suppose that this series converges and let us denote its sum by $V$. Then from the preceding computations we have

$$E\left(\int_T u_t D_t \left(\sum_{n=0}^N I_n(g_n)\right)\mu(dt)\right) = E(V\sum_{n=0}^N I_n(g_n))$$

for all $N \geq 0$. So we get

$$|E(\int_T u_t D_t F \mu(dt))| \leq \|V\|_2 \|F\|_2,$$

for any random variable $F$ with a finite Wiener chaos expansion. By a density argument, this relation holds for any random variable $F$ in $\mathbb{D}^{1,2}$, and by Definition 1.3.1 we conclude that $u$ belongs to $\mathrm{Dom}\,\delta$.    □

From Proposition 1.3.7 it is clear that the class $\mathrm{Dom}\,\delta$ of Skorohod integrable processes coincides with the subspace of $L^2(T \times \Omega)$ formed by the processes that satisfy the following condition:

$$E(\delta(u)^2) = \sum_{n=0}^\infty (n+1)!\|\tilde{f}_n\|^2_{L^2(T^{n+1})} < \infty. \tag{1.53}$$

The space $\mathbb{D}^{1,2}(L^2(T))$, denoted by $\mathbb{L}^{1,2}$, coincides with the class of processes $u \in L^2(T \times \Omega)$ such that $u(t) \in \mathbb{D}^{1,2}$ for almost all $t$, and there exists a measurable version of the two-parameter process $D_s u_t$ verifying $E \int_T \int_T (D_s u_t)^2 \mu(ds)\mu(dt) < \infty$. This space is included in $\mathrm{Dom}\,\delta$ by Proposition 1.3.1. We recall that $\mathbb{L}^{1,2}$ is a Hilbert space with the norm

$$\|u\|^2_{1,2,L^2(T)} = \|u\|^2_{L^2(T\times\Omega)} + \|Du\|^2_{L^2(T^2\times\Omega)}.$$

Note that $\mathbb{L}^{1,2}$ is isomorphic to $L^2(T; \mathbb{D}^{1,2})$.

If $u$ and $v$ are two processes in the space $\mathbb{L}^{1,2}$, then Equation (1.45) can be written as

$$E(\delta(u)\delta(v)) = \int_T E(u_t v_t)\mu(dt) + \int_T \int_T E(D_s u_t D_t v_s)\mu(ds)\mu(dt). \tag{1.54}$$

Suppose that $T = [0, \infty)$ and that $\mu$ is the Lebesgue measure. Then, if both processes are adapted to the filtration generated by the Brownian motion, by Corollary 1.2.1 we have that $D_s u_t = 0$ for almost all $(s,t)$ such that $s > t$, since $\mathcal{F}_t = \mathcal{F}_{[0,t]}$. Consequently, the second summand in Eq. (1.54) is equal to zero, and we recover the usual isometry property of the Itô integral.

We could ask in which sense the Skorohod integral can be interpreted as an integral. Suppose that $u$ is a smooth elementary process of the form

$$u(t) = \sum_{j=1}^n F_j h_j(t), \tag{1.55}$$

where the $F_j$ are smooth random variables, and the $h_j$ are elements of $L^2(T)$. Equation (1.44) can be written as

$$\int_T u_t dW_t = \sum_{j=1}^{n} F_j \int_T h_j(t) dW_j - \sum_{j=1}^{n} \int_T D_t F_j h_j(t) \mu(dt). \qquad (1.56)$$

We see here that the Skorohod integral of a smooth elementary process can be decomposed into two parts, one that can be considered as a path-wise integral, and another that involves the derivative operator. We remark that if for every $j$ the function $h_j$ is an indicator $\mathbf{1}_{A_j}$ of a set $A_j \in \mathcal{B}_0$, and $F_j$ is $\mathcal{F}_{A_j^c}$-measurable, then by Corollary 1.2.1, the second summand of Eq. (1.44) vanishes and the Skorohod integral of $u$ is just the first summand of (1.44).

Proposition 1.3.2 can be reformulated as follows.

**Proposition 1.3.8** *Suppose that $u \in \mathbb{L}^{1,2}$. Assume that for almost all $t$ the process $\{D_t u_s, s \in T\}$ is Skorohod integrable, and there is a version of the process $\{\int_T D_t u_s dW_s, t \in T\}$ which is in $L^2(T \times \Omega)$. Then $\delta(u) \in \mathbb{D}^{1,2}$, and we have*

$$D_t(\delta(u)) = u_t + \int_T D_t u_s dW_s. \qquad (1.57)$$

The next result characterizes the family of stochastic processes that can be written as $DF$ for some random variable $F$.

**Proposition 1.3.9** *Suppose that $u \in L^2(T \times \Omega)$. There exists a random variable $F \in \mathbb{D}^{1,2}$ such that $DF = u$ if and only if the kernels $f_n$ appearing in the integral decomposition (1.51) of $u$ are symmetric functions of all the variables.*

*Proof:* The condition is obviously necessary. To show the sufficiency, define

$$F = \sum_{n=0}^{\infty} \frac{1}{n+1} I_{n+1}(f_n).$$

Clearly, this series converges in $\mathbb{D}^{1,2}$ and $DF = u$. $\qquad \square$

**Proposition 1.3.10** *Every process $u \in L^2(T \times \Omega)$ has a unique orthogonal decomposition $u = DF + u^0$, where $F \in \mathbb{D}^{1,2}$, $E(F) = 0$, and $E(\langle DG, u^0 \rangle_H) = 0$ for all $G$ in $\mathbb{D}^{1,2}$. Furthermore, $u^0$ is Skorohod integrable and $\delta(u^0) = 0$.*

*Proof:* The elements of the form $DF$, $F \in \mathbb{D}^{1,2}$, constitute a closed subspace of $L^2(T \times \Omega)$ by Proposition 1.3.9. Therefore, any process $u \in L^2(T \times \Omega)$ has a unique orthogonal decomposition $u = DF + u^0$, where $F \in \mathbb{D}^{1,2}$, and $u^0 \perp DG$ for all $G$ in $\mathbb{D}^{1,2}$. From $E(\langle DG, u^0 \rangle_H) = 0$ for all $G$ in $\mathbb{D}^{1,2}$, it is clear that $u^0$ is Skorohod integrable and $\delta(u^0) = 0$. $\qquad \square$

## 1.3.3   The Itô stochastic integral as a particular case of the Skorohod integral

It is not difficult to construct processes $u$ that are Skorohod integrable (they belong to Dom $\delta$) and do not belong to the space $\mathbb{L}^{1,2}$. The next result provides a simple method for constructing processes of this type.

**Lemma 1.3.2** *Let $A$ belong to $\mathcal{B}_0$, and let $F$ be a square integrable random variable that is measurable with respect to the $\sigma$-field $\mathcal{F}_{A^c}$. Then the process $F\mathbf{1}_A$ is Skorohod integrable and*

$$\delta(F\mathbf{1}_A) = FW(A).$$

*Proof:*   Suppose first that $F$ belongs to the space $\mathbb{D}^{1,2}$. In that case using (1.48) and Corollary 1.2.1, we have

$$\delta(F\mathbf{1}_A) = FW(A) - \int_T D_t F\mathbf{1}_A(t)\mu(dt) = FW(A).$$

Then, the general case follows by a limit argument, using the fact that $\delta$ is closed.                                                                                □

Notice that Lemma 1.3.2 is a particular case of Proposition 1.3.4 because if $F$ is in $L^2(\Omega, \mathcal{F}_{A^c}, P)$, then $F \in \mathbb{D}^{\mathbf{1}_A,2}$ and $D^{\mathbf{1}_A}F = 0$.

Using this lemma we can show that the operator $\delta$ is an extension of the Itô integral in the case of the Brownian motion. Let $W = \{W^i(t); 0 \le t \le 1, 1 \le i \le d\}$ be a $d$-dimensional Brownian motion. We denote by $L^2_a$ the closed subspace of $L^2([0,1] \times \Omega; \mathbb{R}^d) \cong L^2(T \times \Omega)$ (we recall that here $T = [0,1] \times \{1,\ldots,d\}$) formed by the adapted processes.

In this context we have the following proposition.

**Proposition 1.3.11** $L^2_a \subset \text{Dom } \delta$, *and the operator $\delta$ restricted to $L^2_a$ coincides with the Itô integral, that is,*

$$\delta(u) = \sum_{i=1}^{d} \int_0^1 u^i_t dW^i_t.$$

*Proof:*   Suppose that $u$ is an elementary adapted process of the form

$$u_t = \sum_{j=1}^{n} F_j \mathbf{1}_{(t_j, t_{j+1}]}(t),$$

where $F_j \in L^2(\Omega, \mathcal{F}_{t_j}, P; \mathbb{R}^d)$, and $0 \le t_1 < \cdots < t_{n+1} \le 1$ (here $\mathcal{F}_t = \mathcal{F}_{[0,t]}$). Then from Lemma 1.3.2 we obtain $u \in \text{Dom } \delta$ and

$$\delta(u) = \sum_{i=1}^{d} \sum_{j=1}^{n} F_j^i (W^i(t_{j+1}) - W^i(t_j)). \tag{1.58}$$

We know that any process $u \in L_a^2$ can be approximated in the norm of $L^2(T \times \Omega)$ by a sequence $u^n$ of elementary adapted processes. Then by (1.58) $\delta(u^n)$ is equal to the Itô integral of $u^n$ and it converges in $L^2(\Omega)$ to the Itô integral of $u$. Since $\delta$ is closed we deduce that $u \in \mathrm{Dom}\,\delta$, and $\delta(u)$ is equal to the Itô integral of $u$. □

More generally, any type of adapted stochastic integral with respect to a multiparameter Gaussian white noise $W$ can be considered as a Skorohod integral (see, for instance, Nualart and Zakai [264]).

Let $W = \{W(t), t \in [0,1]\}$ be a one-dimensional Brownian motion. We are going to introduce a class of processes which are differentiable in the future and it contains $L_a^2$. The Skorohod integral is well defined in this class and possesses properties similar to those of the Itô integral (see Chapter 3).

Let $\mathbb{L}^{1,2,f}$ be the closure of $\mathcal{S}_H$ with respect to the seminorm

$$\|u\|_{1,2,f}^2 = E\left(\int_0^1 u_t^2 dt\right) + E\left(\int_{s \leq t} (D_s u_t)^2 ds dt\right),$$

and let $\mathbb{L}^F$ be defined as the closure of $\mathcal{S}_H$ with respect to the seminorm

$$\|u\|_F^2 = \|u\|_{1,2,f}^2 + E\left(\int_{r \vee s \leq t} (D_r D_s u_t)^2 ds dt\right).$$

**Remarks:**

**1.** $\mathbb{L}^F$ coincides with the class of processes $u \in \mathbb{L}^{1,2,f}$ such that $\{D_s u_t \mathbf{1}_{[0,s]}(t), t \in [0,1]\}$ belongs to $\mathbb{D}^{1,2}(L^2([0,1]^2))$.

**2.** If $u \in \mathbb{L}^{1,2,f}$, then $\int_a^b u_t dt \in \mathbb{D}^{1_{[b,1]},2}$ for any $0 \leq a < b \leq 1$.

**Proposition 1.3.12** $L_a^2 \subset \mathbb{L}^F$, and for any $u \in L_a^2$ we have $D_s u_t = 0$ if $t \geq s$, and

$$\|u\|_F^2 = E\left(\int_0^1 u_t^2 dt\right). \tag{1.59}$$

*Proof:* Let $u$ be an elementary adapted process of the form (1.58). Then $u \in \mathbb{L}^F$, and $D_s u_t = 0$ if $t \geq s$. The result follows because these processes are dense in $L_a^2$. □

**Proposition 1.3.13** $\mathbb{L}^F \subset \mathrm{Dom}\delta$ and for all $u \in \mathbb{L}^F$ we have

$$E\left(\delta(u)^2\right) \leq 2\|u\|_F^2.$$

*Proof:* If $u \in \mathcal{S}_H$, then by (1.54) we have

$$\begin{aligned} E\left(\delta(u)^2\right) &= E\left(\int_0^1 u_t^2 dt\right) + E\left(\int_0^1 \int_0^1 D_s u_t D_t u_s ds dt\right) \\ &= E\left(\int_0^1 u_t^2 dt\right) + 2E\left(\int_0^1 \int_0^t D_s u_t D_t u_s ds dt\right). \end{aligned} \tag{1.60}$$

Using the duality between the operators $\delta$ and $D$ and applying again (1.54) yields

$$
\begin{aligned}
E\left(\int_0^1 \int_0^t D_s u_t D_t u_s ds dt\right) &= E\left(\int_0^1 u_t \left(\int_0^t D_t u_s dW_s\right) dt\right) \\
&\leq \frac{1}{2} E\left(\int_0^1 u_t^2 dt\right) \\
&\quad + \frac{1}{2} E\left(\int_0^1 \left(\int_0^t D_t u_s dW_s\right)^2 dt\right). \quad (1.61)
\end{aligned}
$$

Moreover, (1.54) yields

$$
E\left(\int_0^1 \left(\int_0^t D_t u_s dW_s\right)^2 dt\right) = E\left(\int_0^1 \int_0^t (D_t u_s)^2 ds dt\right)
$$
$$
+ E\left(\int_0^1 \int_0^t \int_0^t (D_r D_t u_s)^2 dr ds dt\right). \quad (1.62)
$$

Substituting (1.61) and (1.62) into (1.60) we obtain the inequality (1.59). Finally, the general case follow by a density argument.  □

### 1.3.4  Stochastic integral representation of Wiener functionals

Suppose that $W = \{W(t), t \in [0, 1]\}$ is a one-dimensional Brownian motion. We have seen in Section 1.1.3 that any square integrable random variable $F$, measurable with respect to $W$, can be written as

$$
F = E(F) + \int_0^1 \phi(t) dW_t,
$$

where the process $\phi$ belongs to $L_a^2$. When the variable $F$ belongs to the space $\mathbb{D}^{1,2}$, it turns out that the process $\phi$ can be identified as the optional projection of the derivative of $F$. This is called the Clark-Ocone representation formula:

**Proposition 1.3.14** *Let $F \in \mathbb{D}^{1,2}$, and suppose that $W$ is a one-dimensional Brownian motion. Then*

$$
F = E(F) + \int_0^1 E(D_t F | \mathcal{F}_t) dW_t. \quad (1.63)
$$

*Proof:*  Suppose that $F = \sum_{n=0}^\infty I_n(f_n)$. Using (1.39) and (1.40) we deduce

$$
\begin{aligned}
E(D_t F | \mathcal{F}_t) &= \sum_{n=1}^\infty n E(I_{n-1}(f_n(\cdot, t)) | \mathcal{F}_t) \\
&= \sum_{n=1}^\infty n I_{n-1}\left(f_n(t_1, \ldots, t_{n-1}, t) \mathbf{1}_{\{t_1 \vee \cdots \vee t_{n-1} \leq t\}}\right).
\end{aligned}
$$

Set $\phi_t = E(D_t F | \mathcal{F}_t)$. We can compute $\delta(\phi)$ using the above expression for $\phi$ and (1.52), and we obtain

$$\delta(\phi) = \sum_{n=1}^{\infty} I_n(f_n) = F - E(F),$$

which shows the desired result because $\delta(\phi)$ is equal to the Itô stochastic integral of $\phi$. $\qquad\square$

As a consequence of this integral representation, and applying the Hölder, Burkholder, and Jensen inequalities, we deduce the following inequality for $F \in \mathbb{D}^{1,p}$ and $p \geq 2$ in the case $T = [0,1]$:

$$E(|F|^p) \leq C_p[|E(F)|^p + E(\int_0^1 |D_t F|^p dt)].$$

### 1.3.5  Local properties

In this subsection we will show that the divergence and derivative operators verify a local property. The local property of the Skorohod integral is analogous to that of the Itô integral.

**Proposition 1.3.15** *Let $u \in \mathbb{D}^{1,2}(H)$ and $A \in \mathcal{F}$, such that $u = 0$ on $A$. Then $\delta(u) = 0$ a.s. on $A$.*

*Proof:*  Let $F$ be a smooth random variable of the form

$$F = f(W(h_1), \dots, W(h_n))$$

with $f \in C_0^{\infty}(\mathbb{R}^n)$. We want to show that

$$\delta(u) \mathbf{1}_{\{\|u\|_H = 0\}} = 0$$

a.s. Suppose that $\phi : \mathbb{R} \to \mathbb{R}$ is an infinitely differentiable function such that $\phi \geq 0$, $\phi(0) = 1$ and its support is included in the interval $[-1, 1]$. Define the function $\phi_\epsilon(x) = \phi(\frac{x}{\epsilon})$ for all $\epsilon > 0$. We will use (see Exercise 1.3.3) the fact that the product $F\phi_\epsilon(\|u\|_H^2)$ belongs to $\mathbb{D}^{1,2}$. Then by the duality relation (1.42) we obtain

$$
\begin{aligned}
E\left(\delta(u)\phi_\epsilon\left(\|u\|_H^2\right) F\right) &= E\left(\langle u, D[F\phi_\epsilon(\|u\|_H^2)]\rangle_H\right) \\
&= E\left(\phi_\epsilon\left(\|u\|_H^2\right) \langle u, DF\rangle_H\right) \\
&\quad + 2E\left(F\phi_\epsilon'\left(\|u\|_H^2\right) \langle u, D^u u\rangle_H\right).
\end{aligned}
$$

We claim that the above expression converges to zero as $\epsilon$ tends to zero. In fact, first observe that the random variables

$$V_\epsilon = \phi_\epsilon\left(\|u\|_H^2\right) \langle u, DF\rangle_H + 2F\phi_\epsilon'\left(\|u\|_H^2\right) \langle u, D^u u\rangle_H$$

converge a.s. to zero as $\epsilon \downarrow 0$, since $\|u\|_H = 0$ implies $V_\epsilon = 0$. Second, we can apply the Lebesgue dominated convergence theorem because we have

$$\left|\phi_\epsilon\left(\|u\|_H^2\right)\langle u, DF\rangle_H\right| \leq \|\phi\|_\infty \|u\|_H \|DF\|_H,$$
$$\left|\phi'_\epsilon\left(\|u\|_H^2\right)\langle u, D^u u\rangle_H\right|$$
$$\leq \sup_x |x\phi'_\epsilon(x)| \|Du\|_{H\otimes H} \leq \|\phi'\|_\infty \|Du\|_{H\otimes H}.$$

The proof is now complete.    $\square$

Notice that the local property of the divergence has been established for $H$-valued random variables in the space $\mathbb{D}^{1,2}(H)$. We do not know if it holds for an arbitrary variable $u$ in the domain of $\delta$, although we know that in the Brownian case the local property holds in the subspace $L_a^2$, because as we have seen $\delta$ coincides there with the Itô integral. As an extension of this result we will prove in Proposition 1.3.17 below that the local property of $\delta$ holds in the space $\mathbb{L}^F$.

The next result shows that the operator $D$ is local in the space $\mathbb{D}^{1,1}$.

**Proposition 1.3.16** *Let $F$ be a random variable in the space $\mathbb{D}^{1,1}$ such that $F = 0$ a.s. on some set $A \in \mathcal{F}$. Then $DF = 0$ a.s. on $A$.*

*Proof:*    We can assume that $F \in \mathbb{D}^{1,1} \cap L^\infty(\Omega)$, replacing $F$ by $\arctan(F)$. We want to show that $\mathbf{1}_{\{F=0\}} DF = 0$ a.s. Consider a function $\phi : \mathbb{R} \to \mathbb{R}$ such as that in the proof of Proposition 1.3.15. Set

$$\psi_\epsilon(x) = \int_{-\infty}^x \phi_\epsilon(y) dy.$$

By the chain rule $\psi_\epsilon(F)$ belongs to $\mathbb{D}^{1,1}$ and $D\psi_\epsilon(F) = \phi_\epsilon(F)DF$. Let $u$ be a smooth elementary process of the form

$$u = \sum_{j=1}^n F_j h_j,$$

where $F_j \in \mathcal{S}_b$ and $h_j \in H$. Observe that the duality relation (1.42) holds for $F$ in $\mathbb{D}^{1,1} \cap L^\infty(\Omega)$ and for a process $u$ of this type. Note that the class of such processes $u$ is total in $L^1(\Omega; H)$ in the sense that if $v \in L^1(\Omega; H)$ satisfies $E(\langle v, u\rangle_H) = 0$ for all $u$ in the class, then $v \equiv 0$. Then we have

$$\begin{aligned}|E(\phi_\epsilon(F)\langle DF, u\rangle_H)| &= |E(\langle D(\psi_\epsilon(F)), u\rangle_H)| \\ &= |E(\psi_\epsilon(F)\delta(u))| \leq \epsilon\|\phi\|_\infty E(|\delta(u)|).\end{aligned}$$

Letting $\epsilon \downarrow 0$, we obtain

$$E\left(\mathbf{1}_{\{F=0\}}\langle DF, u\rangle_H\right) = 0,$$

which implies the desired result.    $\square$

**Proposition 1.3.17** *Suppose that* $W = \{W(t), t \in [0,1]\}$ *is a one-dimensional Brownian motion. Let* $u \in \mathbb{L}^F$ *and* $A \in \mathcal{F}$, *such that* $u_t(\omega) = 0$ *a.e. on the product space* $[0,T] \times A$. *Then* $\delta(u) = 0$ *a.s. on* $A$.

*Proof:*  Let $u \in \mathbb{L}^F$. Consider the sequence of processes $\tilde{u}^n$ defined in (1.17). As in Lemma 1.1.3 the operator $P_n$ defined by $P_n(u) = \tilde{u}^n$ has norm bounded by 1 from $\mathbb{L}^F$ to $\mathbb{L}^F$. By Proposition 1.3.13 $\delta(\tilde{u}^n)$ converges in $L^2(\Omega)$ to $\delta(u)$ as $n$ tends to infinity. On the other hand, applying Proposition 1.3.4 we have

$$\delta(\tilde{u}^n) = \sum_{i=1}^{2^n-1} 2^n \left( \int_{(i-1)2^{-n}}^{i2^{-n}} u(s)ds \right) \left( W_{(i+1)2^{-n}} - W_{i2^{-n}} \right)$$
$$- \int_{i2^{-n}}^{(i+1)2^{-n}} \int_{(i-1)2^{-n}}^{i2^{-n}} D_s u_t \, dt \, ds.$$

and by the local property of the operator $D$ in the space $\mathbb{L}^{1,2,f}$ (see Exercise 1.3.12) we deduce that this expression is zero on the set $\int_0^1 u_t^2 dt = 0$, which completes the proof of the proposition.  □

We can localize the domains of the operators $D$ and $\delta$ as follows. If $\mathbb{L}$ is a class of random variables (or processes) we denote by $\mathbb{L}_{\mathrm{loc}}$ the set of random variables $F$ such that there exists a sequence $\{(\Omega_n, F_n), n \geq 1\} \subset \mathcal{F} \times \mathbb{L}$ with the following properties:

(i) $\Omega_n \uparrow \Omega$ , a.s.

(ii) $F = F_n$ a.s. on $\Omega_n$.

If $F \in \mathbb{D}^{1,p}_{\mathrm{loc}}$, $p \geq 1$, and $(\Omega_n, F_n)$ localizes $F$ in $\mathbb{D}^{1,p}$, then $DF$ is defined without ambiguity by $DF = DF_n$ on $\Omega_n$, $n \geq 1$. More generally, the iterated derivative $D^k$ is well defined by localization in the space $\mathbb{D}^{k,p}_{\mathrm{loc}}$. Moreover, for any $h \in H$ the operator $D^h$ is also local (see Exercise 1.3.12) and it has a local domain $\mathbb{D}^{h,p}_{\mathrm{loc}}$, $p \geq 1$.

Then, if $u \in \mathbb{D}^{1,2}_{\mathrm{loc}}(H)$, the divergence $\delta(u)$ is defined as a random variable determined by the conditions

$$\delta(u)|_{\Omega_n} = \delta(u^n)|_{\Omega_n} \qquad \text{for all} \qquad n \geq 1,$$

where $(\Omega_n, u_n)$ is a localizing sequence for $u$.

Although the local property of the divergence operator has not been proved in its domain, we can localize the divergence as follows. Suppose that $\{(\Omega_n, u^n), n \geq 1\}$ is a localizing sequence for $u$ in $(\mathrm{Dom}\delta)_{\mathrm{loc}}$. If $\delta(u^n) = \delta(u^m)$ a.s. on $\Omega_n$ for all $m \geq n$, then, the divergence $\delta(u)$ is the random variable determined by the conditions $\delta(u)|_{\Omega_n} = \delta(u^n)|_{\Omega_n}$ for all $n \geq 1$, but it may depend on the localizing sequence.

**Examples:**    Let $W = \{W(t), t \in [0, 1]\}$ be a one-dimensional Brownian motion. The processes

$$u_t = \frac{W_t}{|W_1|}, \quad v_t = \exp(W_t^4)$$

belong to $\mathbb{L}_{\mathrm{loc}}^{1,2}$. In fact, the sequence $(\Omega_n, u^n)$ with $\Omega_n = \{|W_1| > \frac{1}{n}\}$ and $u_t^n = \frac{W_t}{|W_1| \vee (1/n)}$ localizes the process $u$ in $\mathbb{L}^{1,2}$. On the other hand, if we take $\Omega_n = \{\sup_{t \in [0,1]} |W_t| < n\}$ and $v_t^n = \exp(W_t^4 \wedge n)$, we obtain a localizing sequence for the process $v$ (see also Exercise 1.3.10).

The following proposition asserts that the Skorohod integral defined by localization in the space $\mathbb{L}_{\mathrm{loc}}^F$ thanks to Proposition 1.3.17 is an extension of the Itô stochastic integral.

**Proposition 1.3.18** *Let $W = \{W_t, t \in [0, 1]\}$ be a one-dimensional Brownian motion and consider an adapted process $u$ such that $\int_0^1 u_t^2 dt < \infty$ a. s. Then, $u$ belongs to $\mathbb{L}_{\mathrm{loc}}^F$ and $\delta(u)$ coincides with the Itô stochastic integral $\int_0^1 u_t dW_t$.*

*Proof:*    For any integer $k \geq 1$ consider an infinitely differentiable function $\varphi_k : \mathbb{R} \to \mathbb{R}$ such that $\varphi_k(x) = 1$ if $|x| \leq k$, $\varphi_k(x) = 0$ if $|x| \geq k + 1$ and $|\varphi_k(x)| \leq 1$ for all $x$. Define

$$u_t^k = u_t \varphi_k\left(\int_0^t u_s^2 ds\right)$$

and

$$\Omega_k = \left\{\int_0^1 u_s^2 ds \leq k\right\}.$$

Then we have $\Omega_k \uparrow \Omega$ a.s., $u = u^k$ on $[0, 1] \times \Omega_k$, and $u^k \in L_a^2$ because $u^k$ is adapted and

$$\int_0^1 \left(u_t^k\right)^2 dt = \int_0^1 u_t^2 \varphi_k\left(\int_0^t u_s^2 ds\right) dt \leq k + 1.$$

Then, the result follows because on $L_a^2$ the Skorohod integral is an extension of the Itô integral.    $\square$

The following lemma is helpful in the application of the analysis on the Wiener space. It allows us to transform measurability properties with respect to $\sigma$-fields generated by variables of the first chaos into analytical conditions.

**Lemma 1.3.3** *Let $G$ be a random variable in $\mathbb{D}_{\mathrm{loc}}^{1,2}$. Given a closed subspace $K$ of $H$, we denote by $\mathcal{F}_K$ the $\sigma$-field generated by the Gaussian random variables $\{W(h), h \in K\}$. Let $A \in \mathcal{F}_K$. Suppose that $\mathbf{1}_A G$ is $\mathcal{F}_K$-measurable. Then $DG \in K$, a.s., in $A$.*

*Proof:*    Since we can approximate $G$ by $\varphi_n(G)$, where $\varphi_n \in C_b^\infty(\mathbb{R})$, $\varphi_n(x) = x$ for $|x| \leq n$, it is sufficient to prove the result for $G \in \mathbb{D}_{\text{loc}}^{1,2} \cap L^2(\Omega)$. Let $h \in H$ be an element orthogonal to $K$. Then $E(G \mid \mathcal{F}_K)$ belongs to $\mathbb{D}^{h,2}$ and $D^h E(G \mid \mathcal{F}_K) = 0$. However, $G \in \mathbb{D}_{\text{loc}}^{h,2}$ and $G = E(G \mid \mathcal{F}_K)$ a.s. on $A$. From the local property of $D^h$ it follows that $D^h G = 0$ a.s. on $A$. Then it remains to choose a countable and dense set of elements $h$ in the orthogonal complement of $K$, and we obtain that $DG \in K$ a.s. on $A$.    $\square$

The following lemma shows that an Itô integral is differentiable if and only if its integrand is differentiable (see [279]).

**Lemma 1.3.4** *Let $W = \{W(t), t \in [0,1]\}$ be a one-dimensional Brownian motion. Consider a square integrable adapted process $u = \{u_t, t \in [0,1]\}$, and set $X_t = \int_0^t u_s dW_s$. Then the process $u$ belongs to the space $\mathbb{L}^{1,2}$ if and only if $X_1$ belongs to $\mathbb{D}^{1,2}$. In this case the process $X$ belongs to $\mathbb{L}^{1,2}$, and we have*

$$\int_0^t E(|D_s X_t|^2) ds = \int_0^t E(u_s^2) ds + \int_0^t \int_0^s E(|D_r u_s|^2) dr ds, \qquad (1.64)$$

*for all $t \in [0,1]$.*

*Proof:*    Suppose first that $u \in \mathbb{L}^{1,2}$. Then the process $u$ verifies the hypothesis of Proposition 1.3.8 of the Skorohod integral. In fact, the process $\{D_t u_s, s \in [t,1]\}$ is Skorohod integrable because it is adapted and square integrable. Moreover,

$$E\left(\int_0^1 \left|\int_t^1 D_t u_s dW_s\right|^2 dt\right) = \int_0^1 \int_t^1 E(|D_t u_s|^2) ds dt < \infty,$$

due to the isometry of the Itô integral. Consequently, by Proposition 1.3.8 we obtain that $X_t$ belongs to $\mathbb{D}^{1,2}$ for any $t$ and

$$D_s X_t = u_s \mathbf{1}_{\{s \leq t\}} + \int_s^t D_s u_r dW_r. \qquad (1.65)$$

Taking the expectation of the square of the above expression, we get (1.64) and $X$ belongs to $\mathbb{L}^{1,2}$.

Conversely, suppose that $X_1$ belongs to $\mathbb{D}^{1,2}$. For each $N$ we denote by $u_t^N$ the projection of $u_t$ on $\mathcal{P}_N = \mathcal{H}_0 \oplus \cdots \oplus \mathcal{H}_N$. Set $X_t^N = \int_0^t u_s^N dW_s$. Then $X_t^N$ is the projection of $X_t$ on $\mathcal{P}_{N+1}$. Hence, $X_1^N$ converges to $X_1$ in the topology of the space $\mathbb{D}^{1,2}$. Then the result follows from the inequality

$$
\begin{aligned}
\int_0^1 E(|D_s X_1^N|^2) ds &= \int_0^1 E(|u_s^N|^2) ds + \int_0^1 \int_0^s E(|D_r u_s^N|^2) dr ds \\
&\geq \int_0^1 \int_0^s E(|D_r u_s^N|^2) dr ds \\
&= E\left(\|Du^N\|_{L^2([0,1]^2)}^2\right).
\end{aligned}
$$

$\square$

## *Exercises*

**1.3.1** Show the isometry property (1.54) using the Wiener series expansion of the process $u$.

**1.3.2** Let $\mathcal{R}$ be the class of processes of the form

$$u = \sum_{i=1}^{n} F_i \mathbf{1}_{A_i},$$

where $A_i \in \mathcal{B}_0$, and $F_i \in L^2(\Omega, \mathcal{F}_{A_i^c}, P)$. Show that Dom $\delta$ coincides with the closed hull of $\mathcal{R}$ for the norm $\|u\|_{L^2(T \times \Omega)} + \|\delta(u)\|_2$.

**1.3.3** Let $F$ be a smooth random variable of the form

$$F = f(W(h_1), \dots, W(h_n)),$$

where $f \in C_0^\infty(\mathbb{R}^n)$. Let $g$ be in $C_0^\infty(\mathbb{R})$, and let $u \in \mathbb{L}^{1,2}$. Show that $Fg(\|u\|_H^2)$ belongs to the space $\mathbb{D}^{1,2}$ and

$$D\left(Fg(\|u\|_H^2)\right) = DFg(\|u\|_H^2) + 2Fg'(\|u\|_H^2)D_\cdot^u.$$

**1.3.4** Let $F \in \mathbb{D}^{1,2}$ be a random variable such that $E(|F|^{-2}) < \infty$. Then $P\{F > 0\}$ is zero or one.

    *Hint:* Using the duality relation, compute $E(\varphi_\epsilon(F)\delta(u))$, where $u$ is an arbitrary bounded element in the domain of $\delta$ and $\varphi_\epsilon$ is an approximation of the sign function.

**1.3.5** Show that the random variable $F = \mathbf{1}_{\{W(h)>0\}}$ does not belong to $\mathbb{D}^{1,2}$. Prove that it belongs to $\mathbb{D}_{\text{loc}}^{1,2}$, and $DF = 0$.

**1.3.6** Show the following differentiation rule (see Ocone and Pardoux [272, Lemma 2.3]) . Let $F = (F^1, \dots, F^k)$ be a random vector whose components belong to $\mathbb{D}_{\text{loc}}^{1,2}$. Consider a measurable process $u = \{u(x), x \in \mathbb{R}^k\}$ which can be localized by processes with continuously differentiable paths, such that for any $x \in \mathbb{R}^k$, $u(x) \in \mathbb{D}_{\text{loc}}^{1,2}$ and the derivative $Du(x)$ has a continuous version as an $H$-valued process. Suppose that for any $a > 0$ we have

$$E\left(\sup_{|x| \le a} \left[|u(x)|^2 + \|Du(x)\|_H^2\right]\right) < \infty,$$

$$\|\sup_{|x| \le a} |\nabla u(x)|\|_\infty < \infty.$$

Then the composition $G = u(F)$ belongs to $\mathbb{D}_{\text{loc}}^{1,2}$, and we have

$$DG = \sum_{i=1}^{k} \partial_i u(F)DF^i + (Du)(F).$$

*Hint:* Approximate the composition $u(F)$ by the integral

$$\int_{\mathbb{R}^k} u(x)\psi_\epsilon(F-x)dx,$$

where $\psi_\epsilon$ is an approximation of the identity.

**1.3.7** Suppose that $H = L^2(T)$. Let $\delta^k$ be the adjoint of the operator $D^k$. That is, a multiparameter process $u \in L^2(T^k \times \Omega)$ belongs to the domain of $\delta^k$ if and only if there exists a random variable $\delta^k(u)$ such that

$$E(F\delta^k(u)) = E(\langle u, D^k F\rangle_{L^2(T^k)})$$

for all $F \in \mathbb{D}^{k,2}$. Show that a process $u \in L^2(T^k \times \Omega)$ with an expansion

$$u_t = \sum_{n=0}^{\infty} I_n(f_n(\cdot, t)), \quad t \in T^k,$$

belongs to the domain of $\delta^k$ if and only if the series

$$\delta^k(u) = \sum_{n=0}^{\infty} I_{n+k}(f_n)$$

converges in $L^2(\Omega)$.

**1.3.8** Let $u \in L^2(T^k \times \Omega)$. Show that there exists a random variable $F \in \mathbb{D}^{k,2}$ such that $u = D^k F$ if and only if $u_t = \sum_{n=0}^{\infty} I_n(f_n(\cdot, t))$ and the kernels $f_n \in L^2(T^{n+k})$ are symmetric functions of all their variables. Show that every process $u \in L^2(T^k \times \Omega)$ admits a unique decomposition $u = D^k F + u^0$, where $F \in \mathbb{D}^{k,2}$ and $\delta^k(u^0) = 0$.

**1.3.9** Let $\{W_t, t \in [0,1]\}$ be a one-dimensional Brownian motion. Using Exercise 1.2.6 find the Wiener chaos expansion of the random variables

$$F_1 = \int_0^1 (t^3 W_t^3 + 2t W_t^2)dW_t, \quad F_2 = \int_0^1 t e^{W_t} dW_t.$$

**1.3.10** Suppose that $H = L^2(T)$. Let $u \in \mathbb{L}^{1,2}$ and $F \in \mathbb{D}^{1,2}$ be two elements such that $P(F = 0) = 0$ and $E\left(\int_T |u_t D_s F|^2 \mu(ds)\mu(dt)\right) < \infty$. Show that the process $\frac{u_t}{|F|}$ belongs to $\mathbb{L}^{1,2}_{\text{loc}}$, and compute its derivative and its Skorohod integral.

**1.3.11** In the particular case $H = L^2(T)$, deduce the estimate (1.47) from Equation (1.53) and the inequality

$$\|\tilde{f}_n\|_{L^2(T^{n+1})} \le \|f_n\|_{L^2(T^{n+1})}.$$

**1.3.12** Let $F \in \mathbb{D}^{h,p}$, $p \geq 1$, be such that $F = 0$ a.s. on $A \in \mathcal{F}$. Show that $D^h F = 0$ a.s. on $A$. As a consequence, deduce the local property of the operator $D$ on the space $\mathbb{L}^{1,2,f}$.

**1.3.13** Using Clark-Ocone formula (1.63) find the stochastic integral representation of the following random variables:
  (i) $F = W_1^3$,
  (ii) $F = \exp(2W_1)$,
  (iii) $F = \sup_{0 \leq t \leq 1} W_t$.

# 1.4   The Ornstein-Uhlenbeck semigroup

In this section we describe the main properties of the Ornstein-Uhlenbeck semigroup and, in particular, we show the hypercontractivity property.

## 1.4.1   The semigroup of Ornstein-Uhlenbeck

We assume that $W = \{W(h), h \in H\}$ is an isonormal Gaussian process associated to the Hilbert space $H$ defined in a complete probability space $(\Omega, \mathcal{F}, P)$, and that $\mathcal{F}$ is generated by $W$. We recall that $J_n$ denotes the orthogonal projection on the $n$th Wiener chaos.

**Definition 1.4.1** *The Ornstein-Uhlenbeck semigroup is the one-parameter semigroup $\{T_t, t \geq 0\}$ of contraction operators on $L^2(\Omega)$ defined by*

$$T_t(F) = \sum_{n=0}^{\infty} e^{-nt} J_n F, \qquad (1.66)$$

*for any $F \in L^2(\Omega)$ .*

There is an alternative procedure for introducing this semigroup. Suppose that the process $W' = \{W'(h), h \in H\}$ is an independent copy of $W$. We will assume that $W$ and $W'$ are defined on the product probability space $(\Omega \times \Omega', \mathcal{F} \otimes \mathcal{F}', P \times P')$. For any $t > 0$ we consider the process $Z = \{Z(h), h \in H\}$ defined by

$$Z(h) = e^{-t} W(h) + \sqrt{1 - e^{-2t}} W'(h), \qquad h \in H.$$

This process is Gaussian, with zero mean and with the same covariance function as $W$. In fact, we have

$$E(Z(h_1)Z(h_2)) = e^{-2t}\langle h_1, h_2 \rangle_H + (1 - e^{-2t})\langle h_1, h_2 \rangle_H = \langle h_1, h_2 \rangle_H.$$

Let $W : \Omega \to \mathbb{R}^H$ and $W' : \Omega' \to \mathbb{R}^H$ be the canonical mappings associated with the processes $\{W(h), h \in H\}$ and $\{W'(h), h \in H\}$, respectively. Given a random variable $F \in L^2(\Omega)$, we can write $F = \psi_F \circ W$, where $\psi_F$ is a measurable mapping from $\mathbb{R}^H$ to $\mathbb{R}$, determined $P \circ W^{-1}$ a.s. As a consequence, the random variable $\psi_F(Z(\omega, \omega')) = \psi_F(e^{-t}W(\omega) + \sqrt{1 - e^{-2t}}W'(\omega'))$ is well defined $P \times P'$ a.s. Then, for any $t > 0$ we put

$$T_t(F) = E'(\psi_F(e^{-t}W + \sqrt{1 - e^{-2t}}W')), \qquad (1.67)$$

where $E'$ denotes mathematical expectation with respect to the probability $P'$. Equation (1.67) is called Mehler's formula. We are going to check the equivalence between (1.66) and (1.67). First we will see that both definitions give rise to a linear contraction operator on $L^2(\Omega)$. This is clear for the definition (1.66). On the other hand, (1.67) defines a linear contraction operator on $L^p(\Omega)$ for any $p \geq 1$ because we have

$$
\begin{aligned}
E(|T_t(F)|^p) &= E(|E'(\psi_F(e^{-t}W + \sqrt{1 - e^{-2t}}W'))|^p) \\
&\leq E(E'(|\psi_F(e^{-t}W + \sqrt{1 - e^{-2t}}W')|^p)) = E(|F|^p).
\end{aligned}
$$

So, to show that (1.66) is equal to (1.67) on $L^2(\Omega)$, it suffices to check that both definitions coincide when $F = \exp(W(h) - \frac{1}{2}\|h\|_H^2)$, $h \in H$. We have

$$
E'\left(\exp\left(e^{-t}W(h) + \sqrt{1 - e^{-2t}}W'(h) - \frac{1}{2}\|h\|_H^2\right)\right)
$$
$$
= \exp\left(e^{-t}W(h) - \frac{1}{2}e^{-2t}\|h\|_H^2\right) = \sum_{n=0}^{\infty} e^{-nt}\|h\|_H^n H_n\left(\frac{W(h)}{\|h\|_H}\right)
$$
$$
= \sum_{n=0}^{\infty} \frac{e^{-nt}}{n!} I_n(h^{\otimes n}).
$$

On the other hand,

$$
\begin{aligned}
T_t(F) &= T_t\left(\sum_{n=0}^{\infty} \frac{1}{n!} I_n(h^{\otimes n})\right) \\
&= \sum_{n=0}^{\infty} \frac{e^{-nt}}{n!} I_n(h^{\otimes n}),
\end{aligned}
$$

which yields the desired equality.

The operators $T_t$ verify the following properties:

(i) $T_t$ is nonnegative (i.e., $F \geq 0$ implies $T_t(F) \geq 0$).

(ii) $T_t$ is symmetric:

$$E(GT_t(F)) = E(FT_t(G)) = \sum_{n=0}^{\infty} e^{-nt} E(J_n(F)J_n(G)).$$

**Example 1.4.1** *The classical Ornstein-Uhlenbeck (O.U.) process on the real line $\{X_t, t \in \mathbb{R}\}$ is defined as a Gaussian process with zero mean and covariance function given by $K(s,t) = \beta e^{-\alpha|s-t|}$, where $\alpha, \beta > 0$ and $s, t \in \mathbb{R}$. This process is Markovian and stationary, and these properties characterize the form of the covariance function, assuming that $K$ is continuous.*

*It is easy to check that the transition probabilities of the Ornstein-Uhlenbeck process $X_t$ are the normal distributions*

$$P(X_t \in dy | X_s = x) = N(xe^{-\alpha(t-s)}, \beta(1 - e^{-2\alpha(t-s)})).$$

*In fact, for all $s < t$ we have*

$$\begin{aligned} E(X_t | X_s) &= e^{-\alpha(t-s)} X_s, \\ E((X_t - E(X_t | X_s))^2) &= \beta(1 - e^{-2\alpha(t-s)}). \end{aligned}$$

*Also, the standard normal law $\nu = N(0, \beta)$ is an invariant measure for the Markov semigroup associated with the O.U. process.*

*Consider the semigroup of operators on $L^2(\mathbb{R}, \mathcal{B}(\mathbb{R}), \nu)$ determined by the stationary transition probabilities of the O.U. process (with $\alpha, \beta = 1$). This semigroup is a particular case of the Ornstein-Uhlenbeck semigroup introduced in Definition 1.4.1, if we take $(\Omega, \mathcal{F}, P) = (\mathbb{R}, \mathcal{B}(\mathbb{R}), \nu)$, $H = \mathbb{R}$, and $W(t)(x) = tx$ for any $t \in \mathbb{R}$. In fact, if $\{X_s, s \in \mathbb{R}\}$ is a real-valued O.U. process, for any bounded measurable function $f$ on $\mathbb{R}$ we have for $t \geq 0$ and $s \in \mathbb{R}$*

$$\begin{aligned} \int_{\mathbb{R}} f(y) P(X_{s+t} \in dy | X_s = x) &= \int_{\mathbb{R}} f(y) N(e^{-t}x, 1 - e^{-2t})(dy) \\ &= \int_{\mathbb{R}} f(e^{-t}x + \sqrt{1 - e^{-2t}}y)\nu(dy) \\ &= (T_t f)(x). \end{aligned}$$

*Let $W$ be a Brownian measure on the real line. That is, $\{W(B), B \in \mathcal{B}(\mathbb{R})\}$ is a centered Gaussian family such that*

$$E(W(B_1)W(B_2)) = \int_{\mathbb{R}} \mathbf{1}_{B_1 \cap B_2}(x) dx.$$

*Then the process*

$$X_t = \sqrt{2\alpha\beta} \int_{-\infty}^{t} e^{-\alpha(t-u)} dW_u$$

*has the law of an Ornstein-Uhlenbeck process of parameters $\alpha, \beta$. Furthermore, the process $X_t$ satisfies the stochastic differential equation*

$$dX_t = \sqrt{2\alpha\beta} dW_t - \alpha X_t dt.$$

Consider now the case where $H = L^2(T, \mathcal{B}, \mu)$ and $\mu$ is a $\sigma$-finite atomless measure. Using the above ideas we are going to introduce an Ornstein-Uhlenbeck process parametrized by $H$. To do this we consider a Brownian measure $B$ on $T \times \mathbb{R}$, defined on some probability space $(\widetilde{\Omega}, \widetilde{\mathcal{F}}, \widetilde{P})$ and with intensity equal to $2\mu(dt)dx$. Then we define

$$X_t(h) = \int_{-\infty}^{t} \int_T h(\tau) e^{-(t-s)} B(d\tau, ds). \tag{1.68}$$

It is easy to check that $X_t(h)$ is a Gaussian zero-mean process with covariance function given by

$$\widetilde{E}(X_{t_1}(h_1) X_{t_2}(h_2)) = e^{-|t_1 - t_2|} \langle h_1, h_2 \rangle_H.$$

Consequently, we have the following properties:

(i) For any $h \in H$, $\{X_t(h), t \in \mathbb{R}\}$ is a real-valued Ornstein-Uhlenbeck process with parameters $\alpha = 1$ and $\beta = \|h\|_H^2$.

(ii) For any $t \geq 0$, $\{X_t(h), h \in H\}$ has the same law as $\{W(h), h \in H\}$.

Therefore, for any random variable $F \in L^0(\Omega)$ we can define the composition $F(X_t)$. That is, $F(X_t)$ is short notation for $\psi_F(X_t)$, where $\psi_F$ is the mapping from $\mathbb{R}^H$ to $\mathbb{R}$ determined by $\psi_F(W) = F$. Let $\widetilde{\mathcal{F}}_t$ denote the $\sigma$-field generated by the random variables $B(G)$, where $G$ is a measurable and bounded subset of $T \times (-\infty, t]$. The following result establishes the relationship between the process $X_t(h)$ and the Ornstein-Uhlenbeck semigroup.

**Proposition 1.4.1** *For any $t \geq 0$, $s \in \mathbb{R}$, and for any integrable random variable $F$ we have*

$$\widetilde{E}(F(X_{s+t})|\widetilde{\mathcal{F}}_s) = (T_t F)(X_s). \tag{1.69}$$

*Proof:* Without loss of generality we may assume that $F$ is a smooth random variable of the form

$$F = f(W(h_1), \ldots, W(h_n)),$$

where $f \in C_p^\infty(\mathbb{R}^n)$, $h_1, \ldots, h_n \in H$, $1 \leq i \leq n$. In fact, the set $\mathcal{S}$ of smooth random variables is dense in $L^1(\Omega)$, and both members of Eq. (1.69) are continuous in $L^1(\Omega)$. We are going to use the decomposition $X_{s+t} = X_{s+t} - e^{-t} X_s + e^{-t} X_s$. Note that

(i) $\{e^{-t} X_s(h), h \in H\}$ is $\widetilde{\mathcal{F}}_s$-measurable, and

(ii) the Gaussian family $\{X_{s+t}(h) - e^{-t} X_s(h), h \in H\}$ has the same law as $\{\sqrt{1 - e^{-2t}} W(h), h \in H\}$, and is independent of $\widetilde{\mathcal{F}}_s$.

Therefore, we have

$$
\begin{aligned}
\tilde{E}(F(X_{s+t})|\tilde{\mathcal{F}}_s) &= \tilde{E}(f(X_{s+t}(h_1),\ldots,X_{s+t}(h_n))|\tilde{\mathcal{F}}_s) \\
&= \tilde{E}\Big(f(X_{s+t}(h_1) - e^{-t}X_s(h_1) + e^{-t}X_s(h_1)), \\
&\qquad \ldots, X_{s+t}(h_n) - e^{-t}X_s(h_n) + e^{-t}X_s(h_n))|\tilde{\mathcal{F}}_s\Big) \\
&= E'\Big(f\Big(\sqrt{1-e^{-2t}}W'(h_1) + e^{-t}X_s(h_1), \\
&\qquad \ldots, \sqrt{1-e^{-2t}}W'(h_n) + e^{-t}X_s(h_n)\Big)\Big) \\
&= (T_t F)(X_s),
\end{aligned}
$$

where $W'$ is an independent copy of $W$, and $E'$ denotes the mathematical expectation with respect to $W'$.  $\square$

Consider, in particular, the case of the Brownian motion. That means $\Omega = C_0([0,1])$, and $P$ is the Wiener measure. In that case, $T = [0,1]$, and the process defined by (1.68) can be written as

$$
X_t(h) = \int_0^1 h(\tau) X_t(d\tau),
$$

where $X_t(\tau) = \int_{-\infty}^t \int_0^\tau e^{-(t-s)} W(d\sigma, ds)$, and $W$ is a two-parameter Wiener process on $[0,1] \times \mathbb{R}$ with intensity $2dtdx$. We remark that the stochastic process $\{X_t(\cdot), t \in \mathbb{R}\}$ is a stationary Gaussian continuous Markov process with values on $C_0([0,1])$, which has the Wiener measure as invariant measure.

### 1.4.2  The generator of the Ornstein-Uhlenbeck semigroup

In this section we will study the properties of the infinitesimal generator of the Ornstein-Uhlenbeck semigroup. Let $F \in L^2(\Omega)$ be a square integrable random variable. We define the operator $L$ as follows:

$$
LF = \sum_{n=0}^\infty -n J_n F,
$$

provided this series converges in $L^2(\Omega)$. The domain of this operator will be the set

$$
\text{Dom}\,L = \{F \in L^2(\Omega), F = \sum_{n=0}^\infty I_n(f_n) : \sum_{n=1}^\infty n^2 \|J_n F\|_2^2 < \infty\}.
$$

In particular, $\text{Dom}\,L \subset \mathbb{D}^{1,2}$. Note that $L$ is an unbounded symmetric operator on $L^2(\Omega)$. That is, $E(FLG) = E(GLF)$ for all $F, G \in \text{Dom}\,L$. The next proposition tells us that $L$ coincides with the infinitesimal generator of the Ornstein-Uhlenbeck semigroup $\{T_t, t \geq 0\}$ introduced in Definition 1.4.1. In particular, $L$ is self-adjoint and (hence) closed.

**Proposition 1.4.2** *The operator $L$ coincides with the infinitesimal generator of the Ornstein-Uhlenbeck semigroup $\{T_t, t \geq 0\}$.*

*Proof:* We have to show that $F$ belongs to the domain of $L$ if and only if the limit $\lim_{t \downarrow 0} \frac{1}{t}(T_t F - F)$ exists in $L^2(\Omega)$ and, in this case, this limit is equal to $LF$. Assume first that $F \in \operatorname{Dom} L$. Then

$$
E\left( \left| \frac{1}{t}(T_t F - F) - LF \right|^2 \right) = \sum_{n=0}^{\infty} \left[ \frac{1}{t}(e^{-nt} - 1) + n \right]^2 E(|J_n F|^2),
$$

which converges to zero as $t \downarrow 0$. In fact, for any $n$ the expression $\frac{1}{t}(e^{-nt} - 1) + n$ tends to zero, and moreover $|\frac{1}{t}(e^{-nt} - 1)| \leq n$.

Conversely, suppose that $\lim_{t \downarrow 0} \frac{1}{t}(T_t F - F) = G$ in $L^2(\Omega)$. Then we have that

$$
J_n G = \lim_{t \downarrow 0} \frac{1}{t}(T_t J_n F - J_n F) = -n J_n F.
$$

Therefore, $F$ belongs to the domain of $L$, and $LF = G$. □

The next proposition explains the relationship between the operators $D$, $\delta$, and $L$.

**Proposition 1.4.3** $\delta D F = -LF$, *that is, for $F \in L^2(\Omega)$ the statement $F \in \operatorname{Dom} L$ is equivalent to $F \in \operatorname{Dom} \delta D$ (i.e., $F \in \mathbb{D}^{1,2}$ and $DF \in \operatorname{Dom} \delta$), and in this case $\delta D F = -LF$.*

*Proof:* Suppose first that $F \in \mathbb{D}^{1,2}$ and that $DF$ belongs to $\operatorname{Dom} \delta$. Let $G$ be a random variable in the $n$th chaos $\mathcal{H}_n$. Then, applying Proposition 1.2.2 we have

$$
E(G \delta D F) = E(\langle DG, DF \rangle_H) = n^2(n-1)! \, \langle g, f_n \rangle_{H^{\otimes n}} = nE(G J_n F).
$$

So, $J_n \delta D F = n J_n F$, which implies $F \in \operatorname{Dom} L$ and $\delta D F = -LF$.

Conversely, if $F \in \operatorname{Dom} L$, then $F \in \mathbb{D}^{1,2}$ and for any $G \in \mathbb{D}^{1,2}$, $G = \sum_{n=0}^{\infty} I_n(g_n)$, we have

$$
E(\langle DG, DF \rangle_H) = \sum_{n=1}^{\infty} nE(J_n G J_n F) = -E(GLF).
$$

Therefore, $DF \in \operatorname{Dom} \delta$, and $\delta D F = -LF$. □

We are going to show that the operator $L$ behaves as a second-order differential operator when it acts on smooth random variables.

**Proposition 1.4.4** *It holds that $\mathcal{S} \subset \mathrm{Dom}\, L$, and for any $F \in \mathcal{S}$ of the form $F = f(W(h_1), \ldots, W(h_n))$, $f \in C_p^\infty(\mathbb{R}^n)$, we have*

$$
\begin{aligned}
LF \;=\; & \sum_{i,j=1}^{n} \partial_i \partial_j f(W(h_1), \ldots, W(h_n)) \langle h_i, h_j \rangle_H \\
& - \sum_{i=1}^{n} \partial_i f(W(h_1), \ldots, W(h_n)) W(h_i).
\end{aligned}
\tag{1.70}
$$

*Proof:*    We know that $F$ belongs to $\mathbb{D}^{1,2}$ and that

$$
DF = \sum_{i=1}^{n} \partial_i f(W(h_1), \ldots, W(h_n)) h_i.
$$

Consequently, $DF \in \mathcal{S}_H \subset \mathrm{Dom}\, \delta$ and by Eq. (1.44) we obtain

$$
\begin{aligned}
\delta DF \;=\; & \sum_{i=1}^{n} \partial_i f(W(h_1), \ldots, W(h_n)) W(h_i) \\
& - \sum_{i,j=1}^{n} \partial_i \partial_j f(W(h_1), \ldots, W(h_n)) \langle h_i, h_j \rangle_H.
\end{aligned}
$$

Now the result follows from Proposition 1.4.3.    □

More generally, we can prove the following result.

**Proposition 1.4.5** *Suppose that $F = (F^1, \ldots, F^m)$ is a random vector whose components belong to $\mathbb{D}^{2,4}$. Let $\varphi$ be a function in $C^2(\mathbb{R}^m)$ with bounded first and second partial derivatives. Then $\varphi(F) \in \mathrm{Dom}\, L$, and*

$$
L(\varphi(F)) = \sum_{i,j=1}^{m} \partial_i \partial_j \varphi(F) \langle DF^i, DF^j \rangle_H + \sum_{i=1}^{m} \partial_i \varphi(F) LF^i.
$$

*Proof:*    Approximate $F$ by smooth random variables in the norm $\| \cdot \|_{2,4}$, and $\varphi$ by functions in $C_p^\infty(\mathbb{R}^m)$, and use the continuity of the operator $L$ in the norm $\| \cdot \|_{2,2}$.    □

We can define on $\mathcal{S}$ the norm

$$
\|F\|_L = \left[ E(F^2) + E(|LF|^2) \right]^{\frac{1}{2}}.
$$

Notice that $\mathrm{Dom}\, L = \mathbb{D}^{2,2}$ and that the norms $\| \cdot \|_L$ and $\| \cdot \|_{2,2}$ coincide. In fact,

$$
\begin{aligned}
E(F^2) + E(|LF|^2) \;=\; & \sum_{n=0}^{\infty} (n^2 + 1) \|J_n F\|_2^2 \\
=\; & E(F^2) + E(\|DF\|_H^2) + E(\|D^2 F\|_{H \otimes H}^2).
\end{aligned}
$$

Similarly, the space $\mathbb{D}^{1,2}$ can be characterized as the domain in $L^2(\Omega)$ of the operator $C = -\sqrt{-L}$ defined by

$$CF = \sum_{n=0}^{\infty} -\sqrt{n} J_n F.$$

As in the case of the operator $L$, we can show that $C$ is the infinitesimal generator of a semigroup of operators (the *Cauchy semigroup*) given by

$$Q_t F = \sum_{n=0}^{\infty} e^{-\sqrt{n}t} J_n F.$$

Observe that $\mathrm{Dom}\, C = \mathbb{D}^{1,2}$, and for any $F \in \mathrm{Dom}\, C$ we have

$$E((CF)^2) = \sum_{n=1}^{\infty} n\|J_n F\|_2^2 = E(\|DF\|_H^2).$$

### 1.4.3 Hypercontractivity property and the multiplier theorem

We have seen that $T_t$ is a contraction operator on $L^p(\Omega)$ for any $p \geq 1$. Actually, these operators verify a *hypercontractivity property*, which is due to Nelson [235]. In the next theorem this property will be proved using Itô's formula, according to Neveu's approach (cf. [236]).

**Theorem 1.4.1** *Let $p > 1$ and $t > 0$, and set $q(t) = e^{2t}(p-1) + 1 > p$. Suppose that $F \in L^p(\Omega)$. Then*

$$\|T_t F\|_{q(t)} \leq \|F\|_p.$$

*Proof:*    Put $q = q(t)$, and let $q'$ be the conjugate of $q$. Taking into account the duality between $L^q(\Omega)$ and $L^{q'}(\Omega)$, it suffices to show that $|E((T_t F)G)| \leq \|F\|_p \|G\|_{q'}$ for any $F \in L^p(\Omega)$ and for any $G \in L^{q'}(\Omega)$. With the operator $T_t$ nonnegative (which implies $|T_t F| \leq T_t(|F|)$), we may assume that $F$ and $G$ are nonnegative. By an approximation argument it suffices to suppose that there exist real numbers $a \leq b$ such that $0 < a \leq F, G \leq b < \infty$. Also we may restrict our study to the case where $F = f(W(h_1), \ldots, W(h_n))$ and $G = g(W(h_1), \ldots, W(h_n))$ for some measurable functions $f, g$ such that $0 < a \leq f, g \leq b < \infty$ and orthonormal elements $h_1, \ldots, h_n \in H$.

Let $\{\beta_t, 0 \leq t \leq 1\}$ and $\{\xi_t, 0 \leq t \leq 1\}$ be two independent Brownian motions. Consider orthonormal functions $\phi_1, \ldots, \phi_n \in L^2([0,1])$. By (1.67) we can write

$$E((T_t F)G) = E\left( f\left( e^{-t} \int_0^1 \phi_1 d\beta + \sqrt{1 - e^{-2t}} \int_0^1 \phi_1 d\xi, \right.\right.$$

$$\left.\left. \ldots, e^{-t} \int_0^1 \phi_n d\beta + \sqrt{1 - e^{-2t}} \int_0^1 \phi_n d\xi \right) g\left( \int_0^1 \phi_1 d\beta, \ldots, \int_0^1 \phi_n d\beta \right) \right).$$

In this way we can reduce our problem to show the following inequality:

$$E(XY) \leq \|X\|_p \|Y\|_{q'},$$

where $0 < a \leq X, Y \leq b < \infty$, and $X, Y$ are random variables measurable with respect to the $\sigma$-fields generated by the Brownian motions $\eta_s = e^{-t}\beta_s + \sqrt{1 - e^{-2t}}\xi_s$ and $\beta_s$, respectively. These random variables will have integral representations of the following kind:

$$X^p = E(X^p) + \int_0^1 \varphi_s d\eta_s, \quad Y^{q'} = E(Y^{q'}) + \int_0^1 \psi_s d\beta_s.$$

Appling Itô's formula to the bounded positive martingales

$$M_s = E(X^p) + \int_0^s \varphi_u d\eta_u \quad \text{and} \quad N_s = E(Y^{q'}) + \int_0^s \psi_u d\beta_u,$$

and to the function $f(x, y) = x^\alpha y^\gamma$, $\alpha = \frac{1}{p}$, $\gamma = \frac{1}{q'}$, we obtain

$$
\begin{aligned}
XY &= \|X\|_p \|Y\|_{q'} \\
&+ \int_0^1 (\alpha M_s^{\alpha-1} N_s^\gamma dM_s + \gamma M_s^\alpha N_s^{\gamma-1} dN_s) + \int_0^1 \frac{1}{2} M_s^\alpha N_s^\gamma A_s ds,
\end{aligned}
$$

where

$$A_s = \alpha(\alpha - 1)M_s^{-2}\varphi_s^2 + \gamma(\gamma - 1)N_s^{-2}\psi_s^2 + 2\alpha\gamma M_s^{-1} N_s^{-1}\varphi_s \psi_s e^{-t}.$$

Taking expectations, we get

$$E(XY) = \|X\|_p \|Y\|_{q'} + \frac{1}{2}\int_0^1 E(M_s^\alpha N_s^\gamma A_s) ds.$$

Therefore, it suffices to show that $A_s \leq 0$. Note that $\alpha(\alpha - 1) = \frac{1}{p}(\frac{1}{p} - 1) < 0$. Thus, $A_s$ will be negative if

$$\alpha(\alpha - 1)\gamma(\gamma - 1) - (\alpha\gamma e^{-t})^2 \geq 0.$$

Finally,

$$(\alpha - 1)(\gamma - 1) - \gamma\alpha e^{-2t} = \frac{1}{pq}(p - 1 - (q - 1)e^{-2t}) = 0,$$

which achieves the proof.  $\square$

As a consequence of the hypercontractivity property it can be shown that for any $1 < p < q < \infty$ the norms $\|\cdot\|_p$ and $\|\cdot\|_q$ are equivalent on any Wiener chaos $\mathcal{H}_n$. In fact, let $t > 0$ such that $q = 1 + e^{2t}(p - 1)$. Then for every $F \in \mathcal{H}_n$ we have

$$e^{-nt}\|F\|_q = \|T_t F\|_q \leq \|F\|_p.$$

In addition, for each $n \geq 1$ the operator $J_n$ is bounded in $L^p(\Omega)$ for any $1 < p < \infty$, and

$$\|J_n F\|_p \leq \begin{cases} (p-1)^{\frac{n}{2}} \|F\|_p & \text{if } p > 2 \\ (p-1)^{-\frac{n}{2}} \|F\|_p & \text{if } p < 2. \end{cases}$$

In fact, suppose first that $p > 2$, and let $t > 0$ be such that $p - 1 = e^{2t}$. Using the hypercontractivity property with the exponents $p$ and $2$, we obtain

$$\|J_n F\|_p = e^{nt} \|T_t J_n F\|_p \leq e^{nt} \|J_n F\|_2 \leq e^{nt} \|F\|_2 \leq e^{nt} \|F\|_p. \qquad (1.71)$$

If $p < 2$, we use a duality argument:

$$\begin{aligned} \|J_n F\|_p &= \sup_{\|G\|_q \leq 1} E((J_n F)G) \\ &\leq \|F\|_p \sup_{\|G\|_q \leq 1} \|J_n G\|_q \leq e^{nt} \|F\|_p, \end{aligned}$$

where $q$ is the conjugate of $p$, and $q - 1 = e^{2t}$.

We are going to use the hypercontractivity property to show a multiplier theorem (see Meyer [225] and Watanabe [343]) that will be useful in proving Meyer's inequalities.

Recall that we denote by $\mathcal{P}$ the class of polynomial random variables. That means that a random variable $F$ belongs to $\mathcal{P}$ if it is of the form

$$F = p(W(h_1), \ldots, W(h_n)),$$

where $h_1, \ldots, h_n$ are elements of $H$ and $p$ is a polynomial of $n$ variables. The set $\mathcal{P}$ is dense in $L^p(\Omega)$ for all $p \geq 1$ (see Exercise 1.1.7).

Consider a sequence of real numbers $\{\phi(n), n \geq 0\}$ with $\phi(0) = 0$. This sequence determines a linear operator $T_\phi : \mathcal{P} \to \mathcal{P}$ defined by

$$T_\phi F = \sum_{n=0}^{\infty} \phi(n) J_n F, \qquad F \in \mathcal{P}.$$

We remark that the operators $T_t$, $Q_t$, $L$, and $C$ are of this type, the corresponding sequences being $e^{-nt}$, $e^{-\sqrt{n}t}$, $-n$, $-\sqrt{n}$, respectively. We are interested in the following question: For which sequences is the operator $T_\phi$ bounded in $L^p(\Omega)$ for $p > 1$? Theorem 1.4.2 will give an answer to this problem. The proof of the multiplier theorem is based on the following technical lemma.

**Lemma 1.4.1** *Let $p > 1$ and $F \in \mathcal{P}$. Then for any integer $N \geq 1$ there exists a constant $K$ (depending on $p$ and $N$) such that*

$$\|T_t(I - J_0 - J_1 - \cdots - J_{N-1})(F)\|_p \leq K e^{-Nt} \|F\|_p$$

*for all $t > 0$.*

*Proof:*     Assume first that $p > 2$. Choose $t_0$ such that $p = e^{2t_0} + 1$. Then, by Nelson's hypercontractivity theorem (Theorem 1.4.1) we have, for all $t \geq t_0$,

$$
\|T_{t_0} T_{t-t_0}(I - J_0 - J_1 - \cdots - J_{N-1})(F)\|_p^2
$$
$$
\leq \|T_{t-t_0}(I - J_0 - J_1 - \cdots - J_{N-1})(F)\|_2^2
$$
$$
= \| \sum_{n=N}^{\infty} e^{-n(t-t_0)} J_n F \|_2^2 = \sum_{n=N}^{\infty} e^{-2n(t-t_0)} \| J_n F \|_2^2
$$
$$
\leq e^{-2N(t-t_0)} \|F\|_2^2 \leq e^{-2N(t-t_0)} \|F\|_p^2,
$$

and this proves the desired inequality with $K = e^{Nt_0}$. For $t < t_0$, the inequality can be proved by the following direct argument, using (1.71):

$$
\|T_t(I - J_0 - J_1 - \cdots - J_{N-1})(F)\|_p
$$
$$
\leq \sum_{n=0}^{N-1} \|J_n F\|_p + \|F\|_p \leq \sum_{n=0}^{N-1} e^{nt_0} \|F\|_p + \|F\|_p
$$
$$
\leq \left( Ne^{2Nt_0} + e^{Nt_0} \right) e^{-Nt} \|F\|_p.
$$

For $p = 2$ the inequality is immediate, and for $1 < p < 2$ it can be obtained by duality (see Exercise 1.4.5).   □

The following is the multiplier theorem.

**Theorem 1.4.2** *Consider a sequence of real numbers $\{\phi(n), n \geq 0\}$ such that $\phi(0) = 0$ and $\phi(n) = \sum_{k=0}^{\infty} a_k n^{-k}$ for $n \geq N$ and for some $a_k \in \mathbb{R}$ such that $\sum_{k=0}^{\infty} |a_k| N^{-k} < \infty$. Then the operator*

$$
T_\phi(F) = \sum_{n=0}^{\infty} \phi(n) J_n F
$$

*is bounded in $L^p(\Omega)$ for any $1 < p < \infty$.*

Notice that the assumptions of this theorem are equivalent to saying that there exists a function $h(x)$ analytic near the origin such that $\phi(n) = h(n^{-1})$ for $n \geq N$.

*Proof:*     Define

$$
T_\phi = \sum_{n=0}^{N-1} \phi(n) J_n + \sum_{n=N}^{\infty} \phi(n) J_n = T_\phi^{(1)} + T_\phi^{(2)}.
$$

We know that $T_\phi^{(1)}$ is bounded in $L^p(\Omega)$ because the operators $J_n$ are bounded in $L^p(\Omega)$ for each fixed $n$. We have

$$\left\| T_\phi^{(2)} F \right\|_p = \left\| \sum_{n=N}^\infty \left( \sum_{k=0}^\infty a_k n^{-k} \right) J_n F \right\|_p$$

$$\leq \sum_{k=0}^\infty |a_k| \left\| \sum_{n=N}^\infty n^{-k} J_n F \right\|_p. \qquad (1.72)$$

Now, using the equality

$$n^{-k} = \left( \int_0^\infty e^{-nt} dt \right)^k = \int_{[0,\infty)^k} e^{-n(t_1 + \cdots + t_k)} dt_1 \cdots dt_k$$

we obtain

$$\sum_{n=N}^\infty n^{-k} J_n F = \int_{[0,\infty)^k} T_{t_1 + \cdots + t_k} (I - J_0 - \cdots - J_{N-1})(F) dt_1 \cdots dt_k.$$

Applying Lemma 1.4.1 yields

$$\left\| \sum_{n=N}^\infty n^{-k} J_n F \right\|_p$$

$$\leq \int_{[0,\infty)^k} \left\| T_{t_1 + \cdots + t_k} (I - J_0 - \cdots - J_{N-1})(F) \right\|_p dt_1 \cdots dt_k$$

$$\leq K \|F\|_p \int_{[0,\infty)^k} e^{-N(t_1 + \cdots + t_k)} dt_1 \cdots dt_k$$

$$= K N^{-k} \|F\|_p, \qquad (1.73)$$

where the constant $K$ depends only on $p$ and $N$. Substituting (1.73) into (1.72) we obtain

$$\left\| T_\phi^{(2)} F \right\|_p \leq K \sum_{k=0}^\infty |a_k| N^{-k} \|F\|_p,$$

which allows us to complete the proof. $\qquad \square$

For example, the operator $T_\phi = (I - L)^{-\alpha}$ defined by the sequence $\phi(n) = (1 + n)^{-\alpha}$, where $\alpha > 0$, is bounded in $L^p(\Omega)$, for $1 < p < \infty$, because $h(x) = \left( \frac{x}{x+1} \right)^\alpha$ is analytic in a neibourhood of the origin. Actually this operator is a contraction in $L^p(\Omega)$ for any $1 \leq p < \infty$ (see Exercise 1.4.8).

The following commutativity relationship holds for a multiplier operators $T_\phi$.

**Lemma 1.4.2** *Consider a sequence of real numbers $\{\phi(n), n \geq 0\}$ and the associated linear operator $T_\phi$ from $\mathcal{P}$ into $\mathcal{P}$. Define $T_{\phi+} = \sum_{n=0}^{\infty} \phi(n + 1)J_n$. Then for any $F \in \mathcal{P}$ it holds that*

$$DT_\phi(F) = T_{\phi+}D(F). \tag{1.74}$$

*Proof:*　Without loss of generality we can assume that $F$ belongs to the $n$th Wiener chaos $\mathcal{H}_n$, $n \geq 0$. In that case we have

$$DT_\phi(F) = D(\phi(n)F) = \phi(n)DF = T_{\phi+}D(F).$$

$\square$

## Exercises

**1.4.1** Let $W = \{W_t, t \geq 0\}$ be a standard Brownian motion. Check that the process

$$Y_t = \sqrt{\beta}e^{-\alpha t}W(e^{2\alpha t}), \qquad t \in \mathbb{R},$$

has the law of an Ornstein-Uhlenbeck process with parameters $\alpha, \beta$.

**1.4.2** Suppose that $(\Omega, \mathcal{F}, P)$ is the classical Wiener space (that is, $\Omega = C_0([0, 1])$ and $P$ is the Wiener measure). Let $\{T_t, t \geq 0\}$ be the Ornstein-Uhlenbeck semigroup given by

$$(T_tF)(u) = \int_\Omega F(e^{-t}u + \sqrt{1 - e^{-2t}}\omega)P(d\omega),$$

for all $F \in L^2(\Omega)$. Consider a Brownian measure $W$ on $[0, 1] \times \mathbb{R}_+$, defined on some probability space $(\widetilde{\Omega}, \widetilde{\mathcal{F}}, \widetilde{P})$, and with Lebesgue measure as control measure. Then $W(s, t) = W([0, s] \times [0, t])$, $(s, t) \in [0, 1] \times \mathbb{R}_+$, is a two-parameter Wiener process that possesses a continuous version. Define

$$X(t, \tau) = e^{-t}W(\tau, e^{2t}), \qquad t \in \mathbb{R}, \quad \tau \in [0, 1].$$

Compute the covariance function of $X$. Show that $X_t = X(t, \cdot)$ is a $\Omega$-valued stationary continuous Markov process on the probability space $(\widetilde{\Omega}, \widetilde{\mathcal{F}}, \widetilde{P})$ such that it admits $T_t$ as semigroup of operators.

*Hint:* Use the arguments of Proposition 1.4.1's proof to show that

$$\widetilde{E}\left(F(X_{s+t})|\widetilde{\mathcal{F}}_{e^{2s}}\right) = (T_tF)(X_s)$$

for all $t \geq 0$, $s \in \mathbb{R}$, $F \in L^2(\Omega)$, where $\widetilde{\mathcal{F}}_t$, $t \geq 0$, is the $\sigma$-field generated by the random variables $\{W(\tau, \sigma), 0 \leq \sigma \leq t, \tau \in [0, 1]\}$.

**1.4.3** For any $0 < \epsilon < 1$ put $F_{1-\epsilon} = \sum_{n=0}^{\infty}(1 - \epsilon)^n J_n F$ and $F^\epsilon = \frac{1}{\epsilon}[F_{1-\epsilon} - F]$. Show that $LF$ exists if and only if $F^\epsilon$ converges in $L^2(\Omega)$ as $\epsilon \downarrow 0$, and in this case $LF = \lim_{\epsilon \downarrow 0} F^\epsilon$.

**1.4.4** Set $F = \exp(W(h) - \frac{1}{2}\|h\|_H^2)$, $h \in H$. Show that $LF = -(W(h) - \|h\|_H^2)F$.

**1.4.5** Complete the proof of Lemma 1.4.1 in the case $1 < p < 2$, using a duality argument.

**1.4.6** Using the Gaussian formula (A.1), show that the multiplier theorem (Theorem 1.4.2) is still valid for Hilbert-valued random variables.

**1.4.7** Show that the operator $L$ is local in the domain $\text{Dom } L$. That is, $LF\mathbf{1}_{\{F=0\}} = 0$ for any random variable $F$ in $\text{Dom } L$.

**1.4.8** Show that the operator $(I - L)^{-\alpha}$ is a contraction in $L^p(\Omega)$ for any $1 \le p < \infty$, where $\alpha > 0$.
   *Hint*: Use the equation $(1+n)^{-\alpha} = \Gamma(\alpha)^{-1} \int_0^\infty e^{-(n+1)\alpha} t^{\alpha-1} dt$.

**1.4.9** Show that if $F \in \mathbb{D}^{1,2}$ and $G$ is a square integrable random variable such that $E(G) = 0$, then

$$E(FG) = E\left(\langle DF, DC^{-2}G\rangle_H\right).$$

## 1.5   Sobolev spaces and the equivalence of norms

In this section we establish Meyer's inequalities, following the method of Pisier [285]. Let $V$ be a Hilbert space. We recall that the spaces $\mathbb{D}^{k,p}(V)$, for any integer $k \ge 1$ and any real number $p \ge 1$ have been defined as the completion of the family of $V$-valued smooth random variables $\mathcal{S}_V$ with respect to the norm $\|\cdot\|_{k,p,V}$ defined in (1.37).
   Consider the intersection

$$\mathbb{D}^\infty(V) = \cap_{p\ge 1} \cap_{k\ge 1} \mathbb{D}^{k,p}(V).$$

Then $\mathbb{D}^\infty(V)$ is a complete, countably normed, metric space. We will write $\mathbb{D}^\infty(\mathbb{R}) = \mathbb{D}^\infty$. For every integer $k \ge 1$ and any real number $p \ge 1$ the operator $D$ is continuous from $\mathbb{D}^{k,p}(V)$ into $\mathbb{D}^{k-1,p}(H \otimes V)$. Consequently, $D$ is a continuous linear operator from $\mathbb{D}^\infty(V)$ into $\mathbb{D}^\infty(H \otimes V)$. Moreover, if $F$ and $G$ are random variables in $\mathbb{D}^\infty$, then the scalar product $\langle DF, DG\rangle_H$ is also in $\mathbb{D}^\infty$. The following result can be easily proved by approximating the components of the random vector $F$ by smooth random variables.

**Proposition 1.5.1** *Suppose that* $F = (F^1, \ldots, F^m)$ *is a random vector whose components belong to* $\mathbb{D}^\infty$. *Let* $\varphi \in C_p^\infty(\mathbb{R}^m)$. *Then* $\varphi(F) \in \mathbb{D}^\infty$, *and we have*

$$D(\varphi(F)) = \sum_{i=1}^m \partial_i\varphi(F)DF^i.$$

In particular, we deduce that $\mathbb{D}^\infty$ is an algebra. We will see later that $L$ is a continuous operator from $\mathbb{D}^\infty$ into $\mathbb{D}^\infty$ and that the operator $\delta$ is continuous from $\mathbb{D}^\infty(H)$ into $\mathbb{D}^\infty$. To show these results we will need Meyer's inequalities, which provide the equivalence between the $p$ norm of $CF$ and that of $\|DF\|_H$ for $p > 1$ (we recall that $C$ is the operator defined by $C = -\sqrt{-L}$). This equivalence of norms will follow from the fact that the operator $DC^{-1}$ is bounded in $L^p(\Omega)$ for any $p > 1$, and this property will be proved using the approach by Pisier [285] based on the boundedness in $L^p$ of the Hilbert transform. We recall that the Hilbert transform of a function $f \in C_0^\infty(\mathbb{R})$ is defined by

$$Hf(x) = \int_{\mathbb{R}} \frac{f(x+t) - f(x-t)}{t} dt.$$

The transformation $H$ is bounded in $L^p(\mathbb{R})$ for any $p > 1$ (see Dunford and Schwarz [87], Theorem XI.7.8).

Consider the function $\varphi : [-\frac{\pi}{2}, 0) \cup (0, \frac{\pi}{2}] \to \mathbb{R}_+$ defined by

$$\varphi(\theta) = \frac{1}{\sqrt{2}} |\pi \log \cos^2 \theta|^{-\frac{1}{2}} \operatorname{sign} \theta. \tag{1.75}$$

Notice that when $\theta$ is close to zero this function tends to infinity as $\frac{1}{\sqrt{2\pi\theta}}$. Suppose that $\{W'(h), h \in H\}$ is an independent copy of the Gaussian process $\{W(h), h \in H\}$. We will assume as in Section 1.4 that $W$ and $W'$ are defined in the product probability space $(\Omega \times \Omega', \mathcal{F} \otimes \mathcal{F}', P \times P')$. For any $\theta \in \mathbb{R}$ we consider the process $W_\theta = \{W_\theta(h), h \in H\}$ defined by

$$W_\theta(h) = W(h) \cos \theta + W'(h) \sin \theta, \qquad h \in H.$$

This process is Gaussian, with zero mean and with the same covariance function as $\{W(h), h \in H\}$. Let $W : \Omega \to \mathbb{R}^H$ and $W' : \Omega' \to \mathbb{R}^H$ be the canonical mappings associated with the processes $\{W(h), h \in H\}$ and $\{W'(h), h \in H\}$, respectively. Given a random variable $F \in L^0(\Omega, \mathcal{F}, P)$, we can write $F = \psi_F \circ W$, where $\psi_F$ is a measurable mapping from $\mathbb{R}^H$ to $\mathbb{R}$, determined $P \circ W^{-1}$ a.s. As a consequence, the random variable $\psi_F(W_\theta) = \psi_F(W \cos \theta + W' \sin \theta)$ is well defined $P \times P'$ a.s. We set

$$R_\theta F = \psi_F(W_\theta). \tag{1.76}$$

We denote by $E'$ the mathematical expectation with respect to the probability $P'$, and by $D'$ the derivative operator with respect to the Gaussian process $W'(h)$. With these notations we can write the following expression for the operator $D(-C)^{-1}$.

**Lemma 1.5.1** *For every $F \in \mathcal{P}$ such that $E(F) = 0$ we have*

$$D(-C)^{-1}F = \int_{-\frac{\pi}{2}}^{\frac{\pi}{2}} E'(D'(R_\theta F))\varphi(\theta)d\theta. \tag{1.77}$$

*Proof:*    Suppose that $F = p(W(h_1), \ldots, W(h_n))$, where $h_1, \ldots, h_n \in H$ and $p$ is a polynomial in $n$ variables. We have

$$R_\theta F = p(W(h_1)\cos\theta + W'(h_1)\sin\theta, \ldots, W(h_n)\cos\theta + W'(h_n)\sin\theta),$$

and therefore

$$D'(R_\theta F) = \sum_{i=1}^n \partial_i p(W(h_1)\cos\theta + W'(h_1)\sin\theta,$$
$$\ldots, W(h_n)\cos\theta + W'(h_n)\sin\theta)\sin\theta h_i(s) = \sin\theta R_\theta(DF).$$

Consequently, using Mehler's formula (1.67) we obtain

$$E'(D'(R_\theta F)) = \sin\theta E'(R_\theta(DF)) = \sin\theta T_t(DF),$$

where $t > 0$ is such that $\cos\theta = e^{-t}$. This implies

$$E'(D'(R_\theta F)) = \sum_{n=0}^\infty \sin\theta(\cos\theta)^n J_n DF.$$

Note that since $F$ is a polynomial random variable the above series is actually the sum of a finite number of terms. By Exercise 1.5.3, the right-hand side of (1.77) can be written as

$$\sum_{n=0}^\infty \left( \int_{-\frac{\pi}{2}}^{\frac{\pi}{2}} \sin\theta(\cos\theta)^n \varphi(\theta)d\theta \right) J_n DF = \sum_{n=0}^\infty \frac{1}{\sqrt{n+1}} J_n DF.$$

Finally, applying the commutativity relationship (1.74) to the multiplication operator defined by the sequence $\phi(n) = \frac{1}{\sqrt{n}}$, $n \geq 1$, $\phi(0) = 0$, we get

$$T_{\phi^+}DF = DT_\phi F = D(-C)^{-1}F,$$

and the proof of the lemma is complete.    □

   Now with the help of the preceding equation we can show that the operator $DC^{-1}$ is bounded from $L^p(\Omega)$ into $L^p(\Omega; H)$ for any $p > 1$. Henceforth $c_p$ and $C_p$ denote generic constants depending only on $p$, which can be different from one formula to another.

**Proposition 1.5.2** *Let $p > 1$. There exists a finite constant $c_p > 0$ such that for any $F \in \mathcal{P}$ with $E(F) = 0$ we have*

$$\|DC^{-1}F\|_p \leq c_p\|F\|_p.$$

*Proof:*   Using (1.77) we can write

$$E\left(\|DC^{-1}F\|_H^p\right)$$

$$= E\left(\left\|\int_{-\frac{\pi}{2}}^{\frac{\pi}{2}} E'(D'(R_\theta F))\varphi(\theta)d\theta\right\|_H^p\right)$$

$$= \alpha_p^{-1}EE'\left(\left|W'\left(\int_{-\frac{\pi}{2}}^{\frac{\pi}{2}} E'(D'(R_\theta F))\varphi(\theta)d\theta\right)\right|^p\right),$$

where $\alpha_p = E(|\xi|^p)$ with $\xi$ an $N(0,1)$ random variable. We recall that by Exercise 1.2.6 (Stroock's formula) for any $G \in L^2(\Omega', \mathcal{F}', P')$ the Gaussian random variable

$$W'(E'(D'G))$$

is equal to the projection $J_1'G$ of $G$ on the first Wiener chaos. Therefore, we obtain that

$$E\left(\|DC^{-1}F\|_H^p\right)$$

$$= \alpha_p^{-1}EE'\left(\left|\int_{-\frac{\pi}{2}}^{\frac{\pi}{2}} J_1'R_\theta F\varphi(\theta)d\theta\right|^p\right)$$

$$= \alpha_p^{-1}EE'\left(\left|J_1'\left(\text{p.v.}\int_{-\frac{\pi}{2}}^{\frac{\pi}{2}} R_\theta F\varphi(\theta)d\theta\right)\right|^p\right)$$

$$\leq c_p EE'\left(\left|\text{p.v.}\int_{-\frac{\pi}{2}}^{\frac{\pi}{2}} R_\theta F\varphi(\theta)d\theta\right|^p\right),$$

for some constant $c_p > 0$ (where the abbreviation p.v. stands for principal value). Notice that the function $R_\theta F\varphi(\theta)$ might not belong to $L^1(-\frac{\pi}{2}, \frac{\pi}{2})$ because, unlike the term $J_1'R_\theta F$, the function $R_\theta F$ may not balance the singularity of $\varphi(\theta)$ at the origin. For this reason we have to introduce the principal value integral

$$\text{p.v.}\int_{-\frac{\pi}{2}}^{\frac{\pi}{2}} R_\theta F\varphi(\theta)d\theta = \lim_{\epsilon \downarrow 0}\int_{\epsilon \leq |\theta| \leq \frac{\pi}{2}} R_\theta F\varphi(\theta)d\theta,$$

which can be expressed as a convergent integral in the following way:

$$\int_0^{\frac{\pi}{2}} [R_\theta F\varphi(\theta) + R_{-\theta}F\varphi(-\theta)]d\theta = \int_0^{\frac{\pi}{2}} \frac{[R_\theta F - R_{-\theta}F]}{\sqrt{2\pi|\log \cos^2 \theta|}}d\theta.$$

For any $\xi \in \mathbb{R}$ we define the process

$$\overline{R}_\xi(h) = (W(h)\cos\xi + W'(h)\sin\xi, -W(h)\sin\xi + W'(h)\cos\xi).$$

The law of this process is the same as that of $\{(W(h), W'(h)), h \in H\}$. On the other hand, $\overline{R}_\xi R_\theta F = R_{\xi+\theta} F$, where we set

$$\overline{R}_\xi G((W(h_1), W'(h_1)), \dots, (W(h_n), W'(h_n))) = G(\overline{R}_\xi(h_1), \dots, \overline{R}_\xi(h_n)).$$

Therefore, we get

$$\left\| \text{p.v.} \int_{-\frac{\pi}{2}}^{\frac{\pi}{2}} R_\theta F \varphi(\theta) d\theta \right\|_p = \left\| \overline{R}_\xi \left[ \text{p.v.} \int_{-\frac{\pi}{2}}^{\frac{\pi}{2}} R_\theta F \varphi(\theta) d\theta \right] \right\|_p$$

$$= \left\| \text{p.v.} \int_{-\frac{\pi}{2}}^{\frac{\pi}{2}} R_{\xi+\theta} F \varphi(\theta) d\theta \right\|_p ,$$

where $\| \cdot \|_p$ denotes the $L^p$ norm with respect to $P \times P'$. Integration with respect to $\xi$ yields

$$E\left( \|DC^{-1}F\|_H^p \right) \le c_p E E' \left( \int_{-\frac{\pi}{2}}^{\frac{\pi}{2}} \left| \text{p.v.} \int_{-\frac{\pi}{2}}^{\frac{\pi}{2}} R_{\xi+\theta} F \varphi(\theta) d\theta \right|^p d\xi \right). \quad (1.78)$$

Furthermore, there exists a bounded continuous function $\widetilde{\varphi}$ and a constant $c > 0$ such that

$$\varphi(\theta) = \widetilde{\varphi}(\theta) + \frac{c}{\theta},$$

on $[-\frac{\pi}{2}, \frac{\pi}{2}]$. Consequently, using the $L^p$ boundedness of the Hilbert transform, we see that the right-hand side of (1.78) is dominated up to a constant by

$$E E' \left( \int_{-\frac{\pi}{2}}^{\frac{\pi}{2}} |R_\theta F|^p d\theta \right) = \pi \|F\|_p^p.$$

In fact, the term $\widetilde{\varphi}(\theta)$ is easy to treat. On the other hand, to handle the term $\frac{1}{\theta}$ it suffices to write

$$\int_{-\frac{\pi}{2}}^{\frac{\pi}{2}} \left| \int_{-\frac{\pi}{2}}^{\frac{\pi}{2}} \frac{R_{\xi+\theta}F - R_{\xi-\theta}F}{\theta} d\theta \right|^p d\xi$$

$$\le c_p \left( \int_{\mathbb{R}} \left| \int_{\mathbb{R}} \frac{\widetilde{R}_{\xi+\theta}F - \widetilde{R}_{\xi-\theta}F}{\theta} d\theta \right|^p d\xi \right.$$

$$+ \int_{-\frac{\pi}{2}}^{\frac{\pi}{2}} \int_{[-2\pi, -\frac{\pi}{2}] \cup [\frac{\pi}{2}, 2\pi]} \left| \frac{R_{\xi+\theta}F}{\theta} \right|^p d\theta d\xi \right)$$

$$\le c_p' \left( \int_{\mathbb{R}} |\widetilde{R}_\theta F|^p d\theta + \int_{-\frac{\pi}{2}}^{\frac{\pi}{2}} \int_{-2\pi}^{2\pi} |R_{\xi+\theta}F|^p d\theta d\xi \right),$$

where $\widetilde{R}_\theta F = \mathbf{1}_{[-\frac{3\pi}{2}, \frac{3\pi}{2}]}(\theta) R_\theta F$.     $\square$

**Proposition 1.5.3** *Let $p > 1$. Then there exist positive and finite constants $c_p$ and $C_p$ such that for any $F \in \mathcal{P}$ we have*

$$c_p \|DF\|_{L^p(\Omega;H)} \leq \|CF\|_p \leq C_p \|DF\|_{L^p(\Omega;H)}. \tag{1.79}$$

*Proof:*    We can assume that the random variable $F$ has zero expectation. Set $G = CF$. Then, using Proposition 1.5.2, we have

$$\|DF\|_{L^p(\Omega;H)} = \|DC^{-1}G\|_{L^p(\Omega;H)} \leq c_p \|G\|_p = c_p \|CF\|_p,$$

which shows the left inequality. We will prove the right inequality using a duality argument. Let $F, G \in \mathcal{P}$. Set $\widetilde{G} = C^{-1}(I - J_0)(G)$, and denote the conjugate of $p$ by $q$. Then we have

$$
\begin{aligned}
|E(GCF)| &= |E((I - J_0)(G)CF)| = |E(CFC\widetilde{G})| = |E(\langle DF, D\widetilde{G} \rangle_H)| \\
&\leq \|DF\|_{L^p(\Omega;H)} \|D\widetilde{G}\|_{L^q(\Omega;H)} \leq c_q \|DF\|_{L^p(\Omega;H)} \|C\widetilde{G}\|_q \\
&= c_q \|DF\|_{L^p(\Omega;H)} \|(I - J_0)(G)\|_q \leq c_q' \|DF\|_{L^p(\Omega;H)} \|G\|_q.
\end{aligned}
$$

Taking the supremum with respect to $G \in \mathcal{P}$ with $\|G\|_q \leq 1$, we obtain

$$\|CF\|_p \leq c_q' \|DF\|_{L^p(\Omega;H)}.$$

$\square$

Now we can state Meyer's inequalities in the general case.

**Theorem 1.5.1** *For any $p > 1$ and any integer $k \geq 1$ there exist positive and finite constants $c_{p,k}$ and $C_{p,k}$ such that for any $F \in \mathcal{P}$,*

$$
\begin{aligned}
c_{p,k} E\left( \|D^k F\|_{H^{\otimes k}}^p \right) &\leq E\left( |C^k F|^p \right) \\
&\leq C_{p,k} \left[ E\left( \|D^k F\|_{H^{\otimes k}}^p \right) + E(|F|^p) \right]. \tag{1.80}
\end{aligned}
$$

*Proof:*    The proof will be done by induction on $k$. The case $k = 1$ is included in Proposition 1.5.3. Suppose that the left-hand side of (1.80) holds for $1, \dots, k$. Consider two families of independent random variables, with the identical distribution $N(0,1)$, defined in the probability space $([0,1], \mathcal{B}([0,1]), \lambda)$ ($\lambda$ is the Lebesgue measure) $\{\gamma_\alpha(s), s \in [0,1], \alpha \in \mathbb{N}_*^k\}$, where $\mathbb{N}_* = \{1, 2, \dots\}$ and $\{\gamma_i(s), s \in [0,1], i \geq 1\}$. Suppose that $F = p(W(h_1), \dots, W(h_n))$, where the $h_i$'s are orthonormal elements of $H$. We fix a complete orthonormal system $\{e_i, i \geq 1\}$ in $H$ which contains the $h_i$'s. We set $D_i(F) = \langle DF, e_i \rangle_H$ and $D_\alpha^k(F) = D_{\alpha_1} D_{\alpha_2} \cdots D_{\alpha_k}(F)$ for any multiindex $\alpha = (\alpha_1, \dots, \alpha_k)$. With these notations, using the Gaussian formula (A.1) and Proposition 1.5.3, we can write

$$
E\left(\left\|D^{k+1}F\right\|_{H^{\otimes(k+1)}}^{p}\right)
$$

$$
= E\left(\left|\sum_{i=1}^{\infty}\sum_{\alpha\in\mathbb{N}_{*}^{k}}\left(D_{i}D_{\alpha}^{k}F\right)^{2}\right|^{\frac{p}{2}}\right)
$$

$$
= A_{p}^{-1}\int_{0}^{1}\int_{0}^{1}E\left(\left|\sum_{i=1}^{\infty}\sum_{\alpha\in\mathbb{N}_{*}^{k}}D_{i}D_{\alpha}^{k}F\gamma_{\alpha}(t)\gamma_{i}(s)\right|^{p}\right)dtds
$$

$$
\leq \int_{0}^{1}E\left(\left|\sum_{i=1}^{\infty}\left[D_{i}\left(\sum_{\alpha\in\mathbb{N}_{*}^{k}}D_{\alpha}^{k}F\gamma_{\alpha}(t)\right)\right]^{2}\right|^{\frac{p}{2}}\right)dt
$$

$$
\leq c_{p}\int_{0}^{1}E\left(\left|C\left(\sum_{\alpha\in\mathbb{N}_{*}^{N}}D_{\alpha}^{k}F\gamma_{\alpha}(t)\right)\right|^{p}\right)dt
$$

$$
\leq c_{p}'E\left(\left|\sum_{\alpha\in\mathbb{N}_{*}^{k}}\left(CD_{\alpha}^{k}F\right)^{2}\right|^{\frac{p}{2}}\right).
$$

Consider the operator

$$
R_{k}(F)=\sum_{n=k}^{\infty}\sqrt{1-\frac{k}{n}}J_{n}F,\quad F\in\mathcal{P}.
$$

By Theorem 1.4.2 this operator is bounded in $L^{p}(\Omega)$, and using the induction hypothesis we can write

$$
E\left(\left|\sum_{\alpha\in\mathbb{N}_{*}^{k}}\left(CD_{\alpha}^{k}F\right)^{2}\right|^{\frac{p}{2}}\right)
= E\left(\left|\sum_{\alpha\in\mathbb{N}_{*}^{k}}\left(D_{\alpha}^{k}CR_{k}F\right)^{2}\right|^{\frac{p}{2}}\right)
$$

$$
= E\left(\left\|D^{k}CR_{k}F\right\|_{H^{\otimes k}}^{p}\right)
$$

$$
\leq c_{p,k}E\left(\left|C^{k+1}R_{k}F\right|^{p}\right)
$$

$$
\leq c_{p,k}E\left(\left|C^{k+1}F\right|^{p}\right)
$$

for some constant $c_{p,k}>0$.

We will prove by induction the right inequality in (1.80) for $F\in\mathcal{P}$ satisfying $(J_{0}+J_{1}+\cdots+J_{k-1})(F)=0$. The general case would follow easily (Exercise 1.5.1). Suppose that this holds for $k$. Applying Proposition

1.5.3 and the Gaussian formula (A.1), we have

$$
E\left(\left|C^{k+1}F\right|^{p}\right) \leq c_{p}E\left(\|DC^{k}F\|_{H}^{p}\right) = c_{p}E\left(\left|\sum_{i=1}^{\infty}\left(D_{i}C^{k}F\right)^{2}\right|^{\frac{p}{2}}\right)
$$

$$
= A_{p}^{-1}c_{p}\int_{0}^{1}E\left(\left|\sum_{i=1}^{\infty}\left(D_{i}C^{k}F\right)\gamma_{i}(s)\right|^{p}\right)ds.
$$

Consider the operator defined by

$$
R_{k,1} = \sum_{n=2}^{\infty}\left(\sqrt{\frac{n}{n-1}}\right)^{k}J_{n}.
$$

Using the commutativity relationship (1.74), our induction hypothesis, and the Gaussian formula (A.1), we can write

$$
\int_{0}^{1}E\left(\left|\sum_{i=1}^{\infty}\left(D_{i}C^{k}F\right)\gamma_{i}(s)\right|^{p}\right)ds
$$

$$
= \int_{0}^{1}E\left(\left|\sum_{i=1}^{\infty}\left(C^{k}D_{i}R_{k,1}F\right)\gamma_{i}(s)\right|^{p}\right)ds
$$

$$
\leq C_{p,k}\int_{0}^{1}E\left(\left\|D^{k}\left(\sum_{i=1}^{\infty}\left(D_{i}R_{k,1}F\right)\gamma_{i}(s)\right)\right\|_{H^{\otimes k}}^{p}\right)ds
$$

$$
= C_{p,k}\int_{0}^{1}E\left(\left|\sum_{\alpha\in\mathbb{N}_{*}^{k}}\left(\sum_{i=1}^{\infty}\left(D_{\alpha}^{k}D_{i}R_{k,1}F\right)\gamma_{i}(s)\right)^{2}\right|^{\frac{p}{2}}\right)ds
$$

$$
= C_{p,k}A_{p}^{-1}\int_{0}^{1}\int_{0}^{1}E\left(\left|\sum_{\alpha\in\mathbb{N}_{*}^{k}}\sum_{i=1}^{\infty}\left(D_{\alpha}^{k}D_{i}R_{N,1}F\right)\gamma_{i}(s)\gamma_{\alpha}(t)\right|^{p}\right)dsdt.
$$

Finally, if we introduce the operator

$$
R_{k,2} = \sum_{n=0}^{\infty}\left(\frac{n+1+k}{n+k}\right)^{\frac{k}{2}}J_{n},
$$

we obtain, by applying the commutativity relationship, the Gaussian formula (A.1), and the boundedness in $L^p(\Omega)$ of the operator $R_{k,2}$, that

$$\int_0^1 \int_0^1 E\left(\left|\sum_{\alpha\in\mathbb{N}_*^k}\sum_{i=1}^\infty (D_\alpha^k D_i R_{k,1} F)\,\gamma_i(s)\gamma_\alpha(t)\right|^p\right) ds\,dt$$

$$= \int_0^1 \int_0^1 E\left(\left|\sum_{\alpha\in\mathbb{N}_*^k}\sum_{i=1}^\infty (R_{k,2} D_\alpha^k D_i F)\,\gamma_i(s)\gamma_\alpha(t)\right|^p\right) ds\,dt$$

$$\leq C_{p,k} \int_0^1 \int_0^1 E\left(\left|\sum_{\alpha\in\mathbb{N}_*^k}\sum_{i=1}^\infty (D_\alpha^k D_i F)\,\gamma_i(s)\gamma_\alpha(t)\right|^p\right) ds\,dt$$

$$= C_{p,k} A_p E\left(\left|\sum_{\alpha\in\mathbb{N}_*^k}\sum_{i=1}^\infty (D_\alpha^k D_i F)^2\right|^{\frac{p}{2}}\right)$$

$$= C_{p,k} A_p E\left(\|D^{k+1}F\|_{H^{\otimes(k+1)}}^p\right),$$

which completes the proof of the theorem. $\qquad\square$

The inequalities (1.80) also hold for polynomial random variables taking values in a separable Hilbert space (see Execise 1.5.5). One of the main applications of Meyer's inequalities is the following result on the continuity of the operator $\delta$. Here we consider $\delta$ as the adjoint of the derivative operator $D$ on $L^p(\Omega)$.

**Proposition 1.5.4** *The operator $\delta$ is continuous from $\mathbb{D}^{1,p}(H)$ into $L^p(\Omega)$ for all $p > 1$.*

*Proof:*    Let $q$ be the conjugate of $p$. For any $u$ in $\mathbb{D}^{1,p}(H)$ and any polynomial random variable $G$ with $E(G) = 0$ we have

$$E(\delta(u)G) = E(\langle u, DG\rangle_H) = E(\langle \tilde{u}, DG\rangle_H) + E(\langle E(u), DG\rangle_H),$$

where $\tilde{u} = u - E(u)$. Notice that the second summand in the above expression can be bounded by a constant times $\|u\|_{L^p(\Omega;H)}\|G\|_q$. So we can assume $E(u) = E(DG) = 0$. Then we have, using Exercise 1.4.9

$$\begin{aligned}|E(\delta(u)G)| &= |E(\langle u, DG\rangle_H)| = |E(\langle Du, DC^{-2}DG\rangle_{H\otimes H})|\\ &\leq \|Du\|_{L^p(\Omega;H\otimes H)}\|DC^{-2}DG\|_{L^q(\Omega;H\otimes H)}\\ &\leq c_p\|Du\|_{L^p(\Omega;H\otimes H)}\|D^2 C^{-2} RG\|_{L^q(\Omega;H\otimes H)}\\ &\leq c_p'\|Du\|_{L^p(\Omega;H\otimes H)}\|G\|_q,\end{aligned}$$

where

$$R = \sum_{n=2}^\infty \frac{n}{n-1} J_n,$$

and we have used Meyer's inequality and the boundedness in $L^q(\Omega)$ of the operator $R$. So, we have proved that $\delta$ is continuous from $\mathbb{D}^{1,p}(H)$ into $L^p(\Omega)$.    $\square$

Consider the set $\mathcal{P}_H$ of $H$-valued polynomial random variables. We have the following result:

**Lemma 1.5.2** *For any process $u \in \mathcal{P}_H$ and for any $p > 1$, we have*

$$\|C^{-1}\delta(u)\|_p \le c_p \|u\|_{L^p(\Omega;H)}.$$

*Proof:*    Let $G \in \mathcal{P}$ with $E(G) = 0$ and $u \in \mathcal{P}_H$. Using Proposition 1.5.3 we can write

$$
\begin{aligned}
|E(C^{-1}\delta(u)\, G)| &= |E(\langle u, DC^{-1}G \rangle_H)| \\
&\le \|u\|_{L^p(\Omega;H)} \|DC^{-1}G\|_{L^q(\Omega;H)} \\
&\le c_p \|u\|_{L^p(\Omega;H)} \|G\|_{L^q(\Omega)},
\end{aligned}
$$

where $q$ is the conjugate of $p$. This yields the desired estimation.    $\square$

As a consequence, the operator $D(-L)^{-1}\delta$ is bounded from $L^p(\Omega; H)$ into $L^p(\Omega; H)$. In fact, we can write

$$D(-L)^{-1}\delta = [DC^{-1}][C^{-1}\delta].$$

Using Lemma 1.5.2 we can show the following result:

**Proposition 1.5.5** *Let $F$ be a random variable in $\mathbb{D}^{k,\alpha}$ with $\alpha > 1$. Suppose that $D^i F$ belongs to $L^p(\Omega; H^{\otimes i})$ for $i = 0, 1, \ldots, k$ and for some $p > \alpha$. Then $F \in \mathbb{D}^{k,p}$, and there exists a sequence $G_n \in \mathcal{P}$ that converges to $F$ in the norm $\|\cdot\|_{k,p}$.*

*Proof:*    We will prove the result only for $k = 1$; a similar argument can be used for $k > 1$. We may assume that $E(F) = 0$. We know that $\mathcal{P}_H$ is dense in $L^p(\Omega; H)$. Hence, we can find a sequence of $H$-valued polynomial random variables $\eta_n$ that converges to $DF$ in $L^p(\Omega; H)$. Without loss of generality we may assume that $J_k \eta_n \in \mathcal{P}_H$ for all $k \ge 1$. Note that $-L^{-1}\delta D = (I - J_0)$ on $\mathbb{D}^{1,\alpha}$. Consider the decomposition $\eta_n = DG_n + u_n$ given by Proposition 1.3.10. Notice that $G_n \in \mathcal{P}$ because $G_n = -L^{-1}\delta(\eta_n)$ and $\delta(u_n) = 0$. Using the boundedness in $L^p$ of the operator $C^{-1}\delta$ (which implies that of $L^{-1}\delta$ by Exercise 1.4.8), we obtain that $F - G_n = L^{-1}\delta(\eta_n - DF)$ converges to zero in $L^p(\Omega)$ as $n$ tends to infinity. On the other hand,

$$\|DF - DG_n\|_{L^p(\Omega;H)} = \|DL^{-1}\delta(\eta_n - DF)\|_{L^p(\Omega;H)} \le c_p \|\eta_n - DF\|_{L^p(\Omega;H)};$$

hence, $\|DG_n - DF\|_H$ converges to zero in $L^p(\Omega)$ as $n$ tends to infinity. So the proof of the proposition is complete.    $\square$

**Corollary 1.5.1** *The class $\mathcal{P}$ is dense in $\mathbb{D}^{k,p}$ for all $p > 1$ and $k \ge 1$.*

As a consequence of the above corollary, Theorem 1.5.1 holds for random variables in $\mathbb{D}^{k,p}$, and the operator $(-C)^k = (-L)^{\frac{k}{2}}$ is continuous from $\mathbb{D}^{k,p}$ into $L^p$. Thus, $L$ is a continuous operator on $\mathbb{D}^\infty$.

The following proposition is a Hölder inequality for the $\|\cdot\|_{k,p}$ norms.

**Proposition 1.5.6** *Let $F \in \mathbb{D}^{k,p}$, $G \in \mathbb{D}^{k,q}$ for $k \in \mathbb{N}^*$, $1 < p, q < \infty$ and let $r$ be such that $\frac{1}{p} + \frac{1}{q} = \frac{1}{r}$. Then, $FG \in \mathbb{D}^{k,r}$ and*

$$\|FG\|_{k,r} \le c_{p,q,k} \|F\|_{k,p} \|G\|_{k,q}.$$

*Proof:* Suppose that $F, G \in \mathcal{P}$. By Leipnitz rule (see Exercise 1.2.13) we can write

$$D^k(FG) = \sum_{i=0}^{k} \binom{k}{i} \|D^i F\|_{H^{\otimes i}} \|D^{k-i}G\|_{H^{\otimes(k-i)}}.$$

Hence, by Hölder's inequality

$$
\begin{aligned}
\|FG\|_{k,r} &\le \sum_{j=0}^{k} \sum_{i=0}^{j} \binom{j}{i} \left\| \|D^i F\|_{H^{\otimes i}} \right\|_p \left\| \|D^{j-i}G\|_{H^{\otimes(j-i)}} \right\|_q \\
&\le c_{p,q,k} \|F\|_{k,p} \|G\|_{k,q}.
\end{aligned}
$$

$\square$

We will now introduce the continuous family of Sobolev spaces defined by Watanabe (see [343]). For any $p > 1$ and $s \in \mathbb{R}$ we will denote by $\|\|\cdot\|\|_{s,p}$ the seminorm

$$\|\|F\|\|_{s,p} = \left\| (I - L)^{\frac{s}{2}} F \right\|_p,$$

where $F$ is a polynomial random variable. Note that $(I-L)^{\frac{s}{2}} F = \sum_{n=0}^{\infty} (1+n)^{\frac{s}{2}} J_n F$.

These seminorms have the following properties:

(i) $\|\|F\|\|_{s,p}$ is increasing in both coordinates $s$ and $p$. The monotonicity in $p$ is clear and in $s$ follows from the fact that the operators $(I - L)^{\frac{s}{2}}$ are contractions in $L^p$ for all $s < 0$, $p > 1$ (see Exercise 1.4.8).

(ii) The seminorms $\|\|\cdot\|\|_{s,p}$ are compatible, in the sense that for any sequence $F_n$ in $\mathcal{P}$ converging to zero in the norm $\|\|\cdot\|\|_{s,p}$, and being a Cauchy sequence in another norm $\|\|\cdot\|\|_{s',p'}$, it also converges to zero in the norm $\|\|\cdot\|\|_{s',p'}$.

For any $p > 1$, $s \in \mathbb{R}$, we define $\mathbb{D}^{s,p}$ as the completion of $\mathcal{P}$ with respect to the norm $\|\|\cdot\|\|_{s,p}$.

**Remarks:**

**1.** $|||F|||_{0,p} = ||F||_{0,p} = ||F||_p$, and $\mathbb{D}^{0,p} = L^p(\Omega)$. For $k = 1, 2, \ldots$ the seminorms $|||\cdot|||_{k,p}$ and $||\cdot||_{k,p}$ are equivalent due to Meyer's inequalities. In fact, we have

$$|||F|||_{k,p} = ||(I-L)^{\frac{k}{2}} F||_p \leq |E(F)| + \left\| R(-L)^{\frac{k}{2}} F \right\|_p,$$

where $R = \sum_{n=1}^{\infty} \left(\frac{n+1}{n}\right)^{\frac{k}{2}} J_n$. By Theorem 1.4.2 this operator is bounded in $L^p(\Omega)$ for all $p > 1$. Hence, applying Theorem 1.5.1 we obtain

$$
\begin{aligned}
|||F|||_{k,p} &\leq c_{k,p} \left( ||F||_p + \left\| (-L)^{\frac{k}{2}} F \right\|_p \right) \\
&\leq c'_{k,p} \left( ||F||_p + \left\| D^k F \right\|_{L^p(\Omega; H^{\otimes k})} \right) \leq c''_{k,p} ||F||_{k,p} .
\end{aligned}
$$

In a similar way one can show the converse inequality (Exercise 1.5.9). Thus, by Corollary 1.5.1 the spaces $\mathbb{D}^{k,p}$ coincide with those defined using the derivative operator.

**2.** From properties (i) and (ii) we have $\mathbb{D}^{s,p} \subset \mathbb{D}^{s',p'}$ if $p' \leq p$ and $s' \leq s$.

**3.** For $s > 0$ the operator $(I-L)^{-\frac{s}{2}}$ is an isometric isomorphism (in the norm $|||\cdot|||_{s,p}$) between $L^p(\Omega)$ and $\mathbb{D}^{s,p}$ and between $\mathbb{D}^{-s,p}$ and $L^p(\Omega)$ for all $p > 1$. As a consequence, the dual of $\mathbb{D}^{s,p}$ is $\mathbb{D}^{-s,q}$ where $\frac{1}{p} + \frac{1}{q} = 1$. If $s < 0$ the elements of $\mathbb{D}^{s,p}$ may not be ordinary random variables and they are interpreted as distributions on the Gaussian space or generalized random variables. Set $\mathbb{D}^{-\infty} = \cup_{s,p} \mathbb{D}^{s,p}$. The space $\mathbb{D}^{-\infty}$ is the dual of the space $\mathbb{D}^{\infty}$ which is a countably normed space.

   The interest of the space $\mathbb{D}^{-\infty}$ is that it contains the composition of Schwartz distributions with smooth and nondegenerate random variables, as we shall show in the next chapter. An example of a distribution random variable is the compostion $\delta_0(W(h))$ (see Exercise 1.5.6).

**4.** Suppose that $V$ is a real separable Hilbert space. We can define the Sobolev spaces $\mathbb{D}^{s,p}(V)$ of $V$-valued functionals as the completion of the class $\mathcal{P}_V$ of $V$-valued polynomial random variables with respect to the seminorm $|||\cdot|||_{s,p,V}$ defined in the same way as before. The above properties are still true for $V$-valued functionals. If $F \in \mathbb{D}^{s,p}(V)$ and $G \in \mathbb{D}^{-s,q}(V)$, where $\frac{1}{p} + \frac{1}{q} = 1$, then we denote the pairing $\langle F, G \rangle$ by $E(\langle F, G \rangle_V)$.

**Proposition 1.5.7** *Let $V$ be a real separable Hilbert space. For every $p > 1$ and $s \in \mathbb{R}$, the operator $D$ is continuous from $\mathbb{D}^{s,p}(V)$ to $\mathbb{D}^{s-1,p}(V \otimes H)$ and the operator $\delta$ (defined as the adjoint of $D$) is continuous from $\mathbb{D}^{s,p}(V \otimes H)$ into $\mathbb{D}^{s-1,p}(V)$. That is, for all $p > 1$ and $s \in \mathbb{R}$, we have*

$$|||\delta(u)|||_{s-1,p} \leq c_{s,p} |||u|||_{s,p,H} .$$

*Proof:*    For simplicity we assume that $V = \mathbb{R}$. Let us prove first the continuity of $D$. For any $F \in \mathcal{P}$ we have

$$(I - L)^{\frac{s}{2}} DF = DR(I - L)^{\frac{s}{2}} F,$$

where

$$R = \sum_{n=1}^{\infty} \left( \frac{n}{n+1} \right)^{\frac{s}{2}} J_n.$$

By Theorem 1.4.2 the operator $R$ is bounded in $L^p(\Omega)$ for all $p > 1$, and we obtain

$$
\begin{aligned}
\||DF\||_{s+1,p,H} &= \left\| (I - L)^{\frac{s}{2}} DF \right\|_{L^p(\Omega;H)} = \left\| DR(I - L)^{\frac{s}{2}} F \right\|_{L^p(\Omega;H)} \\
&\leq \left\| R(I - L)^{\frac{s}{2}} F \right\|_{1,p} \leq c_p \left\| \left| R(I - L)^{\frac{s}{2}} F \right| \right\|_{1,p} \\
&= c_p \left\| (I - L)^{\frac{1}{2}} R(I - L)^{\frac{s}{2}} F \right\|_p = c_p \left\| R(I - L)^{\frac{s+1}{2}} F \right\|_p \\
&\leq c_p' \left\| (I - L)^{\frac{s+1}{2}} F \right\|_p = c_p \||F\||_{s+1,p}.
\end{aligned}
$$

The continuity of the operator $\delta$ follows by a duality argument. In fact, for any $u \in \mathbb{D}^{s,p}(H)$ we have

$$
\begin{aligned}
\||\delta(u)\||_{s-1,p} &= \sup_{\||F\||_{1-s,q} \leq 1} |E\left( \langle u, DF \rangle_H \right)| \leq \||u\||_{s,p,H} \||DF\||_{-s,q,H} \\
&\leq c_{s,p} \||u\||_{s,p,H}.
\end{aligned}
$$

□

Proposition 1.5.7 allows us to generalize Lemma 1.2.3 in the following way:

**Lemma 1.5.3** *Let $\{F_n, n \geq 1\}$ be a sequence of random variables converging to $F$ in $L^p(\Omega)$ for some $p > 1$. Suppose that $\sup_n \||F_n\||_{s,p} < \infty$ for some $s > 0$. Then Then $F$ belongs to $\mathbb{D}^{s,p}$.*

*Proof:*    We know that

$$\sup_n \left\| (I - L)^{\frac{s}{2}} F_n \right\|_p < \infty.$$

Let $q$ be the conjugate of $p$. There exists a subsequence $\{F_{n(i)}, i \geq 1\}$ such that $(I - L)^{\frac{s}{2}} F_{n(i)}$ converges weakly in $\sigma(L^p, L^q)$ to some element $G$. Then for any polynomial random variable $Y$ we have

$$
\begin{aligned}
E\left( F(I - L)^{\frac{s}{2}} Y \right) &= \lim_n E\left( F_{n(i)}(I - L)^{\frac{s}{2}} Y \right) \\
&= \lim_n E\left( (I - L)^{\frac{s}{2}} F_{n(i)} Y \right) = E(GY).
\end{aligned}
$$

Thus, $F = (I - L)^{-\frac{s}{2}} G$, and this implies that $F \in \mathbb{D}^{s,p}$.    □

The following proposition provides a precise estimate for the norm $p$ of the divergence operator.

**Proposition 1.5.8** *Let $u$ be an element of $\mathbb{D}^{1,p}(H)$, $p > 1$. Then we have*

$$\|\delta(u)\|_p \leq c_p \left( \|E(u)\|_H + \|Du\|_{L^p(\Omega;H\otimes H)} \right).$$

*Proof:*    From Proposition 1.5.7 we know that $\delta$ is continuous from $\mathbb{D}^{1,p}(H)$ into $L^p(\Omega)$. This implies that

$$\|\delta(u)\|_p \leq c_p \left( \|u\|_{L^p(\Omega;H)} + \|Du\|_{L^p(\Omega;H\otimes H)} \right).$$

On the other hand, we have

$$\|u\|_{L^p(\Omega;H)} \leq \|E(u)\|_H + \|u - E(u)\|_{L^p(\Omega;H)},$$

and

$$
\begin{aligned}
\|u - E(u)\|_{L^p(\Omega;H)} &= \left\| (I-L)^{-\frac{1}{2}} RCu \right\|_{L^p(\Omega;H)} \leq c_p \|Cu\|_{L^p(\Omega;H)} \\
&\leq c_p' \|Du\|_{L^p(\Omega;H\otimes H)},
\end{aligned}
$$

where $R = \sum_{n=1}^{\infty} (1 + \frac{1}{n})^{\frac{1}{2}} J_n$.    $\square$

## Exercises

**1.5.1** Complete the proof of Meyer's inequality (1.80) without the condition $(J_0 + \cdots + J_{N-1})(F) = 0$.

**1.5.2** Derive the right inequality in (1.80) from the left inequality by means of a duality argument.

**1.5.3** Show that

$$\int_0^{\frac{\pi}{2}} \frac{\sin\theta \cos^n\theta}{\sqrt{\pi|\log\cos^2\theta|}} d\theta = \frac{1}{\sqrt{2(n+1)}}.$$

*Hint:* Change the variables, substituting $\cos\theta = y$ and $y = \exp(-\frac{x^2}{2})$.

**1.5.4** Let $W = \{W_t, t \in [0,1]\}$ be a Brownian motion. For every $0 < \gamma < \frac{1}{2}$ and $p = 2, 3, 4, \ldots$ such that $\gamma < \frac{1}{2} - \frac{1}{2p}$, we define the random variable

$$\|W\|_{p,\gamma}^{2p} = \int_{[0,1]^2} \frac{|W_s - W_t|^{2p}}{|s-t|^{1+2p\gamma}} ds dt.$$

Show that $\|W\|_{p,\gamma}^{2p}$ belongs to $\mathbb{D}^\infty$ (see Airault and Malliavin [3]).

**1.5.5** Using the Gaussian formula (A.1), extend Theorem 1.5.1 to a polynomial random variable with values on a separable Hilbert space $V$ (see Sugita [323]).

**1.5.6** Let $p_\epsilon(x)$ be the density of the normal distribution $N(0, \epsilon)$, for any $\epsilon > 0$. Fix $h \in H$. Using Stroock's formula (see Exercise 1.2.6) and the expression of the derivatives of $p_\epsilon(x)$ in terms of Hermite polynomials, show the following chaos expansion:

$$p_\epsilon(W(h)) = \sum_{m=0}^{\infty} \frac{(-1)^m \; I_{2m}(h^{\otimes 2m})}{\sqrt{2\pi} \; 2^m \; m! \left(\| h \|_H^2 + \epsilon\right)^{m + \frac{1}{2}}} \; .$$

Letting $\epsilon$ tend to zero in the above expression, find the chaos expansion of $\delta_0(W(h))$ and deduce that $\delta_0(W(h))$ belongs to the negative Sobolev space $\mathbb{D}^{-\alpha,2}$ for any $\alpha > \frac{1}{2}$, and also that $\delta_0(W(h))$ is not in $\mathbb{D}^{-\frac{1}{2},2}$.

**1.5.7** (See Sugita [325]) Let $F$ be a smooth functional of a Gaussian process $\{W(h), h \in H\}$. Let $\{W'(h), h \in H\}$ be an independent copy of $\{W(h), h \in H\}$.
  a) Prove the formula

$$D(T_t F) = \frac{e^{-t}}{\sqrt{1 - e^{-2t}}} E'(D'(F(e^{-t}W + \sqrt{1 - e^{-2t}}W')))$$

for all $t > 0$, where $D'$ denotes the derivative operator with respect to $W'$.
  b) Using part a), prove the inequality

$$E(\|D(T_t F)\|_H^p) \leq c_p \left( \frac{e^{-t}}{\sqrt{1 - e^{-2t}}} \right)^p E(|F|^p),$$

for all $p > 1$.
  c) Applying part b), show that the operator $(-L)^k T_t$ is bounded in $L^p$ and that $T_t$ is continuous from $L^p$ into $\mathbb{D}^{k,p}$, for all $k \geq 1$ and $p > 1$.

**1.5.8** Prove Proposition 1.5.7 for $k > 1$.

**1.5.9** Prove that $\||F|\|_{k,p} \leq c_{k,p} \|F\|_{k,p}$ for all $p > 1$, $k \in \mathbb{N}$ and $F \in \mathcal{P}$.

## Notes and comments

**[1.1]**    The notion of Gaussian space or the isonormal Gaussian process was introduced by Segal [303], and the orthogonal decomposition of the space of square integrable functionals of the Wiener process is due to Wiener [349]. We are interested in results on Gaussian families $\{W(h), h \in H\}$ that depend only on the covariance function, that is, on the underlying Hilbert space $H$. One can always associate to the Hilbert space $H$ an abstract Wiener space (see Gross [128]), that is, a Gaussian measure $\mu$ on a Banach space $\Omega$ such that $H$ is injected continuously into $\Omega$ and

$$\int_\Omega \exp(it\langle y, x \rangle)\mu(dy) = \frac{1}{2}\|x\|_H^2$$

for any $x \in \Omega^* \subset H$. In this case the probability space has a nice topological structure, but most of the notions introduced in this chapter are not related to this structure. For this reason we have chosen an arbitrary probability space as a general framework.

For the definition and properties of multiple stochastic integrals with respect to a Gaussian measure we have followed the presentation provided by Itô in [153]. The stochastic integral of adapted processes with respect to the Brownian motion originates in Itô [152]. In Section 1.1.3 we described some elementary facts about the Itô integral. For a complete exposition of this subject we refer to the monographs by Ikeda and Watanabe [146], Karatzas and Shreve [164], and Revuz and Yor [292].

[1.2]    The derivative operator and its representation on the chaotic development has been used in different frameworks. In the general context of a Fock space the operator $D$ coincides with the annihilation operator studied in quantum probability.

The notation $D_t F$ for the derivative of a functional of a Gaussian process has been taken from the work of Nualart and Zakai [263].

The bilinear form $(F, G) \to E(\langle DF, DG \rangle_H)$ on the space $\mathbb{D}^{1,2}$ is a particular type of a Dirichlet form in the sense of Fukushima [113]. In this sense some of the properties of the operator $D$ and its domain $\mathbb{D}^{1,2}$ can be proved in the general context of a Dirichlet form, under some additional hypotheses. This is true for the local property and for the stability under Lipschitz maps. We refer to Bouleau and Hirsch [46] and to Ma and Röckner [205] for monographs on this theory.

In [324] Sugita provides a characterization of the space $\mathbb{D}^{1,2}$ in terms of differentiability properties. More precisely, in the case of the Brownian motion, a random variable $F \in L^2(\Omega)$ belongs to $\mathbb{D}^{1,2}$ if and only if the following two conditions are satisfied:

(i) $F$ is ray absolutely continuous (RAC). This means that for any $h \in H$ there exists a version of the process $\{F(\omega + t \int_0^\cdot h_s ds), t \in \mathbb{R}\}$ that is absolutely continuous.

(ii) There exists a random vector $DF \in L^2(\Omega; H)$ such that for any $h \in H$, $\frac{1}{t}[F(\omega + t \int_0^\cdot h_s ds) - F(\omega)]$ converges in probability to $\langle DF, h \rangle_H$ as $t$ tends to zero.

In Lemma 2.1.5 of Chapter 2 we will show that properties (i) and (ii) hold for any random variable $F \in \mathbb{D}^{1,p}$, $p > 1$. Proposition 1.2.6 is due to Sekiguchi and Shiota [305].

[1.3]    The generalization of the stochastic integral with respect to the Brownian motion to nonadapted processes was introduced by Skorohod in [315], obtaining the isometry formula (1.54), and also by Hitsuda in [136, 135]. The identification of the Skorohod integral as the adjoint of the derivative operator has been proved by Gaveau and Trauber [116].

We remark that in [290] (see also Kusuoka [178]) Ramer has also introduced this type of stochastic integral, independently of Skorohod's work, in connection with the study of nonlinear transformations of the Wiener measure.

One can show that the iterated derivative operator $D^k$ is the adjoint of the multiple Skorohod integral $\delta^k$, and some of the properties of the Skorohod integral can be extended to multiple integrals (see Nualart and Zakai [264]).

Formula (1.63) was first proved by Clark [68], where $F$ was assumed to be Fréchet differentiable and to satisfy some technical conditions. In [269] Ocone extends this result to random variables $F$ in the space $\mathbb{D}^{1,2}$. Clark's representation theorem has been extended by Karatzas et al. [162] to random variables in the space $\mathbb{D}^{1,1}$.

The spaces $\mathbb{L}^{1,2,f}$ and $\mathbb{L}^F$ of random variables differentiable in future times were introduced by Alòs and Nualart in [10]. These spaces lead to a stochastic calculus which generalizes both the classical Itô calculus and the Skorohod calculus (see Chapter 3).

**[1.4]**   For a complete presentation of the hypercontractivity property and its relation with the Sobolev logarithmic inequality, we refer to the Saint Flour course by Bakry [15]. The multiplier theorem proved in this section is due to Meyer [225], and the proof given here has been taken from Watanabe [343].

**[1.5]**   The Sobolev spaces of Wiener functionals have been studied by different authors. In [172] Krée and Krée proved the continuity of the divergence operator in $L^2$.

The equivalence between the the norms $\|D^k F\|_p$ and $\|(-L)^{\frac{k}{2}}\|_p$ for any $p > 1$ was first established by Meyer [225] using the Littlewood-Payley inequalities. In finite dimension the operator $DC^{-1}$ is related to the Riesz transform. Using this idea, Gundy [129] gives a probabilistic proof of Meyer's inequalities which is based on the properties of the three-dimensional Bessel process and Burkholder inequalities for martingales. On the other hand, using the boundedness in $L^p$ of the Hilbert transform, Pisier [285] provides a short analytical proof of the fact that the operator $DC^{-1}$ is bounded in $L^p$. We followed Pisier's approach in Section 1.5.

In [343] Watanabe developed the theory of distributions on the Wiener space that has become a useful tool in the analysis of regularity of probability densities.

# 2
# Regularity of probability laws

In this chapter we apply the techniques of the Malliavin calculus to study the regularity of the probability law of a random vector defined on a Gaussian probability space. We establish some general criteria for the absolute continuity and regularity of the density of such a vector. These general criteria will be applied to the solutions of stochastic differential equations and stochastic partial differential equations driven by a space-time white noise.

## 2.1 Regularity of densities and related topics

This section is devoted to study the regularity of the law of a random vector $F = (F^1, \ldots, F^m)$, which is measurable with respect to an underlying isonormal Gaussian process $\{W(h), h \in H\}$. Using the duality between the operators $D$ and $\delta$ we first derive an explicit formula for the density of a one-dimensional random variable and we deduce some estimates. Then we establish a criterion for absolute continuity for a random vector under the assumption that its Malliavin matrix is invertible a.s. An alternative approach, due to Bouleau and Hirsch, is presented in the third part of this section. This approach is based on a criterion for absolute continuity in finite dimension and it then uses a limit argument. The criterion obtained in this way is stronger than that obtained by integration by parts, in that it requires weaker regularity hypotheses on the random vector.

We later introduce the notion of smooth and nondegenerate random vector by the condition that the inverse of the determinant of the Malliavin matrix has moments of all orders. We show that smooth and nondegenerate random vectors have infinitely differentiable densities. Two different proofs of this result are given. First we show by a direct argument the local smoothness of the density under more general hypotheses. Secondly, we derive the smoothness of the density from the properties of the composition of a Schwartz tempered distribution with a smooth and nondegenerated random vector.

We also study some properties of the topological support of the law of a random vector. The last part of this section is devoted to the regularity of the law of the supremum of a continuous process.

## 2.1.1  Computation and estimation of probability densities

As in the previous chapter, let $W = \{W(h), h \in H\}$ be an isonormal Gaussian process associated to a separable Hilbert space $H$ and defined on a complete probability space $(\Omega, \mathcal{F}, P)$. Assume also that $\mathcal{F}$ is generated by $W$.

The integration-by-parts formula leads to the following explicit expression for the density of a one-dimensional random variable.

**Proposition 2.1.1** *Let $F$ be a random variable in the space $\mathbb{D}^{1,2}$. Suppose that $\frac{DF}{\|DF\|_H^2}$ belongs to the domain of the operator $\delta$ in $L^2(\Omega)$. Then the law of $F$ has a continuous and bounded density given by*

$$p(x) = E\left[ \mathbf{1}_{\{F>x\}} \delta\left( \frac{DF}{\|DF\|_H^2} \right) \right]. \qquad (2.1)$$

*Proof:*    Let $\psi$ be a nonnegative smooth function with compact support, and set $\varphi(y) = \int_{-\infty}^{y} \psi(z)dz$. We know that $\varphi(F)$ belongs to $\mathbb{D}^{1,2}$, and making the scalar product of its derivative with $DF$ obtains

$$\langle D(\varphi(F)), DF \rangle_H = \psi(F)\|DF\|_H^2.$$

Using the duality relationship between the operators $D$ and $\delta$ (see (1.42)), we obtain

$$
\begin{aligned}
E[\psi(F)] &= E\left[ \left\langle D(\varphi(F)), \frac{DF}{\|DF\|_H^2} \right\rangle_H \right] \\
&= E\left[ \varphi(F)\delta\left( \frac{DF}{\|DF\|_H^2} \right) \right]. \qquad (2.2)
\end{aligned}
$$

By an approximation argument, Equation (2.2) holds for $\psi(y) = \mathbf{1}_{[a,b]}(y)$, where $a < b$. As a consequence, we apply Fubini's theorem to get

$$
\begin{aligned}
P(a \leq F \leq b) &= E\left[\left(\int_{-\infty}^{F} \psi(x)dx\right) \delta\left(\frac{DF}{\|DF\|_H^2}\right)\right] \\
&= \int_a^b E\left[\mathbf{1}_{\{F>x\}} \delta\left(\frac{DF}{\|DF\|_H^2}\right)\right] dx,
\end{aligned}
$$

which implies the desired result. $\qquad\square$

We note that sufficient conditions for $\frac{DF}{\|DF\|_H^2} \in \mathrm{Dom}\,\delta$ are that $F$ is in $\mathbb{D}^{2,4}$ and that $E(\|DF\|_H^{-8}) < \infty$ (see Exercise 2.1.1). On the other hand, Equation (2.1) still holds under the hypotheses $F \in \mathbb{D}^{1,p}$ and $\frac{DF}{\|DF\|_H^2} \in \mathbb{D}^{1,p'}(H)$ for some $p, p' > 1$. We will see later that the property $\|DF\|_H > 0$ a.s. (assuming that $F$ is in $\mathbb{D}_{\mathrm{loc}}^{1,1}$) is sufficient for the existence of a density.

From expression (2.1) we can deduce estimates for the density. Fix $p$ and $q$ such that $\frac{1}{p} + \frac{1}{q} = 1$. By Hölder's inequality we obtain

$$
p(x) \leq (P(F > x))^{1/q} \left\|\delta\left(\frac{DF}{\|DF\|_H^2}\right)\right\|_p.
$$

In the same way, taking into account the relation $E[\delta(DF/\|DF\|_H^2)] = 0$ we can deduce the inequality

$$
p(x) \leq (P(F < x))^{1/q} \left\|\delta\left(\frac{DF}{\|DF\|_H^2}\right)\right\|_p.
$$

As a consequence, we obtain

$$
p(x) \leq (P(|F| > |x|))^{1/q} \left\|\delta\left(\frac{DF}{\|DF\|_H^2}\right)\right\|_p, \tag{2.3}
$$

for all $x \in \mathbb{R}$. Now using the $L^p(\Omega)$ estimate of the operator $\delta$ established in Proposition 1.5.8 we obtain

$$
\left\|\delta\left(\frac{DF}{\|DF\|_H^2}\right)\right\|_p \leq c_p \left(\left\|E\left(\frac{DF}{\|DF\|_H^2}\right)\right\|_H + \left\|D\left(\frac{DF}{\|DF\|_H^2}\right)\right\|_{L^p(\Omega;H\otimes H)}\right). \tag{2.4}
$$

We have

$$
D\left(\frac{DF}{\|DF\|_H^2}\right) = \frac{D^2 F}{\|DF\|_H^2} - 2\frac{\langle D^2 F, DF \otimes DF\rangle_{H\otimes H}}{\|DF\|_H^4},
$$

and, hence,

$$
\left\|D\left(\frac{DF}{\|DF\|_H^2}\right)\right\|_{H\otimes H} \leq \frac{3\left\|D^2 F\right\|_{H\otimes H}}{\|DF\|_H^2}. \tag{2.5}
$$

Finally, from the inequalities (2.3), (2.4) and (2.5) we deduce the following estimate.

**Proposition 2.1.2** *Let* $q$, $\alpha$, $\beta$ *be three positive real numbers such that* $\frac{1}{q} + \frac{1}{\alpha} + \frac{1}{\beta} = 1$. *Let* $F$ *be a random variable in the space* $\mathbb{D}^{2,\alpha}$, *such that* $E(\|DF\|_H^{-2\beta}) < \infty$. *Then the density* $p(x)$ *of* $F$ *can be estimated as follows*

$$p(x) \;\leq\; c_{q,\alpha,\beta}\left(P(|F| > |x|)\right)^{1/q}$$
$$\times \left( E(\|DF\|_H^{-1}) + \|D^2 F\|_{L^\alpha(\Omega; H \otimes H)} \left\| \|DF\|_H^{-2} \right\|_\beta \right). \quad (2.6)$$

Let us apply the preceding proposition to a Brownian martingale.

**Proposition 2.1.3** *Let* $W = \{W(t), t \in [0,T]\}$ *be a Brownian motion and let* $u = \{u(t), t \in [0,T]\}$ *be an adapted process verifying the following hypotheses:*

*(i)* $E\left(\int_0^T u(t)^2 dt\right) < \infty$, $u(t)$ *belongs to the space* $\mathbb{D}^{2,2}$ *for each* $t \in [0,T]$, *and*

$$\lambda := \sup_{s,t \in [0,T]} E(|D_s u_t|^p) + \sup_{r,s \in [0,T]} E\left(\left(\int_0^T |D^2_{r,s} u_t|^p dt\right)^{\frac{p}{2}}\right) < \infty,$$

*for some* $p > 3$.

*(ii)* $|u(t)| \geq \rho > 0$ *for some constant* $\rho$.

Set $M_t = \int_0^t u(s) dW_s$, *and denote by* $p_t(x)$ *the probability density of* $M_t$. *Then for any* $t > 0$ *we have*

$$p_t(x) \leq \frac{c}{\sqrt{t}} P(|M_t| > |x|)^{\frac{1}{q}}, \quad (2.7)$$

*where* $q > \frac{p}{p-3}$ *and the constant* $c$ *depends on* $\lambda$, $\rho$ *and* $p$.

*Proof:*    Fix $t \in (0,T]$. We will apply Proposition 2.1.2 to the random variable $M_t$. We claim that $M_t \in \mathbb{D}^{2,2}$. In fact, note first that by Lemma 1.3.4 $M_t \in \mathbb{D}^{1,2}$ and for $s < t$

$$D_s M_t = u_s + \int_s^t D_s u_r dW_r. \quad (2.8)$$

For almost all $s$, the process $\{D_s u_r, r \in [0,T]\}$ is adapted and belongs to $\mathbb{L}^{1,2}$. Hence, by Lemma 1.3.4 $\int_s^t D_s u_r dW_r$ belongs to $\mathbb{D}^{1,2}$ and

$$D_\theta \left( \int_s^t D_s u_r dW_r \right) = D_s u_\theta + \int_{s \vee \theta}^t D_\theta D_s u_r dW_r. \quad (2.9)$$

From (2.8) and (2.9) we deduce for any $\theta, s \leq t$

$$D_\theta D_s M_t = D_\theta u_s + D_s u_\theta + \int_{s \vee \theta}^t D_\theta D_s u_r dW_r. \tag{2.10}$$

We will take $\alpha = p$ in Proposition 2.1.2. Using Hölder's and Burkholder's inequalities we obtain from (2.10)

$$E(\|D^2 M_t\|_{H \otimes H}^p) \leq c_p \lambda t^p \ .$$

Set

$$\sigma(t) := \|DM_t\|_H^2 = \int_0^t \left( u_s + \int_s^t D_s u_r dW_r \right)^2 ds.$$

We have the following estimates for any $h \leq 1$

$$
\begin{aligned}
\sigma(t) &\geq \int_{t(1-h)}^t \left( u_s + \int_s^t D_s u_r dW_r \right)^2 ds \\
&\geq \int_{t(1-h)}^t \frac{u_s^2}{2} ds - \int_{t(1-h)}^t \left( \int_s^t D_s u_r dW_r \right)^2 ds \\
&\geq \frac{th\rho^2}{2} - I_h(t),
\end{aligned}
$$

where

$$I_h(t) = \int_{t(1-h)}^t \left( \int_s^t D_s u_r dW_r \right)^2 ds.$$

Choose $h = \frac{4}{t\rho^2 y}$, and notice that $h \leq 1$ provided $y \geq a := \frac{4}{t\rho^2}$. We have

$$P\left( \sigma(t) \leq \frac{1}{y} \right) \leq P\left( I_h(t) \geq \frac{1}{y} \right) \leq y^{\frac{p}{2}} E(|I_h(t)|^{\frac{p}{2}}). \tag{2.11}$$

Using Burkholder' inequality for square integrable martingales we get the following estimate

$$
\begin{aligned}
E(|I_h(t)|^{\frac{p}{2}}) &\leq c_p(th)^{\frac{p}{2}-1} \int_{t(1-h)}^t E\left( \left( \int_s^t (D_s u_r)^2 dr \right)^{\frac{p}{2}} \right) ds \\
&\leq c_p' \sup_{s,r \in [0,t]} E(|D_s u_r|^p)(th)^p. \tag{2.12}
\end{aligned}
$$

Consequently, for $0 < \gamma < \frac{p}{2}$ we obtain, using (2.11) and (2.12),

$$
\begin{aligned}
E(\sigma(t)^{-\gamma}) &= \int_0^\infty \gamma y^{\gamma-1} P\left(\sigma(t)^{-1} > y\right) dy \\
&\leq a^\gamma + \gamma \int_a^\infty y^{\gamma-1} P\left(\sigma(t) < \frac{1}{y}\right) dy \\
&\leq \left(\frac{4}{t\rho^2}\right)^\gamma + \gamma \int_{\frac{4}{t\rho^2}}^\infty E(|I_h(t)|^{\frac{p}{2}}) y^{\gamma-1+\frac{p}{2}} dy \\
&\leq c\left(t^{-\gamma} + \int_{\frac{4}{t\rho^2}}^\infty y^{\gamma-1-\frac{p}{2}} dy\right) \leq c'\left(t^{-\gamma} + t^{\frac{p}{2}-\gamma}\right). \quad (2.13)
\end{aligned}
$$

Substituting (2.13) in Equation (2.6) with $\alpha = p$, $\beta < \frac{p}{2}$, and with $\gamma = \frac{1}{2}$ and $\gamma = \beta$, we get the desired estimate. □

Applying Tchebychev and Burkholder's inequalities, from (2.7) we deduce the following inequality for any $\theta > 1$

$$
p_t(x) \leq \frac{c|x|^{-\frac{\theta}{q}}}{\sqrt{t}} \left(E\left(\left(\int_0^t u_s^2 ds\right)^{\frac{\theta}{2}}\right)\right)^{\frac{1}{q}}.
$$

**Corollary 2.1.1** *Under the conditions of Proposition 2.1.3, if the process $u$ satisfies $|u_t| \leq M$ for some constant $M$, then*

$$
p_t(x) \leq \frac{c}{\sqrt{t}} \exp\left(-\frac{|x|^2}{qM^2 t}\right).
$$

*Proof:* It suffices to apply the martingale exponential inequality (A.5). □

### 2.1.2  A criterion for absolute continuity based on the integration-by-parts formula

We recall that $C_b^\infty(\mathbb{R}^m)$ denotes the class of functions $f : \mathbb{R}^m \to \mathbb{R}$ that are bounded and possess bounded derivatives of all orders, and we write $\partial_i = \frac{\partial}{\partial x_i}$. We start with the following lemma of real analysis (cf. Malliavin [207]).

**Lemma 2.1.1** *Let $\mu$ be a finite measure on $\mathbb{R}^m$. Assume that for all $\varphi \in C_b^\infty(\mathbb{R}^m)$ the following inequality holds:*

$$
\left|\int_{\mathbb{R}^m} \partial_i \varphi d\mu\right| \leq c_i \|\varphi\|_\infty, \qquad 1 \leq i \leq m, \quad (2.14)
$$

*where the constants $c_i$ do not depend on $\varphi$. Then $\mu$ is absolutely continuous with respect to the Lebesgue measure.*

*Proof:*    If $m = 1$ there is a simple proof of this result. Fix $a < b$, and consider the function $\varphi$ defined by

$$\varphi(x) = \begin{cases} 0 & \text{if } x \leq a \\ \frac{x-a}{b-a} & \text{if } a < x < b \\ 1 & \text{if } x \geq b. \end{cases}$$

Although this function is not infinitely differentiable, we can approximate it by functions of $C_b^\infty(\mathbb{R})$ in such a way that Eq. (2.14) still holds. In this form we get $\mu([a, b]) \leq c_1(b - a)$, which implies the absolute continuity of $\mu$.

For an arbitrary value of $m$, Malliavin [207] gives a proof of this lemma that uses techniques of harmonic analysis. Following a remark in Malliavin's paper, we are going to give a different proof and show that the density of $\mu$ belongs to $L^{\frac{m}{m-1}}$ if $m > 1$. Consider an approximation of the identity $\{\psi_\epsilon, \epsilon > 0\}$ on $\mathbb{R}^m$. Take, for instance,

$$\psi_\epsilon(x) = (2\pi\epsilon)^{-\frac{m}{2}} \exp(-\frac{|x|^2}{2\epsilon}).$$

Let $c_M(x)$, $M \geq 1$, be a sequence of functions of the space $C_0^\infty(\mathbb{R}^m)$ such that $0 \leq c_M \leq 1$ and

$$c_M(x) = \begin{cases} 1 & \text{if } |x| \leq M \\ 0 & \text{if } |x| \geq M + 1. \end{cases}$$

We assume that the partial derivatives of $c_M$ of all orders are bounded uniformly with respect to $M$. Then the functions

$$c_M(x)(\psi_\epsilon * \mu)(x) = c_M(x) \int_{\mathbb{R}^m} \psi_\epsilon(x - y)\mu(dy)$$

belong to $C_0^\infty(\mathbb{R}^m)$.

The Gagliardo-Nirenberg inequality says that for any function $f$ in the space $C_0^\infty(\mathbb{R}^m)$ one has

$$\|f\|_{L^{\frac{m}{m-1}}} \leq \prod_{i=1}^m \|\partial_i f\|_{L^1}^{1/m}.$$

An elementary proof of this inequality can be found in Stein [317, p. 129]. Applying this inequality to the functions $c_M(\psi_\epsilon * \mu)$, we obtain

$$\|c_M(\psi_\epsilon * \mu)\|_{L^{\frac{m}{m-1}}} \leq \prod_{i=1}^m \|\partial_i(c_M(\psi_\epsilon * \mu))\|_{L^1}^{\frac{1}{m}}. \tag{2.15}$$

Equation (2.14) implies that the mapping $\varphi \hookrightarrow \int_{\mathbb{R}^m} \partial_i\varphi\, d\mu$, defined on $C_0^\infty(\mathbb{R}^m)$, is a signed measure, which will be denoted by $\nu_i$, $1 \leq i \leq m$.

Then we have

$$
\begin{aligned}
\|\partial_i(c_M(\psi_\epsilon * \mu))\|_{L^1} &\leq \int_{\mathbb{R}^m} c_M(x) \left| \int_{\mathbb{R}^m} \partial_i \psi_\epsilon(x-y)\mu(dy) \right| dx \\
&+ \int_{\mathbb{R}^m} |\partial_i c_M(x)| \left( \int_{\mathbb{R}^m} \psi_\epsilon(x-y)\mu(dy) \right) dx \\
&\leq \int_{\mathbb{R}^m} \left| \int_{\mathbb{R}^m} \psi_\epsilon(x-y)\nu_i(dy) \right| dx \\
&+ \int_{\mathbb{R}^m} |\partial_i c_M(x)| \left( \int_{\mathbb{R}^m} \psi_\epsilon(x-y)\mu(dy) \right) dx \leq K,
\end{aligned}
$$

where $K$ is a constant not depending on $M$ and $\epsilon$. Consequently, the family of functions $\{c_M(\psi_\epsilon * \mu), M \geq 1, \epsilon > 0\}$ is bounded in $L^{\frac{m}{m-1}}$. We use the weak compactness of the unit ball of $L^{\frac{m}{m-1}}$ to deduce the desired result. □

Suppose that $F = (F^1, \ldots, F^m)$ is a random vector whose components belong to the space $\mathbb{D}^{1,1}_{loc}$. We associate to $F$ the following random symmetric nonnegative definite matrix:

$$
\gamma_F = (\langle DF^i, DF^j \rangle_H)_{1 \leq i,j \leq m}.
$$

This matrix will be called the *Malliavin matrix* of the random vector $F$. The basic condition for the absolute continuity of the law of $F$ will be that the matrix $\gamma_F$ is invertible a.s. The first result in this direction follows.

**Theorem 2.1.1** *Let $F = (F^1, \ldots, F^m)$ be a random vector verifying the following conditions:*

*(i) $F^i \in \mathbb{D}^{2,p}_{loc}$ for all $i, j = 1, \ldots, m$, for some $p > 1$.*

*(ii) The matrix $\gamma_F$ is invertible a.s.*

*Then the law of $F$ is absolutely continuous with respect to the Lebesgue measure on $\mathbb{R}^m$.*

*Proof:*    We will assume that $F^i \in \mathbb{D}^{2,p}$ for each $i$. Fix a test function $\varphi \in C_b^\infty(\mathbb{R}^m)$. From Proposition 1.2.3, we know that $\varphi(F)$ belongs to the space $\mathbb{D}^{1,p}$ and that

$$
D(\varphi(F)) = \sum_{i=1}^m \partial_i \varphi(F) DF^i.
$$

Hence,

$$
\langle D(\varphi(F)), DF^j \rangle_H = \sum_{i=1}^m \partial_i \varphi(F) \gamma_F^{ij};
$$

therefore,

$$
\partial_i \varphi(F) = \sum_{j=1}^m \langle D(\varphi(F)), DF^j \rangle_H (\gamma_F^{-1})^{ji}. \tag{2.16}
$$

The inverse of $\gamma_F$ may not have moments, and for this reason we need a localizing argument.

For any integer $N \geq 1$ we consider a function $\Psi_N \in C_0^\infty(\mathbb{R}^m \otimes \mathbb{R}^m)$ such that $\Psi_N \geq 0$ and

(a) $\Psi_N(\sigma) = 1$   if $\sigma \in K_N$,

(b) $\Psi_N(\sigma) = 0$   if $\sigma \notin K_{N+1}$, where

$$K_N = \{\sigma \in \mathbb{R}^m \otimes \mathbb{R}^m : |\sigma^{ij}| \leq N \quad \text{for all} \quad i, j, \quad \text{and} \quad |\det \sigma| \geq \frac{1}{N}\}.$$

Note that $K_N$ is a compact subset of $GL(m) \subset \mathbb{R}^m \otimes \mathbb{R}^m$. Multiplying (2.16) by $\Psi_N(\gamma_F)$ yields

$$E[\Psi_N(\gamma_F)\partial_i\varphi(F)] = \sum_{j=1}^m E[\Psi_N(\gamma_F)\langle D(\varphi(F)), DF^j\rangle_H (\gamma_F^{-1})^{ji}].$$

Condition (i) implies that $\Psi_N(\gamma_F)(\gamma_F^{-1})^{ji}DF^j$ belongs to $\mathbb{D}^{1,p}(H)$. Consequently, we use the continuity of the operator $\delta$ from $\mathbb{D}^{1,p}(H)$ into $L^p(\Omega)$ (Proposition 1.5.4) and the duality relationship (1.42) to obtain

$$\left|E[\Psi_N(\gamma_F)\partial_i\varphi(F)]\right| = \left|E\left[\varphi(F)\sum_{j=1}^m \delta\left(\Psi_N(\gamma_F)(\gamma_F^{-1})^{ji}DF^j\right)\right]\right|$$

$$\leq E\left(\left|\sum_{j=1}^m \delta\left(\Psi_N(\gamma_F)(\gamma_F^{-1})^{ji}DF^j\right)\right|\right)\|\varphi\|_\infty.$$

Therefore, by Lemma 2.1.1 the measure $[\Psi_N(\gamma_F) \cdot P] \circ F^{-1}$ is absolutely continuous with respect to the Lebesgue measure on $\mathbb{R}^m$. Thus, for any Borel set $A \subset \mathbb{R}^m$ with zero Lebesgue measure we have

$$\int_{F^{-1}(A)} \Psi_N(\gamma_F)dP = 0.$$

Letting $N$ tend to infinity and using hypothesis (ii), we obtain the equality $P(F^{-1}(A)) = 0$, thereby proving that the probability $P \circ F^{-1}$ is absolutely continuous with respect to the Lebesgue measure on $\mathbb{R}^m$.    $\square$

Notice that if we only assume condition (i) in Theorem 2.1.1 and if no nondegeneracy condition on the Malliavin matrix is made, then we deduce that the measure $(\det(\gamma_F) \cdot P) \circ F^{-1}$ is absolutely continuous with respect to the Lebesgue measure on $\mathbb{R}^m$. In other words, the random vector $F$ has an absolutely continuous law conditioned by the set $\{\det(\gamma_F) > 0\}$; that is,

$$P\{F \in B, \det(\gamma_F) > 0\} = 0$$

for any Borel subset $B$ of $\mathbb{R}^m$ of zero Lebesgue measure.

### 2.1.3  Absolute continuity using Bouleau and Hirsch's approach

In this section we will present the criterion for absolute continuity obtained by Bouleau and Hirsch [46]. First we introduce some results in finite dimension, and we refer to Federer [96, pp. 241–245] for the proof of these results. We denote by $\lambda^n$ the Lebesgue measure on $\mathbb{R}^n$.

Let $\varphi$ be a measurable function from $\mathbb{R}$ to $\mathbb{R}$. Then $\varphi$ is said to be *approximately differentiable* at $a \in \mathbb{R}$, with an approximate derivative equal to $b$, if

$$\lim_{\eta \to 0} \frac{1}{\eta} \lambda^1 \{ x \in [a - \eta, a + \eta] : |\varphi(x) - \varphi(a) - (x - a)b| > \epsilon |x - a| \} = 0$$

for all $\epsilon > 0$. We will write $b = \operatorname{ap} \varphi'(a)$. The following property is an immediate consequence of the above definition.

(a) If $\varphi = \tilde{\varphi}$ a.e. and $\varphi$ is differentiable a.e., then $\tilde{\varphi}$ is approximately differentiable a.e. and $\operatorname{ap} \tilde{\varphi}' = \varphi'$ a.e.

If $\varphi$ is a measurable function from $\mathbb{R}^n$ to $\mathbb{R}$, we will denote by $\operatorname{ap} \partial_i \varphi$ the approximate partial derivative of $\varphi$ with respect to the $i$th coordinate. We will also denote by

$$\operatorname{ap} \nabla \varphi = (\operatorname{ap} \partial_1 \varphi, \dots, \operatorname{ap} \partial_n \varphi)$$

the approximate gradient of $\varphi$. Then we have the following result:

**Lemma 2.1.2** *Let $\varphi : \mathbb{R}^n \to \mathbb{R}^m$ be a measurable function, with $m \le n$, such that the approximate derivatives $\operatorname{ap} \partial_j \varphi_i$, $1 \le i \le m$, $1 \le j \le n$, exist for almost every $x \in \mathbb{R}^n$ with respect to the Lebesgue measure on $\mathbb{R}^n$. Then we have*

$$\int_{\varphi^{-1}(B)} \det[\langle \operatorname{ap} \nabla \varphi_j, \operatorname{ap} \nabla \varphi_k \rangle]_{1 \le j, k \le m} d\lambda^n = 0 \qquad (2.17)$$

*for any Borel set $B \subset \mathbb{R}^m$ with zero Lebesgue measure.*

Notice that the conclusion of Lemma 2.1.2 is equivalent to saying that

$$(\det[\langle \operatorname{ap} \nabla \varphi_j, \operatorname{ap} \nabla \varphi_k \rangle] \cdot \lambda^n) \circ \varphi^{-1} \ll \lambda^m.$$

We will also make use of linear transformations of the underlying Gaussian process $\{W(h), h \in H\}$. Fix an element $g \in H$ and consider the translated Gaussian process $\{W^g(h), h \in H\}$ defined by $W^g(h) = W(h) + \langle h, g \rangle_H$.

**Lemma 2.1.3** *The process $W^g$ has the same law (that is, the same finite dimensional distributions) as $W$ under a probability $Q$ equivalent to $P$ given by*

$$\frac{dQ}{dP} = \exp(-W(g) - \frac{1}{2} \|g\|_H^2).$$

*Proof:*    Let $f : \mathbb{R}^n \to \mathbb{R}$ be a bounded Borel function, and let $e_1, \ldots, e_n$ be orthonormal elements of $H$. Then we have

$$E\left[f(W^g(e_1), \ldots, W^g(e_n)) \exp\left(-W(g) - \frac{1}{2}\|g\|_H^2\right)\right]$$

$$= E\left[f(W^g(e_1), \ldots, W^g(e_n))\right.$$

$$\left. \times \exp\left(-\sum_{i=1}^n \langle e_i, g\rangle_H W(e_i) - \frac{1}{2}\sum_{i=1}^n \langle e_i, g\rangle_H^2\right)\right]$$

$$= \int_{\mathbb{R}^n} f(x_1 + \langle g, e_1\rangle_H, \ldots, x_n + \langle g, e_n\rangle_H)$$

$$\times \exp\left(-\frac{1}{2}\sum_{i=1}^n |x_i + \langle g, e_i\rangle_H|^2\right) dx$$

$$= E[f(W(e_1), \ldots, W(e_n))].$$

$\square$

Now consider a random variable $F \in L^0(\Omega)$. We can write $F = \psi_F \circ W$, where $\psi_F$ is a measurable mapping from $\mathbb{R}^H$ to $\mathbb{R}$ that is uniquely determined except on a set of measure zero for $P \circ W^{-1}$. By the preceding lemma on the equivalence between the laws of $W$ and $W^g$, we can define the shifted random variable $F^g = \psi_F \circ W^g$. Then the following result holds.

**Lemma 2.1.4** *Let $F$ be a random variable in the space $\mathbb{D}^{1,p}$, $p > 1$. Fix two elements $h, g \in H$. Then there exists a version of the process $\{\langle DF, h\rangle_H^{sh+g}, s \in \mathbb{R}\}$ such that for all $a < b$ we have*

$$F^{bh+g} - F^{ah+g} = \int_a^b \langle DF, h\rangle_H^{sh+g} ds \tag{2.18}$$

*a.s. Consequently, there exists a version of the process $\{F^{th+g}, t \in \mathbb{R}\}$ that has absolutely continuous paths with respect to the Lebesgue measure on $\mathbb{R}$, and its derivative is equal to $\langle DF, h\rangle_H^{th+g}$.*

*Proof:*    The proof will be done in two steps.

*Step 1:*    First we will show that $F^{th+g} \in L^q(\Omega)$ for all $q \in [1, p)$ with an $L^q$ norm uniformly bounded with respect to $t$ if $t$ varies in some bounded interval. In fact, let us compute

$$E(|F^{th+g}|^q) = E\left(|F|^q \exp\left\{tW(h) + W(g) - \frac{1}{2}\|th + g\|_H^2\right\}\right)$$

$$\leq (E(|F|^p))^{\frac{q}{p}} \left(E\left[\exp\left\{\frac{p}{p-q}(tW(h) + W(g))\right\}\right]\right)^{1-\frac{q}{p}}$$

$$\times e^{-\frac{1}{2}\|th+g\|_H^2}$$

$$= (E(|F|^p))^{\frac{q}{p}} \exp\left(\frac{q}{2(p-q)}\|th + g\|_H^2\right) < \infty. \tag{2.19}$$

*Step 2:* Suppose first that $F$ is a smooth functional of the form $F = f(W(h_1), \ldots, W(h_k))$. In this case the mapping $t \to F^{th+g}$ is continuously differentiable and

$$\frac{d}{dt}(F^{th+g}) = \sum_{i=1}^{k} \partial_i f(W(h_1) + t\langle h, h_1 \rangle_H + \langle g, h_1 \rangle_H,$$

$$\ldots, W(h_k) + t\langle h, h_k \rangle_H + \langle g, h_k \rangle_H)\langle h, h_i \rangle_H = \langle DF, h \rangle_H^{th+g}.$$

Now suppose that $F$ is an arbitrary element in $\mathbb{D}^{1,p}$, and let $\{F_k, k \geq 1\}$ be a sequence of smooth functionals such that as $k$ tends to infinity $F_k$ converges to $F$ in $L^p(\Omega)$ and $DF_k$ converges to $DF$ in $L^p(\Omega; H)$. By taking suitable subsequences, we can also assume that these convergences hold almost everywhere. We know that for any $k$ and any $a < b$ we have

$$F_k^{bh+g} - F_k^{ah+g} = \int_a^b \langle DF_k, h \rangle_H^{sh+g} ds. \tag{2.20}$$

For any $t \in \mathbb{R}$ the random variables $F_k^{th+g}$ converge almost surely to $F^{th+g}$ as $k$ tends to infinity. On the other hand, the sequence of random variables $\int_a^b \langle DF_k, h \rangle_H^{sh+g} ds$ converges in $L^1(\Omega)$ to $\int_a^b \langle DF, h \rangle_H^{sh+g} ds$ as $k$ tends to infinity. In fact, using Eq. (2.19) with $q = 1$, we obtain

$$E\left( \left| \int_a^b \langle DF_k, h \rangle_H^{sh+g} ds - \int_a^b \langle DF, h \rangle_H^{sh+g} ds \right| \right)$$

$$\leq E\left( \int_a^b |\langle DF_k, h \rangle_H^{sh+g} - \langle DF, h \rangle_H^{sh+g}| ds \right)$$

$$\leq (E(|D^h F_k - D^h F|^p))^{\frac{1}{p}} (b - a)$$

$$\times \sup_{t \in [a,b]} \exp\left( \frac{1}{2(p-1)} \|th + g\|_H^2 \right).$$

In conclusion, by taking the limit of both sides of Eq. (2.20) as $k$ tends to infinity, we obtain (2.18). This completes the proof. ☐

Here is a useful consequence of Lemma 2.1.4.

**Lemma 2.1.5** *Let $F$ be a random variable in the space $\mathbb{D}^{1,p}$ for some $p > 1$. Fix $h \in H$. Then, a.s. we have*

$$\lim_{\epsilon \to 0} \frac{1}{\epsilon} \int_0^\epsilon (F^{th} - F) dt = \langle DF, h \rangle_H. \tag{2.21}$$

*Proof:* By Lemma 2.1.4, for almost all $(\omega, x) \in \Omega \times \mathbb{R}$ we have

$$\lim_{\epsilon \to 0} \frac{1}{\epsilon} \int_x^{x+\epsilon} (F^{yh}(\omega) - F(\omega)) dy = \langle DF(\omega), h \rangle_H^{xh}. \tag{2.22}$$

Hence, there exists an $x \in \mathbb{R}$ for which (2.22) holds a.s. Finally, if we consider the probability $Q$ defined by

$$\frac{dQ}{dP} = \exp(-xW(h) - \frac{x^2}{2}\|h\|_H^2)$$

we obtain that (2.21) holds $Q$ a.s. This completes the proof. □

Now we can state the main result of this section.

**Theorem 2.1.2** *Let $F = (F^1, \ldots, F^m)$ be a random vector satisfying the following conditions:*

(i) *$F^i$ belongs to the space $\mathbb{D}_{\mathrm{loc}}^{1,p}$, $p > 1$, for all $i = 1, \ldots, m$.*

(ii) *The matrix $\gamma_F = (\langle DF^i, DF^j \rangle)_{1 \leq i,j \leq m}$ is invertible a.s.*

*Then the law of $F$ is absolutely continuous with respect to the Lebesgue measure on $\mathbb{R}^m$.*

*Proof:*   We may assume by a localization argument that $F^k$ belongs to $\mathbb{D}^{1,p}$ for $k = 1, \ldots, m$. Fix a complete orthonormal system $\{e_i, i \geq 1\}$ in the Hilbert space $H$. For any natural number $n \geq 1$ we define

$$\varphi^{n,k}(t_1, \ldots, t_n) = (F^k)^{t_1 e_1 + \cdots + t_n e_n},$$

for $1 \leq k \leq m$. By Lemma 2.1.4, if we fix the coordinates $t_1, \ldots, t_{i-1}$, $t_{i+1}, \ldots, t_n$, the process $\{\varphi^{n,k}(t_1, \ldots, t_n), t_i \in \mathbb{R}\}$ has a version with absolutely continuous paths. So, for almost all $t$ the function $\varphi^{n,k}(t_1, \ldots, t_n)$ has an approximate partial derivative with respect to the $i$th coordinate, and moreover,

$$\mathrm{ap}\partial_i \varphi^{n,k}(t) = \langle DF^k, e_i \rangle_H^{t_1 e_1 + \cdots + t_n e_n}.$$

Consequently, we have

$$\langle \mathrm{ap}\nabla\varphi^{n,k}, \mathrm{ap}\nabla\varphi^{n,j} \rangle = (\sum_{i=1}^{n} \langle DF^k, e_i \rangle_H \langle DF^j, e_i \rangle_H)^{t_1 e_1 + \cdots + t_n e_n}. \quad (2.23)$$

Let $B$ be a Borel subset of $\mathbb{R}^m$ of zero Lebesgue measure. Then, Lemma 2.1.2 applied to the function $\varphi^n = (\varphi^{n,1}, \ldots, \varphi^{n,m})$ yields, for almost all $\omega$, assuming $n \geq m$

$$\int_{(\varphi^n)^{-1}(B)} \det[\langle \mathrm{ap}\nabla\varphi^{n,k}, \mathrm{ap}\nabla\varphi^{n,j} \rangle]dt_1 \ldots dt_n = 0.$$

Set $G = \{t \in \mathbb{R}^n : F^{t_1 e_1 + \cdots + t_n e_n}(\omega) \in B\}$. Taking expectations in the above expression and using (2.23), we deduce

$$
\begin{aligned}
0 &= E \int_G \left( \det(\sum_{i=1}^n \langle DF^k, e_i \rangle_H \langle DF^j, e_i \rangle_H) \right)^{t_1 e_1 + \cdots + t_n e_n} dt_1 \cdots dt_n \\
&= \int_{\mathbb{R}^n} E \left\{ \det(\sum_{i=1}^n \langle DF^k, e_i \rangle_H \langle DF^j, e_i \rangle_H) \mathbf{1}_{F^{-1}(B)} \right. \\
&\qquad \left. \times \exp(\sum_{i=1}^n (t_i W(e_i) - \frac{1}{2} t_i^2)) \right\} dt_1 \cdots dt_n.
\end{aligned}
$$

Consequently,

$$
\mathbf{1}_{F^{-1}(B)} \det(\sum_{i=1}^n \langle DF^k, e_i \rangle_H \langle DF^j, e_i \rangle_H) = 0
$$

almost surely, and letting $n$ tend to infinity yields

$$
\mathbf{1}_{F^{-1}(B)} \det(\langle DF^k DF^j \rangle_H) = 0,
$$

almost surely. Therefore, $P(F^{-1}(B)) = 0$, and the proof of the theorem is complete. □

As in the remark after the proof of Theorem 2.1.1, if we only assume condition (i) in Theorem 2.1.2, then the measure $(\det(\langle DF^k, DF^j \rangle_H) \cdot P) \circ F^{-1}$ is absolutely continuous with respect to the Lebesgue measure on $\mathbb{R}^m$.

The following result is a version of Theorem 2.1.2 for one-dimensional random variables. The proof we present here, which has been taken from [266], is much shorter than the proof of Theorem 2.1.2. It even works for $p = 1$.

**Theorem 2.1.3** *Let $F$ be a random variable of the space $\mathbb{D}^{1,1}_{loc}$, and suppose that $\|DF\|_H > 0$ a.s. Then the law of $F$ is absolutely continuous with respect to the Lebesgue measure on $\mathbb{R}$.*

*Proof:*   By the standard localization argument we may assume that $F$ belongs to the space $\mathbb{D}^{1,1}$. Also, we can assume that $|F| < 1$. We have to show that for any measurable function $g : (-1, 1) \to [0, 1]$ such that $\int_{-1}^1 g(y) dy = 0$ we have $E(g(F)) = 0$. We can find a sequence of continuously differentiable functions with bounded derivatives $g^n : (-1, 1) \to [0, 1]$ such that as $n$ tends to infinity $g^n(y)$ converges to $g(y)$ for almost all $y$ with respect to the measure $P \circ F^{-1} + \lambda^1$. Set

$$
\psi^n(y) = \int_{-1}^y g^n(x) dx
$$

and

$$\psi(y) = \int_{-1}^{y} g(x)dx.$$

By the chain rule, $\psi^n(F)$ belongs to the space $\mathbb{D}^{1,1}$ and we have $D[\psi^n(F)] = g^n(F)DF$. We have that $\psi^n(F)$ converges to $\psi(F)$ a.s. as $n$ tends to infinity, because $g^n$ converges to $g$ a.e. with respect to the Lebesgue measure. This convergence also holds in $L^1(\Omega)$ by dominated convergence. On the other hand, $D\psi^n(F)$ converges a.s. to $g(F)DF$ because $g^n$ converges to $g$ a.e. with respect to the law of $F$. Again by dominated convergence, this convergence holds in $L^1(\Omega; H)$. Observe that $\psi(F) = 0$ a.s. Now we use the property that the operator $D$ is closed to deduce that $g(F)DF = 0$ a.s. Consequently, $g(F) = 0$ a.s., which completes the proof of the theorem.  □

As in the case of Theorems 2.1.1 and 2.1.2, the proof of Theorem 2.1.3 yields the following result:

**Corollary 2.1.2** *Let $F$ be a random variable in $\mathbb{D}_{loc}^{1,1}$. Then the measure $(\|DF\|_H \cdot P) \circ F^{-1}$ is absolutely continuous with respect to the Lebesgue measure.*

This is equivalent to saying that the random variable $F$ has an absolutely continuous law conditioned by the set $\{\|DF\|_H > 0\}$; this means that

$$P\{F \in B, \|DF\|_H > 0\} = 0$$

for any Borel subset of $\mathbb{R}$ of zero Lebesgue measure.

### 2.1.4  Smoothness of densities

In order to derive the smoothness of the density of a random vector we will impose the nondegeneracy condition given in the following definition.

**Definition 2.1.1** *We will say that a random vector $F = (F^1, \ldots, F^m)$ whose components are in $\mathbb{D}^\infty$ is nondegenerate if the Malliavin matrix $\gamma_F$ is invertible a.s. and*

$$(\det \gamma_F)^{-1} \in \cap_{p \geq 1} L^p(\Omega).$$

We aim to cover some examples of random vectors whose components are not in $\mathbb{D}^\infty$ and satisfy a local nondegenerary condition. In these examples, the density of the random vector will be smooth only on an open subset of $\mathbb{R}^m$. To handle these example we introduce the following definition.

**Definition 2.1.2** *We will say that a random vector $F = (F^1, \ldots, F^m)$ whose components are in $\mathbb{D}^{1,2}$ is locally nondegenerate in an open set $A \subset \mathbb{R}^m$ if there exist elements $u_A^j \in \mathbb{D}^\infty(H)$, $j = 1, \ldots, m$ and an $m \times m$ random matrix $\gamma_A = (\gamma_A^{ij})$ such that $\gamma_A^{ij} \in \mathbb{D}^\infty$, $|\det \gamma_A|^{-1} \in L^p(\Omega)$ for all $p \geq 1$, and $\langle DF^i, u_A^j \rangle_H = \gamma_A^{ij}$ on $\{F \in A\}$ for any $i, j = 1, \ldots, m$.*

Clearly, a nondegenerate random vector is also locally nondegenerate in $\mathbb{R}^m$, and we can take $u_{\mathbb{R}^m}^j = DF^j$, and $\gamma_A = \gamma_F$.

We need the following preliminary lemma.

**Lemma 2.1.6** *Suppose that $\gamma$ is an $m \times m$ random matrix that is invertible a.s. and such that $|\det \gamma|^{-1} \in L^p(\Omega)$ for all $p \geq 1$. Suppose that the entries $\gamma^{ij}$ of $\gamma$ are in $\mathbb{D}^\infty$. Then $(\gamma^{-1})^{ij}$ belongs to $\mathbb{D}^\infty$ for all $i, j$, and*

$$D\left(\gamma^{-1}\right)^{ij} = -\sum_{k,l=1}^m \left(\gamma^{-1}\right)^{ik} \left(\gamma^{-1}\right)^{lj} D\gamma^{kl}. \tag{2.24}$$

*Proof:*    First notice that $\{\det \gamma > 0\}$ has probability zero or one (see Exercise 1.3.4). We will assume that $\det \gamma > 0$ a.s. For any $\epsilon > 0$ define

$$\gamma_\epsilon^{-1} = \frac{\det \gamma}{\det \gamma + \epsilon} \gamma^{-1}.$$

Note that $(\det \gamma + \epsilon)^{-1}$ belongs to $\mathbb{D}^\infty$ because it can be expressed as the composition of $\det \gamma$ with a function in $C_p^\infty(\mathbb{R})$. Therefore, the entries of $\gamma_\epsilon^{-1}$ belong to $\mathbb{D}^\infty$. Furthermore, for any $i, j$, $\left(\gamma_\epsilon^{-1}\right)^{ij}$ converges in $L^p(\Omega)$ to $\left(\gamma^{-1}\right)^{ij}$ as $\epsilon$ tends to zero. Then, in order to check that the entries of $\gamma^{-1}$ belong to $\mathbb{D}^\infty$, it suffices to show (taking into account Lemma 1.5.3) that the iterated derivatives of $\left(\gamma_\epsilon^{-1}\right)^{ij}$ are bounded in $L^p(\Omega)$, uniformly with respect to $\epsilon$, for any $p \geq 1$. This boundedness in $L^p(\Omega)$ holds, from the Leibnitz rule for the operator $D^k$ (see Exercise 1.2.13), because $(\det \gamma)\gamma^{-1}$ belongs to $\mathbb{D}^\infty$, and on the other hand, $(\det \gamma + \epsilon)^{-1}$ has bounded $\|\cdot\|_{k,p}$ norms for all $k, p$, due to our hypotheses.

Finally, from the expression $\gamma_\epsilon^{-1}\gamma = \frac{\det \gamma}{\det \gamma + \epsilon} I$, we deduce Eq. (2.24) by first applying the derivative operator $D$ and then letting $\epsilon$ tend to zero. $\square$

For a locally nondegenerate random vector the following integration-by-parts formula plays a basic role. For any multiindex $\alpha \in \{1, \ldots, m\}^k$, $k \geq 1$ we will denote by $\partial_\alpha$ the partial derivative $\frac{\partial^k}{\partial x_{\alpha_1} \cdots \partial x_{\alpha_k}}$.

**Proposition 2.1.4** *Let $F = (F^1, \ldots, F^m)$ be a locally nondegenerate random vector in an open set $A \subset \mathbb{R}^m$ in the sense of Definition 2.1.2. Let $G \in \mathbb{D}^\infty$ and let $\varphi$ be a function in the space $C_p^\infty(\mathbb{R}^m)$. Suppose that $G = 0$ on the set $\{F \notin A\}$. Then for any multiindex $\alpha \in \{1, \ldots, m\}^k$, $k \geq 1$, there exists an element $H_\alpha \in \mathbb{D}^\infty$ such that*

$$E\left[\partial_\alpha \varphi(F) G\right] = E\left[\varphi(F) H_\alpha\right]. \tag{2.25}$$

*Moreover, the elements $H_\alpha$ are recursively given by*

$$H_{(i)} = \sum_{j=1}^m \delta\left(G\left(\gamma_A^{-1}\right)^{ij} u_A^j\right), \tag{2.26}$$

$$H_\alpha = H_{\alpha_k}(H_{(\alpha_1, \ldots, \alpha_{k-1})}), \tag{2.27}$$

*and for $1 \leq p < q < \infty$ we have*

$$\|H_\alpha\|_p \leq c_{p,q} \left\|\gamma_A^{-1} u\right\|_{k,2^k-1_r}^k \|G\|_{k,q}, \qquad (2.28)$$

*where $\frac{1}{p} = \frac{1}{q} + \frac{1}{r}$.*

*Proof:* By the chain rule (Proposition 1.2.3) we have on $\{F \in A\}$

$$\langle D(\varphi(F)), u_A^j \rangle_H = \sum_{i=1}^m \partial_i \varphi(F) \langle DF^i, u_A^j \rangle_H = \sum_{i=1}^m \partial_i \varphi(F) \gamma_A^{ij},$$

and, consequently,

$$\partial_i \varphi(F) = \sum_{j=1}^m \langle D(\varphi(F)), u_A^j \rangle_H (\gamma_A^{-1})^{ji}.$$

Taking into account that $G$ vanishes on the set $\{F \notin A\}$, we obtain

$$G \partial_i \varphi(F) = \sum_{j=1}^m G \langle D(\varphi(F)), u_A^j \rangle_H (\gamma_A^{-1})^{ji}.$$

Finally, taking expectations and using the duality relationship between the derivative and the divergence operators we get

$$E\left[\partial_i \varphi(F)G\right] = E\left[\varphi(F)H_{(i)}\right],$$

where $H_{(i)}$ equals to the right-hand side of Equation (2.26). Equation (2.27) follows by recurrence.

Using the continuity of the operator $\delta$ from $\mathbb{D}^{1,p}(H)$ into $L^p(\Omega)$ and the Hölder inequality for the $\|\cdot\|_{p,k}$ norms (Proposition 1.5.6) we obtain

$$\|H_\alpha\|_p \leq c_p \left\| H_{(\alpha_1,\ldots,\alpha_{k-1})} \sum_{j=1}^m (\gamma_A^{-1})^{\alpha_k j} u^j \right\|_{1,p}$$

$$\leq c_p \left\| H_{(\alpha_1,\ldots,\alpha_{k-1})} \right\|_{1,q} \left\| (\gamma_A^{-1} u)^{\alpha_k} \right\|_{1,r}.$$

This implies (2.28) for $k = 1$, and the general case follows by recurrence. $\square$

If $F$ is nondegenerate then Equation (2.25) holds for any $G \in \mathbb{D}^\infty$, and we replace in this equation $\gamma_A$ and $u_A^j$ by $\gamma_F$ and $DF^j$, respectively. In that case, the element $H_\alpha$ depends only on $F$ and $G$ and we denote it by $H_\alpha(F, G)$. Then, formulas (2.25) to (2.28) are tranformed into

$$E\left[\partial_\alpha \varphi(F)G\right] = E\left[\varphi(F)H_\alpha(F, G)\right], \qquad (2.29)$$

where

$$H_{(i)}(F,G) = \sum_{j=1}^{m} \delta\left(G\left(\gamma_F^{-1}\right)^{ij} DF^j\right), \qquad (2.30)$$

$$H_{\alpha}(F,G) = H_{\alpha_k}(H_{(\alpha_1,\ldots,\alpha_{k-1})}(F,G)), \qquad (2.31)$$

and

$$\|H_{\alpha}(F,G)\|_p \le c_{p,q} \left\|\gamma_F^{-1} DF\right\|_{k,2^{k-1}r}^{k} \|G\|_{k,q}. \qquad (2.32)$$

As a consequence, there exists constants $\beta, \gamma > 1$ and integers $n, m$ such that

$$\|H_{\alpha}(F,G)\|_p \le c_{p,q} \left\|\det \gamma_F^{-1}\right\|_{\beta}^{m} \|DF\|_{k,\gamma}^{n} \|G\|_{k,q}. $$

Now we can state the local criterion for smoothness of densities which allows us to show the smoothness of the density for random variables that are not necessarily in the space $\mathbb{D}^{\infty}$.

**Theorem 2.1.4** *Let $F = (F^1, \ldots, F^m)$ be a locally nondegenerate random vector in an open set $A \subset \mathbb{R}^m$ in the sense of Definition 2.1.2. Then $F$ possesses an infinitely differentiable density on the open set $A$.*

*Proof:* Fix $x_0 \in A$, and consider an open ball $B_{\delta}(x_0)$ of radius $\delta < \frac{1}{2}d(x_0, A^c)$. Let $\delta < \delta' < d(x_0, A^c)$. Consider a function $\psi \in C^{\infty}(\mathbb{R}^m)$ such that $0 \le \psi(x) \le 1$, $\psi(x) = 1$ on $B_{\delta}(x_0)$, and $\psi(x) = 0$ on the complement of $B_{\delta'}(x_0)$. Equality (2.25) applied to the multiindex $\alpha = (1, 2, \ldots, m)$ and to the random variable $G = \psi(F)$ yields, for any function $\varphi$ in $C_p^{\infty}(\mathbb{R}^m)$

$$E\left[\psi(F)\partial_{\alpha}\varphi(F)\right] = E[\varphi(F)H_{\alpha}].$$

Notice that

$$\varphi(F) = \int_{-\infty}^{F^1} \cdots \int_{-\infty}^{F^m} \partial_{\alpha}\varphi(x) dx.$$

Hence, by Fubini's theorem we can write

$$E\left[\psi(F)\partial_{\alpha}\varphi(F)\right] = \int_{\mathbb{R}^m} \partial_{\alpha}\varphi(x) E\left[\mathbf{1}_{\{F > x\}} H_{\alpha}\right] dx. \qquad (2.33)$$

We can take as $\partial_{\alpha}\varphi$ any function in $C_0^{\infty}(\mathbb{R}^m)$. Then Equation (2.33) implies that on the ball $B_{\delta}(x_0)$ the random vector $F$ has a density given by

$$p(x) = E\left[\mathbf{1}_{\{F > x\}} H_{\alpha}\right].$$

Moreover, for any multiindex $\beta$ we have

$$\begin{aligned} E\left[\psi(F)\partial_{\beta}\partial_{\alpha}\varphi(F)\right] &= E[\varphi(F)H_{\beta}(H_{\alpha})] \\ &= \int_{\mathbb{R}^m} \partial_{\alpha}\varphi(x) E\left[\mathbf{1}_{\{F > x\}} H_{\beta}(H_{\alpha})\right] dx. \end{aligned}$$

Hence, for any $\xi \in C_0^\infty(B_\delta(x_0))$

$$\int_{\mathbb{R}^m} \partial_\beta \xi(x) p(x) dx = \int_{\mathbb{R}^m} \xi(x) E\left[\mathbf{1}_{\{F>x\}} H_\beta(H_\alpha)\right] dx.$$

Therefore $p(x)$ is infinitely differentiable in the ball $B_\delta(x_0)$, and for any multiindex $\beta$ we have

$$\partial_\beta p(x) = (-1)^{|\beta|} E\left[\mathbf{1}_{\{F>x\}} H_\beta(H_\alpha)\right].$$

□

We denote by $\mathcal{S}(\mathbb{R}^m)$ the space of all infinitely differentiable functions $f : \mathbb{R}^m \to \mathbb{R}$ such that for any $k \geq 1$, and for any multiindex $\beta \in \{1,\ldots,m\}^j$ one has $\sup_{x \in \mathbb{R}^m} |x|^k |\partial_\beta f(x)| < \infty$ (Schwartz space).

**Proposition 2.1.5** *Let $F = (F^1,\ldots,F^m)$ be a nondegenerate random vector in the sense of Definition 2.1.1. Then the density of $F$ belongs to the space $\mathcal{S}(\mathbb{R}^m)$, and*

$$p(x) = E\left[\mathbf{1}_{\{F>x\}} H_{(1,2,\ldots,m)}(F, 1)\right]. \tag{2.34}$$

*Proof:*    The proof of Theorem 2.1.4 implies, taking $G = 1$, that $F$ possesses an infinitely differentiable density and (2.34) holds. Moreover, for any multiindex $\beta$

$$\partial_\beta p(x) = (-1)^{|\beta|} E\left[\mathbf{1}_{\{F>x\}} H_\beta(H_{(1,2,\ldots,m)}(F, 1))\right].$$

In order to show that the density belongs to $\mathcal{S}(\mathbb{R}^m)$ we have to prove that for any multiindex $\beta$ and for any $k \geq 1$ and for all $j = 1,\ldots,m$

$$\sup_{x \in \mathbb{R}^m} x_j^{2k} |E\left[\mathbf{1}_{\{F>x\}} H_\beta(H_{(1,2,\ldots,m)}(F, 1))\right]| < \infty.$$

If $x_j > 0$ we have

$$x_j^{2k} |E\left[\mathbf{1}_{\{F>x\}} H_\beta(H_{(1,2,\ldots,m)}(F, 1))\right]|$$
$$\leq E\left[|F^j|^{2k} |H_\beta(H_{(1,2,\ldots,m)}(F, 1))|\right] < \infty.$$

If $x_j < 0$ then we use the alternative expression for the density

$$p(x) = E\left[\prod_{i \neq j} \mathbf{1}_{\{x^i < F^i\}} \mathbf{1}_{\{x^j > F^j\}} H_{(1,2,\ldots,m)}(F, 1)\right],$$

and we deduce a similar estimate.

□

### 2.1.5   Composition of tempered distributions with nondegenerate random vectors

Let $F$ be an $m$-dimensional random vector. The probability density of $F$ at $x \in \mathbb{R}^m$ can be formally defined as the generalized expectation $E(\delta_x(F))$, where $\delta_x$ denotes the Dirac function at $x$. The expression $E(\delta_x(F))$ can be interpreted as the coupling $\langle \delta_x(F), 1 \rangle$, provided we show that $\delta_x(F)$ is an element of $\mathbb{D}^{-\infty}$. The Dirac function $\delta_x$ is a measure, and more generally we will see that we can define the composition $T(F)$ of a Schwartz distribution $T \in \mathcal{S}'(\mathbb{R}^m)$ with a nondegenerate random vector, and the composition will belong to $\mathbb{D}^{-\infty}$. Furthermore, the diferentiability of the mapping $x \to \delta_x(F)$ from $\mathbb{R}^m$ into some Sobolev space $\mathbb{D}^{-k,p}$ provides an alternative proof of the smoothness of the density of $F$.

Consider the following sequence of seminorms in the space $\mathcal{S}(\mathbb{R}^m)$:

$$\|\phi\|_{2k} = \left\| (1 + |x|^2 - \Delta)^k \phi \right\|_\infty, \quad \phi \in \mathcal{S}(\mathbb{R}^m), \tag{2.35}$$

for $k \in \mathbb{Z}$. Let $\mathcal{S}_{2k}$, $k \in \mathbb{Z}$, be the completion of $\mathcal{S}(\mathbb{R}^m)$ by the seminorm $\|\cdot\|_{2k}$. Then we have

$$\mathcal{S}_{2k+2} \subset \mathcal{S}_{2k} \subset \cdots \subset \mathcal{S}_2 \subset \mathcal{S}_0 \subset \mathcal{S}_{-2} \subset \cdots \subset \mathcal{S}_{-2k} \subset \mathcal{S}_{-2k-2},$$

and $\mathcal{S}_0 = \widehat{C}(\mathbb{R}^m)$ is the space of continuous functions on $\mathbb{R}^m$ which vanish at infinity. Moreover, $\cap_{k \geq 1} \mathcal{S}_{2k} = \mathcal{S}(\mathbb{R}^m)$ and $\cup_{k \geq 1} \mathcal{S}_{-2k} = \mathcal{S}'(\mathbb{R}^m)$.

**Proposition 2.1.6** *Let $F = (F^1, \ldots, F^m)$ be a nondegenerate random vector in the sense of Definition 2.1.1. For any $k \in \mathbb{N}$ and $p > 1$, there exists a constant $c(p, k, F)$ such that for any $\phi \in \mathcal{S}(\mathbb{R}^m)$ we have*

$$\|\phi(F)\|_{-2k,p} \leq c(p, k, F) \|\phi\|_{-2k}.$$

*Proof:*   Let $\psi = (1 + |x|^2 - \Delta)^{-k} \phi \in \mathcal{S}(\mathbb{R}^m)$. By Proposition 2.1.4 for any $G \in \mathbb{D}^\infty$ there exists $R_{2k}(G) \in \mathbb{D}^\infty$ such that

$$E\left[\phi(F)G\right] = E\left[(1 + |x|^2 - \Delta)^k \psi(F)G\right] = E\left[\psi(F) R_{2k}(G)\right].$$

Therefore, using (2.35) and (2.28) with $q$ such that $\frac{1}{p} + \frac{1}{q} = 1$, yields

$$|E\left[\phi(F)G\right]| \leq \|\psi\|_\infty E\left[|R_{2k}(G)|\right] \leq c(p, k, F) \|\phi\|_{-2k} \|G\|_{2k,q}.$$

Finally, it suffices to use the fact that

$$\||\phi(F)|\|_{-2k,p} = \sup \left\{ |E\left[\phi(F)G\right]|, G \in \mathbb{D}^{2k,q}, \|G\|_{2k,q} \leq 1 \right\}.$$

$\square$

**Corollary 2.1.3** *Let $F$  be a nondegenerate random vector. For any $k \in \mathbb{N}$ and $p > 1$ we can uniquely extend the mapping  $\phi \to \phi(F)$ to a continuous linear mapping from $\mathcal{S}_{-2k}$ into $\mathbb{D}^{-2k,p}$.*

As a consequence of the above Corollary, we can define the composition of a Schwartz distribution $T \in \mathcal{S}'(\mathbb{R}^m)$ with the nondegenerate random vector $F$, as a generalized random variable $T(F) \in \mathbb{D}^{-\infty}$. Actually,

$$T(F) \in \cup_{k=1}^{\infty} \cap_{p>1} \mathbb{D}^{-2k,p}.$$

For $k = 0$, $\phi(F)$ coincides with the usual composition of the continuous function $\phi \in \mathcal{S}_0 = \widehat{C}(\mathbb{R}^m)$ and the random vector $F$.

For any $x \in \mathbb{R}^m$, the Dirac function $\delta_x$ belongs to $\mathcal{S}_{-2k}$, where $k = \left[\frac{m}{2}\right] + 1$, and the mapping $x \to \delta_x$ is $2j$ continuously differentiable from $\mathbb{R}^m$ to $\mathcal{S}_{-2k-2j}$, for any $j \in \mathbb{N}$. Therefore, for any nondegenerate random vector $F$, the composition $\delta_x(F)$ belongs to $\mathbb{D}^{-2k,p}$ for any $p > 1$, and the mapping $x \to \delta_x(F)$ is $2j$ continuously differentiable from $\mathbb{R}^m$ to $\mathbb{D}^{-2k-2j,p}$, for any $j \in \mathbb{N}$. This implies that for any $G \in \mathbb{D}^{2k+2j,p}$ the mapping $x \to \langle \delta_x(F), G \rangle$ belongs to $C^{2j}(\mathbb{R}^m)$.

**Lemma 2.1.7** Let $k = \left[\frac{m}{2}\right] + 1$ and $p > 1$. If $f \in C_0(\mathbb{R}^m)$, then for any $G \in \mathbb{D}^{2k,q}$

$$\int_{\mathbb{R}^m} f(x) \langle \delta_x(F), G \rangle \, dx = E\left[f(F)G\right].$$

*Proof:*    We have

$$f = \int_{\mathbb{R}^m} f(x) \delta_x dx,$$

where the integral is $\mathcal{S}_{-2k}$-valued and in the sense of Bochner. Thus, approximating the integral by Riemann sums we obtain

$$f(F) = \int_{\mathbb{R}^m} f(x) \delta_x(F) dx,$$

in $\mathbb{D}^{-2k,p}$. Finally, multiplying by $G$ and taking expectations we get the result.    $\square$

This lemma and previous remarks imply that for any $G \in \mathbb{D}^{2k+2j,p}$, the measure

$$\mu_G(B) = E\left[\mathbf{1}_{\{F \in B\}} G\right], \quad B \in \mathcal{B}(\mathbb{R}^m)$$

has a density $p_G(x) = \langle \delta_x(F), G \rangle \in C^{2j}(\mathbb{R}^m)$. In particular, $\langle \delta_x(F), 1 \rangle$ is the density of $F$ and it will be infinitely differentiable.

## 2.1.6  Properties of the support of the law

Given a random vector $F : \Omega \to \mathbb{R}^m$, the topological support of the law of $F$ is defined as the set of points $x \in \mathbb{R}^m$ such that $P(|x - F| < \varepsilon) > 0$ for all $\varepsilon > 0$. The following result asserts the connectivity property of the support of a smooth random vector.

**Proposition 2.1.7** *Let $F = (F^1, \ldots, F^m)$ be a random vector whose components belong to $\mathbb{D}^{1,p}$ for some $p > 1$. Then, the topological support of the law of $F$ is a closed connected subset of $\mathbb{R}^m$.*

*Proof:*   If the support of $F$ is not connected, it can be decomposed as the union of two nonempty disjoint closed sets $A$ and $B$.

For each integer $M \geq 2$ let $\psi_M : \mathbb{R}^m \to \mathbb{R}$ be an infinitely differentiable function such that $0 \leq \psi_M \leq 1$, $\psi_M(x) = 0$ if $|x| \geq M$, $\psi_M(x) = 1$ if $|x| \leq M - 1$, and $\sup_{x,M} |\nabla \psi_M(x)| < \infty$.

Set $A_M = A \cap \{|x| \leq M\}$ and $B_M = B \cap \{|x| \leq M\}$. For $M$ large enough we have $A_M \neq \emptyset$ and $B_M \neq \emptyset$, and there exists an infinitely differentiable function $f_M$ such that $0 \leq f_M \leq 1$, $f_M = 1$ in a neighborhood of $A_M$, and $f_M = 0$ in a neighborhood of $B_M$.

The sequence $(f_M \psi_M)(F)$ converges a.s. and in $L^p(\Omega)$ to $\mathbf{1}_{\{F \in A\}}$ as $M$ tends to infinity. On the other hand, we have

$$
D\left[(f_M \psi_M)(F)\right] = \sum_{i=1}^{m} \left[(\psi_M \partial_i f_M)(F)DF^i + (f_M \partial_i \psi_M)(F)DF^i\right]
$$

$$
= \sum_{i=1}^{m} (f_M \partial_i \psi_M)(F)DF^i.
$$

Hence,

$$
\sup_M \|D\left[(f_M \psi_M)(F)\right]\|_H \leq \sum_{i=1}^{m} \sup_M \|\partial_i \psi_M\|_\infty \|DF^i\|_H \in L^p(\Omega).
$$

By Lemma 1.5.3 we get that $\mathbf{1}_{\{F \in A\}}$ belongs to $\mathbb{D}^{1,p}$, and by Proposition 1.2.6 this is contradictory because $0 < P(F \in A) < 1$.   $\square$

As a consequence, the support of the law of a random variable $F \in \mathbb{D}^{1,p}$, $p > 1$ is a closed interval. The next result provides sufficient conditions for the density of $F$ to be nonzero in the interior of the support.

**Proposition 2.1.8** *Let $F \in \mathbb{D}^{1,p}$, $p > 2$, and suppose that $F$ possesses a density $p(x)$ which is locally Lipschitz in the interior of the support of the law of $F$. Let $a$ be a point in the interior of the support of the law of $F$. Then $p(a) > 0$.*

*Proof:*     Suppose $p(a) = 0$. Set $r = \frac{2p}{p+2} > 1$. From Proposition 1.2.6 we know that $\mathbf{1}_{\{F > a\}} \notin \mathbb{D}^{1,r}$ because $0 < P(F > a) < 1$. Fix $\epsilon > 0$ and set

$$
\varphi_\epsilon(x) = \int_{-\infty}^{x} \frac{1}{2\epsilon} \mathbf{1}_{[a-\epsilon, a+\epsilon]}(y) dy.
$$

Then $\varphi_\epsilon(F)$ converges to $\mathbf{1}_{\{F > a\}}$ in $L^r(\Omega)$ as $\epsilon \downarrow 0$. Moreover, $\varphi_\epsilon(F) \in \mathbb{D}^{1,r}$ and

$$
D(\varphi_\epsilon(F)) = \frac{1}{2\epsilon} \mathbf{1}_{[a-\epsilon, a+\epsilon]}(F)DF.
$$

We have

$$E\left(\|D(\varphi_\epsilon(F))\|_H^r\right) \leq (E(\|DF\|_H^p)^{\frac{2}{p+2}} \left(\frac{1}{(2\epsilon)^2} \int_{a-\epsilon}^{a+\epsilon} p(x)dx\right)^{\frac{p}{p+2}}.$$

The local Lipschitz property of $p$ implies that $p(x) \leq K|x - a|$, and we obtain

$$E\left(\|D(\varphi_\epsilon(F))\|_H^r\right) \leq (E(\|DF\|_H^p)^{\frac{2}{p+2}} 2^{-r} K^{\frac{p}{p+2}}.$$

By Lemma 1.5.3 this implies $\mathbf{1}_{\{F>a\}} \in \mathbb{D}^{1,r}$, resulting in a contradiction.
□

Sufficient conditions for the density of $F$ to be continuously differentiable are given in Exercise 2.1.8.

The following example shows that, unlike the one-dimensional case, in dimension $m > 1$ the density of a nondegenerate random vector may vanish in the interior of the support.

**Example 2.1.1** *Let $h_1$ and $h_2$ be two orthonormal elements of $H$. Define $X = (X_1, X_2)$, where*

$$\begin{aligned}
X_1 &= \arctan W(h_1), \\
X_2 &= \arctan W(h_2).
\end{aligned}$$

*Then, $X_i \in \mathbb{D}^\infty$ and*

$$DX_i = (1 + W(h_i)^2)^{-1} h_i,$$

*for $i = 1, 2$, and*

$$\det \gamma_X = \left[(1 + W(h_1)^2)(1 + W(h_2)^2)\right]^{-2}.$$

*The support of the law of the random vector $X$ is the rectangle $\left[-\frac{\pi}{2}, \frac{\pi}{2}\right]^2$, and the density of $X$ is strictly positive in the interior of the support. Now consider the vector $Y = (Y_1, Y_2)$ given by*

$$\begin{aligned}
Y_1 &= (X_1 + \frac{3\pi}{2})\cos(2X_2 + \pi), \\
Y_2 &= (X_1 + \frac{3\pi}{2})\sin(2X_2 + \pi).
\end{aligned}$$

*We have that $Y_i \in \mathbb{D}^\infty$ for $i = 1, 2$, and*

$$\det \gamma_Y = 4(X_1 + \frac{3\pi}{2})^2 \left[(1 + W(h_1)^2)(1 + W(h_2)^2)\right]^{-2}.$$

*This implies that $Y$ is a nondegenerate random vector. Its support is the set $\{(x,y) : \pi^2 \leq x^2 + y^2 \leq 4\pi^2\}$, and the density of $Y$ vanishes on the points $(x, y)$ in the support such that $\pi < y < 2\pi$ and $x = 0$.*

For a nondegenerate random vector when the density vanishes, then all its partial derivatives also vanish.

**Proposition 2.1.9** *Let $F = (F^1, \ldots, F^m)$ be a nondegenerate random vector in the sense of Definition 2.1.1 and denote its density by $p(x)$. Then $p(x) = 0$ implies $\delta_\alpha p(x) = 0$ for any multiindex $\alpha$.*

*Proof:*    Suppose that $p(x) = 0$. For any nonnegative random variable $G \in \mathbb{D}^\infty$, $\langle \delta_x(F), G \rangle \geq 0$ because this is the density of the measure $\mu_G(B) = E\left[\mathbf{1}_{\{F \in B\}} G\right]$, $B \in \mathcal{B}(\mathbb{R}^m)$. Fix a complete orthonormal system $\{e_i, i \geq 1\}$ in $H$. For each $n \geq 1$ the function $\varphi : \mathbb{R}^n \to \mathbb{C}$ given by

$$\varphi(t) = \left\langle \delta_x(F), \exp\left( i \sum_{j=1}^n t^j W(e_j) \right) \right\rangle$$

is nonnegative definite and continuous. Thus, there exists a measure $\nu_n$ on $\mathbb{R}^n$ such that

$$\varphi(t) = \int_{\mathbb{R}^n} e^{i\langle t, x \rangle} d\nu_n(x).$$

Note that $\nu_n(\mathbb{R}^n) = \langle \delta_x(F), 1 \rangle = p(x) = 0$. So, this measure is zero and we get that $\langle \delta_x(F), G \rangle = 0$ for any polynomial random variable $G \in \mathcal{P}$. This implies that $\delta_x(F) = 0$ as an element of $\mathbb{D}^{-\infty}$.

For any multiindex $\alpha$ we have

$$\partial_\alpha p(x) = \partial_\alpha \langle \delta_x(F), 1 \rangle = \langle (\partial_\alpha \delta_x)(F), 1 \rangle.$$

Hence, it suffices to show that $(\partial_\alpha \delta_x)(F)$ vanishes. Suppose first that $\alpha = \{i\}$. We can write

$$D(\delta_x(F)) = \sum_{i=1}^m (\partial_i \delta_x)(F) DF^i$$

as elements of $\mathbb{D}^{-\infty}$, which implies

$$(\partial_i \delta_x)(F) = \sum_{j=1}^m \langle D(\delta_x(F)), DF^j \rangle_H (\gamma_F^{-1})^{ji} = 0$$

because $D(\delta_x(F)) = 0$. The general case follows by recurrence.    □

## 2.1.7    Regularity of the law of the maximum of continuous processes

In this section we present the application of the Malliavin calculus to the absolute continuity and smoothness of the density for the supremum of a continuous process. We assume that the $\sigma$-algebra of the underlying probability space $(\Omega, \mathcal{F}, P)$ is generated by an isonormal Gaussian process $W = \{W(h), h \in H\}$. Our first result provides sufficient conditions for the differentiability of the supremum of a continuous process.

**Proposition 2.1.10** *Let* $X = \{X(t), t \in S\}$ *be a continuous process parametrized by a compact metric space* $S$. *Suppose that*

(i) $E(\sup_{t \in S} X(t)^2) < \infty$;

(ii) *for any* $t \in S$, $X(t) \in \mathbb{D}^{1,2}$, *the* $H$-*valued process* $\{DX(t), t \in S\}$ *possesses a continuous version, and* $E(\sup_{t \in S} \|DX(t)\|_H^2) < \infty$.

*Then the random variable* $M = \sup_{t \in S} X(t)$ *belongs to* $\mathbb{D}^{1,2}$.

*Proof:*    Consider a countable and dense subset $S_0 = \{t_n, n \geq 1\}$ in $S$. Define $M_n = \sup\{X(t_1), \ldots, X(t_n)\}$. The function $\varphi_n : \mathbb{R}^n \to \mathbb{R}$ defined by $\varphi_n(x_1, \ldots, x_n) = \max\{x_1, \ldots, x_n\}$ is Lipschitz. Therefore, from Proposition 1.2.4 we deduce that $M_n$ belongs to $\mathbb{D}^{1,2}$. The sequence $M_n$ converges in $L^2(\Omega)$ to $M$. Thus, by Lemma 1.2.3 it suffices to see that the sequence $DM_n$ is bounded in $L^2(\Omega; H)$. In order to evaluate the derivative of $M_n$, we introduce the following sets:

$$A_1 = \{M_n = X(t_1)\},$$

$$\ldots$$

$$A_k = \{M_n \neq X(t_1), \ldots, M_n \neq X(t_{k-1}), M_n = X(t_k)\}, \quad 2 \leq k \leq n.$$

By the local property of the operator $D$, on the set $A_k$ the derivatives of the random variables $M_n$ and $X(t_k)$ coincide. Hence, we can write

$$DM_n = \sum_{k=1}^{n} \mathbf{1}_{A_k} DX(t_k).$$

Consequently,

$$E(\|DM_n\|_H^2) \leq E\left(\sup_{t \in S} \|DX(t)\|_H^2\right) < \infty,$$

and the proof is complete.    □

We can now establish the following general criterion of absolute continuity.

**Proposition 2.1.11** *Let* $X = \{X(t), t \in S\}$ *be a continuous process parametrized by a compact metric space* $S$ *verifying the hypotheses of Proposition 2.1.10. Suppose that* $\|DX(t)\|_H \neq 0$ *on the set* $\{t : X(t) = M\}$. *Then the law of* $M = \sup_{t \in S} X(t)$ *is absolutely continuous with respect to the Lebesgue measure.*

*Proof:*     By Theorem 2.1.3 it suffices to show that a.s. $DM = DX(t)$ on the set $\{t : X(t) = M\}$. Thus, if we define the set

$$G = \{\text{there exists } t \in S : DX(t) \neq DM, \text{ and } X(t) = M\},$$

then $P(G) = 0$. Let $S_0 = \{t_n, n \geq 1\}$ be a countable and dense subset of $S$. Let $H_0$ be a countable and dense subset of the unit ball of $H$. We can write

$$G \subset \bigcup_{s \in S_0, r \in \mathbb{Q}, r > 0, k \geq 1, h \in H_0} G_{s,r,k,h},$$

where

$$G_{s,r,k,h} = \{\langle DX(t) - DM, h \rangle_H > \frac{1}{k} \text{ for all } t \in B_r(s)\} \cap \{\sup_{t \in B_r(s)} X_t = M\}.$$

Here $B_r(s)$ denotes the open ball with center $s$ and radius $r$. Because it is a countable union, it suffices to check that $P(G_{s,r,k,h}) = 0$ for fixed $s, r, k, h$. Set $M' = \sup\{X(t), t \in \overline{B_r(s)}\}$ and $M'_n = \sup\{X(t_i), 1 \leq i \leq n, t_i \in \overline{B_r(s)}\}$. By Lemma 1.2.3, $DM'_n$ converges to $DM'$ in the weak topology of $L^2(\Omega; H)$ as $n$ tends to infinity, but on the set $G_{s,r,k,h}$ we have

$$\langle DM'_n - DM', h \rangle_H \geq \frac{1}{k}$$

for all $n \geq 1$. This implies that $P(G_{s,r,k,h}) = 0$.     $\square$

Consider the case of a continuous Gaussian process $X = \{X(t), t \in S\}$ with covariance function $K(s,t)$, and suppose that the Gaussian space $\mathcal{H}_1$ is the closed span of the random variables $X(t)$. We can choose as Hilbert space $H$ the closed span of the functions $\{K(t, \cdot), t \in S\}$ with the scalar product

$$\langle K(t, \cdot), K(s, \cdot) \rangle_H = K(t, s),$$

that is, $H$ is the reproducing kernel Hilbert space (RKHS) (see [13]) associated with the process $X$. The space $H$ contains all functions of the form $\varphi(t) = E(YX(t))$, where $Y \in \mathcal{H}_1$. Then, $DX(t) = K(t, \cdot)$ and $\|DX(t)\|_H = K(t, t)$. As a consequence, the criterion of the above proposition reduces to $K(t, t) \neq 0$ on the set $\{t : X(t) = M\}$.

Let us now discuss the differentiability of the density of $M = \sup_{t \in S} X(t)$. If $S = [0, 1]$ and the process $X$ is a Brownian motion, then the law of $M$ has the density

$$p(x) = \frac{2}{\sqrt{2\pi}} e^{-\frac{x^2}{2}} \mathbf{1}_{[0,\infty)}(x).$$

Indeed, the reflection principle (see [292, Proposition III.3.7]) implies that $P\{\sup_{t \in [0,1]} X(t) > a\} = 2P\{X(1) > a\}$ for all $a > 0$. Note that $p(x)$ is infinitely differentiable in $(0, +\infty)$.

Consider now the case of a two-parameter Wiener process on the unit square $W = \{W(z), z \in [0,1]^2\}$. That is, $S = T = [0,1]^2$ and $\mu$ is the Lebesgue measure. Set $M = \sup_{z \in [0,1]^2} W(z)$. The explicit form of the density of $M$ is unknown. We will show that the density of $M$ is infinitely differentiable in $(0, +\infty)$, but first we will show some preliminary results.

**Lemma 2.1.8** *With probability one the Wiener sheet $W$ attains its maximum on $[0,1]^2$ on a unique random point $(S, T)$.*

*Proof:*    We want to show that the set

$$G = \left\{ \omega : \sup_{z \in [0,1]^2} W(z) = W(z_1) = W(z_2) \quad \text{for} \quad \text{some} \quad z_1 \neq z_2 \right\}$$

has probability zero. For each $n \geq 1$ we denote by $\mathcal{R}_n$ the class of dyadic rectangles of the form $[(j-1)2^{-n}, j2^{-n}] \times [(k-1)2^{-n}, k2^{-n}]$, with $1 \leq j, k \leq 2^n$. The set $G$ is included in the countable union

$$\bigcup_{n \geq 1} \bigcup_{R_1, R_2 \in \mathcal{R}_n, R_1 \cap R_2 = \emptyset} \left\{ \sup_{z \in R_1} W(z) = \sup_{z \in R_2} W(z) \right\}.$$

Finally, it suffices to check that for each $n \geq 1$ and for any couple of disjoint rectangles $R_1, R_2$ with sides parallel to the axes, $P\{\sup_{z \in R_1} W(z) = \sup_{z \in R_2} W(z)\} = 0$ (see Exercise 2.1.7). $\qquad \square$

**Lemma 2.1.9** *The random variable $M = \sup_{z \in [0,1]^2} W(z)$ belongs to $\mathbb{D}^{1,2}$ and $D_z M = \mathbf{1}_{[0,S] \times [0,T]}(z)$, where $(S, T)$ is the point where the maximum is attained.*

*Proof:*    We introduce the approximation of $M$ defined by

$$M_n = \sup\{W(z_1), \dots, W(z_n)\},$$

where $\{z_n, n \geq 1\}$ is a countable and dense subset of $[0,1]^2$. It holds that

$$D_z M_n = \mathbf{1}_{[0,S_n] \times [0,T_n]}(z),$$

where $(S_n, T_n)$ is the point where $M_n = W(S_n, T_n)$. We know that the sequence of derivatives $DM_n$ converges to $DM$ in the weak topology of $L^2([0,1]^2 \times \Omega)$. On the other hand, $(S_n, T_n)$ converges to $(S, T)$ almost surely. This implies the result. $\qquad \square$

As an application of Theorem 2.1.4 we can prove the regularity of the density of $M$.

**Proposition 2.1.12** *The random variable $M = \sup_{z \in [0,1]^2} W(z)$ possesses an infinitely differentiable density on $(0, +\infty)$.*

*Proof:*     Fix $a > 0$ and set $A = (a, +\infty)$. By Theorem 2.1.4 it suffices to show that $M$ is locally nondegenerate in $A$ in the sense of Definition 2.1.2. Define the following random variables:

$$T_a = \inf\{t : \sup_{\{0 \leq x \leq 1, 0 \leq y \leq t\}} W(x, y) > a\}$$

and

$$S_a = \inf\{s : \sup_{\{0 \leq x \leq s, 0 \leq y \leq 1\}} W(x, y) > a\}.$$

We recall that $S_a$ and $T_a$ are stopping times with respect to the one-parameter filtrations $\mathcal{F}_s^1 = \sigma\{W(x, y) : 0 \leq x \leq s, 0 \leq y \leq 1\}$ and $\mathcal{F}_t^2 = \sigma\{W(x, y) : 0 \leq x \leq 1, 0 \leq y \leq t\}$.

Note that $(S_a, T_a) \leq (S, T)$ on the set $\{M > a\}$. Hence, by Lemma 2.1.9 it holds that $D_z M(\omega) = 1$ for almost all $(z, \omega)$ such that $z \leq (S_a(\omega), T_a(\omega))$ and $M(\omega) > a$.

For every $0 < \gamma < \frac{1}{2}$ and $p > 2$ such that $\frac{1}{2p} < \gamma < \frac{1}{2} - \frac{1}{2p}$, we define the Hölder seminorm on $C_0([0, 1])$,

$$\|f\|_{p,\gamma} = \left( \int_{[0,1]^2} \frac{|f(x) - f(y)|^{2p}}{|x - y|^{1 + 2p\gamma}} dx dy \right)^{\frac{1}{2p}}.$$

We denote by $\mathcal{H}_{p,\gamma}$ the Banach space of continuous functions on $[0, 1]$ vanishing at zero and having a finite $(p, \gamma)$ norm.

We define two families of random variables:

$$Y^1(\sigma) = \int_{[0,\sigma]^2} \frac{\|W(s, \cdot) - W(s', \cdot)\|_{p,\gamma}^{2p}}{|s - s'|^{1 + 2p\gamma}} ds ds'$$

and

$$Y^2(\tau) = \int_{[0,\tau]^2} \frac{\|W(\cdot, t) - W(\cdot, t')\|_{p,\gamma}^{2p}}{|t - t'|^{1 + 2p\gamma}} dt dt',$$

where $\sigma, \tau \in [0, 1]$. Set $Y(\sigma, \tau) = Y^1(\sigma) + Y^2(\tau)$.

We claim that there exists a constant $R$, depending on $a$, $p$, and $\gamma$, such that

$$Y(\sigma, \tau) \leq R \quad \text{implies} \quad \sup_{z \in [0,\sigma] \times [0,1] \cup [0,1] \times [0,\tau]} W_z \leq a. \tag{2.36}$$

In order to show this property, we first apply Garsia, Rodemich, and Rumsey's lemma (see Appendix, Lemma A.3.1) to the $\mathcal{H}_{p,\gamma}$-valued function $s \hookrightarrow W(s, \cdot)$. From this lemma, and assuming $Y^1(\sigma) < R$, we deduce

$$\|W(s, \cdot) - W(s', \cdot)\|_{p,\gamma}^{2p} \leq c_{p,\gamma} R |s - s'|^{2p\gamma - 1}$$

for all $s, s' \in [0, \sigma]$. Hence,

$$\|W(s, \cdot)\|_{p,\gamma}^{2p} \leq c_{p,\gamma} R$$

for all $s \in [0, \sigma]$. Applying the same lemma to the real-valued function $t \hookrightarrow W(s, t)$ ($s$ is now fixed), we obtain

$$|W(s, t) - W(s, t')|^{2p} \leq c_{p,\gamma}^2 R |t - t'|^{2p\gamma - 1}$$

for all $t, t' \in [0, 1]$. Hence,

$$\sup_{0 \leq s \leq \sigma, 0 \leq t \leq 1} |W(s, t)| \leq c_{p,\gamma}^{1/p} R^{\frac{1}{2p}}.$$

Similarly, we can prove that

$$\sup_{0 \leq s \leq 1, 0 \leq t \leq \tau} |W(s, t)| \leq c_{p,\gamma}^{1/p} R^{\frac{1}{2p}},$$

and it suffices to choose $R$ in such a way that $c_{p,\gamma}^{1/p} R^{\frac{1}{2p}} < a$.

Now we introduce the stochastic process $u_A(s, t)$ and the random variable $\gamma_A$ that will verify the conditions of Definition 2.1.2.

Let $\psi : \mathbb{R}_+ \to \mathbb{R}_+$ be an infinitely differentiable function such that $\psi(x) = 0$ if $x > R$, $\psi(x) = 1$ if $x < \frac{R}{2}$, and $0 \leq \psi(x) \leq 1$. Then we define

$$u_A(s, t) = \psi(Y(s, t))$$

and

$$\gamma_A = \int_{[0,1]^2} \psi(Y(s, t)) ds dt.$$

On the set $\{M > a\}$ we have

(1) $\psi(Y(s, t)) = 0$ if $(s, t) \notin [0, S_a] \times [0, T_a]$. Indeed, if $\psi(Y(s, t)) \neq 0$, then $Y(s, t) \leq R$ (by definition of $\psi$) and by (2.36) this would imply $\sup_{z \in [0,s] \times [0,1] \cup [0,1] \times [0,t]} W_z \leq a$, and, hence, $s \leq S_a$, $t \leq T_a$, which is contradictory.

(2) $D_{s,t} M = 1$ if $(s, t) \in [0, S_a] \times [0, T_a]$, as we have proven before.

Consequently, on $\{M > a\}$ we obtain

$$\begin{aligned} \langle DM, u_A \rangle_H &= \int_{[0,1]^2} D_{s,t} M \psi(Y(s, t)) ds dt \\ &= \int_{[0,S_a] \times [0,T_a]} \psi(Y(s, t)) ds dt = \gamma_A. \end{aligned}$$

We have $\gamma_A \in \mathbb{D}^\infty$ and $u_A \in \mathbb{D}^\infty(H)$ because the variables $Y^1(s)$ and $Y^2(t)$ are in $\mathbb{D}^\infty$ (see Exercise 1.5.4 and [3]). So it remains to prove that $\gamma_A^{-1}$ has moments of all orders. We have

$$\int_{[0,1]^2} \psi(Y(s,t)) ds dt \geq \int_{[0,1]^2} \mathbf{1}_{\{Y(s,t) < \frac{R}{2}\}} ds dt$$

$$= \lambda^2 \{(s,t) \in [0,1]^2 : Y^1(s) + Y^2(t) < \frac{R}{2}\}$$

$$\geq \lambda^1 \{s \in [0,1] : Y^1(s) < \frac{R}{4}\}$$

$$\times \lambda^1 \{t \in [0,1] : Y^2(t) < \frac{R}{4}\}$$

$$= (Y^1)^{-1}(\frac{R}{4})(Y^2)^{-1}(\frac{R}{4}).$$

Here we have used the fact that the stochastic processes $Y^1$ and $Y^2$ are continuous and increasing. Finally for any $\epsilon$ we can write

$$P((Y^1)^{-1}(\frac{R}{4}) < \epsilon) = P(\frac{R}{4} < Y^1(\epsilon))$$

$$\leq P\left(\int_{[0,\epsilon]^2} \frac{\|W(s,\cdot) - W(s',\cdot)\|_{p,\gamma}^{2p}}{|s - s'|^{1+2p\gamma}} ds ds' > \frac{R}{4}\right)$$

$$\leq (\frac{4}{R})^p E\left(\left|\int_{[0,\epsilon]^2} \frac{\|W(s,\cdot) - W(s',\cdot)\|_{p,\gamma}^{2p}}{|s - s'|^{1+2p\gamma}} ds ds'\right|^p\right)$$

$$\leq C\epsilon^{2p}$$

for some constant $C > 0$. This completes the proof of the theorem.    $\square$

## Exercises

**2.1.1** Show that if $F$ is a random variable in $\mathbb{D}^{2,4}$ such that $E(\|DF\|^{-8}) < \infty$, then $\frac{DF}{\|DF\|^2} \in \text{Dom}\,\delta$ and

$$\delta\left(\frac{DF}{\|DF\|_H^2}\right) = -\frac{LF}{\|DF\|_H^2} - 2\frac{\langle DF \otimes DF, D^2 F\rangle_{H \otimes H}}{\|DF\|_H^4}.$$

*Hint:* Show first that $\frac{DF}{\|DF\|_H^2 + \epsilon}$ belongs to $\text{Dom}\,\delta$ for any $\epsilon > 0$ using Proposition 1.3.3, and then let $\epsilon$ tend to zero.

**2.1.2** Let $u = \{u_t, t \in [0,1]\}$ be an adapted continuous process belonging to $\mathbb{L}^{1,2}$ and such that $\sup_{s,t\in[0,1]} E[|D_s u_t|^2] < \infty$. Show that if $u_1 \neq 0$ a.s., then the random variable $F = \int_0^1 u_s dW_s$ has an absolutely continuous law.

**2.1.3** Suppose that $F$ is a random variable in $\mathbb{D}^{1,2}$, and let $h$ be an element of $H$ such that $\langle DF, h\rangle_H \neq 0$ a.s. and $\frac{h}{\langle DF,h\rangle_H}$ belongs to the domain of $\delta$.

Show that $F$ possesses a continuous and bounded density given by

$$f(x) = E\left(\mathbf{1}_{\{F > x\}}\delta\left(\frac{h}{\langle DF, h\rangle_H}\right)\right).$$

**2.1.4** Let $F$ be a random variable in $\mathbb{D}^{1,2}$ such that $G_k\frac{DF}{\|DF\|_H^2}$ belongs to Dom $\delta$ for any $k = 0, \ldots, n$, where $G_0 = 1$ and

$$G_k = \delta\left(G_{k-1}\frac{DF}{\|DF\|_H^2}\right)$$

if $1 \le k \le n + 1$. Show that $F$ has a density of class $C^n$ and

$$f^{(k)}(x) = (-1)^k E\left[\mathbf{1}_{\{F > x\}}G_{k+1}\right],$$

$0 \le k \le n$.

**2.1.5** Let $F \ge 0$ be a random variable in $\mathbb{D}^{1,2}$ such that $\frac{DF}{\|DF\|_H^2} \in \mathrm{Dom}\,\delta$. Show that the density $f$ of $F$ verifies

$$\|f\|_p \le \|\delta\left(\frac{DF}{\|DF\|_H^2}\right)\|_q (E(F))^{\frac{1}{p}}$$

for any $p > 1$, where $q$ is the conjugate of $p$.

**2.1.6** Let $W = \{W_t, t \ge 0\}$ be a standard Brownian motion, and consider a random variable $F$ in $\mathbb{D}^{1,2}$. Show that for all $t \ge 0$, except for a countable set of times, the random variable $F + W_t$ has an absolutely continuous law (see [218]).

**2.1.7** Let $W = \{W(s,t), (s,t) \in [0,1]^2\}$ be a two-parameter Wiener process. Show that for any pair of disjoint rectangles $R_1, R_2$ with sides parallel to the axes we have

$$P\{\sup_{z \in R_1} W(z) = \sup_{z \in R_2} W(z)\} = 0.$$

*Hint:* Fix a rectangle $[a, b] \subset [0, 1]^2$. Show that the law of the random variable $\sup_{z \in [a,b]} W(z)$ conditioned by the $\sigma$-field generated by the family $\{W(s,t), s \le a_1\}$ is absolutely continuous.

**2.1.8** Let $F \in \mathbb{D}^{3,\alpha}$, $\alpha > 4$, be a random variable such that $E(\|DF\|_H^{-p}) < \infty$ for all $p \ge 2$. Show that the density $p(x)$ of $F$ is continuously differentiable, and compute $p'(x)$.

**2.1.9** Let $F = (F^1, \ldots, F^m)$ be a random vector whose components belong to the space $\mathbb{D}^\infty$. We denote by $\gamma_F$ the Malliavin matrix of $F$. Suppose that $\det \gamma_F > 0$ a.s. Show that the density of $F$ is lower semicontinuous.

*Hint:* The density of $F$ is the nondecreasing limit as $N$ tends to infinity of the densities of the measures $[\Psi_N(\gamma_F) \cdot P] \circ F^{-1}$ introduced in the proof of Theorem 2.1.1.

**2.1.10** Let $F = (W(h_1) + W(h_2))e^{-W(h_2)^2}$, where $h_1$, $h_2$ are orthonormal elements of $H$. Show that $F \in \mathbb{D}^\infty$, $\|DF\|_H > 0$ a.s., and the density of $F$ has a lower semicontinuous version satisfying $p(0) = +\infty$ (see [197]).

**2.1.11** Show that the random variable $F = \int_0^1 t^2 \arctan(W_t)dt$, where $W$ is a Brownian motion, has a $C^\infty$ density.

**2.1.212** Let $W = \{W(s,t), (s,t) \in [0,1]^2\}$ be a two-parameter Wiener process. Show that the density of $\sup_{(s,t)\in[0,1]^2} W(s,t)$ is strictly positive in $(0, +\infty)$.
    *Hint:* Apply Proposition 2.1.8.

## 2.2   Stochastic differential equations

In this section we discuss the existence, uniqueness, and smoothness of solutions to stochastic differential equations. Suppose that $(\Omega, \mathcal{F}, P)$ is the canonical probability space associated with a $d$-dimensional Brownian motion $\{W^i(t), t \in [0,T], 1 \le i \le d\}$ on a finite interval $[0,T]$. This means $\Omega = C_0([0,T]; \mathbb{R}^d)$, $P$ is the $d$-dimensional Wiener measure, and $\mathcal{F}$ is the completion of the Borel $\sigma$-field of $\Omega$ with respect to $P$. The underlying Hilbert space here is $H = L^2([0,T]; \mathbb{R}^d)$.
    Let $A_j, B : [0,T] \times \mathbb{R}^m \to \mathbb{R}^m$, $1 \le j \le d$, be measurable functions satisfying the following globally Lipschitz and boundedness conditions:

**(h1)** $\sum_{j=1}^d |A_j(t,x) - A_j(t,y)| + |B(t,x) - B(t,y)| \le K|x - y|$,     for any
    $x, y \in \mathbb{R}^m$, $t \in [0,T]$;

**(h2)** $t \to A_j(t,0)$ and $t \to B(t,0)$ are bounded on $[0,T]$.

We denote by $X = \{X(t), t \in [0,T]\}$ the solution of the following $m$-dimensional stochastic differential equation:

$$X(t) = x_0 + \sum_{j=1}^d \int_0^t A_j(s, X(s))dW_s^j + \int_0^t B(s, X(s))ds, \qquad (2.37)$$

where $x_0 \in \mathbb{R}^m$ is the initial value of the process $X$. We will show that there is a unique continuous solution to this equation, such that for all $t \in [0,T]$ and for all $i = 1, \ldots, m$ the random variable $X^i(t)$ belongs to the space $\mathbb{D}^{1,p}$ for all $p \ge 2$. Furthermore, if the coefficients are infinitely differentiable in the space variable and their partial derivatives of all orders are uniformly bounded, then $X^i(t)$ belongs to $\mathbb{D}^\infty$.
    From now on we will use the convention of summation over repeated indices.

## 2.2.1 Existence and uniqueness of solutions

Here we will establish an existence and uniqueness result for equations that are generalizations of (2.37). This more general type of equation will be satisfied by the iterated derivatives of the process $X$.

Let $V = \{V(t), 0 \le t \le T\}$ be a continuous and adapted $M$-dimensional stochastic process such that

$$\beta_p = \sup_{0 \le t \le T} E(|V(t)|^p) < \infty$$

for all $p \ge 2$. Suppose that

$$\sigma : \mathbb{R}^M \times \mathbb{R}^m \to \mathbb{R}^m \otimes \mathbb{R}^d \qquad \text{and} \qquad b : \mathbb{R}^M \times \mathbb{R}^m \to \mathbb{R}^m$$

are measurable functions satisfying the following conditions, for a positive constant $K$:

**(h3)** $|\sigma(x, y) - \sigma(x, y')| + |b(x, y) - b(x, y')| \le K|y - y'|$, for any $x \in \mathbb{R}^M$, $y, y' \in \mathbb{R}^m$;

**(h4)** the functions $x \to \sigma(x, 0)$ and $x \to b(x, 0)$ have at most polynomial growth order (i.e., $|\sigma(x, 0)| + |b(x, 0)| \le K(1 + |x|^\nu)$ for some integer $\nu \ge 0$).

With these assumptions, we have the next result.

**Lemma 2.2.1** *Consider a continuous and adapted $m$-dimensional process $\alpha = \{\alpha(t), 0 \le t \le T\}$ such that $d_p = E(\sup_{0 \le t \le T} |\alpha(t)|^p) < \infty$ for all $p \ge 2$. Then there exists a unique continuous and adapted $m$-dimensional process $Y = \{Y(t), 0 \le t \le T\}$ satisfying the stochastic differential equation*

$$Y(t) = \alpha(t) + \int_0^t \sigma_j(V(s), Y(s))dW_s^j + \int_0^t b(V(s), Y(s))ds. \qquad (2.38)$$

*Moreover,*

$$E\left(\sup_{0 \le t \le T} |Y(t)|^p\right) \le C_1$$

*for any $p \ge 2$, where $C_1$ is a positive constant depending on $p, T, K, \beta_{p\nu}, m$, and $d_p$.*

*Proof:* Using Picard's iteration scheme, we introduce the processes $Y_0(t) = \alpha(t)$ and

$$Y_{n+1}(t) = \alpha(t) + \int_0^t \sigma_j(V(s), Y_n(s))dW_s^j + \int_0^t b(V(s), Y_n(s))ds \qquad (2.39)$$

if $n \geq 0$. By a recursive argument one can show that $Y_n$ is a continuous and adapted process such that

$$E\left(\sup_{0 \leq t \leq T} |Y_n(t)|^p\right) < \infty \qquad (2.40)$$

for any $p \geq 2$. Indeed, applying Doob's maximal inequality (A.2) and Burkholder's inequality (A.4) for $m$-dimensional martingales, and making use of hypotheses (h3) and (h4), we obtain

$$
\begin{aligned}
& E\left(\sup_{0 \leq t \leq T} |Y_{n+1}(t)|^p\right) \\
& \leq c_p\left[d_p + E\left(\left|\int_0^T \sigma_j(V(s), Y_n(s))dW_s^j\right|^p\right)\right. \\
& \qquad \left. + E\left(\left(\int_0^T |b(V(s), Y_n(s))|\, ds\right)^p\right)\right] \\
& \leq c_p\left[d_p + c_p' K^p T^{p-1} \int_0^T (1 + E(|V(s)|^{\nu p}) + E(|Y_n(s)|^p))\, ds\right] \\
& \leq c_p\left[d_p + c_p' K^p T^p\left(1 + \beta_{\nu p} + \sup_{0 \leq t \leq T} E(|Y_n(t)|^p)\right)\right],
\end{aligned}
$$

where $c_p$ and $c_p'$ are constants depending only on $p$. Thus, Eq. (2.40) holds.

Again applying Doob's maximal inequality, Burkholder's inequality, and condition (h3), we obtain, for any $p \geq 2$,

$$E\left(\sup_{0 \leq t \leq T} |Y_{n+1}(t) - Y_n(t)|^p\right) \leq c_p K^p T^{p-1} \int_0^T E\left(|Y_n(s) - Y_{n-1}(s)|^p\right) ds.$$

It follows inductively that the preceding expression is bounded by

$$\frac{1}{n!}(c_p K^p T^{p-1})^{n+1} \sup_{0 \leq s \leq T} E(|Y_1(s)|^p).$$

Consequently, we have

$$\sum_{n=0}^{\infty} E\left(\sup_{0 \leq t \leq T} |Y_{n+1}(t) - Y_n(t)|^p\right) < \infty,$$

which implies the existence of a continuous process $Y$ satisfying (2.38) and such that $E(\sup_{0 \leq t \leq T} |Y(t)|^p) \leq C_1$ for all $p \geq 2$. The uniqueness of the solution is derived by means of a similar method.                □

As a consequence, taking $V(t) = t$ in the Lemma 2.2.1 produces the following result.

**Corollary 2.2.1** *Assume that the coefficients $A_j$ and $B$ of Eq. (2.37) are globally Lipschitz and have linear growth (conditions (h1) and (h2)). Then there exists a unique continuous solution $X = \{X(t), t \in [0,T]\}$ to Eq. (2.37). Moreover,*

$$E\left(\sup_{0 \le t \le T} |X(t)|^p\right) \le C_1$$

*for any $p \ge 2$, where $C_1$ is a positive constant depending on $p, T, K, \nu$, and $x_0$.*

### 2.2.2  Weak differentiability of the solution

We will first consider the case where the coefficients $A_j$ and $B$ of the stochastic differential equation (2.37) are globally Lipschitz functions and have linear growth. Our aim is to show that the coordinates of the solution at each time $t \in [0,T]$ belong to the space $\mathbb{D}^{1,\infty} = \cap_{p \ge 1} \mathbb{D}^{1,p}$. To show this result we will make use of an extension of the chain rule to Lipschitz functions established in Proposition 1.2.4.

We denote by $D_t^j(F)$, $t \in [0,T]$, $j = 1, \dots, d$, the derivative of a random variable $F$ as an element of $L^2([0,T] \times \Omega; \mathbb{R}^d) \simeq L^2(\Omega; H)$. Similarly we denote by $D_{t_1,\dots,t_N}^{j_1,\dots,j_N}(F)$ the $N$th derivative of $F$.

Using Proposition 1.2.4, we can show the following result.

**Theorem 2.2.1** *Let $X = \{X(t), t \in [0,T]\}$ be the solution to Eq. (2.37), where the coefficients are supposed to be globally Lipschitz functions with linear growth (hypotheses (h1) and (h2)). Then $X^i(t)$ belongs to $\mathbb{D}^{1,\infty}$ for any $t \in [0,T]$ and $i = 1, \dots, m$. Moreover,*

$$\sup_{0 \le r \le t} E\left(\sup_{r \le s \le T} |D_r^j X^i(s)|^p\right) < \infty,$$

*and the derivative $D_r^j X^i(t)$ satisfies the following linear equation:*

$$
\begin{aligned}
D_r^j X(t) &= A_j(r, X(r)) + \int_r^t \overline{A}_{k,\alpha}(s) D_r^j(X^k(s)) dW_s^\alpha \\
&\quad + \int_r^t \overline{B}_k(s) D_r^j X^k(s) ds
\end{aligned}
\tag{2.41}
$$

*for $r \le t$ a.e., and*

$$D_r^j X(t) = 0$$

*for $r > t$ a.e., where $\overline{A}_{k,\alpha}(s)$ and $\overline{B}_k(s)$ are uniformly bounded and adapted $m$-dimensional processes.*

*Proof:*  Consider the Picard approximations given by

$$
\begin{aligned}
X_0(t) &= x_0, \\
X_{n+1}(t) &= x_0 + \int_0^t A_j(s, X_n(s)) dW_s^j + \int_0^t B(s, X_n(s)) ds
\end{aligned}
\tag{2.42}
$$

if $n \geq 0$. We will prove the following property by induction on $n$:

(P)       $X_n^i(t) \in \mathbb{D}^{1,\infty}$ for all $i = 1, \ldots, m$, $n \geq 0$, and $t \in [0, T]$; further-more, for all $p > 1$ we have

$$\psi_n(t) := \sup_{0 \leq r \leq t} E\left(\sup_{s \in [r,t]} |D_r X_n(s)|^p\right) < \infty \qquad (2.43)$$

and

$$\psi_{n+1}(t) \leq c_1 + c_2 \int_0^t \psi_n(s)ds, \qquad (2.44)$$

for some constants $c_1, c_2$.

Clearly, (P) holds for $n = 0$. Suppose it is true for $n$. Applying Proposition 1.2.4 to the random vector $X_n(s)$ and to the functions $A_j^i$ and $B^i$, we deduce that the random variables $A_j^i(s, X_n(s))$ and $B^i(s, X_n(s))$ belong to $\mathbb{D}^{1,2}$ and that there exist $m$-dimensional adapted processes $\overline{A}_j^{n,i}(s) = (\overline{A}_{j,1}^{n,i}(s), \ldots, \overline{A}_{j,m}^{n,i}(s))$ and $\overline{B}^{n,i}(s) = (\overline{B}_1^{n,i}(s), \ldots, \overline{B}_m^{n,i}(s))$, uniformly bounded by $K$, such that

$$D_r[A_j^i(s, X_n(s))] = \overline{A}_{j,k}^{n,i}(s)D_r(X_n^k(s))\mathbf{1}_{\{r \leq s\}} \qquad (2.45)$$

and

$$D_r[B^i(s, X_n(s))] = \overline{B}_k^{n,i}(s)D_r(X_n^k(s))\mathbf{1}_{\{r \leq s\}}. \qquad (2.46)$$

In fact, these processes are obtained as the weak limit of the sequences $\{\partial_k[A_j^i * \alpha_m](s, X_n(s)), m \geq 1\}$ and $\{\partial_k[B^i * \alpha_m](s, X_n(s)), m \geq 1\}$, where $\alpha_m$ denotes an approximation of the identity, and it is easy to check the adaptability of the limit. From Proposition 1.5.5 we deduce that the random variables $A_j^i(s, X_n(s))$ and $B^i(s, X_n(s))$ belong to $\mathbb{D}^{1,\infty}$.

Thus the processes $\{D_r^l[A_j^i(s, X_n(s))], s \geq r\}$ and $\{D_r^l[B^i(s, X_n(s))], s \geq r\}$ are square integrable and adapted, and from (2.45) and (2.46) we get

$$|D_r[A_j^i(s, X_n(s))]| \leq K|D_r X_n(s)|, \qquad |D_r[B^i(s, X_n(s))]| \leq K|D_r X_n(s)|. \qquad (2.47)$$

Using Lemma 1.3.4 we deduce that the Itô integral $\int_0^t A_j^i(s, X_n(s))dW_s^j$ belongs to the space $\mathbb{D}^{1,2}$, and for $r \leq t$ we have

$$D_r^l\left[\int_0^t A_j^i(s, X_n(s))dW_s^j\right] = A_l^i(r, X_n(r)) + \int_r^t D_r^l[A_j^i(s, X_n(s))]dW_s^j. \qquad (2.48)$$

On the other hand, $\int_0^t B^i(s, X_n(s))ds \in \mathbb{D}^{1,2}$, and for $r \leq t$ we have

$$D_r^l\left[\int_0^t B^i(s, X_n(s))ds\right] = \int_r^t D_r^l[B^i(s, X_n(s))]ds. \qquad (2.49)$$

From these equalities and Eq. (2.42) we see that $X^i_{n+1}(t) \in \mathbb{D}^{1,\infty}$ for all $t \in [0, T]$, and we obtain

$$E \left( \sup_{r \leq s \leq t} |D^j_r X_{n+1}(s)|^p \right) \leq c_p \left[ \gamma_p + T^{p-1} K^p \int_r^t E \left( |D^j_r X_n(s)|^p \right) ds \right],$$
(2.50)

where

$$\gamma_p = \sup_{n,j} E(\sup_{0 \leq t \leq T} |A_j(t, X_n(t))|^p) < \infty.$$

So (2.43) and (2.44) hold for $n + 1$. From Lemma 2.2.1 we know that

$$E \left( \sup_{s \leq T} |X_n(s) - X(s)|^p \right) \longrightarrow 0$$

as $n$ tends to infinity. By Gronwall's lemma applied to (2.50) we deduce that derivatives of the sequence $X^i_n(t)$ are bounded in $L^p(\Omega; H)$ uniformly in $n$ for all $p \geq 2$. Therefore, from Proposition 1.5.5 we deduce that the random variables $X^i(t)$ belong to $\mathbb{D}^{1,\infty}$. Finally, applying the operator $D$ to Eq. (2.37) and using Proposition 1.2.4, we deduce the linear stochastic differential equation (2.41) for the derivative of $X^i(t)$.          □

If the coefficients of Eq. (2.37) are continuously differentiable, then we can write

$$\overline{A}^i_{k,l}(s) = (\partial_k A^i_l)(s, X(s))$$

and

$$\overline{B}^i_k(s) = (\partial_k B^i)(s, X(s)).$$

In order to prove the existence of higher-order derivatives, we will need the following technical lemma.

Consider adapted and continuous processes $\alpha = \{\alpha(r, t), t \in [r, T]\}$ and $V = \{V_j(t), 0 \leq t \leq T, j = 0, \ldots, d\}$ such that $\alpha$ is $m$-dimensional and $V_j$ is uniformly bounded and takes values on the set of matrices of order $m \times m$. Suppose that the random variables $\alpha^i(r, t)$ and $V^{kl}_j(t)$ belong to $\mathbb{D}^{1,\infty}$ for any $i, j, k, l$, and satisfy the following estimates:

$$\sup_{0 \leq r \leq T} E \left( \sup_{r \leq t \leq T} |\alpha(r, t)|^p \right) < \infty,$$

$$\sup_{0 \leq s \leq T} E \left( \sup_{s \leq t \leq T} |D_s V_j(t)|^p \right) < \infty,$$

$$\sup_{0 \leq s, r \leq T} E \left( \sup_{r \vee s \leq t \leq T} |D_s \alpha(r, t)|^p \right) < \infty,$$

for any $p \geq 2$ and any $j = 0, \ldots, d$.

**Lemma 2.2.2** *Let* $Y = \{Y(t), r \leq t \leq T\}$ *be the solution of the linear stochastic differential equation*

$$Y(t) = \alpha(r,t) + \int_r^t V_j(s)Y(s)dW_s^j + \int_r^t V_0(s)Y(s)ds. \qquad (2.51)$$

*Then* $\{Y^i(t)\}$ *belongs to* $\mathbb{D}^{1,\infty}$ *for any* $i = 1, \ldots, m$, *and the derivative* $D_s Y^i(t)$ *verifies the following linear equation, for* $s \leq t$:

$$
\begin{aligned}
D_s^j Y(t) \;=\;& D_s^j \alpha(r,t) + V_j(s)Y(s)\mathbf{1}_{\{r \leq s \leq t\}} \\
& + \int_r^t [D_s^j V_l(u)Y(u) + V_l(u)D_s^j Y(u)]dW_u^l \\
& + \int_r^t [D_s^j V_0(u)Y(u) + V_0(u)D_s^j Y(u)]du. \qquad (2.52)
\end{aligned}
$$

*Proof:*    The proof can be done using the same technique as the proof of Theorem 2.2.1, and so we will omit the details. The main idea is to observe that Eq. (2.51) is a particular case of (2.38) when the coefficients $\sigma_j$ and $b$ are linear. Consider the Picard approximations defined by the recursive equations (2.39). Then we can show by induction that the variables $Y_n^i(t)$ belong to $\mathbb{D}^{1,\infty}$ and satisfy the equation

$$
\begin{aligned}
D_s^j Y_{n+1}(t) \;=\;& D_s^j \alpha(r,t) + V_j(s)Y_n(s)\mathbf{1}_{\{r \leq s \leq t\}} \\
& + \int_r^t [D_s^j V_l(u)Y_n(u) + V_l(u)D_s^j Y_n(u)]dW_u^l \\
& + \int_r^t [D_s^j V_0(u)Y_n(u) + V_0(u)D_s^j Y_n(u)]du.
\end{aligned}
$$

Finally, we conclude our proof as we did in the proof of Theorem 2.2.1. $\Box$

Note that under the assumptions of Lemma 2.2.2 the solution $Y$ of Eq. (2.51) satisfies the estimates

$$E\left(\sup_{0 \leq t \leq T} |Y(t)|^p\right) < \infty,$$

$$\sup_{0 \leq s \leq t} E\left(\sup_{r \leq t \leq T} |D_s Y(t)|^p\right) < \infty,$$

for all $p \geq 2$.

**Theorem 2.2.2** *Let* $X$ *be the solution of the stochastic differential equation (2.37), and suppose that the coefficients* $A_j^i$ *and* $B^i$ *are infinitely differentiable functions in* $x$ *with bounded derivatives of all orders greater than or equal to one and that the functions* $A_j^i(t,0)$ *and* $B^i(t,0)$ *are bounded. Then* $X^i(t)$ *belongs to* $\mathbb{D}^{\infty}$ *for all* $t \in [0,T]$, *and* $i = 1, \ldots, m$.

*Proof:*     We know from Theorem 2.2.1 that for any $i = 1, \ldots, m$ and any $t \in [0, T]$, the random variable $X^i(t)$ belongs to $\mathbb{D}^{1,p}$ for all $p \geq 2$. Furthermore, the derivative $D_r^j X^i(t)$ verifies the following linear stochastic differential equation:

$$D_r^j X^i(t) = A_j^i(r, X_r) \; + \; \int_r^t (\partial_k A_l^i)(s, X(s)) D_r^j X^k(s) dW_s^l$$

$$+ \; \int_r^t (\partial_k B)(s, X(s)) D_r^j X^k(s) ds. \qquad (2.53)$$

Now we will recursively apply Lemma 2.2.2 to this linear equation. We will denote by $D_{r_1,\ldots,r_N}^{j_1,\ldots,j_N}(X(t))$ the iterated derivative of order $N$. We have to introduce some notation. For any subset $K = \{\epsilon_1 < \cdots < \epsilon_\eta\}$ of $\{1, \ldots, N\}$, we put $j(K) = j_{\epsilon_1}, \ldots, j_{\epsilon_\eta}$ and $r(K) = r_{\epsilon_1}, \ldots, r_{\epsilon_\eta}$. Define

$$\alpha_{l,j_1,\ldots,j_N}^i(s, r_1, \ldots, r_N) \; = \; \sum (\partial_{k_1} \cdots \partial_{k_\nu} A_l^i)(s, X(s))$$

$$\times D_{r(I_1)}^{j(I_1)}[X^{k_1}(s)] \cdots D_{r(I_\nu)}^{j(I_\nu)}[X^{k_\nu}(s)]$$

and

$$\beta_{j_1,\ldots,j_N}^i(s, r_1, \ldots, r_N) \; = \; \sum (\partial_{k_1} \cdots \partial_{k_\nu} B^i)(s, X(s))$$

$$\times D_{r(I_1)}^{j(I_1)}[X^{k_1}(s)] \cdots D_{r(I_\nu)}^{j(I_\nu)}[X^{k_\nu}(s)],$$

where the sums are extended to the set of all partitions $\{1, \ldots, N\} = I_1 \cup \cdots \cup I_\nu$. We also set $\alpha_j^i(s) = A_j^i(s, X(s))$. With these notations we will recursively show the following properties for any integer $N \geq 1$:

(P1)     For any $t \in [0, T]$, $p \geq 2$, and $i = 1, \ldots, m$, $X^i(t)$ belongs to $\mathbb{D}^{N,p}$, and

$$\sup_{r_1,\ldots,r_N \in [0,T]} E \left( \sup_{r_1 \vee \cdots \vee r_N \leq t \leq T} |D_{r_1,\ldots,r_N}(X(t))|^p \right) < \infty.$$

(P2)     The $N$th derivative satisfies the following linear equation:

$$D_{r_1,\ldots,r_N}^{j_1,\ldots,j_N}(X^i(t)) \; = \; \sum_{\epsilon=1}^N \alpha_{j_\epsilon,j_1,\ldots,j_{\epsilon-1},j_{\epsilon+1},\ldots,j_N}^i(r_\epsilon, r_1, \ldots, r_{\epsilon-1}, r_{\epsilon+1}, \ldots, r_N)$$

$$+ \int_{r_1 \vee \cdots \vee r_N}^t \left[ \alpha_{l,j_1,\ldots,j_N}^i(s, r_1, \ldots, r_N) dW_s^l \right.$$

$$\left. + \beta_{j_1,\ldots,j_N}^i(s, r_1, \ldots, r_N) ds \right] \qquad (2.54)$$

if $t \geq r_1 \vee \cdots \vee r_N$, and $D_{r_1,\ldots,r_N}^{j_1,\ldots,j_N}(X(t)) = 0$ if $t < r_1 \vee \cdots \vee r_N$.

We know that these properties hold for $N = 1$ because of Theorem 2.2.1. Suppose that the above properties hold up to the index $N$. Observe that $\alpha_{l,j_1,\ldots,j_N}^i(s, r_1, \ldots, r_N)$ is equal to

$$(\partial_k A_l^i)(s, X(s)) D_{r_1,\ldots,r_N}^{j_1,\ldots,j_N}(X^k(s))$$

(this term corresponds to $\nu = 1$) plus a polynomial function on the derivatives $(\partial_{k_1} \cdots \partial_{k_\nu} A_l^i)(s, X(s))$ with $\nu \geq 2$, and the processes $D_{r(I)}^{j(I)}(X^k(s))$, with $\text{card}(I) \leq N - 1$. Therefore, we can apply Lemma 2.2.2 to $r = r_1 \vee \cdots \vee r_N$, and the processes

$$
\begin{aligned}
Y(t) &= D_{r_1,\ldots,r_N}^{j_1,\ldots,j_N}(X(t)), \quad t \geq r, \\
V_j^{ik}(t) &= (\partial_k A_j^i)(s, X(s)), \quad 1 \leq i, k \leq m, \quad j = 1, \ldots, d,
\end{aligned}
$$

and $\alpha(r, t)$ is equal to the sum of the remaining terms in the right-hand side of Eq. (2.54).

Notice that with the above notations we have

$$
D_r^j\left[\alpha_{l,j_1,\ldots,j_N}^i(t, r_1, \ldots, r_N)\right] = \alpha_{l,j_1,\ldots,j_N,j}^i(t, r_1, \ldots, r_N, r)
$$

and

$$
D_r^j\left[\beta_{j_1,\ldots,j_N}^i(t, r_1, \ldots, r_N)\right] = \beta_{j_1,\ldots,j_N,j}^i(t, r_1, \ldots, r_N, r).
$$

Using these relations and computing the derivative of (2.54) by means of Lemma 2.2.2, we obtain

$$
\begin{aligned}
&D_r^j D_{r_1,\ldots,r_N}^{j_1,\ldots,j_N}(X^i(t)) \\
&= \sum_{\epsilon=1}^N \alpha_{j_\epsilon,j_1,\ldots,j_{\epsilon-1},j_{\epsilon+1},\ldots,j_N,j}^i(r_\epsilon, r_1, \ldots, r_{\epsilon-1}, r_{\epsilon+1}, \ldots, r_N, r) \\
&\quad + \alpha_{j,j_1,\ldots,j_N}^i(r, r_1, \ldots, r_N) \\
&\quad + \int_{r_1 \vee \cdots \vee r_N}^t \left[\alpha_{l,j_1,\ldots,j_N,j}^i(s, r_1, \ldots, r_N, r)dW_s^l \right. \\
&\qquad \left. + \beta_{j_1,\ldots,j_N,j}^i(s, r_1, \ldots, r_N, r)ds\right],
\end{aligned}
$$

which implies that property (P2) holds for $N+1$. The estimates of property (P1) are also easily derived. The proof of the theorem is now complete. $\square$

## Exercises

**2.2.1** Let $\sigma$ and $b$ be continuously differentiable functions on $\mathbb{R}$ with bounded derivatives. Consider the solution $X = \{X_t, t \in [0, T]\}$ of the stochastic differential equation

$$
X_t = x_0 + \int_0^t \sigma(X_s)dW_s + \int_0^t b(X_s)ds.
$$

Show that for $s \leq t$ we have

$$
D_s X_t = \sigma(X_s)\exp\left(\int_0^t \sigma'(X_s)dW_s + \int_0^t [b' - \frac{1}{2}(\sigma')^2](X_s)ds\right).
$$

**2.2.2** (Doss [84]) Suppose that $\sigma$ is a function of class $C^2(\mathbb{R})$ with bounded first and second partial derivatives and that $b$ is Lipschitz continuous. Show that the one-dimensional stochastic differential equation

$$X_t = x_0 + \int_0^t \sigma(X_s)dW_s + \int_0^t [b + \frac{1}{2}\sigma\sigma'](X_s)ds \qquad (2.55)$$

has a solution that can be written in the form $X_t = u(W_t, Y_t)$, where

(i) $u(x, y)$ is the solution of the ordinary differential equation

$$\frac{\partial u}{\partial x} = \sigma(u), \qquad u(0, y) = y;$$

(ii) for each $\omega \in \Omega$, $\{Y_t(\omega), t \geq 0\}$ is the solution of the ordinary differential equation

$$Y_t'(\omega) = f(W_t(\omega), Y_t(\omega)), \qquad Y_0(\omega) = x_0,$$

where $f(x, y) = b(u(x, y))\left(\frac{\partial u}{\partial y}\right)^{-1} = b(u(x, y))\exp(-\int_0^x \sigma'(u(z, y)dz)$.

Using the above representation of the solution to Eq. (2.55), show that $X_t$ belongs to $\mathbb{D}^{1,p}$ for all $p \geq 2$ and compute the derivative $D_s X_t$.

## 2.3 Hypoellipticity and Hörmander's theorem

In this section we introduce nondegeneracy conditions on the coefficients of Eq. (2.37) and show that under these conditions the solution $X(t)$ at any time $t \in (0, T]$ has a (smooth) density. Clearly, if the subspace spanned by $\{A_j(t, y), B(t, y); 1 \leq j \leq d, t \in [0, T], y \in \mathbb{R}^m\}$ has dimension strictly smaller than $m$, then the law of $X(t)$, for all $t \geq 0$, will be singular with respect to the Lebesgue measure. We thus need some kind of nondegeneracy assumption.

### 2.3.1 *Absolute continuity in the case of Lipschitz coefficients*

Let $\{X(t), t \in [0, T]\}$ be the solution of the stochastic differential equation (2.37), where the coefficients are supposed to be globally Lipschitz functions with linear growth. In Theorem 2.2.1 we proved that $X^i(t)$ belongs to $\mathbb{D}^{1,\infty}$ for all $i = 1, \ldots, m$ and $t \in [0, T]$, and we found that the derivative $D_r^j X_t^i$ satisfies the following linear stochastic differential equation:

$$D_r^j X_t^i = A_j^i(r, X_r) + \int_r^t \overline{A}_{k,l}^i(s)D_r^j X_s^k dW_s^l$$

$$+ \int_r^t \overline{B}_k^i(s)D_r^j X_s^k ds. \qquad (2.56)$$

We are going to deduce a simpler expression for the derivative $DX_t^i$. Consider the $m \times m$ matrix-valued process defined by

$$Y_j^i(t) = \delta_j^i + \int_0^t \left[ \overline{A}_{k,l}^i(s) Y_j^k(s) dW_s^l + \overline{B}_k^i(s) Y_j^k(s) ds \right], \qquad (2.57)$$

$i, j = 1, \ldots, m$. If the coefficients of Eq. (2.37) are of class $C^{1+\alpha}$, $\alpha > 0$ (see Kunita [173]), then there is a version of the solution $X(t, x_0)$ to this equation that is continuously differentiable in $x_0$, and $Y(t)$ is the Jacobian matrix $\frac{\partial X}{\partial x_0}(t, x_0)$.

Now consider the $m \times m$ matrix-valued process $Z(t)$ solution to the system

$$
\begin{aligned}
Z_j^i(t) &= \delta_j^i - \int_0^t Z_k^i(s) \overline{A}_{j,l}^k(s) dW_s^l \\
&\quad - \int_0^t Z_k^i(s) \left[ \overline{B}_j^k(s) - \overline{A}_{\alpha,l}^k(s) \overline{A}_{j,l}^\alpha(s) \right] ds.
\end{aligned}
\qquad (2.58)
$$

By means of Itô's formula, one can check that $Z_t Y_t = Y_t Z_t = I$. In fact,

$$
\begin{aligned}
Z_j^i(t) Y_k^j(t) &= \delta_k^i + \int_0^t Z_j^i(s) \overline{A}_{l,\theta}^j(s) Y_k^l(s) dW_s^\theta \\
&\quad + \int_0^t Z_j^i(s) \overline{B}_l^j(s) Y_k^l(s) ds - \int_0^t Z_l^i(s) \overline{A}_{j,\theta}^l(s) Y_k^j(s) dW_s^\theta \\
&\quad - \int_0^t Z_l^i(s) \left[ \overline{B}_j^l(s) - \overline{A}_{\alpha,\theta}^l(s) \overline{A}_{j,\theta}^\alpha(s) \right] Y_k^j(s) ds \\
&\quad - \int_0^t Z_l^i(s) \overline{A}_{j,\theta}^l(s) \overline{A}_{\alpha,\theta}^j(s) Y_k^\alpha(s) ds = \delta_k^i,
\end{aligned}
$$

and similarly for $Y_t Z_t$. As a consequence, for any $t \geq 0$ the matrix $Y_t$ is invertible and $Y_t^{-1} = Z_t$. Then it holds that

$$D_r^j X_t^i = Y_l^i(t) Y^{-1}(r)_k^l A_j^k(r, X_r). \qquad (2.59)$$

Indeed, it is enough to verify that the process $\{Y_l^i(t) Y^{-1}(r)_k^l A_j^k(r, X_r), t \geq r\}$ satisfies Eq. (2.56):

$$
\begin{aligned}
A_j^i(r, X_r) &+ \int_r^t \overline{A}_{k,l}^i(s) \left\{ Y_\alpha^k(s) Y^{-1}(r)_\beta^\alpha A_j^\beta(r, X_r) \right\} dW^l(s) \\
&+ \int_r^t \overline{B}_k^i(s) \left\{ Y_\alpha^k(s) Y^{-1}(r)_\beta^\alpha A_j^\beta(r, X_r) \right\} ds \\
&= A_j^i(r, X_r) + \left[ Y_l^i(t) - Y_l^i(r) \right] Y^{-1}(r)_\theta^l A_j^\theta(r, X_r) \\
&= Y_l^i(t) Y^{-1}(r)_k^l A_j^k(r, X_r).
\end{aligned}
$$

We will denote by

$$Q_t^{ij} = \langle DX_t^i, DX_t^j \rangle_H = \sum_{l=1}^{d} \int_0^t D_r^l X_t^i D_r^l X_t^j dr$$

the Malliavin matrix of the vector $X(t)$. Equation (2.59) allows us to write the following expression for this matrix:

$$Q_t = Y_t C_t Y_t^T, \qquad (2.60)$$

where

$$C_t^{ij} = \sum_{l=1}^{d} \int_0^t Y^{-1}(s)_k^i A_l^k(s, X_s) Y^{-1}(s)_{k'}^j A_l^{k'}(s, X_s) ds. \qquad (2.61)$$

Define both the time-dependent $m \times m$ diffusion matrix

$$\sigma^{ij}(t, x) = \sum_{k=1}^{d} A_k^i(t, x) A_k^j(t, x)$$

and the stopping time

$$S = \inf\{t > 0 : \int_0^t \mathbf{1}_{\{\det \sigma(s, X_s) \neq 0\}} ds > 0\} \wedge T.$$

The following absolute continuity result has been established by Bouleau and Hirsch in [46].

**Theorem 2.3.1** *Let $\{X(t), t \in [0, T]\}$ be the solution of the stochastic differential equation (2.37), where the coefficients are globally Lipschitz functions and of at most linear growth. Then for any $0 < t \leq T$ the law of $X(t)$ conditioned by $\{t > S\}$ is absolutely continuous with respect to the Lebesgue measure on $\mathbb{R}^m$.*

*Proof:*    Taking into account Theorem 2.2.1 and Corollary 2.1.2, it suffices to show that $\det Q_t > 0$ a.s. on the set $\{t > S\}$. In view of expression (2.60) it is sufficient to prove that $\det C_t > 0$ a.s. on this set. Suppose $t > S$. Then there exists a set $G \subset [0, t]$ of positive Lebesgue measure such that for any $s \in G$ and $v \in \mathbb{R}^m$ we have

$$v^T \sigma(s, X_s) v \geq \lambda(s) |v|^2,$$

where $\lambda(s) > 0$. Taking $v = (Y_s^{-1})^T u$ and integrating over $[0, t] \cap G$, we obtain

$$u^T C_t u = \int_0^t u^T Y(s)^{-1} \sigma(s, X_s) (Y(s)^{-1})^T u \, ds \geq k|u|^2,$$

where $k = \int_0^t \mathbf{1}_G(s) \frac{\lambda(s)}{|Y(s)|^2} ds$. Consequently, if $t > S$, the matrix $C_t$ is invertible and the result is proved.    $\square$

### 2.3.2  Absolute continuity under Hörmander's conditions

In this section we assume that the coefficients of Eq. (2.37) are infinitely differentiable with bounded derivatives of all orders and do not depend on the time. Let us denote by $X = \{X(t), t \geq 0\}$ the solution of this equation on $[0, \infty)$. We have seen in Theorem 2.2.2 that in such a case the random variables $X^i(t)$ belong to the space $\mathbb{D}^\infty$. We are going to impose nondegeneracy conditions on the coefficients in such a way that the solution has a smooth density. To introduce these conditions, consider the following vector fields on $\mathbb{R}^m$ associated with the coefficients of Eq. (2.37):

$$A_j = A^i_j(x)\frac{\partial}{\partial x_i}, \quad j = 1, \ldots, d,$$

$$B = B^i(x)\frac{\partial}{\partial x_i}.$$

The covariant derivative of $A_k$ in the direction of $A_j$ is defined as the vector field $A^\nabla_j A_k = A^l_j \partial_l A^i_k \frac{\partial}{\partial x_i}$, and the Lie bracket between the vector fields $A_j$ and $A_k$ is defined by

$$[A_j, A_k] = A^\nabla_j A_k - A^\nabla_k A_j.$$

Set

$$A_0 = \left[ B^i(x) - \frac{1}{2}A^j_l(x)\partial_j A^i_l(x) \right] \frac{\partial}{\partial x_i}$$

$$= B - \frac{1}{2}\sum_{l=1}^{d} A^\nabla_l A_l.$$

The vector field $A_0$ appears when we write the stochastic differential equation (2.37) in terms of the Stratonovich integral instead of the Itô integral:

$$X_t = x_0 + \int_0^t A_j(X_s) \circ dW^j_s + \int_0^t A_0(X_s)ds.$$

Hörmander's condition can be stated as follows:

**(H)**    The vector space spanned by the vector fields

$$A_1, \ldots, A_d, \quad [A_i, A_j], 0 \leq i, j \leq d, \quad [A_i, [A_j, A_k]], 0 \leq i, j, k \leq d, \ldots$$

at point $x_0$ is $\mathbb{R}^m$.

For instance, if $m = d = 1$, $A^1_1(x) = a(x)$, and $A^1_0(x) = b(x)$, then Hörmander's condition means that $a(x_0) \neq 0$ or $a^n(x_0)b(x_0) \neq 0$ for some $n \geq 1$. In this situation we have the following result.

**Theorem 2.3.2** *Assume that Hörmander's condition (H) holds. Then for any $t > 0$ the random vector $X(t)$ has a probability distribution that is absolutely continuous with respect to the Lebesgue measure.*

We will see in the next section that the density of the law of $X_t$ is infinitely differentiable on $\mathbb{R}^m$. This result can be considered as a probabilistic version of Hörmander's theorem on the hypoellipticity of second-order differential operators. Let us discuss this point with some detail. We recall that a differential operator $\mathcal{A}$ on an open set $G$ of $\mathbb{R}^m$ with smooth (i.e., infinitely differentiable) coefficients is called hypoelliptic if, whenever $u$ is a distribution on $G$, $u$ is a smooth function on any open set $G' \subset G$ on which $\mathcal{A}u$ is smooth.

Consider the second-order differential operator

$$\mathcal{A} = \frac{1}{2} \sum_{i=1}^{d} (A_i)^2 + A_0. \tag{2.62}$$

Hörmander's theorem [138] states that if the Lie algebra generated by the vector fields $A_0, A_1, \ldots, A_d$ has full rank at each point of $\mathbb{R}^m$, then the operator $\mathcal{L}$ is hypoelliptic. Notice that this assumtion is stronger than (H).

A straightforward proof of this result using the calculus of pseudo-differential operators can be found in Khon [170]. On the other hand, Oleĭnik and Radkevič [277] have made generalizations of Hörmander's theorem to include operators $\mathcal{L}$, which cannot be written in Hörmander's form (as a sum of squares).

In order to relate the hypoellipticity property with the smoothness of the density of $X_t$, let us consider an infinitely differentiable function $f$ with compact support on $(0, \infty) \times \mathbb{R}^m$. By means of Itô's formula we can write for $t$ large enough

$$0 = E[f(t, X_t)] - E[f(0, X_0)] = E\left[\int_0^t (\frac{\partial}{\partial s} + \mathcal{G})f(s, X_s)ds\right],$$

where

$$\mathcal{G} = \frac{1}{2} \sum_{i,j=1}^{m} (AA^T)^{ij} \frac{\partial^2}{\partial x_i \partial x_j} + \sum_{i=1}^{m} B^i \frac{\partial}{\partial x_i}.$$

Notice that $\mathcal{G} - B = \mathcal{L} - A_0$, where $\mathcal{L}$ is defined in (2.62). Denote by $p_t(dy)$ the probability distribution of $X_t$. We have

$$0 = E\left[\int_0^\infty (\frac{\partial}{\partial s} + \mathcal{G})f(s, X_s)ds\right] = \int_0^\infty \int_{\mathbb{R}^m} (\frac{\partial}{\partial s} + \mathcal{G})f(s, y)p_s(dy)ds.$$

This means that $p_t(dy)$ satisfies the forward Fokker-Planck equation $(-\frac{\partial}{\partial t} + \mathcal{G}^*)p = 0$ (where $\mathcal{G}^*$ denotes the adjoint of the operator $\mathcal{G}$) in the distribution sense. Therefore, the fact that $p_t(dy)$ has a $C^\infty$ density in the variable $y$ is implied by the hypoelliptic character of the operator $\frac{\partial}{\partial t} - \mathcal{G}^*$. Increasing the dimension by one and applying Hörmander's theorem to the operator

$\frac{\partial}{\partial t} - \mathcal{G}^*$, one can deduce its hypoellipticity assuming hypothesis (H) at each point $x_0$ in $\mathbb{R}^m$. We refer to Williams [350] for a more detailed discussion of this subject.

Let us turn to the proof of Theorem 2.3.2. First we carry out some preliminary computations that will explain the role played by the nondegeneracy condition (H). Suppose that $V(x) = V^i(x)\frac{\partial}{\partial x_i}$ is a $C^\infty$ vector field on $\mathbb{R}^m$. The Lie brackets appear when we apply Itô's formula to the process $Y_t^{-1}V(X_t)$, where the process $Y_t^{-1}$ has been defined in (2.58). In fact, we have

$$
\begin{aligned}
Y_t^{-1}V(X_t) \quad =& V(x_0) + \int_0^t Y_s^{-1}[A_k,V](X_s)dW_s^k \\
&+ \int_0^t Y_s^{-1}\left\{[A_0,V] + \frac{1}{2}\sum_{k=1}^d [A_k,[A_k,V]]\right\}(X_s)ds. \quad (2.63)
\end{aligned}
$$

We recall that from (2.58) we have

$$
\begin{aligned}
Y_t^{-1} \quad =& \quad I - \sum_{k=1}^d \int_0^t Y_s^{-1}\partial A_k(X_s)dW_s^k \\
&- \int_0^t Y_s^{-1}\left[\partial B(X_s) - \sum_{k=1}^d \partial A_k(X_s)\partial A_k(X_s)\right]ds,
\end{aligned}
$$

where $\partial A_k$ and $\partial B$ respectively denote the Jacobian matrices $\left(\partial_j A_k^i\right)$ and $\left(\partial_j B^i\right)$, $i,j = 1,\ldots,m$. In order to show Eq. (2.63), we first use Itô's formula:

$$
\begin{aligned}
Y_t^{-1}V(X_t) = V(x_0) \quad +& \int_0^t Y_s^{-1}\sum_{k=1}^d \left(\partial V A_k - \partial A_k V\right)(X_s)dW_s^k \\
+& \int_0^t Y_s^{-1}\left(\partial V B - \partial B V\right)(X_s)ds \\
+& \int_0^t Y_s^{-1}\sum_{k=1}^d (\partial A_k \partial A_k V)(X_s)ds \quad\quad (2.64) \\
+& \frac{1}{2}\int_0^t Y_s^{-1}\sum_{i,j=1}^m \partial_i\partial_j V(X_s)\sum_{k=1}^d A_k^i(X_s)A_k^j(X_s)ds \\
-& \int_0^t Y_s^{-1}\sum_{k=1}^d (\partial A_k \partial V A_k)(X_s)ds.
\end{aligned}
$$

Notice that

$$
\begin{aligned}
\partial V A_k - \partial A_k V &= [A_k,V], \quad \text{and} \\
\partial V B - \partial B V &= [B,V].
\end{aligned}
$$

Additionally, we can write

$$[A_0, V] + \frac{1}{2} \sum_{k=1}^{d} [A_k, [A_k, V]] - [B, V]$$

$$= \left[ -\frac{1}{2} \sum_{k=1}^{d} A_k^\nabla A_k, V \right] + \frac{1}{2} \sum_{k=1}^{d} [A_k, [A_k, V]]$$

$$= \frac{1}{2} \sum_{k=1}^{d} \left\{ -(A_k^\nabla A_k)^\nabla V + V^\nabla (A_k^\nabla A_k) + A_k^\nabla (A_k^\nabla V) \right.$$
$$\left. - A_k^\nabla (V^\nabla A_k) - (A_k^\nabla V)^\nabla A_k + (V^\nabla A_k)^\nabla A_k \right\}$$

$$= \frac{1}{2} \sum_{k=1}^{d} \left\{ - A_k^i \partial_i A_k^l \partial_l V + V^i \partial_i A_k^l \partial_l A_k + V^i A_k^l \partial_i \partial_l A_k \right.$$
$$+ A_k^i \partial_i A_k^l \partial_l V + A_k^i A_k^l \partial_i \partial_l V - A_k^i \partial_i V^l \partial_l A_k$$
$$\left. - A_k^i V^l \partial_i \partial_l A_k - A_k^i \partial_i V^l \partial_l A_k + V^i \partial_i A_k^l \partial_l A_k \right\}$$

$$= \sum_{k=1}^{d} \left\{ V^i \partial_i A_k^l \partial_l A_k + \frac{1}{2} A_k^i A_k^l \partial_i \partial_l V - A_k^i \partial_i V^l \partial_l A_k \right\}.$$

Finally expression (2.63) follows easily from the previous computations.

*Proof of Theorem 2.3.2:*    Fix $t > 0$. Using Theorem 2.1.2 (or Theorem 2.1.1) it suffices to show that the matrix $C_t$ given by (2.61) is invertible with probability one. Suppose that $P\{\det C_t = 0\} > 0$. We want to show that under this assumption condition (H) cannot be satisfied. Let $K_s$ be the random subspace of $\mathbb{R}^m$ spanned by $\{Y_\sigma^{-1} A_k(X_\sigma); 0 \le \sigma \le s, k = 1, \ldots, d\}$. The family of vector spaces $\{K_s, s \ge 0\}$ is increasing. Set $K_{0+} = \cap_{s>0} K_s$. By the Blumenthal zero-one law for the Brownian motion (see Revuz and Yor [292, Theorem III.2.15]), $K_{0+}$ is a deterministic space with probability one. Define the increasing adapted process $\{\dim K_s, s > 0\}$ and the stopping time

$$\tau = \inf\{s > 0 : \dim K_s > \dim K_{0+}\}.$$

Notice that $P\{\tau > 0\} = 1$. For any vector $v \in \mathbb{R}^m$ of norm one we have

$$v^T C_t v = \sum_{k=1}^{d} \int_0^t |v^T Y_s^{-1} A_k(X_s)|^2 ds.$$

As a consequence, by continuity $v^T C_t v = 0$ implies $v^T Y_s^{-1} A_k(X_s) = 0$ for any $s \in [0, t]$ and any $k = 1, \ldots, d$. Therefore, $K_{0+} \ne \mathbb{R}^m$, otherwise $K_s = \mathbb{R}^m$ for any $s > 0$ and any vector $v$ verifying $v^T C_t v = 0$ would be equal to zero, which implies that $C_t$ is invertible a.s., in contradiction with our hypothesis. Let $v$ be a fixed nonzero vector orthogonal to $K_{0+}$. Observe that $v \perp K_s$ if $s < \tau$, that is,

$$v^T Y_s^{-1} A_k(X_s) = 0, \quad \text{for } k = 1, \dots, d \quad \text{and } s < \tau. \tag{2.65}$$

We introduce the following sets of vector fields:

$$
\begin{aligned}
\Sigma_0 &= \{A_1, \dots, A_d\}, \\
\Sigma_n &= \{[A_k, V], k = 1, \dots, d, V \in \Sigma_{n-1}\} \quad \text{if } n \geq 1, \\
\Sigma &= \cup_{n=0}^{\infty} \Sigma_n,
\end{aligned}
$$

and

$$
\begin{aligned}
\Sigma_0' &= \Sigma_0, \\
\Sigma_n' &= \{[A_k, V], k = 1, \dots, d, V \in \Sigma_{n-1}'; \\
&\quad [A_0, V] + \frac{1}{2}\sum_{j=1}^{d}[A_j, [A_j, V]], V \in \Sigma_{n-1}'\} \quad \text{if } n \geq 1, \\
\Sigma' &= \cup_{n=0}^{\infty} \Sigma_n'.
\end{aligned}
$$

We denote by $\Sigma_n(x)$ (resp. $\Sigma_n'(x)$) the subset of $\mathbb{R}^m$ obtained by freezing the variable $x$ in the vector fields of $\Sigma_n$ (resp. $\Sigma_n'$). Clearly, the vector spaces spanned by $\Sigma(x)$ or by $\Sigma'(x)$ coincide, and under Hörmander's condition this vector space is $\mathbb{R}^m$. We will show that for all $n \geq 0$ the vector $v$ is orthogonal to $\Sigma_n'(x_0)$, which is in contradiction with Hörmander's condition. This claim will follow from the following stronger orthogonality property:

$$v^T Y_s^{-1} V(X_s) = 0, \quad \text{for all } s < \tau, V \in \Sigma_n', n \geq 0. \tag{2.66}$$

Indeed, for $s = 0$ we have $Y_0^{-1}V(X_0) = V(x_0)$. Property (2.66) can be proved by induction on $n$. For $n = 0$ it reduces to (2.65). Suppose that it holds for $n - 1$, and let $V \in \Sigma_{n-1}'$. Using formula (2.63) and the induction hypothesis, we obtain

$$
\begin{aligned}
0 &= \int_0^s v^T Y_u^{-1}[A_k, V](X_u) dW_u^k \\
&\quad + \int_0^s v^T Y_u^{-1}\left\{[A_0, V] + \frac{1}{2}\sum_{k=1}^{d}[A_k, [A_k, V]]\right\}(X_u) du
\end{aligned}
$$

for $s < \tau$. If a continuous semimartingale vanishes in a random interval $[0, \tau)$, where $\tau$ is a stopping time, then the quadratic variation of the martingale part and the bounded variation part of the semimartingale must be zero on this interval. As a consequence we obtain

$$v^T Y_s^{-1}[A_k, V](X_s) = 0$$

and

$$v^T Y_s^{-1} \left\{ [A_0, V] + \frac{1}{2} \sum_{k=1}^{d} [A_k, [A_k, V]] \right\} (X_s) = 0,$$

for any $s < \tau$. Therefore (2.66) is true for $n$, and the proof of the theorem is complete. □

## 2.3.3  Smoothness of the density under Hörmander's condition

In this section we will show the following result.

**Theorem 2.3.3** *Assume that $\{X(t), t \geq 0\}$ is the solution to Eq. (2.37), where the coefficients do not depent on the time. Suppose that the coefficients $A_j$, $1 \leq j \leq d$, $B$ are infinitely differentiable with bounded partial derivatives of all orders and that Hörmander's condition (H) holds. Then for any $t > 0$ the random vector $X(t)$ has an infinitely differentiable density.*

From the previous results it suffices to show that $(\det C_t)^{-1}$ has moments of all orders. We need the following preliminary lemmas.

**Lemma 2.3.1** *Let $C$ be a symmetric nonnegative definite $m \times m$ random matrix. Assume that the entries $C^{ij}$ have moments of all orders and that for any $p \geq 2$ there exists $\epsilon_0(p)$ such that for all $\epsilon \leq \epsilon_0(p)$*

$$\sup_{|v|=1} P\{v^T C v \leq \epsilon\} \leq \epsilon^p.$$

*Then $(\det C_t)^{-1} \in L^p(\Omega)$ for all $p$.*

*Proof:*    Let $\lambda = \inf_{|v|=1} v^T C v$ be the smallest eigenvalue of $C$. We know that $\lambda^m \leq \det C$. Thus, it suffices to show that $E(\lambda^{-p}) < \infty$ for all $p \geq 2$. Set $|C| = \left[ \sum_{i,j=1}^{m} (C^{ij})^2 \right]^{\frac{1}{2}}$. Fix $\epsilon > 0$, and let $v_1, \ldots, v_N$ be a finite set of unit vectors such that the balls with their center in these points and radius $\frac{\epsilon^2}{2}$ cover the unit sphere $S^{m-1}$. Then we have

$$
\begin{aligned}
P\{\lambda < \epsilon\} &= P\{\inf_{|v|=1} v^T C v < \epsilon\} \\
&\leq P\{\inf_{|v|=1} v^T C v < \epsilon, |C| \leq \frac{1}{\epsilon}\} + P\{|C| > \frac{1}{\epsilon}\}. \quad (2.67)
\end{aligned}
$$

Assume that $|C| \leq \frac{1}{\epsilon}$ and $v_k^T C v_k \geq 2\epsilon$ for any $k = 1, \ldots, N$. For any unit vector $v$ there exists a $v_k$ such that $|v - v_k| \leq \frac{\epsilon^2}{2}$ and we can deduce the following inequalities:

$$
\begin{aligned}
v^T C v &\geq v_k^T C v_k - |v^T C v - v_k^T C v_k| \\
&\geq 2\epsilon - \left[|v^T C v - v^T C v_k| + |v^T C v_k - v_k^T C v_k|\right] \\
&\geq 2\epsilon - 2|C||v - v_k| \geq \epsilon.
\end{aligned}
$$

As a consequence, (2.67) is bounded by

$$
P\left(\bigcup_{k=1}^{N} \{v_k^T C v_k < 2\epsilon\}\right) + P\{|C| > \frac{1}{\epsilon}\} \leq N(2\epsilon)^{p+2m} + \epsilon^p E(|C|^p)
$$

if $\epsilon \leq \frac{1}{2}\epsilon_0(p + 2m)$. The number $N$ depends on $\epsilon$ but is bounded by a constant times $\epsilon^{-2m}$. Therefore, we obtain $P\{\lambda < \epsilon\} \leq \text{const.}\epsilon^p$ for all $\epsilon \leq \epsilon_1(p)$ and for all $p \geq 2$. Clearly, this implies that $\lambda^{-1}$ has moments of all orders. $\qquad\square$

The next lemma has been proved by Norris in [239], following the ideas of Stroock [320], and is the basic ingredient in the proof of Theorem 2.3.3. The heuristic interpretation of this lemma is as follows: It is well known that if the quadratic variation and the bounded variation component of a continuous semimartingale vanish in some time interval, then the semimartingale vanishes in this interval. (Equation (2.69) provides a quantitative version of this result.) That is, when the quadratic variation or the bounded variation part of a continuous semimartingale is large, then the semimartingale is small with an exponentially small probability.

**Lemma 2.3.2** *Let $\alpha, y \in \mathbb{R}$. Suppose that $\beta(t)$, $\gamma(t) = (\gamma_1(t), \ldots, \gamma_d(t))$, and $u(t) = (u_1(t), \ldots, u_d(t))$ are adapted processes. Set*

$$
\begin{aligned}
a(t) &= \alpha + \int_0^t \beta(s)ds + \int_0^t \gamma_i(s)dW_s^i \\
Y(t) &= y + \int_0^t a(s)ds + \int_0^t u_i(s)dW_s^i,
\end{aligned}
$$

*and assume that there exists $t_0 > 0$ and $p \geq 2$ such that*

$$
c = E\left(\sup_{0 \leq t \leq t_0} (|\beta(t)| + |\gamma(t)| + |a(t)| + |u(t)|)^p\right) < \infty. \tag{2.68}
$$

*Then, for any $q > 8$ and for any $r, \nu > 0$ such that $18r + 9\nu < q - 8$, there exists $\epsilon_0 = \epsilon_0(t_0, q, r, \nu)$ such that for all $\epsilon \leq \epsilon_0$*

$$
P\left\{\int_0^{t_0} Y_t^2 dt < \epsilon^q, \int_0^{t_0} (|a(t)|^2 + |u(t)|^2)dt \geq \epsilon\right\} \leq c\epsilon^{rp} + e^{-\epsilon^{-\nu}}. \tag{2.69}
$$

*Proof:*    Set $\theta_t = |\beta(t)| + |\gamma(t)| + |a(t)| + |u(t)|$. Fix $q > 8$ and $r$, $\nu$ such that $18r + 9\nu < q - 8$. Suppose that $\nu' < \nu$ also satisfies $18r + 9\nu' < q - 8$ Then we define the bounded stopping time

$$T = \inf\left\{s \geq 0: \sup_{0 \leq u \leq s} \theta_u > \epsilon^{-r}\right\} \wedge t_0.$$

We have

$$P\left\{\int_0^{t_0} Y_t^2 dt < \epsilon^q, \int_0^{t_0} (|a(t)|^2 + |u(t)|^2)dt \geq \epsilon\right\} \leq A_1 + A_2,$$

with $A_1 = P\{T < t_0\}$ and

$$A_2 = P\left\{\int_0^{t_0} Y_t^2 dt < \epsilon^q, \int_0^{t_0} (|a(t)|^2 + |u(t)|^2)dt \geq \epsilon, T = t_0\right\}.$$

By the definition of $T$ and condition (2.68), we obtain

$$A_1 \leq P\left\{\sup_{0 \leq s \leq t_0} \theta_s > \epsilon^{-r}\right\} \leq \epsilon^{rp} E\left[\sup_{0 \leq s \leq t_0} \theta_s^p\right] \leq c\epsilon^{rp}.$$

Let us introduce the following notation:

$$A_t = \int_0^t a(s)ds, \qquad M_t = \int_0^t u_i(s)dW_s^i,$$

$$N_t = \int_0^t Y(s)u_i(s)dW_s^i, \quad Q_t = \int_0^t A(s)\gamma_i(s)dW_s^i.$$

Define for any $\rho_i > 0$, $\delta_i > 0$, $i = 1, 2, 3$,

$$B_1 = \left\{\langle N\rangle_T < \rho_1, \sup_{0 \leq s \leq T} |N_s| \geq \delta_1\right\},$$

$$B_2 = \left\{\langle M\rangle_T < \rho_2, \sup_{0 \leq s \leq T} |M_s| \geq \delta_2\right\},$$

$$B_3 = \left\{\langle Q\rangle_T < \rho_3, \sup_{0 \leq s \leq T} |Q_s| \geq \delta_3\right\}.$$

By the exponential martingale inequality (cf. (A.5)),

$$P(B_i) \leq 2\exp\left(-\frac{\delta_i^2}{2\rho_i}\right), \tag{2.70}$$

for $i = 1, 2, 3$. Our aim is to prove the following inclusion:

$$\left\{\int_0^T Y_t^2 dt < \epsilon^q, \int_0^T (|a(t)|^2 + |u(t)|^2)dt \geq \epsilon, T = t_0\right\}$$

$$\subset B_1 \cup B_2 \cup B_3, \tag{2.71}$$

for the particular choices of $\rho_i$ and $\delta_i$:

$$\rho_1 = \epsilon^{-2r+q}, \qquad\qquad \delta_1 = \epsilon^{q_1}, \quad q_1 = \frac{q}{2} - r - \frac{\nu'}{2},$$

$$\rho_2 = 2(2t_0+1)^{\frac{1}{2}}\epsilon^{-2r+\frac{q_1}{2}}, \quad \delta_2 = \epsilon^{q_2}, \quad q_2 = \frac{q}{8} - \frac{5r}{4} - \frac{5\nu'}{8},$$

$$\rho_3 = 36\epsilon^{-2r+2q_2}t_0, \qquad\qquad \delta_3 = \epsilon^{q_3}, \quad q_3 = \frac{q}{8} - \frac{9r}{4} - \frac{9\nu'}{8}.$$

From the inequality (2.70) and the inclusion (2.71) we get

$$\begin{aligned}
A_2 &\leq 2\left(\exp(-\frac{\delta_1^2}{2\rho_1}) + \exp(-\frac{\delta_2^2}{2\rho_2}) + \exp(-\frac{\delta_3^2}{2\rho_3})\right) \\
&\leq 2\left(\exp(-\frac{1}{2}\epsilon^{-\nu'}) + \exp(-\frac{1}{4\sqrt{1+2t_0}}\epsilon^{-\nu'}) + \exp(-\frac{1}{72t_0}\epsilon^{-\nu'})\right) \\
&\leq \exp(-\epsilon^{-\nu})
\end{aligned}$$

for $\epsilon \leq \epsilon_0$, because

$$\begin{aligned}
2q_1 + 2r - q &= -\nu', \\
2q_2 + 2r - \frac{q_1}{2} &= -\nu', \\
2q_3 + 2r - 2q_2 &= -\nu',
\end{aligned}$$

which allows us to complete the proof of the lemma. It remains only to check the inclusion (2.71).

*Proof of (2.71):* Suppose that $\omega \notin B_1 \cup B_2 \cup B_3$, $T(\omega) = t_0$, and $\int_0^T Y_t^2 dt < \epsilon^q$. Then

$$\langle N \rangle_T = \int_0^T Y_t^2 |u_t|^2 dt < \epsilon^{-2r+q} = \rho_1.$$

Then since $\omega \notin B_1$, $\sup_{0\leq s\leq T}\left|\int_0^t Y_s u_s^i dW_s^i\right| < \delta_1 = \epsilon^{q_1}$. Also

$$\sup_{0\leq s\leq T}\left|\int_0^t Y_s a_s ds\right| \leq \left(t_0\int_0^T Y_t^2 a_t^2 dt\right)^{\frac{1}{2}} < t_0^{\frac{1}{2}}\epsilon^{-r+\frac{q}{2}}.$$

Thus,

$$\sup_{0\leq s\leq T}\left|\int_0^t Y_s dY_s\right| < \sqrt{t_0}\epsilon^{-r+\frac{q}{2}} + \epsilon^{q_1}.$$

By Itô's formula $Y_t^2 = y^2 + 2\int_0^t Y_s dY_s + \langle M \rangle_t$, and therefore

$$\begin{aligned}
\int_0^T \langle M \rangle_t dt &= \int_0^T Y_t^2 dt - Ty^2 - 2\int_0^T\left(\int_0^t Y_s dY_s\right)dt \\
&< \epsilon^q + 2t_0\left(\sqrt{t_0}\epsilon^{-r+\frac{q}{2}} + \epsilon^{q_1}\right) < (2t_0+1)\epsilon^{q_1},
\end{aligned}$$

for $\epsilon \leq \epsilon_0$ because $q > q_1$ and $-r + \frac{q}{2} > q_1$. Since $\langle M \rangle_t$ is an increasing process, for any $0 < \gamma < T$ we have

$$\gamma \langle M \rangle_{T-\gamma} < (2t_0 + 1)\epsilon^{q_1},$$

and hence $\langle M \rangle_T < \gamma^{-1}(2t_0 + 1)\epsilon^{q_1} + \gamma\epsilon^{-2r}$. Choosing $\gamma = (2t_0 + 1)^{\frac{1}{2}}\epsilon^{\frac{q_1}{2}}$, we obtain $\langle M \rangle_T < \rho_2$, provided $\epsilon < 1$. Since $\omega \notin B_2$ we get

$$\sup_{0 \leq s \leq T} |M_t| < \delta_2 = \epsilon^{q_2}.$$

Recall that $\int_0^T Y_t^2 dt < \epsilon^q$ so that, by Tchebychev's inequality,

$$\lambda^1\{t \in [0,T] : |Y_t(\omega)| \geq \epsilon^{\frac{q}{3}}\} \leq \epsilon^{\frac{q}{3}},$$

and therefore

$$\lambda^1\{t \in [0,T] : |y + A_t(\omega)| \geq \epsilon^{\frac{q}{3}} + \epsilon^{q_2}\} \leq \epsilon^{\frac{q}{3}}.$$

We can assume that $\epsilon^{\frac{q}{3}} < \frac{t_0}{2}$, provided $\epsilon \leq \epsilon_0(t_0)$. So for each $t \in [0,T]$, there exists $s \in [0,T]$ such that $|s - t| \leq \epsilon^{\frac{q}{3}}$ and $|y + A_s| < \epsilon^{\frac{q}{3}} + \epsilon^{q_2}$. Consequently,

$$|y + A_t| \leq |y + A_s| + \left| \int_s^t a_r dr \right| < (1 + \epsilon^{-r})\epsilon^{\frac{q}{3}} + \epsilon^{q_2}.$$

In particular, $|y| < (1 + \epsilon^{-r})\epsilon^{\frac{q}{3}} + \epsilon^{q_2}$, and for all $t \in [0,T]$ we have

$$|A_t| < 2\left((1 + \epsilon^{-r})\epsilon^{\frac{q}{3}} + \epsilon^{q_2}\right) \leq 6\epsilon^{q_2},$$

because $q_2 < \frac{q}{3} - r$. This implies that

$$\langle Q \rangle_T = \int_0^T A_t^2 |\gamma_t|^2 dt < 36t_0\epsilon^{2q_2 - 2r} = \rho_3.$$

So since $\omega \notin B_3$, we have

$$|Q_T| = \left| \int_0^T A_t \gamma_i(t) dW_i(t) \right| < \delta_3 = \epsilon^{q_3}.$$

Finally, by Itô's formula we obtain

$$
\begin{aligned}
\int_0^T (a_t^2 + |u_t|^2) dt &= \int_0^T a_t dA_t + \langle M \rangle_T \\
&= a_T A_T - \int_0^T A_t \beta_t dt - \int_0^T A_t \gamma_i(t) dW_t^i + \langle M \rangle_T \\
&\leq (1 + t_0) 6\epsilon^{q_2 - r} + \epsilon^{q_3} + 2\sqrt{2t_0 + 1}\epsilon^{-2r + \frac{q_1}{2}} < \epsilon
\end{aligned}
$$

for $\epsilon \leq \epsilon_0$, because $q_2 - r > q_3$, $q_3 > 1$, and $-2r + \frac{q_1}{2} > 1$. $\qquad\square$

Now we can proceed to the proof of Theorem 2.3.3.

*Proof of Theorem 2.3.3:* Fix $t > 0$. We want to show that $E[(\det C_t)^{-p}] < \infty$ for all $p \geq 2$. By Lemma 2.3.1 it suffices to see that for all $p \geq 2$ we have

$$\sup_{|v|=1} P\{v^T C_t v \leq \epsilon\} \leq \epsilon^p$$

for any $\epsilon \leq \epsilon_0(p)$. We recall the following expression for the quadratic form associated to the matrix $C_t$:

$$v^T C_t v = \sum_{j=1}^{d} \int_0^t \left| v^T Y_s^{-1} A_j(X_s) \right|^2 ds.$$

By Hörmander's condition, there exists an integer $j_0 \geq 0$ such that the linear span of the set of vector fields $\bigcup_{j=0}^{j_0} \Sigma'_j(x)$ at point $x_0$ has dimension $m$. As a consequence there exist constants $R > 0$ and $c > 0$ such that

$$\sum_{j=0}^{j_0} \sum_{V \in \Sigma'_j} (v^T V(y))^2 \geq c,$$

for all $v$ and $y$ with $|v| = 1$ and $|y - x_0| < R$.

For any $j = 0, 1, \ldots, j_0$ we put $m(j) = 2^{-4j}$ and we define the set

$$E_j = \left\{ \sum_{V \in \Sigma'_j} \int_0^t \left( v^T Y_s^{-1} V(X_s) \right)^2 ds \leq \epsilon^{m(j)} \right\}.$$

Notice that $\{v^T C_t v \leq \epsilon\} = E_0$ because $m(0) = 1$. Consider the decomposition

$$E_0 \subset (E_0 \cap E_1^c) \cup (E_1 \cap E_2^c) \cup \cdots \cup (E_{j_0-1} \cap E_{j_0}^c) \cup F,$$

where $F = E_0 \cap E_1 \cap \cdots \cap E_{j_0}$. Then for any unit vector $v$ we have

$$P\{v^T C_t v \leq \epsilon\} = P(E_0) \leq P(F) + \sum_{j=0}^{j_0} P(E_j \cap E_{j+1}^c).$$

We are going to estimate each term of this sum. This will be done in two steps.

*Step 1:*    Consider the following stopping time:

$$S = \inf\{\sigma \geq 0 : \sup_{0 \leq s \leq \sigma} |X_s - x_0| \geq R \quad \text{or} \quad \sup_{0 \leq s \leq \sigma} |Y_s^{-1} - I| \geq \frac{1}{2}\} \wedge t.$$

We can write

$$P(F) \leq P\left(F \cap \{S \geq \epsilon^\beta\}\right) + P\{S < \epsilon^\beta\},$$

where $0 < \beta < m(j_0)$. For $\epsilon$ small enough, the intersection $F \cap \{S \geq \epsilon^\beta\}$ is empty. In fact, if $S \geq \epsilon^\beta$, we have

$$\sum_{j=0}^{j_0} \sum_{V \in \Sigma'_j} \int_0^t \left(v^T Y_s^{-1} V(X_s)\right)^2 ds$$

$$\geq \sum_{j=0}^{j_0} \sum_{V \in \Sigma'_j} \int_0^S \left(\frac{v^T Y_s^{-1} V(X_s)}{|v^T Y_s^{-1}|}\right)^2 |v^T Y_s^{-1}|^2 ds \geq \frac{c\epsilon^\beta}{4}, \quad (2.72)$$

because $s < S$ implies $|v^T Y_s^{-1}| \geq 1 - |I - Y_s^{-1}| \geq \frac{1}{2}$. On the other hand, the left-hand side of (2.72) is bounded by $(j_0 + 1)\epsilon^{m(j_0)}$ on the set $F$, and for $\epsilon$ small enough we therefore obtain $F \cap \{S \geq \epsilon^\beta\} = \emptyset$. Moreover, it holds that

$$P\{S < \epsilon^\beta\} \leq P\left\{\sup_{0 \leq s \leq \epsilon^\beta} |X_s - x_0| \geq R\right\}$$

$$+ P\left\{\sup_{0 \leq s \leq \epsilon^\beta} |Y_s^{-1} - I| \geq \frac{1}{2}\right\}$$

$$\leq R^{-q} E\left[\sup_{0 \leq s \leq \epsilon^\beta} |X_s - x_0|^q\right] + 2^q E\left[\sup_{0 \leq s \leq \epsilon^\beta} |Y_s^{-1} - I|^q\right]$$

for any $q \geq 2$. Now using Burkholder's and Hölder's inequalities, we deduce that $P\{S < \epsilon^\beta\} \leq C\epsilon^{\frac{q\beta}{2}}$ for any $q \geq 2$, which provides the desired estimate for $P(F)$.

*Step 2:*   For any $j = 0, \ldots, j_0$ we introduce the following probability:

$$P(E_j \cap E_{j+1}^c) = P\left\{\sum_{V \in \Sigma'_j} \int_0^t \left(v^T Y_s^{-1} V(X_s)\right)^2 ds \leq \epsilon^{m(j)},\right.$$

$$\sum_{V \in \Sigma'_j} \int_0^t \left(v^T Y_s^{-1} V(X_s)\right)^2 ds > \epsilon^{m(j+1)}\right\}$$

$$\leq \sum_{V \in \Sigma'_j} P\left\{\int_0^t \left(v^T Y_s^{-1} V(X_s)\right)^2 ds \leq \epsilon^{m(j)},\right.$$

$$\sum_{k=1}^d \int_0^t \left(v^T Y_s^{-1} [A_k, V](X_s)\right)^2 ds + \int_0^t \left(v^T Y_s^{-1}\left([A_0, V]\right.\right.$$

$$\left.\left. + \frac{1}{2} \sum_{j=1}^d [A_j, [A_j, V]]\right)(X_s)\right)^2 ds > \frac{\epsilon^{m(j+1)}}{n(j)}\right\},$$

where $n(j)$ denotes the cardinality of the set $\Sigma'_j$. Consider the continuous semimartingale $\{v^T Y_s^{-1} V(X_s), s \geq 0\}$. From (2.63) we see that the quadratic variation of this semimartingale is equal to

$$\sum_{k=1}^{d} \int_0^s \left( v^T Y_\sigma^{-1} [A_k, V](X_\sigma) \right)^2 d\sigma,$$

and the bounded variation component is

$$\int_0^s v^T Y_\sigma^{-1} \left\{ [A_0, V] + \frac{1}{2} \sum_{j=1}^{d} [A_j, [A_j, V]] \right\} (X_\sigma) d\sigma.$$

Taking into account that $8m(j+1) < m(j)$, we get the desired estimate from Lemma 2.3.2 applied to the semimartingale $Y_t = v^T Y_s^{-1} V(X_s)$. The proof of the theorem is now complete.                                              □

**Remarks:**

**1.** Note that if the diffusion matrix $\sigma(x) = \sum_{j=1}^{d} A_j(x) A_j^T(x)$ is elliptic at the initial point (that is, $\sigma(x_0) > 0$), then Hörmander's condition (H) holds, and for any $t > 0$ the random variable $X_t$ has an infinitely differentiable density. The interesting applications of Hörmander's theorem appear when $\sigma(x_0)$ is degenerate.

Consider the following elementary example. Let $m = d = 2$, $X_0 = 0$, $B = 0$, and consider the vector fields

$$A_1(x) = \begin{bmatrix} 1 \\ 2x_1 \end{bmatrix} \quad \text{and} \quad A_2(x) = \begin{bmatrix} \sin x_2 \\ x_1 \end{bmatrix}.$$

In this case the diffusion matrix

$$\sigma(x) = \begin{bmatrix} 1 + \sin^2 x_2 & x_1(2 + \sin x_2) \\ x_1(2 + \sin x_2) & 5x_1^2 \end{bmatrix}$$

degenerates along the line $x_1 = 0$. The Lie bracket $[A_1, A_2]$ is equal to $\begin{bmatrix} 2x_1 \cos x_2 \\ 1 - 2\sin x_2 \end{bmatrix}$. Therefore, the vector fields $A_1$ and $[A_1, A_2]$ at $x = 0$ span $\mathbb{R}^2$ and Hörmander's condition holds. So from Theorem 2.3.3 $X(t)$ has a $C^\infty$ density for any $t > 0$.

**2.** The following is a stronger version of Hörmander's condition:

**(H1)** The Lie algebra space spanned by the vector fields $A_1, \ldots, A_d$ at point $x_0$ is $\mathbb{R}^m$.

The proof of Theorem 2.3.3 under this stronger hypothesis can be done using the simpler version of Lemma 2.3.2 stated in Exercise 2.3.4.

## *Exercises*

**2.3.1** Let $W = \{(W^1, W^2), t \geq 0\}$ be a two-dimensional Brownian motion, and consider the process $X = \{X_t, t \geq 0\}$ defined by

$$X_t^1 = W_t^1,$$

$$X_t^2 = \int_0^t W_s^1 dW_s^2.$$

Compute the Malliavin matrix $\gamma_t$ of the vector $X_t$, and show that

$$\det \gamma_t \geq t \int_0^t (W_s^1)^2 ds.$$

Using Lemma 2.3.2 show that $E[|\int_0^t (W_s^1)^2 ds|^{-p}] < \infty$ for all $p \geq 2$, and conclude that for all $t > 0$ the random variable $X_t$ has an infinitely differentiable density. Obtain the same result by applying Theorem 2.3.3 to a stochastic differential equation satisfied by $X(t)$.

**2.3.2** Let $f(s, t)$ be a square integrable symmetric kernel on $[0, 1]$. Set $F = I_2(f)$. Show that the norm of the derivative of $F$ is given by

$$\|DF\|_H^2 = \sum_{n=1}^{\infty} \lambda_n^2 W(e_n)^2,$$

where $\{\lambda_n\}$ and $\{e_n\}$ are the corresponding sequence of eigenvalues and orthogonal eigenvectors of the operator associated with $f$. In the particular case where $\lambda_n = (\pi n)^{-2}$, show that

$$P(\|DF\|_H < \epsilon) \leq \sqrt{2} \exp(-\frac{1}{8\epsilon^2}),$$

and conclude that $F$ has an infinitely differentiable density.

*Hint:* Use Tchebychev's exponential inequality with the function $e^{-\lambda^2 \epsilon^2 x}$ and then optimize over $\lambda$.

**2.3.3** Let $m = 3$, $d = 2$, and $X_0 = 0$, and consider the vector fields

$$A_1(x) = \begin{bmatrix} 1 \\ 0 \\ 0 \end{bmatrix}, \quad A_2(x) = \begin{bmatrix} 0 \\ \sin x_2 \\ x_1 \end{bmatrix}, \quad B(x) = \begin{bmatrix} 0 \\ \frac{1}{2} \sin x_2 \cos x_2 + 1 \\ 1 \end{bmatrix}.$$

Show that the solution to the stochastic differential equation $X(t)$ associated to these coefficients has a $C^\infty$ density for any $t > 0$.

**2.3.4** Prove the following stronger version of Lemma 2.3.2: Let

$$Y(t) = y + \int_0^t a(s)ds + \int_0^t u_i(s)dW_s^i, \quad t \in [0, t_0],$$

be a continuous semimartingale such that $y \in \mathbb{R}$ and $a$ and $u_i$ are adapted processes verifying

$$c := E\left[ \sup_{0 \le t \le t_0} (|a_t| + |u_t|)^p \right] < \infty.$$

Then for any $q, r, \nu > 0$ verifying $q > \nu + 10r + 1$ there exists $\epsilon_0 = \epsilon_0(t_0, q, r, \nu)$ such that for $\epsilon \le \epsilon_0$

$$P\left\{ \int_0^{t_0} Y_t^2 dt < \epsilon^q, \int_0^{t_0} |u(t)|^2 dt \ge \epsilon \right\} \le c\epsilon^{rp} + e^{-\epsilon^{-\nu}}.$$

**2.3.5** (Elworthy formula [90]) Let $X = \{X(t),\ t \in [0,T]\}$ be the solution to the following $d$-dimensional stochastic differential equation:

$$X(t) = x_0 + \sum_{j=1}^{d} \int_0^t A_j(X(s)) dW_s^j + \int_0^t B(X(s)) ds,$$

where the coefficients $A_j$ and $B$ are of class $C^{1+\alpha}$, $\alpha > 0$, with bounded derivatives. We also assume that the $m \times m$ matrix $A$ is invertible and that its inverse has polynomial growth. Show that for any function $\varphi \in C_b^1(\mathbb{R}^d)$ and for any $t > 0$ the following formula holds:

$$E[\partial_i \varphi(X_t)] = \frac{1}{t} E\left[ \varphi(X_t) \int_0^t (A^{-1})_k^j(X_s) Y_i^k(s) dW_s^j \right],$$

where $Y(s)$ denotes the Jacobian matrix $\frac{\partial X_s}{\partial x_0}$ given by (2.57).

  *Hint:* Use the decomposition $D_s X_t = Y(t) Y^{-1}(s) A(X_s)$ and the duality relationship between the derivative operator and the Skorohod (Itô) integral.

## 2.4   Stochastic partial differential equations

In this section we discuss the applications of the Malliavin calculus to establishing the existence and smoothness of densities for solutions to stochastic partial differential equations. First we will treat the case of a hyperbolic system of equations using the techniques of the two-parameter stochastic calculus. Second we will prove a criterion for absolute continuity in the case of the heat equation perturbed by a space-time white noise.

### 2.4.1   Stochastic integral equations on the plane

Suppose that $W = \{W_z = (W_z^1, \ldots, W_z^d), z \in \mathbb{R}_+^2\}$ is a $d$-dimensional, two-parameter Wiener process. That is, $W$ is a $d$-dimensional, zero-mean

Gaussian process with a covariance function given by

$$E[W^i(s_1, t_1)W^j(s_2, t_2)] = \delta_{ij}(s_1 \wedge s_2)(t_1 \wedge t_2).$$

We will assume that this process is defined in the canonical probability space $(\Omega, \mathcal{F}, P)$, where $\Omega$ is the space of all continuous functions $\omega : \mathbb{R}_+^2 \to \mathbb{R}^d$ vanishing on the axes, and endowed with the topology of the uniform convergence on compact sets, $P$ is the law of the process $W$ (which is called the two-parameter, $d$-dimensional Wiener measure), and $\mathcal{F}$ is the completion of the Borel $\sigma$-field of $\Omega$ with respect to $P$. We will denote by $\{\mathcal{F}_z, z \in \mathbb{R}_+^2\}$ the increasing family of $\sigma$-fields such that for any $z$, $\mathcal{F}_z$ is generated by the random variables $\{W(r), r \leq z\}$ and the null sets of $\mathcal{F}$. Here $r \leq z$ stands for $r_1 \leq z_1$ and $r_2 \leq z_2$. Given a rectangle $\Delta = (s_1, s_2] \times (t_1, t_2]$, we will denote by $W(\Delta)$ the increment of $W$ on $\Delta$ defined by

$$W(\Delta) = W(s_2, t_2) - W(s_2, t_1) - W(s_1, t_2) + W(s_1, t_1).$$

The Gaussian subspace of $L^2(\Omega, \mathcal{F}, P)$ generated by $W$ is isomorphic to the Hilbert space $H = L^2(\mathbb{R}_+^2; \mathbb{R}^d)$. More precisely, to any element $h \in H$ we associate the random variable $W(h) = \sum_{j=1}^d \int_{\mathbb{R}_+^2} h_j(z) dW^j(z)$.

A stochastic process $\{Y(z), z \in \mathbb{R}_+^2\}$ is said to be adapted if $Y(z)$ is $\mathcal{F}_z$-measurable for any $z \in \mathbb{R}_+^2$. The Itô stochastic integral of adapted and square integrable processes can be constructed as in the one-parameter case and is a special case of the Skorohod integral:

**Proposition 2.4.1** *Let $L_a^2(\mathbb{R}_+^2 \times \Omega)$ be the space of square integrable and adapted processes $\{Y(z), z \in \mathbb{R}_+^2\}$ such that $\int_{\mathbb{R}_+^2} E(Y^2(z)) dz < \infty$. For any $j = 1, \ldots, d$ there is a linear isometry $I^j : L_a^2(\mathbb{R}_+^2 \times \Omega) \to L^2(\Omega)$ such that*

$$I^j(\mathbf{1}_{(z_1, z_2]}) = W^j((z_1, z_2])$$

*for any $z_1 \leq z_2$. Furthermore, $L_a^2(\mathbb{R}_+^2 \times \Omega; \mathbb{R}^d) \subset \mathrm{Dom}\,\delta$, and $\delta$ restricted to $L_a^2(\mathbb{R}_+^2 \times \Omega; \mathbb{R}^d)$ coincides with the sum of the Itô integrals $I^j$, in the sense that for any $d$-dimensional process $Y \in L_a^2(\mathbb{R}_+^2 \times \Omega; \mathbb{R}^d)$ we have*

$$\delta(Y) = \sum_{j=1}^d I^j(Y^j) = \sum_{j=1}^d \int_{\mathbb{R}_+^2} Y^j(z) dW^j(z).$$

Let $A_j, B : \mathbb{R}^m \to \mathbb{R}^m$, $1 \leq j \leq d$, be globally Lipschitz functions. We denote by $X = \{X(z), z \in \mathbb{R}_+^2\}$ the $m$-dimensional, two-parameter, continuous adapted process given by the following system of stochastic integral equations on the plane:

$$X(z) = x_0 + \sum_{j=1}^d \int_{[0,z]} A_j(X_r) dW_r^j + \int_{[0,z]} B(X_r) dr, \qquad (2.73)$$

where $x_0 \in \mathbb{R}^m$ represents the constant value of the process $X(z)$ on the axes. As in the one-parameter case, we can prove that this system of stochastic integral equations has a unique continuous solution:

**Theorem 2.4.1** *There is a unique m-dimensional, continuous, and adapted process $X$ that satisfies the integral equation (2.73). Moreover,*

$$E \left[ \sup_{r \in [0,z]} |X_r|^p \right] < \infty$$

*for any $p \geq 2$, and any $z \in \mathbb{R}_+^2$.*

*Proof:*     Use the Picard iteration method and two-parameter martingale inequalities (see (A.7) and (A.8)) in order to show the uniform convergence of the approximating sequence.     □

Equation (2.73) is the integral version of the following nonlinear hyperbolic stochastic partial differential equation:

$$\frac{\partial^2 X(s,t)}{\partial s \partial t} = \sum_{j=1}^{d} A_j(X(s,t)) \frac{\partial^2 W^j(s,t)}{\partial s \partial t} + B(X(s,t)).$$

Suppose that $z = (s,t)$ is a fixed point in $\mathbb{R}_+^2$ not on the axes. Then we may look for nondegeneracy conditions on the coefficients of Eq. (2.73) so that the random vector $X(z) = (X^1(z), \ldots, X^m(z))$ has an absolutely continuous distribution with a smooth density.

We will assume that the coefficients $A_j$ and $B$ are infinitely differentiable functions with bounded partial derivatives of all orders. We can show as in the one-parameter case that $X^i(z) \in \mathbb{D}^\infty$ for all $z \in \mathbb{R}_+^2$ and $i = 1, \ldots, m$. Furthermore, the Malliavin matrix $Q_z^{ij} = \langle DX_z^i, DX_z^j \rangle_H$ is given by

$$Q_z^{ij} = \sum_{l=1}^{d} \int_{[0,z]} D_r^l X_z^i D_r^l X_z^j dr, \tag{2.74}$$

where for any $r$, the process $\{D_r^k X_z^i, r \leq z, 1 \leq i \leq m, 1 \leq k \leq d\}$ satisfies the following system of stochastic differential equations:

$$D_r^j X_z^i = A_j^i(X_r) + \int_{[r,z]} \partial_k A_l^i(X_u) D_r^j X_u^k dW_u^l$$

$$+ \int_{[r,z]} \partial_k B^i(X_u) D_r^j X_u^k du. \tag{2.75}$$

Moreover, we can write $D_r^j X_z^i = \xi_l^i(r,z) A_j^l(X_r)$, where for any $r$, the process $\{\xi_j^i(r,z), r \leq z, 1 \leq i,j \leq m\}$ is the solution to the following system of stochastic differential equations:

$$\xi_j^i(r,z) = \delta_j^i + \int_{[r,z]} \partial_k A_l^i(X_u)\xi_j^k(r,u)dW_u^l$$

$$+ \int_{[r,z]} \partial_k B^i(X_u)\xi_j^k(r,u)du. \tag{2.76}$$

However, unlike the one-parameter case, the processes $D_r^j X_z^i$ and $\xi_j^i(r,z)$ cannot be factorized as the product of a function of $z$ and a function of $r$. Furthermore, these processes satisfy two-parameter linear stochastic differential equations and the solution to such equations, even in the case of constant coefficients, are not exponentials, and may take negative values. As a consequence, we cannot estimate expectations such as $E(|\xi_j^i(r,z)|^{-p})$. The behavior of solutions to two-parameter linear stochastic differential equations is analyzed in the following proposition (cf. Nualart [243]).

**Proposition 2.4.2** *Let $\{X(z), z \in \mathbb{R}_+^2\}$ be the solution to the equation*

$$X_z = 1 + \int_{[0,z]} aX_r dW_r, \tag{2.77}$$

*where $a \in \mathbb{R}$ and $\{W(z), z \in \mathbb{R}_+^2\}$ is a two-parameter, one-dimensional Wiener process. Then,*

*(i) there exists an open set $\Delta \subset \mathbb{R}_+^2$ such that*

$$P\{X_z < 0 \quad for\ all \quad z \in \Delta\} > 0;$$

*(ii) $E(|X_z|^{-1}) = \infty$ for any $z$ out of the axes.*

*Proof:*   Let us first consider the deterministic version of Eq. (2.77):

$$g(s,t) = 1 + \int_0^s \int_0^t ag(u,v)dudv. \tag{2.78}$$

The solution to this equation is $g(s,t) = f(ast)$, where

$$f(x) = \sum_{n=0}^{\infty} \frac{x^n}{(n!)^2}.$$

In particular, for $a > 0$, $g(s,t) = I_0(2\sqrt{ast})$, where $I_0$ is the modified Bessel function of order zero, and for $a < 0$, $g(s,t) = J_0(2\sqrt{|a|st})$, where $J_0$ is the Bessel function of order zero. Note that $f(x)$ grows exponentially as $x$ tends to infinity and that $f(x)$ is equivalent to $(\pi\sqrt{|x|})^{-\frac{1}{2}} \cos(2\sqrt{|x|} - \frac{\pi}{4})$ as $x$ tends to $-\infty$. Therefore, we can find an open interval $I = (-\beta, -\alpha)$ with $0 < \alpha < \beta$ such that $f(x) < -\delta < 0$ for all $x \in I$.

In order to show part (i) we may suppose by symmetry that $a > 0$. Fix $N > 0$ and set $\Delta = \{(s,t) : \frac{\alpha}{a} < st < \frac{\beta}{a}, 0 < s, t < N\}$. Then $\Delta$ is an open set contained in the rectangle $T = [0,N]^2$ and such that $f(-ast) < -\delta$ for any $(s,t) \in \Delta$. For any $\epsilon > 0$ we will denote by $X_z^\epsilon$ the solution to the equation

$$X_z^\epsilon = 1 + \int_{[0,z]} a\epsilon X_r^\epsilon dW_r.$$

By Lemma 2.1.3 the process $W^\epsilon(s,t) = W(s,t) - ste^{-1}$ has the law of a two-parameter Wiener process on $T = [0,N]^2$ under the probability $P_\epsilon$ defined by

$$\frac{dP_\epsilon}{dP} = \exp\left(\epsilon^{-1}W(N,N) - \frac{1}{2}\epsilon^{-2}N^2\right).$$

Let $Y_z^\epsilon$ be the solution to the equation

$$Y_z^\epsilon = 1 + \int_{[0,z]} a\epsilon Y_r^\epsilon dW_r^\epsilon = 1 + \int_{[0,z]} a\epsilon Y_r^\epsilon dW_r - \int_{[0,z]} a Y_r^\epsilon dr. \qquad (2.79)$$

It is not difficult to check that

$$K = \sup_{0 < \epsilon \leq 1} \sup_{z \in T} E(|Y_z^\epsilon|^2) < \infty.$$

Then, for any $\epsilon \leq 1$, from Eqs. (2.78) and (2.79) we deduce

$$E\left(\sup_{(s,t) \in T} |Y^\epsilon(s,t) - f(-ast)|^2\right)$$

$$\leq C\left(\int_T E(|Y^\epsilon(x,y) - f(-axy)|^2)dxdy + a^2\epsilon^2 K\right)$$

for some constant $C > 0$. Hence,

$$\lim_{\epsilon \downarrow 0} E\left(\sup_{(s,t) \in T} |Y^\epsilon(s,t) - f(-ast)|^2\right) = 0,$$

and therefore

$$P\{Y_z^\epsilon < 0 \text{ for all } z \in \Delta\} \geq P\left\{\sup_{(s,t) \in \Delta} |Y^\epsilon(s,t) - f(-ast)| \leq \delta\right\}$$

$$\geq P\left\{\sup_{(s,t) \in T} |Y^\epsilon(s,t) - f(-ast)| \leq \delta\right\},$$

which converges to one as $\epsilon$ tends to zero. So, there exists an $\epsilon_0 > 0$ such that

$$P\{Y_z^\epsilon < 0 \text{ for all } z \in \Delta\} > 0$$

for any $\epsilon \leq \epsilon_0$. Then

$$P_\epsilon \{Y_z^\epsilon < 0 \quad \text{for all} \quad z \in \Delta\} > 0$$

because the probabilities $P_\epsilon$ and $P$ are equivalent, and this implies

$$P\{X_z^\epsilon < 0 \quad \text{for all} \quad z \in \Delta\} > 0.$$

By the scaling property of the two-parameter Wiener process, the processes $X^\epsilon(s,t)$ and $X(\epsilon s, \epsilon t)$ have the same law. Therefore,

$$P\{X(\epsilon s, \epsilon t) < 0 \quad \text{for all} \quad (s,t) \in \Delta\} > 0,$$

which gives the desired result with the open set $\epsilon \Delta$ for all $\epsilon \leq \epsilon_0$. Note that one can also take the open set $\{(\epsilon^2 s, t) : (s,t) \in \Delta\}$.

To prove (ii) we fix $(s,t)$ such that $st \neq 0$ and define $T = \inf\{\sigma \geq 0 : X(\sigma, t) = 0\}$. $T$ is a stopping time with respect to the increasing family of $\sigma$-fields $\{\mathcal{F}_{\sigma t}, \sigma \geq 0\}$. From part (i) we have $P\{T < s\} > 0$. Then, applying Itô's formula in the first coordinate, we obtain for any $\epsilon > 0$

$$E[(X(s,t)^2 + \epsilon)^{-\frac{1}{2}}] = E[(X(s \wedge T, t)^2 + \epsilon)^{-\frac{1}{2}}]$$
$$+ \frac{1}{2} E\left[\int_{s \wedge T}^s (2X(x,t)^2 - \epsilon)(X(x,t)^2 + \epsilon)^{-\frac{5}{2}} d\langle X(\cdot, t)\rangle_x\right].$$

Finally, if $\epsilon \downarrow 0$, by monotone convergence we get

$$E(|X(s,t)|^{-1}) = \lim_{\epsilon \downarrow 0} E[(X(s,t)^2 + \epsilon)^{-\frac{1}{2}}] \geq \infty P\{T < s\} = \infty.$$

$\square$

In spite of the technical problems mentioned before, it is possible to show the absolute continuity of the random vector $X_z$ solution of (2.73) under some nondegeneracy conditions that differ from Hörmander's hypothesis.

We introduce the following hypothesis on the coefficients $A_j$ and $B$, which are assumed to be infinitely differentiable with bounded partial derivatives of all orders:

**(P)**    The vector space spanned by the vector fields $A_1, \ldots, A_d, \quad A_i^\nabla A_j,$ $1 \leq i,j \leq d, \quad A_i^\nabla (A_j^\nabla A_k), 1 \leq i,j,k \leq d, \ldots, \quad A_{i_1}(\cdots (A_{i_{n-1}}^\nabla A_{i_n}) \cdots),$ $1 \leq i_1, \ldots, i_n \leq d$, at the point $x_0$ is $\mathbb{R}^m$.

Then we have the following result.

**Theorem 2.4.2** *Assume that condition (P) holds. Then for any point $z$ out of the axes the random vector $X(z)$ has an absolutely continuous probability distribution.*

We remark that condition (P) and Hörmander's hypothesis (H) are not comparable. Consider, for instance, the following simple example. Assume that $m \geq 2$, $d = 1$, $x_0 = 0$, $A_1(x) = (1, x^1, x^2, \ldots, x^{m-1})$, and $B(x) = 0$. This means that $X_z$ is the solution of the differential system

$$\begin{aligned}
dX_z^1 &= dW_z \\
dX_z^2 &= X_z^1 dW_z \\
dX_z^3 &= X_z^2 dW_z \\
&\cdots \\
dX_z^m &= X_z^{m-1} dW_z,
\end{aligned}$$

and $X_z = 0$ if $z$ is on the axes. Then condition (P) holds and, as a consequence, Theorem 2.4.2 implies that the joint distribution of the iterated stochastic integrals $W_z$, $\int_{[0,z]} W \, dW$, ..., $\int_{[0,z]} (\cdots (\int W \, dW) \cdots) dW = \int_{z_1 \leq \cdots \leq z_m} dW(z_1) \cdots dW(z_m)$ possesses a density on $\mathbb{R}^m$. However, Hörmander's hypothesis is not true in this case. Notice that in the one-parameter case the joint distribution of the random variables $W_t$ and $\int_0^t W_s dW_s$ is singular because Itô's formula implies that $W_t^2 - 2 \int_0^t W_s dW_s - t = 0$.

*Proof of Theorem 2.4.2:* The first step will be to show that the process $\xi_j^i(r, z)$ given by system (2.76) has a version that is continuous in the variable $r \in [0, z]$. By means of Kolmogorov's criterion (see the appendix, Section A.3), it suffices to prove the following estimate:

$$E(|\xi(r, z) - \xi(r', z)|^p) \leq C(p, z)|r - r'|^{\frac{p}{2}} \tag{2.80}$$

for any $r, r' \in [0, z]$ and $p > 4$. One can show that

$$\sup_{r \in [0,z]} E\left( \sup_{v \in [r,z]} |\xi(r, v)|^p \right) \leq C(p, z), \tag{2.81}$$

where the constant $C(p, z)$ depends on $p$, $z$ and on the uniform bounds of the derivatives $\partial_k B^i$ and $\partial_k A_l^i$. As a consequence, using Burkholder's and Hölder's inequalities, we can write

$$E(|\xi(r,z) - \xi(r',z)|^p)$$

$$\leq C(p,z)\left\{E\left(\left|\sum_{i,j=1}^{m}\left(\int_{[r\vee r',z]}\left[\partial_k A_l^i(X_v)(\xi_j^k(r,v) - \xi_j^k(r',v))dW_v^l\right.\right.\right.\right.$$

$$\left.\left.\left.\left.+\ \partial_k B^i(X_v)(\xi_j^k(r,v) - \xi_j^k(r',v))dv\right]\right)^2\right|^{\frac{p}{2}}\right)$$

$$+E\left(\left|\sum_{i,j=1}^{m}\left(\int_{[r,z]-[r',z]}\left[\partial_k A_l^i(X_v)\xi_j^k(r,v)dW_v^l\right.\right.\right.\right.$$

$$\left.\left.\left.\left.+\ \partial_k B^i(X_v)\xi_j^k(r,v)dv\right]\right)^2\right|^{\frac{p}{2}}\right)$$

$$+E\left(\left|\sum_{i,j=1}^{m}\left(\int_{[r',z]-[r,z]}\left[\partial_k A_l^i(X_v)\xi_j^k(r',v)dW_v^l\right.\right.\right.\right.$$

$$\left.\left.\left.\left.+\ \ \partial_k B^i(X_v)\xi_j^k(r',v)dv\right]\right)^2\right|^{\frac{p}{2}}\right)\right\}$$

$$\leq C(p,z)\left(|r-r'|^{\frac{p}{2}} + \int_{[r\vee r',z]}E(|\xi(r,v) - \xi(r',v)|^p)dv\right).$$

Using a two-parameter version of Gronwall's lemma (see Exercise 2.4.3) we deduce Eq. (2.80).

In order to prove the theorem, it is enough to show that $\det Q_z > 0$ a.s., where $z = (s,t)$ is a fixed point such that $st \neq 0$, and $Q_z$ is given by (2.74). Suppose that $P\{\det Q_z = 0\} > 0$. We want to show that under this assumption condition (P) cannot be satisfied. For any $\sigma \in (0,s]$ let $K_\sigma$ denote the vector subspace of $\mathbb{R}^m$ spanned by

$$\{A_j(X_{\xi t}); 0 \leq \xi \leq \sigma, j = 1,\dots,d\}.$$

Then $\{K_\sigma, 0 < \sigma \leq s\}$ is an increasing family of subspaces. We set $K_{0+} = \cap_{\sigma>0}K_\sigma$. By the Blumenthal zero-one law, $K_{0+}$ is a deterministic subspace with probability one. Define

$$\rho = \inf\{\sigma > 0 : \dim K_\sigma > \dim K_{0+}\}.$$

Then $\rho > 0$ a.s., and $\rho$ is a stopping time with respect to the increasing family of $\sigma$-fields $\{\mathcal{F}_{\sigma t}, \sigma \geq 0\}$. For any vector $v \in \mathbb{R}^m$ we have

$$v^T Q_z v = \sum_{j=1}^{d}\int_{[0,z]}(v_i\xi_l^i(r,z)A_j^l(X_r))^2 dr.$$

Assume that $v^T Q_z v = 0$. Due to the continuity in $r$ of $\xi_j^i(r,z)$, we deduce $v_i \xi_l^i(r,z) A_j^l(X_r) = 0$ for any $r \in [0,z]$ and for any $j = 1, \ldots, d$. In particular, for $r = (\sigma, t)$ we get $v^T A_j(X_{\sigma t}) = 0$ for any $\sigma \in [0,s]$. As a consequence, $K_{0+} \neq \mathbb{R}^m$. Otherwise $K_\sigma = \mathbb{R}^m$ for all $\sigma \in [0,s]$, and any vector $v$ verifying $v^T Q_z v = 0$ would be equal to zero. So, $Q_z$ would be invertible a.s., which contradicts our assumption. Let $v$ be a fixed nonzero vector orthogonal to $K_{0+}$. We remark that $v$ is orthogonal to $K_\sigma$ if $\sigma < \rho$, that is,

$$v^T A_j(X_{\sigma t}) = 0 \quad \text{for all} \quad \sigma < \rho \quad \text{and} \quad j = 1, \ldots, d. \tag{2.82}$$

We introduce the following sets of vector fields:

$$
\begin{aligned}
\Sigma_0 &= \{A_1, \ldots, A_d\}, \\
\Sigma_n &= \{A_j^\nabla V, j = 1, \ldots, d, V \in \Sigma_{n-1}\}, \quad n \geq 1, \\
\Sigma &= \cup_{n=0}^\infty \Sigma_n.
\end{aligned}
$$

Under property (P), the vector space $\langle \Sigma(x_0) \rangle$ spanned by the vector fields of $\Sigma$ at point $x_0$ has dimension $m$. We will show that the vector $v$ is orthogonal to $\langle \Sigma_n(x_0) \rangle$ for all $n \geq 0$, which contradicts property (P). Actually, we will prove the following stronger orthogonality property:

$$v^T V(X_{\sigma t}) = 0 \quad \text{for all} \quad \sigma < \rho, V \in \Sigma_n \quad \text{and} \quad n \geq 0. \tag{2.83}$$

Assertion (2.83) is proved by induction on $n$. For $n = 0$ it reduces to (2.82). Suppose that it holds for $n - 1$, and let $V \in \Sigma_{n-1}$. The process $\{v^T V(X_{\sigma t}), \sigma \in [0,s]\}$ is a continuous semimartingale with the following integral representation:

$$
\begin{aligned}
v^T V(X_{\sigma t}) = {}& v^T V(x_0) + \int_0^\sigma \int_0^t \Big[ v^T (\partial_k V)(X_{\xi t}) A_j^k (X_{\xi \tau}) dW_{\xi \tau}^j \\
& + v^T (\partial_k V)(X_{\xi t}) B^k (X_{\xi \tau}) d\xi d\tau \\
& + \frac{1}{2} v^T \partial_k \partial_{k'} V(X_{\xi t}) \sum_{l=1}^d A_l^k (X_{\xi \tau}) A_l^{k'} (X_{\xi \tau}) d\xi d\tau \Big].
\end{aligned}
$$

The quadratic variation of this semimartingale is equal to

$$
\sum_{j=1}^d \int_0^\sigma \int_0^t \left( v^T (\partial_k V)(X_{\xi t}) A_j^k (X_{\xi \tau}) \right)^2 d\xi d\tau.
$$

By the induction hypothesis, the semimartingale vanishes in the random interval $[0, \rho)$. As a consequence, its quadratic variation is also equal to zero in this interval, and we have, in particular,

$$v^T(A_j^\nabla V)(X_{\sigma t}) = 0 \quad \text{for all} \quad \sigma < \rho \quad \text{and} \quad j = 1, \ldots, d.$$

Thus, (2.83) holds for $n$. This achieves the proof of the theorem.    □

It can be proved (cf. [256]) that under condition (P), the density of $X_z$ is infinitely differentiable. Moreover, it is possible to show the smoothness of the density of $X_z$ under assumptions that are weaker than condition (P). In fact, one can consider the vector space spanned by the algebra generated by $A_1, \ldots, A_d$ with respect to the operation $\nabla$, and we can also add other generators formed with the vector field $B$. We refer to references [241] and [257] for a discussion of these generalizations.

## 2.4.2 Absolute continuity for solutions to the stochastic heat equation

Suppose that $W = \{W(t, x), t \in [0, T], x \in [0, 1]\}$ is a two-parameter Wiener process defined on a complete probability space $(\Omega, \mathcal{F}, P)$. For each $t \in [0, T]$ we will denote by $\mathcal{F}_t$ the $\sigma$-field generated by the random variables $\{W(s, x), (s, x) \in [0, t] \times [0, 1]\}$ and the $P$-null sets. We say that a random field $\{u(t, x), t \in [0, T], x \in [0, 1]\}$ is adapted if for all $(t, x)$ the random variable $u(t, x)$ is $\mathcal{F}_t$-measurable.

Consider the following parabolic stochastic partial differential equation on $[0, T] \times [0, 1]$:

$$\frac{\partial u}{\partial t} = \frac{\partial^2 u}{\partial x^2} + b(u(t, x)) + \sigma(u(t, x))\frac{\partial^2 W}{\partial t \partial x} \tag{2.84}$$

with initial condition $u(0, x) = u_0(x)$, and Dirichlet boundary conditions $u(t, 0) = u(t, 1) = 0$. We will assume that $u_0 \in C([0, 1])$ satisfies $u_0(0) = u_0(1) = 0$.

It is well known that the associated homogeneous equation (i.e., when $b \equiv 0$ and $\sigma \equiv 0$) has a unique solution given by $v(t, x) = \int_0^1 G_t(x, y)u_0(y)dy$, where $G_t(x, y)$ is the fundamental solution of the heat equation with Dirichlet boundary conditions. The kernel $G_t(x, y)$ has the following explicit formula:

$$G_t(x, y) = \frac{1}{\sqrt{4\pi t}} \sum_{n=-\infty}^{\infty} \left\{ \exp\left(-\frac{(y - x - 2n)^2}{4t}\right) \right.$$
$$\left. - \exp\left(-\frac{(y + x - 2n)^2}{4t}\right) \right\}. \tag{2.85}$$

On the other hand, $G_t(x, y)$ coincides with the probability density at point $y$ of a Brownian motion with variance $\sqrt{2t}$ starting at $x$ and killed if it leaves the iterval $[0, 1]$. This implies that

$$G_t(x, y) \leq \frac{1}{\sqrt{4\pi t}} \exp\left(-\frac{|x - y|^2}{4t}\right). \tag{2.86}$$

Therefore, for any $\beta > 0$ we have

$$\int_0^1 G_t(x,y)^\beta \, dy \leq (4\pi t)^{-\frac{\beta}{2}} \int_{\mathbb{R}} e^{-\frac{\beta|x|^2}{4t}} \, dx = C_\beta t^{\frac{1-\beta}{2}}. \qquad (2.87)$$

Note that the right-hand side of (2.87) is integrable in $t$ near the origin, provided that $\beta < 3$.

Equation (2.84) is formal because the derivative $\frac{\partial^2 W}{\partial t \partial x}$ does not exist, and we will replace it by the following integral equation:

$$\begin{aligned} u(t,x) & = \int_0^1 G_t(x,y) u_0(y) \, dy + \int_0^t \int_0^1 G_{t-s}(x,y) b(u(s,y)) \, dy \, ds \\ & + \int_0^t \int_0^1 G_{t-s}(x,y) \sigma(u(s,y)) W(dy, ds) . \end{aligned} \qquad (2.88)$$

One can define a solution to (2.84) in terms of distributions and then show that such a solution exists if and only if (2.88) holds. We refer to Walsh [342] for a detailed discussion of this topic. We can state the following result on the integral equation (2.88).

**Theorem 2.4.3** *Suppose that the coefficients $b$ and $\sigma$ are globally Lipschitz functions. Then there is a unique adapted process $u = \{u(t,x), t \in [0,T], x \in [0,1]\}$ such that*

$$E\left( \int_0^T \int_0^1 u(t,x)^2 \, dx \, dt \right) < \infty,$$

*and satisfies (2.88). Moreover, the solution $u$ satisfies*

$$\sup_{(t,x) \in [0,T] \times [0,1]} E(|u(t,x)|^p) < \infty \qquad (2.89)$$

*for all $p \geq 2$.*

*Proof:*    Consider the Picard iteration scheme defined by

$$u_0(t,x) = \int_0^1 G_t(x,y) u_0(y) \, dy$$

and

$$\begin{aligned} u_{n+1}(t,x) & = u_0(t,x) + \int_0^t \int_0^1 G_{t-s}(x,y) b(u_n(s,y)) \, dy \, ds \\ & + \int_0^t \int_0^1 G_{t-s}(x,y) \sigma(u_n(s,y)) W(dy, ds), \qquad (2.90) \end{aligned}$$

$n \geq 0$. Using the Lipschitz condition on $b$ and $\sigma$ and the isometry property of the stochastic integral with respect to the two-parameter Wiener process (see the Appendix, Section A.3), we obtain

$$E(|u_{n+1}(t,x) - u_n(t,x)|^2)$$

$$\leq 2E\left(\left(\int_0^t \int_0^1 G_{t-s}(x,y)|u_n(s,y) - u_{n-1}(s,y)|dyds\right)^2\right)$$

$$+2E\left(\int_0^t \int_0^1 G_{t-s}(x,y)^2|u_n(s,y) - u_{n-1}(s,y)|^2 dyds\right)$$

$$\leq 2(T+1)\int_0^t \int_0^1 G_{t-s}(x,y)^2 E\left(|u_n(s,y) - u_{n-1}(s,y)|^2\right) dyds.$$

Now we apply (2.87) with $\beta = 2$, and we obtain

$$E(|u_{n+1}(t,x) - u_n(t,x)|^2)$$

$$\leq C_T \int_0^t \int_0^1 E(|u_n(s,y) - u_{n-1}(s,y)|^2)(t-s)^{-\frac{1}{2}} dyds.$$

Hence,

$$E(|u_{n+1}(t,x) - u_n(t,x)|^2)$$

$$\leq C_T^2 \int_0^t \int_0^s \int_0^1 E(|u_n(r,z) - u_{n-1}(r,z)|^2)(s-r)^{-\frac{1}{2}}(t-s)^{-\frac{1}{2}} dzdrds$$

$$= C_T' \int_0^t \int_0^1 E(|u_n(r,z) - u_{n-1}(r,z)|^2) dzdr.$$

Iterating this inequality yields

$$\sum_{n=0}^{\infty} \sup_{t \in [0,T]} \int_0^1 E(|u_{n+1}(t,x) - u_n(t,x)|^2) dx < \infty.$$

This implies that the sequence $u_n(t,x)$ converges in $L^2([0,1] \times \Omega)$, uniformly in time, to a stochastic process $u(t,x)$. The process $u(t,x)$ is adapted and satisfies (2.88). Uniqueness is proved by the same argument.

Let us now show (2.89). Fix $p > 6$. Applying Burkholder's inequality for stochastic integrals with respect to the Brownian sheet (see (A.8)) and the boundedness of the function $u_0$ yields

$$E(|u_{n+1}(t,x)|^p) \leq c_p \left(\|u_0\|_\infty^p\right.$$

$$+E\left(\left(\int_0^t \int_0^1 G_{t-s}(x,y)\,|b(u_n(s,y))|\,dyds\right)^p\right)$$

$$\left.+E\left(\left(\int_0^t \int_0^1 G_{t-s}(x,y)^2 \sigma(u_n(s,y))^2 dyds\right)^{\frac{p}{2}}\right)\right).$$

Using the linear growth condition of $b$ and $\sigma$ we can write

$$E\left(|u_{n+1}(t,x)|^p\right) \leq C_{p,T}\left(1 + E\left(\left(\int_0^t \int_0^1 G_{t-s}(x,y)^2 u_n(s,y)^2 dyds\right)^{\frac{p}{2}}\right)\right).$$

Now we apply Hölder's inequality and (2.87) with $\beta = \frac{2p}{p-2} < 3$, and we obtain

$$\begin{aligned}
E\left(|u_{n+1}(t,x)|^p\right) &\leq C_{p,T}\left(1 + \left(\int_0^t \int_0^1 G_{t-s}(x,y)^{\frac{2p}{p-2}} dyds\right)^{\frac{p-2}{2}}\right.\\
&\qquad \times \left.\int_0^t \int_0^1 E(|u_n(s,y)|^p)dyds\right)\\
&\leq C'_{p,T}\left(1 + \int_0^t \int_0^1 E(|u_n(s,y)|^p)dyds\right),
\end{aligned}$$

and we conclude using Gronwall's lemma.  $\square$

The next proposition tells us that the trajectories of the solution to the Equation (2.88) are $\alpha$-Hölder continuous for any $\alpha < \frac{1}{4}$. For its proof we need the following technical inequalities.

(a) Let $\beta \in (1,3)$. For any $x \in [0,1]$ and $t,h \in [0,T]$ we have

$$\int_0^t \int_0^1 |G_{s+h}(x,y) - G_s(x,y)|^\beta dyds \leq C_{T,\beta} h^{\frac{3-\beta}{2}}, \qquad (2.91)$$

(b) Let $\beta \in (\frac{3}{2},3)$. For any $x,y \in [0,1]$ and $t \in [0,T]$ we have

$$\int_0^t \int_0^1 |G_s(x,z) - G_s(y,z)|^\beta dzds \leq C_{T,\beta}|x-y|^{3-\beta}. \qquad (2.92)$$

**Proposition 2.4.3** *Fix $\alpha < \frac{1}{4}$. Let $u_0$ be a $2\alpha$-Hölder continuous function such that $u_0(0) = u_0(1) = 0$. Then, the solution $u$ to Equation (2.88) has a version with $\alpha$-Hölder continuous paths.*

*Proof:*    We first check the regularity of the first term in (2.88). Set $G_t(x,u_0) := \int_0^1 G_t(x,y)u_0(y)dy$. The semigroup property of $G$ implies

$$G_t(x,u_0) - G_s(x,u_0) = \int_0^1 \int_0^1 G_s(x,y)G_{t-s}(y,z)[u_0(z) - u_0(y)]dzdy.$$

Hence, using (2.86) we get

$$\begin{aligned}
|G_t(x,u_0) - G_s(x,u_0)| &\leq C\int_0^1 \int_0^1 G_s(x,y)G_{t-s}(y,z)|z-y|^{2\alpha}dzdy\\
&\leq C'\int_0^1 G_s(x,y)|t-s|^\alpha dy \leq C'|t-s|^\alpha.
\end{aligned}$$

On the other hand, from (2.85) we can write

$$G_t(x, y) = \psi_t(y - x) - \psi_t(y + x),$$

where $\psi_t(x) = \frac{1}{\sqrt{4\pi t}} \sum_{n=-\infty}^{+\infty} e^{-(x-2n)/4t}$. Notice that $\sup_{x \in [0,1]} \int_0^1 \psi_t(z - x)dz \leq C$. We can write

$$
\begin{aligned}
G_t(x, u_0) - G_t(y, u_0) &= \int_0^1 [\psi_t(z - x) - \psi_t(z - y)] u_0(z)dz \\
&\quad - \int_0^1 [\psi_t(z + x) - \psi_t(z + y)] u_0(z)dz \\
&= A_1 + B_1.
\end{aligned}
$$

It suffices to consider the term $A_1$, because $B_1$ can be treated by a similar method. Let $\eta = y - x > 0$. Then, using the Hölder continuity of $u_0$ and the fact that $u_0(0) = u_1(0) = 1$ we obtain

$$
\begin{aligned}
|A_1| &\leq \int_0^{1-\eta} \psi_t(z - x) |u_0(z) - u_0(z + \eta)| \, dz \\
&\quad + \int_{1-\eta}^1 \psi_t(z - x) |u_0(z)| \, dz + \int_0^\eta \psi_t(z - y) |u_0(z)| \, dz \\
&\leq C\eta^{2\alpha} + C \int_{1-\eta}^1 \psi_t(z - x)(1 - z)^{2\alpha} dz + C \int_0^\eta \psi_t(z - y) z^{2\alpha} dz \\
&\leq C'\eta^{2\alpha}.
\end{aligned}
$$

Set

$$U(t, x) = \int_0^t \int_0^1 G_{t-s}(x, y)\sigma(u(s, y))W(dy, ds).$$

Applying Burkholder's and Hölder's inequalities (see (A.8)), we have for any $p > 6$

$$
\begin{aligned}
E(|U(t, x) - U(t, y)|^p) \\
\leq C_p E \left( \left| \int_0^t \int_0^1 |G_{t-s}(x, z) - G_{t-s}(y, z)|^2 |\sigma(u(s, z))|^2 dzds \right|^{\frac{p}{2}} \right) \\
\leq C_{p,T} \left( \int_0^t \int_0^1 |G_{t-s}(x, z) - G_{t-s}(y, z)|^{\frac{2p}{p-2}} dzds \right)^{\frac{p-2}{2}},
\end{aligned}
$$

because $\int_0^T \int_0^1 E(|\sigma(u(s, z))|^p)dzds < \infty$. From (2.92) with $\beta = \frac{2p}{p-2}$, we know that this is bounded by $C|x - y|^{\frac{p-6}{2}}$.

On the other hand, for $t > s$ we can write

$$E(|U(t,x) - U(s,x)|^p)$$

$$\leq C_p \left\{ E\left( \left| \int_0^s \int_0^1 |G_{t-\theta}(x,y) - G_{s-\theta}(x,y)|^2 |\sigma(u(\theta,y))|^2 dy d\theta \right|^{\frac{p}{2}} \right) \right.$$

$$\left. + E\left( \left| \int_s^t \int_0^1 |G_{t-\theta}(x,y)|^2 |\sigma(u(\theta,y))|^2 dy d\theta \right|^{\frac{p}{2}} \right) \right\}$$

$$\leq C_{p,T} \left\{ \left| \int_0^s \int_0^1 |G_{t-\theta}(x,y) - G_{s-\theta}(x,y)|^{\frac{2p}{p-2}} dy d\theta \right|^{\frac{p-2}{2}} \right.$$

$$\left. + \left| \int_0^s \int_0^1 G_{t-\theta}(x,y)^{\frac{2p}{p-2}} dy d\theta \right|^{\frac{p-2}{2}} \right\}.$$

Using (2.91) we can bound the first summand by $C_p |t-s|^{\frac{p-6}{4}}$. From (2.87) the second summand is bounded by

$$\int_0^{t-s} \int_0^1 G_\theta(x,y)^{\frac{2p}{p-2}} dy d\theta \leq C_p \int_0^{t-s} \theta^{-\frac{p+2}{2(p-2)}} d\theta$$

$$= C_p' |t-s|^{\frac{p-6}{2(p-2)}}.$$

As a consequence,

$$E(|U(t,x) - U(s,y)|^p) \leq C_{p,T} \left( |x-y|^{\frac{p-6}{2}} + |t-s|^{\frac{p-6}{4}} \right),$$

and we conclude using Kolmogorov's continuity criterion. In a similar way we can handle that the term

$$V(t,x) = \int_0^t \int_0^1 G_{t-s}(x,y) b(u(s,y)) dy ds.$$

$\square$

In order to apply the criterion for absolute continuity, we will first show that the random variable $u(t,x)$ belongs to the space $\mathbb{D}^{1,2}$.

**Proposition 2.4.4** *Let $b$ and $\sigma$ be Lipschitz functions. Then $u(t,x) \in \mathbb{D}^{1,2}$, and the derivative $D_{s,y}u(t,x)$ satisfies*

$$D_{s,y}u(t,x) = G_{t-s}(x,y)\sigma(u(s,y))$$

$$+ \int_s^t \int_0^1 G_{t-\theta}(x,\eta) B_{\theta,\eta} D_{s,y}u(\theta,\eta) d\eta d\theta$$

$$+ \int_s^t \int_0^1 G_{t-\theta}(x,\eta) S_{\theta,\eta} D_{s,y}u(\theta,\eta) W(d\theta,d\eta)$$

*if $s < t$, and $D_{s,y}u(t,x) = 0$ if $s > t$, where $B_{\theta,\eta}$ and $S_{\theta,\eta}$, $(\theta,\eta) \in [0,T] \times [0,1]$, are adapted and bounded processes.*

**Remarks:**    If the coefficients $b$ and $\sigma$ are functions of class $C^1$ with bounded derivatives, then $B_{\theta,\eta} = b'(u(\theta,\eta))$ and $S_{\theta,\eta} = \sigma'(u(\theta,\eta))$.

*Proof:*    Consider the Picard approximations $u_n(t,x)$ introduced in (2.90). Suppose that $u_n(t,x) \in \mathbb{D}^{1,2}$ for all $(t,x) \in [0,T] \times [0,1]$ and

$$\sup_{(t,x) \in [0,T] \times [0,1]} E\left( \int_0^t \int_0^1 |D_{s,y} u_n(t,x)|^2 dy ds \right) < \infty. \tag{2.93}$$

Applying the operator $D$ to Eq. (2.90), we obtain that $u_{n+1}(t,x) \in \mathbb{D}^{1,2}$ and that

$$\begin{aligned}
D_{s,y} u_{n+1}(t,x) &= G_{t-s}(x,y)\sigma(u_n(s,y)) \\
&+ \int_s^t \int_0^1 G_{t-\theta}(x,\eta) B_{\theta,\eta}^n D_{s,y} u_n(\theta,\eta) d\eta d\theta \\
&+ \int_s^t \int_0^1 G_{t-\theta}(x,\eta) S_{\theta,\eta}^n D_{s,y} u_n(\theta,\eta) W(d\theta, d\eta),
\end{aligned}$$

where $B_{\theta,\eta}^n$ and $S_{\theta,\eta}^n$, $(\theta,\eta) \in [0,T] \times [0,1]$, are adapted processes, uniformly bounded by the Lipschitz constants of $b$ and $\sigma$, respectively. Note that

$$E\left( \int_0^T \int_0^1 G_{t-s}(x,y)^2 \sigma(u_n(s,y))^2 dy ds \right)$$

$$\leq C_1 \left( 1 + \sup_{t \in [0,T], x \in [0,1]} E(u_n(t,x)^2) \right) \leq C_2,$$

for some constants $C_1, C_2 > 0$. Hence

$$\begin{aligned}
&E\left( \int_0^t \int_0^1 |D_{s,y} u_{n+1}(t,x)|^2 dy ds \right) \\
&\leq \; C_3 \left( 1 + E\left( \int_0^t \int_0^1 \int_s^t \int_0^1 G_{t-\theta}(x,\eta)^2 |D_{s,y} u_n(\theta,\eta)|^2 d\eta d\theta dy ds \right) \right) \\
&\leq \; C_4 \left( 1 + \int_0^t \sup_{\eta \in [0,1]} \int_s^t \int_0^1 (t-\theta)^{-\frac{1}{2}} E(|D_{s,y} u_n(\theta,\eta)|^2) d\theta dy ds \right).
\end{aligned}$$

Let

$$V_n(t) = \sup_{x \in [0,1]} E\left( \int_0^t \int_0^1 |D_{s,y} u_n(t,x)|^2 dy ds \right).$$

Then

$$\begin{aligned}
V_{n+1}(t) &\leq \; C_4 \left( 1 + \int_0^t V_n(\theta)(t-\theta)^{-\frac{1}{2}} d\theta \right) \\
&\leq \; C_5 \left( 1 + \int_0^t \int_0^\theta V_{n-1}(u)(t-\theta)^{-\frac{1}{2}}(\theta-u)^{-\frac{1}{2}} du d\theta \right) \\
&\leq \; C_6 \left( 1 + \int_0^t V_{n-1}(u) du \right) < \infty,
\end{aligned}$$

due to (2.93). By iteration this implies that

$$\sup_{t\in[0,T], x\in[0,1]} V_n(t) < C,$$

where the constant $C$ does not depend on $n$. Taking into account that $u_n(t,x)$ converges to $u(t,x)$ in $L^p(\Omega)$ for all $p \geq 1$, we deduce that $u(t,x) \in \mathbb{D}^{1,2}$, and $Du_n(t,x)$ converges to $Du(t,x)$ in the weak topology of $L^2(\Omega; H)$ (see Lemma 1.2.3). Finally, applying the operator $D$ to both members of Eq. (2.88), we deduce the desired result.     □

The main result of this section is the following;

**Theorem 2.4.4** *Let* $b$ *and* $\sigma$ *be globally Lipschitz functions. Assume that* $\sigma(u_0(y)) \neq 0$ *for some* $y \in (0,1)$. *Then the law of* $u(t,x)$ *is absolutely continuous for any* $(t,x) \in (0,T] \times (0,1)$.

*Proof:*     Fix $(t,x) \in (0,T] \times (0,1)$. According to the general criterion for absolute continuity (Theorem 2.1.3), we have to show that

$$\int_0^t \int_0^1 |D_{s,y}u(t,x)|^2 dy ds > 0 \qquad (2.94)$$

a.s. There exists an interval $[a,b] \subset (0,1)$ and a stopping time $\tau > 0$ such that $\sigma(u(s,y)) \geq \delta > 0$ for all $y \in [a,b]$ and $0 \leq s \leq \tau$. Then a sufficient condition for (2.94) is

$$\int_a^b D_{s,y}u(t,x)dy > 0 \quad \text{for all} \quad 0 \leq s \leq \tau, \qquad (2.95)$$

a.s. for some $b \geq a$. We will show (2.95) only for the case where $s = 0$. The case where $s > 0$ can be treated by similar arguments, restricting the study to the set $\{s < \tau\}$. On the other hand, one can show using Kolmogorov's continuity criterion that the process $\{D_{s,y}u(t,x), s \in [0,t], y \in [0,1]\}$ possesses a continuous version, and this implies that it suffices to consider the case $s = 0$.

The process

$$v(t,x) = \int_a^b D_{0,y}u(t,x)dy$$

is the unique solution of the following linear stochastic parabolic equation:

$$\begin{aligned} v(t,x) &= \int_a^b G_t(x,y)\sigma(u_0(y))dy + \int_0^t \int_0^1 G_{t-s}(x,y)B_{s,y}v(s,y)dsdy \\ &\quad + \int_0^t \int_0^1 G_{t-s}(x,y)S_{s,y}v(s,y)W(ds,dy). \end{aligned} \qquad (2.96)$$

We are going to prove that the solution to this equation is strictly positive at $(t,x)$. By the comparison theorem for stochastic parabolic equations (see

Exercise 2.4.5) it suffices to show the result when the initial condition is $\delta \mathbf{1}_{[a,b]}$, and by linearity we can take $\delta = 1$. Moreover, for any constant $c > 0$ the process $e^{ct}v(t,x)$ satisfies the same equation as $v$ but with $B_{s,y}$ replaced by $B_{s,y} + c$. Hence, we can assume that $B_{s,y} \geq 0$, and by the comparison theorem it suffices to prove the result with $B \equiv 0$.

Suppose that $a \leq x < 1$ (the case where $0 < x \leq a$ would be treated by similar arguments). Let $d > 0$ be such that $x \leq b + d < 1$. We divide $[0, t]$ into $m$ smaller intervals $[\frac{k-1}{m}t, \frac{kt}{m}]$, $1 \leq k \leq m$. We also enlarge the interval $[a, b]$ at each stage $k$, until by stage $k = m$ it covers $[a, b + d]$. Set

$$\alpha = \frac{1}{2} \inf_{m \geq 1} \inf_{1 \leq k \leq m} \inf_{y \in [a, b + \frac{kd}{m}]} \int_a^{b + \frac{d(k-1)}{m}} G_{\frac{t}{m}}(y, z) dz,$$

and note that $\alpha > 0$. For $k = 1, 2, \ldots, m$ we define the set

$$E_k = \left\{ v(\frac{kt}{m}, y) \geq \alpha^k \mathbf{1}_{[a, b + \frac{kd}{m}]}(y), \forall y \in [0, 1] \right\}.$$

We claim that for any $\delta > 0$ there exists $m_0 \geq 1$ such that if $m \geq m_0$ then

$$P(E_{k+1}^c | E_1 \cap \cdots \cap E_k) \leq \frac{\delta}{m} \tag{2.97}$$

for all $0 \leq k \leq m - 1$. If this is true, then we obtain

$$
\begin{aligned}
P\{v(t, x) > 0\} &\geq P\left\{ v(t, y) \geq \alpha^m \mathbf{1}_{[a, b+d]}(y), \forall y \in [0, 1] \right\} \\
&\geq P(E_m | E_{m-1} \cap \cdots \cap E_1) \\
&\quad \times P(E_{m-1} | E_{m-2} \cap \cdots \cap E_1) \ldots P(E_1) \\
&\geq \left( 1 - \frac{\delta}{m} \right)^m \geq 1 - \delta,
\end{aligned}
$$

and since $\delta$ is arbitrary we get $P\{v(t, x) > 0\} = 1$. So it only remains to check Eq. (2.97). We have for $s \in [\frac{tk}{m}, \frac{t(k+1)}{m}]$

$$
\begin{aligned}
v(s, y) &= \int_0^1 G_{\frac{t}{m}}(y, z)v(\frac{kt}{m}, z) dz \\
&\quad + \int_{\frac{t}{m}}^s \int_0^1 G_{s-\theta}(y, z)S_{\theta,z}v(\theta, z)W(d\theta, dz).
\end{aligned}
$$

Again by the comparison theorem (see Exercise 2.4.5) we deduce that on the set $E_1 \cap \cdots \cap E_k$ the following inequalities hold

$$v(s, y) \geq w(s, y) \geq 0$$

for all $(s, y) \in [\frac{tk}{m}, \frac{t(k+1)}{m}] \times [0, 1]$, where $w = \{w(s, y), (s, y) \in [\frac{tk}{m}, \frac{t(k+1)}{m}] \times [0, 1]\}$ is the solution to

$$
\begin{aligned}
w(s, y) &= \int_0^1 G_{\frac{t}{m}}(y, z) \alpha^k \mathbf{1}_{[a, b + \frac{kd}{m}]}(z) dz \\
&+ \int_{\frac{tk}{m}}^s \int_0^1 G_{s-\theta}(y, z) S_{\theta, z} w(\theta, z) W(d\theta, dz).
\end{aligned}
$$

Hence,

$$
\begin{aligned}
P(E_{k+1} | E_1 &\cap \cdots \cap E_k) \\
&\geq P\left\{ w(\frac{(k+1)t}{m}, y) \geq \alpha^{k+1}, \forall y \in [a, b + \frac{(k+1)d}{m}] \right\}. \quad (2.98)
\end{aligned}
$$

On the set $E_k$ and for $y \in [a, b + \frac{(k+1)d}{m}]$, it holds that

$$
\int_a^{b + \frac{kd}{m}} G_{\frac{t}{m}}(y, z) dz \geq 2\alpha.
$$

Thus, from (2.98) we obtain that

$$
\begin{aligned}
P(E_{k+1}^c | E_1 \cap \cdots \cap E_k) &\leq P\left( \sup_{y \in [a, b + \frac{(k+1)d}{m}]} |\Phi_{k+1}(y)| > \alpha | E_1 \cap \cdots \cap E_k \right) \\
&\leq \alpha^{-p} E\left( \sup_{y \in [0, 1]} |\Phi_{k+1}(y)|^p | E_1 \cap \cdots \cap E_k \right),
\end{aligned}
$$

for any $p \geq 2$, where

$$
\Phi_{k+1}(y) = \int_{\frac{tk}{m}}^{\frac{t(k+1)}{m}} \int_0^1 G_{\frac{t(k+1)}{m} - s}(y, z) S_{s, z} \frac{w(s, z)}{\alpha^k} W(ds, dz).
$$

Applying Burkholder's inequality and taking into account that $S_{s, z}$ is uniformly bounded we obtain

$$
\begin{aligned}
E\left( |\Phi_{k+1}(y_1) - \Phi_{k+1}(y_2)|^p | E_1 \cap \cdots \cap E_k \right) \\
\leq CE\left( \left| \int_0^{\frac{t}{m}} \int_0^1 (G_s(y_1, z) - G_s(y_2, z))^2 \alpha^{-2k} \right. \right. \\
\left. \left. \left( w(\frac{t(k+1)}{m} - s, z) \right)^2 ds dz \right|^{\frac{p}{2}} | E_1 \cap \cdots \cap E_k \right).
\end{aligned}
$$

Note that $\sup_{k \geq 1, z \in [0, 1], s \in [\frac{tk}{m}, \frac{t(k+1)}{m}]} \alpha^{-2kq} E\left( w(s, z)^{2q} | E_1 \cap \cdots \cap E_k \right)$ is bounded by a constant not depending on $m$ for all $q \geq 2$. As a consequence, Hölder's inequality and Eq. (2.68) yield for $p > 6$

$$
\begin{aligned}
& E\left(|\Phi_{k+1}(y_1) - \Phi_{k+1}(y_2)|^p | E_1 \cap \cdots \cap E_k\right) \\
& \leq C\left(\frac{t}{m}\right)^{\frac{1}{\eta}} \left(\int_0^{\frac{t}{m}} \int_0^1 |G_s(y_1, z) - G_s(y_2, z)|^{3\eta} ds dz\right)^{\frac{p}{3\eta}} \\
& \leq Cm^{-\frac{1}{\eta}} |x - y|^{\frac{p(1-\eta)}{\eta}},
\end{aligned}
$$

where $\frac{2}{3} \vee \frac{2}{p} < \eta < 1$. Now from (A.11) we get

$$
E\left(\sup_{y \in [0,1]} |\Phi_{k+1}(y)|^p | E_1 \cap \cdots \cap E_k\right) \leq Cm^{-\frac{1}{\eta}},
$$

which concludes the proof of (2.97). $\qquad\qquad\qquad\qquad\qquad\qquad$ □

## *Exercises*

**2.4.1** Prove Proposition 2.4.1.

*Hint:* Use the same method as in the proof of Proposition 1.3.11.

**2.4.2** Let $\{X_z, z \in \mathbb{R}_+^2\}$ be the two-parameter process solution to the linear equation

$$
X_z = 1 + \int_{[0,z]} a X_r dW_r.
$$

Find the Wiener chaos expansion of $X_z$.

**2.4.3** Let $\alpha, \beta : \mathbb{R}_+^2 \to \mathbb{R}$ be two measurable and bounded functions. Let $f : \mathbb{R}_+^2 \to \mathbb{R}$ be the solution of the linear equation

$$
f(z) = \alpha(z) + \int_{[0,z]} \beta(r) f(r) dr.
$$

Show that for any $z = (s, t)$ we have

$$
|f(z)| \leq \sup_{r \in [0,z]} |\alpha(r)| \sum_{m=0}^{\infty} (m!)^{-2} \sup_{r \in [0,z]} |\beta(r)|^m (st)^m.
$$

**2.4.4** Prove Eqs. (2.91) and (2.92).

*Hint:* It suffices to consider the term $\frac{1}{\sqrt{4\pi t}} e^{-\frac{|x-y|^2}{4t}}$ in the series expansion of $G_t(x, y)$. Then, for the proof of (2.92) it is convenient to majorize by the integral over $[0, t] \times \mathbb{R}$ and make the change of variables $z = (x - y)\xi$, $s = (x - y)^2 \eta$. For (2.91) use the change of variables $s = hu$ and $y = \sqrt{h}z$.

**2.4.5** Consider the pair of parabolic stochastic partial differential equations

$$
\frac{\partial u^i}{\partial t} = \frac{\partial^2 u^i}{\partial x^2} + f_i(u^i(t,x)) B(t,x) + g(u^i(t,x)) G(t,x) \frac{\partial^2 W}{\partial t \partial x}, \quad i = 1, 2,
$$

where $f_i$, $g$ are Lipschitz functions, and $B$ and $G$ are measurable, adapted, and bounded random fields. The initial conditions are $u^i(0,x) = \varphi_i(x)$. Then $\varphi_1 \leq \varphi_2$ ($f_1 \leq f_2$) implies $u_1 \leq u_2$.

*Hint:* Let $\{e_i, i \geq 1\}$ be a complete orthonormal system on $L^2([0,1])$. Projecting the above equations on the first $N$ vectors produces a stochastic partial differential equation driven by the $N$ independent Brownian motions defined by

$$W^i(t) = \int_0^1 e_i(x)W(t,dx), \quad i = 1,\ldots,N.$$

In this case we can use Itô's formula to get the inequality, and in the general case one uses a limit argument (see Donati-Martin and Pardoux [83] for the details).

**2.4.6** Let $u = \{u(t,x), t \in [0,T], x \in [0,1]\}$ be an adapted process such that $\int_0^T \int_0^1 E(u_{s,y}^2)dyds < \infty$. Set

$$Z_{t,x} = \int_0^t \int_0^1 G_{t-s}(x,y)u_{s,y}dW_{s,y}.$$

Show the following maximal inequality

$$E\left(\sup_{0 \leq t \leq T} |Z_{t,x}|^p\right)$$
$$\leq C_{p,T} \int_0^T \int_0^1 E\left(\left(\int_0^t \int_0^1 G_{t-s}(x,y)^2(t-s)^{-2\alpha}u_{s,y}^2 dyds\right)^{\frac{p}{2}}\right)dxdt,$$

where $\alpha < \frac{1}{4}$ and $p > \frac{3}{2\alpha}$.
   *Hint:* Write

$$Z_{t,x} = \frac{\sin \pi \alpha}{\pi} \int_0^t \int_0^1 G_{t-s}(x,y)(t-s)^{\alpha-1}Y_{s,y}dyds,$$

where

$$Y_{s,y} = \int_0^s \int_0^1 G_{s-\theta}(y,z)(s-\theta)^{-\alpha}u_{\theta,z}dW_{\theta,z},$$

and apply Hölder and Burholder's inequalities.

## Notes and comments

[2.1]     The use of the integration-by-parts formula to deduce the existence and regularity of densities is one of the basic applications of the Malliavin calculus, and it has been extensively developed in the literature. The starting point of these applications was the paper by Malliavin [207] that exhibits a probabilistic proof of Hörmander's theorem. Stroock

[318], Bismut [38], Watanabe [343], and others, have further developed the technique Malliavin introduced. The absolute continuity result stated in Theorem 2.1.1 is based on Shigekawa's paper [307].

Bouleau and Hirsch [46] introduced an alternative technique to deal with the problem of the absolute continuity, and we described their approach in Section 2.1.2. The method of Bouleau and Hirsch works in the more general context of a Dirichlet form, and we refer to reference [47] for a complete discussion of this generalization. The simple proof of Bouleau and Hirsch criterion's for absolute continuity in dimension one stated in Theorem 2.1.3 is based on reference [266]. For another proof of a similar criterion of absolute continuity, we refer to the note of Davydov [77].

The approach to the smoothness of the density based on the notion of distribution on the Wiener space was developed by Watanabe [343] and [144]. The main ingredient in this approach is the fact that the composition of a Schwartz distribution with a nondegenerate random vector is well defined as a distribution on the Wiener space (i.e., as an element of $\mathbb{D}^{-\infty}$). Then we can interpret the density $p(x)$ of a nondegenerate random vector $F$ as the expectation $E[\delta_x(F)]$, and from this representation we can deduce that $p(x)$ is infinitely differentiable.

The connected property of the topological support of the law of a smooth random variable was first proved by Fang in [95]. For further works on the properties on the positivity of the density of a random vector we refer to [63]. On the other hand, general criterion on the positivity of the density using technique of Malliavin calculus can be deduced (see [248]).

The fact that the supremum of a continuous process belongs to $\mathbb{D}^{1,2}$ (Proposition 2.1.10) has been proved in [261]. Another approach to the differentiability of the supremum based on the derivative of Banach-valued functionals is provided by Bouleau and Hirsch in [47]. The smoothness of the density of the Wiener sheet's supremum has been established in [107]. By a similar argument one can show that the supremum of the fractional Brownian motion has a smooth density in $(0, +\infty)$ (see [190]). In the case of a Gaussian process parametrized by a compact metric space $S$, Ylvisaker [352], [353] has proved by a direct argument that the supremum has a bounded density provided the variance of the process is equal to 1. See also [351, Theorem 2.1].

**[2.2]**    The weak differentibilility of solutions to stochastic differential equations with smooth coefficients can be proved by several arguments. In [146] Ikeda and Watanabe use the approximation of the Wiener process by means of polygonal paths. They obtain a sequence of finite-difference equations whose solutions are smooth functionals that converge to the diffusion process in the topology of $\mathbb{D}^\infty$. Stroock's approach in [320] uses an iterative family of Hilbert-valued stochastic differential equations. We have used the Picard iteration scheme $X_n(t)$. In order to show that the limit $X(t)$ belongs to the space $\mathbb{D}^\infty$, it suffices to show the convergence in $L^p$, for any

$p \geq 2$, and the boundedness of the derivatives $D^N X_n(t)$ in $L^p(\Omega; H^{\otimes N})$, uniformly in $n$.

In the one-dimensional case, Doss [84] has proved that a stochastic differential equation can be solved path-wise – it can be reduced to an ordinary differential equation (see Exercise 2.2.2). This implies that the solution in this case is not only in the space $\mathbb{D}^{1,p}$ but, assuming the coefficients are of class $C^1(\mathbb{R})$, that it is Fréchet differentiable on the Wiener space $C_0([0, T])$. In the multidimensional case the solution might not be a continuous functional of the Wiener process. The simplest example of this situation is Lévy's area (cf. Watanabe [343]). However, it is possible to show, at least if the coefficients have compact support (Üstünel and Zakai [337]), that the solution is $H$-continuously differentiable. The notion of $H$-continuous differentiability will be introduced in Chapter 4 and it requires the existence and continuity of the derivative along the directions of the Cameron-Martin space.

[2.3]     The proof of Hörmander's theorem using probabilistic methods was first done by Malliavin in [207]. Different approaches were developed after Malliavin's work. In [38] Bismut introduces a direct method for proving Hörmander's theorem, based on integration by parts on the Wiener space. Stroock [319, 320] developed the Malliavin calculus in the context of a symmetric diffusion semigroup, and a general criteria for regularity of densities was provided by Ikeda and Watanabe [144, 343]. The proof we present in this section has been inspired by the work of Norris [239]. The main ingredient is an estimation for continuous semimartingales (Lemma 2.3.2), which was first proved by Stroock [320]. Ikeda and Watanabe [144] prove Hörmander's theorem using the following estimate for the tail of the variance of the Brownian motion:

$$P\left(\int_0^1 \left(W_t - \int_0^1 W_s ds\right)^2 dt < \epsilon\right) \leq \sqrt{2}\exp(-\frac{1}{2^7\epsilon}).$$

In [186] Kusuoka and Stroock derive  Gaussian exponential bounds for the density $p_t(x_0, \cdot)$ of the diffusion $X_t(x_0)$ starting at $x_0$ under hypoellipticity conditions. In  [166] Kohatsu-Higa introduced in the notion of uniformly elliptic random vector and obtained Gaussian lower bound estimates for the density of a such a vector. The results are applied to the solution to the stochastic heat equation. Further applications to the potential theory for two-parameter diffusions are given in [76].

Malliavin calculus can be applied to study the asymptotic behavior of the fundamental solution to the heat equation (see Watanabe [344], Ben Arous, Léandre [26], [27]). More generally, it can be used to analyze the asymptotic behavior of the solution stochastic partial differential equations like the stochastic heat equation (see [167]) and stochastic differential equations with two parameters (see [168]).

On the other hand, the stochastic calculus of variations can be used to show hypoellipticity (existence of a smooth density) under conditions that are strictly weaker than Hörmander's hypothesis. For instance, in [24] the authors allow the Lie algebra condition to fail exponentially fast on a submanifold of $\mathbb{R}^m$ of dimension less than $m$ (see also [106]).

In addition to the case of a diffusion process, Malliavin calculus has been applied to show the existence and smoothness of densities for different types of Wiener functionals. In most of the cases analytical methods are not available and the Malliavin calculus is a suitable approach. The following are examples of this type of application:

(i) Bell and Mohammed [23] considered stochastic delay equations. The asymptotic behaviour of the density of the solution when the variance of the noise tends to zero is analized in [99].

(ii) Stochastic differential equations with coefficients depending on the past of the solution have been analyzed by Kusuoka and Stroock [187] and by Hirsch [134].

(iii) The smoothness of the density in a filtering problem has been discussed in Bismut and Michel [43], Chaleyat-Maurel and Michel [61], and Kusuoka and Stroock [185]. The general problem of the existence and smoothness of conditional densities has been considered by Nualart and Zakai [266].

(iv) The application of the Malliavin calculus to diffusion processes with boundary conditions has been developed in the works of Bismut [40] and Cattiaux [60].

(v) Existence and smoothness of the density for solutions to stochastic differential equations, including a stochastic integral with respect to a Poisson measure, have been considered by Bichteler and Jacod [36], and by Bichteler et al. [35], among others.

(vi) Absolute continuity of probability laws in infinite-dimensional spaces have been studied by Moulinier [232], Mazziotto and Millet [220], and Ocone [271].

(vii) Stochastic Volterra equations have been considered by Rovira and Sanz-Solé in [295].

Among other applications of the integration-by-parts formula on the Wiener space, not related with smoothness of probability laws, we can mention the following problems:

(i) time reversal of continuous stochastic processes (see Föllmer [109], Millet et al. [229], [230]),

(ii) estimation of oscillatory integrals (see Ikeda and Shigekawa [143], Moulinier [233], and Malliavin [209]),

(iii) approximation of local time of Brownian martingales by the normalized number of crossings of the regularized process (see Nualart and Wschebor [262]),

(iv) the relationship between the independence of two random variables $F$ and $G$ on the Wiener space and the almost sure orthogonality of their derivatives. This subject has been developed by Üstünel and Zakai [333], [334].

The Malliavin calculus leads to the development of the potential theory on the Wiener space. The notion of $c_{p,r}$ capacities and the associated quasisure analysis were introduced by Malliavin in [208]. One of the basic results of this theory is the regular disintegration of the Wiener measure by means of the coarea measure on submanifolds of the Wiener space with finite codimension (see Airault and Malliavin [3]). In [2] Airault studies the differential geometry of the submanifold $F = c$, where $F$ is a smooth nondegenerate variable on the Wiener space.

[2.4]    The Malliavin calculus is a helpful tool for analyzing the regularity of probability distributions for solutions to stochastic integral equations and stochastic partial differential equations. For instance, the case of the solution $\{X(z), z \in \mathbb{R}^2_+\}$ of two-parameter stochastic differential equations driven by the Brownian sheet, discussed in Section 2.4.1, has been studied by Nualart and Sanz [256], [257]. Similar methods can be applied to the analysis of the wave equation perturbed by a two-parameter white noise (cf. Carmona and Nualart [59], and Léandre and Russo [194]).

The application of Malliavin calculus to the absolute continuity of the solution to the heat equation perturbed by a space-time white noise has been taken from Pardoux and Zhang [282]. The arguments used in the last part of the proof of Theorem 2.4.4 are due to Mueller [234]. The smoothness of the density in this example has been studied by Bally and Pardoux [19]. As an application of the $L^p$ estimates of the density obtained by means of Malliavin calculus (of the type exhibited in Exercise 2.1.5), Bally et al. [18] prove the existence of a unique strong solution for the white noise driven heat equation (2.84) when the coefficient $b$ is measurable and locally bounded, and satisfies a one-sided linear growth condition, while the diffusion coefficient $\sigma$ does not vanish, has a locally Lipschitz derivative, and satisfies a linear growth condition. Gyöngy [130] has generalized this result to the case where $\sigma$ is locally Lipschitz.

The smoothness of the density of the vector $(u(t, x_1), \ldots, u(t, x_n))$, where $u(t, x)$ is the solution of a two-dimensional non-linear stochastic wave equation driven by Gaussian noise that is white in time and correlated in the space variable, has been derived in [231]. These equations were studied by

Dalang and Frangos in [75]. The absolute continuity of the law and the smoothness of the density for the three-dimensional non-linear stochastic wave equation has been considered in [288] and [289], following an approach to construct a solution for these equations developed by Dalang in [77].

The smoothness of the density of the projection onto a finite-dimensional subspace of the solution at time $t > 0$ of the two-dimensional Navier-Stokes equation forced by a finite-dimensional Gaussian white noise has been established by Mattingly and Pardoux in [219] (see also [132]).

# 3

# Anticipating stochastic calculus

As we have seen in Chapter 2, the Skorohod integral is an extension of the Itô integral that allows us to integrate stochastic processes that are not necessarily adapted to the Brownian motion. The adaptability assumption is replaced by some regularity condition. It is possible to develop a stochastic calculus for the Skorohod integral which is similar in some aspects to the classical Itô calculus. In this chapter we present the fundamental facts about this stochastic calculus, and we also discuss other approaches to the problem of constructing stochastic integrals for nonadapted processes (approximation by Riemann sums, development in a basis of $L^2([0,1])$, substitution methods). The last section discusses noncausal stochastic differential equations formulated using anticipating stochastic integrals.

## 3.1  Approximation of stochastic integrals

In order to define the stochastic integral $\int_0^1 u_t dW_t$ of a not necessarily adapted process $u = \{u_t, t \in [0,1]\}$ with respect to the Brownian motion $W$, one could use the following heuristic approach. First approximate $u$ by a sequence of step processes $u^n$, then define the stochastic integral of each process $u^n$ as a finite sum of the increments of the Brownian motion multiplied by the values of the process in each interval, and finally try to check if the sequence of integrals converges in some topology. What happens is that different approximations by step processes will produce different types of integrals. In this section we discuss this approach, and in particular we study two types of approximations, one leading to the Skorohod integral, and a second one that produces a Stratonovich-type stochastic integral.

## 3.1.1  Stochastic integrals defined by Riemann sums

In this section we assume that $\{W(t), t \in [0,1]\}$ is a one-dimensional Brownian motion, defined in the canonical probability space $(\Omega, \mathcal{F}, P)$.

We denote by $\pi$ an arbitrary partition of the interval $[0,1]$ of the form $\pi = \{0 = t_0 < t_1 < \cdots < t_n = 1\}$. We have to take limits (in probability, or in $L^p(\Omega)$, $p \geq 1$) of families of random variables $S_\pi$, depending on $\pi$, as the norm of $\pi$ (defined as $|\pi| = \sup_{0 \leq i \leq n-1}(t_{i+1} - t_i)$) tends to zero. Notice first that this convergence is equivalent to the convergence along any sequence of partitions whose norms tend to zero. In most of the cases it suffices to consider increasing sequences, as the next technical lemma explains.

**Lemma 3.1.1** *Let $S_\pi$ be a family of elements of some complete metric space $(V, d)$ indexed by the class of all partitions of $[0,1]$. Suppose that for any fixed partition $\pi_0$ we have*

$$\lim_{|\pi| \to 0} d(S_{\pi \vee \pi_0}, S_\pi) = 0, \tag{3.1}$$

*where $\pi \vee \pi_0$ denotes the partition induced by the union of $\pi$ and $\pi_0$. Then the family $S_\pi$ converges to some element $S$ if and only if for any increasing sequence of partitions $\{\pi(k), k \geq 1\}$ of $[0,1]$, such that $|\pi(k)| \to 0$, the sequence $S_{\pi(k)}$ converges to $S$ as $k$ tends to infinity.*

*Proof:*  Clearly, the convergence of the family $S_\pi$ implies the convergence of any sequence $S_{\pi(k)}$ with $|\pi(k)| \to 0$ to the same limit. Conversely, suppose that $S_{\tilde{\pi}(k)} \to S$ for any increasing sequence $\tilde{\pi}(k)$ with $|\tilde{\pi}(k)| \to 0$, but there exists an $\epsilon > 0$ and a sequence $\pi(k)$ with $|\pi(k)| \to 0$ such that $d(S_{\pi(k)}, S) > \epsilon$ for all $k$. Then we fix $k_0$ and by (3.1) we can find a $k_1$ such that $k_1 > k_0$ and

$$d(S_{\pi(k_0) \vee \pi(k_1)}, S_{\pi(k_1)}) < \frac{\epsilon}{2}.$$

Next we choose $k_2 > k_1$ large enough so that

$$d(S_{\pi(k_0) \vee \pi(k_1) \vee \pi(k_2)}, S_{\pi(k_2)}) < \frac{\epsilon}{2},$$

and we continue recursively. Set $\tilde{\pi}(n) = \pi(k_0) \vee \pi(k_1) \vee \cdots \vee \pi(k_n)$. Then after the $n$th step we have

$$d(S_{\tilde{\pi}(n)}, S) \geq d(S_{\pi(k_n)}, S) - d(S_{\tilde{\pi}(n)}, S_{\pi(k_n)}) > \frac{\epsilon}{2}.$$

Then $\tilde{\pi}(n)$ is an increasing sequence of partitions such that the sequence of norms $|\tilde{\pi}(n)|$ tends to zero but $d(S_{\tilde{\pi}(n)}, S) > \frac{\epsilon}{2}$, which completes the proof by contradiction.  □

Consider a measurable process $u = \{u_t, t \in [0,1]\}$ such that $\int_0^1 |u_t| dt < \infty$ a.s. For any partition $\pi$ we introduce the following step process:

$$u^\pi(t) = \sum_{i=0}^{n-1} \frac{1}{t_{i+1} - t_i} \left( \int_{t_i}^{t_{i+1}} u_s ds \right) \mathbf{1}_{(t_i, t_{i+1}]}(t). \tag{3.2}$$

If $E\left( \int_0^1 |u_t| dt \right) < \infty$ we define the step process

$$\widehat{u}^\pi(t) = \sum_{i=0}^{n-1} \frac{1}{t_{i+1} - t_i} \left( \int_{t_i}^{t_{i+1}} E(u_s | \mathcal{F}_{[t_i, t_{i+1}]^c}) ds \right) \mathbf{1}_{(t_i, t_{i+1}]}(t). \tag{3.3}$$

We recall that $\mathcal{F}_{[t_i, t_{i+1}]}^c$ denotes the $\sigma$-field generated by the increments $W_t - W_s$, where the interval $(s, t]$ is disjoint with $[t_i, t_{i+1}]$.

The next lemma presents in which topology the step processes $u^\pi$ and $\widehat{u}^\pi$ are approximations of the process $u$.

**Lemma 3.1.2** *Suppose that $u$ belongs to $L^2([0,1] \times \Omega)$. Then, the processes $u^\pi$ and $\widehat{u}^\pi$ converge to the process $u$ in the norm of the space $L^2([0,1] \times \Omega)$ as $|\pi|$ tends to zero. Furthermore, these convergences also hold in $\mathbb{L}^{1,2}$ whenever $u \in \mathbb{L}^{1,2}$.*

*Proof:*    The convergence $u^\pi \to u$ in $L^2([0,1] \times \Omega)$ as $|\pi|$ tends to zero can be proved as in Lemma 1.1.3, but for the convergence of $\widehat{u}^\pi$ we need a different argument.

One can show that the families $u^\pi$ and $\widehat{u}^\pi$ satisfy condition (3.1) with $V = L^2([0,1] \times \Omega)$ (see Exercise 3.1.1). Consequently, by Lemma 3.1.1 it suffices to show the convergence along any fixed increasing sequence of partitions $\pi(k)$ such that $|\pi(k)|$ tends to zero. In the case of the family $u^\pi$, we can regard $u^\pi$ as the conditional expectation of the variable $u$, in the probability space $[0,1] \times \Omega$, given the product $\sigma$-field of the finite algebra of parts of $[0,1]$ generated by $\pi$ times $\mathcal{F}$. Then the convergence of $u^\pi$ to $u$ in $L^2([0,1] \times \Omega)$ along a fixed increasing sequence of partitions follows from the martingale convergence theorem. For the family $\widehat{u}^\pi$ the argument of the proof is as follows.

Let $\pi(k)$ be an increasing sequence of partitions such that $|\pi(k)| \to 0$. Set $\pi(k) = \{0 = t_0^k < t_1^k < \cdots < t_{n_k}^k = 1\}$. For any $k$ we consider the $\sigma$-field $\mathcal{G}^k$ of parts of $[0,1] \times \Omega$ generated by the sets $(t_i^k, t_{i+1}^k] \times F$, where $0 \le i \le n_k - 1$ and $F \in \mathcal{F}_{[t_i^k, t_{i+1}^k]^c}$. Then notice that $\widehat{u}^{\pi(k)} = \widetilde{E}(u | \mathcal{G}^k)$, where $\widetilde{E}$ denotes the mathematical expectation in the probability space $[0,1] \times \Omega$. By the martingale convergence theorem, $\widehat{u}^{\pi(k)}$ converges to some element $\bar{u}$ in $L^2([0,1] \times \Omega)$. We want to show that $u = \bar{u}$. The difference $v = u - \bar{u}$ is orthogonal to $L^2([0,1] \times \Omega, \mathcal{G}^k)$ for every $k$. Consequently, for any fixed $k \ge 1$, such a process $v$ satisfies $\int_{I \times F} v(t, \omega) dt dP = 0$ for any $F \in \mathcal{F}_{[t_i^k, t_{i+1}^k]^c}$ and for any interval $I \subset [t_i^k, t_{i+1}^k]$ in $\pi(m)$ with $m \ge k$. Therefore,

$E(v(t)|\mathcal{F}_{[t_i^k,t_{i+1}^k]^c}) = 0$ for all $(t,\omega)$ almost everywhere in $[t_i^k, t_{i+1}^k] \times \Omega$. Therefore, for almost all $t$, with respect to the Lebesgue measure, the above conditional expectation is zero for any $i, k$ such that $t \in [t_i^k, t_{i+1}^k]$. This implies that $v(t,\omega) = 0$ a.s., for almost all $t$, and the proof of the first part of the lemma is complete.

In order to show the convergence in $\mathbb{L}^{1,2}$ we first compute the derivatives of the processes $u^\pi$ and $\widehat{u}^\pi$ using Proposition 1.2.8:

$$D_r u^\pi(t) = \sum_{i=0}^{n-1} \frac{1}{t_{i+1} - t_i} \left( \int_{t_i}^{t_{i+1}} D_r u_s ds \right) \mathbf{1}_{(t_i,t_{i+1}]}(t),$$

and

$$D_r \widehat{u}^\pi(t) = \sum_{i=0}^{n-1} \frac{1}{t_{i+1} - t_i} \left( \int_{t_i}^{t_{i+1}} E(D_r u_s | \mathcal{F}_{[t_i,t_{i+1}]^c}) ds \right) \times \mathbf{1}_{(t_i,t_{i+1}]}(t) \mathbf{1}_{(t_i,t_{i+1}]^c}(r).$$

Then, the same arguments as in the first part of the proof will give the desired convergence. □

Now consider the Riemann sums associated to the preceding approximations:

$$S^\pi = \sum_{i=0}^{n-1} \frac{1}{t_{i+1} - t_i} \left( \int_{t_i}^{t_{i+1}} u_s ds \right) (W(t_{i+1}) - W(t_i))$$

and

$$\widehat{S}^\pi = \sum_{i=0}^{n-1} \frac{1}{t_{i+1} - t_i} \left( \int_{t_i}^{t_{i+1}} E(u_s | \mathcal{F}_{[t_i,t_{i+1}]^c}) ds \right) (W(t_{i+1}) - W(t_i)).$$

Notice that from Lemma 1.3.2 the processes $\widehat{u}^\pi$ are Skorohod integrable for any process $u$ in $L^2([0,1] \times \Omega)$ and that

$$\widehat{S}^\pi = \delta(\widehat{u}^\pi).$$

On the other hand, for the process $u^\pi$ to be Skorohod integrable we need some additional conditions. For instance, if $u \in \mathbb{L}^{1,2}$, then $u^\pi \in \mathbb{L}^{1,2} \subset \text{Dom}\,\delta$, and we have

$$\delta(u^\pi) = S^\pi - \sum_{i=0}^{n-1} \frac{1}{t_{i+1} - t_i} \int_{t_i}^{t_{i+1}} \int_{t_i}^{t_{i+1}} D_s u_t ds dt. \qquad (3.4)$$

In conclusion, from Lemma 3.1.2 we deduce the following results:

(i) Let $u \in L^2([0,1] \times \Omega)$. If the family $\widehat{S}^\pi$ converges in $L^2(\Omega)$ to some limit, then $u$ is Skorohod integrable and this limit is equal to $\delta(u)$.

(ii) Let $u \in \mathbb{L}^{1,2}$. Then both families $\widehat{S}^{\pi} = \delta(\widehat{u}^{\pi})$ and $\delta(u^{\pi})$ converge in $L^2(\Omega)$ to $\delta(u)$.

Let us now discuss the convergence of the family $S^{\pi}$. Notice that

$$S^{\pi} = \int_0^1 u_t W_t^{\pi} \, dt,$$

where

$$W_t^{\pi} = \sum_{i=0}^{n-1} \frac{W(t_{i+1}) - W(t_i)}{t_{i+1} - t_i} \mathbf{1}_{(t_i, t_{i+1}]}(t). \tag{3.5}$$

**Definition 3.1.1** *We say that a measurable process $u = \{u_t, 0 \le t \le 1\}$ such that $\int_0^1 |u_t| dt < \infty$ a.s. is Stratonovich integrable if the family $S^{\pi}$ converges in probability as $|\pi| \to 0$, and in this case the limit will be denoted by $\int_0^1 u_t \circ dW_t$.*

From (3.4) we see that for a given process $u$ to be Stratonovich integrable it is not sufficient that $u \in \mathbb{L}^{1,2}$. In fact, the second summand in (3.4) can be regarded as an approximation of the trace of the kernel $D_s u_t$ in $[0,1]^2$, and this trace is not well defined for an arbitrary square integrable kernel. Let us introduce the following definitions:

Let $X \in \mathbb{L}^{1,2}$ and $1 \le p \le 2$. We denote by $D^+ X$ (resp. $D^- X$) the element of $L^p([0,1] \times \Omega)$ satisfying

$$\lim_{n \to \infty} \int_0^1 \sup_{s < t \le (s + \frac{1}{n}) \wedge 1} E(|D_s X_t - (D^+ X)_s|^p) ds = 0 \tag{3.6}$$

(resp.

$$\lim_{n \to \infty} \int_0^1 \sup_{(s - \frac{1}{n}) \vee 0 \le t < s} E(|D_s X_t - (D^- X)_s|^p) ds = 0). \tag{3.7}$$

We denote by $\mathbb{L}_{p+}^{1,2}$ (resp. $\mathbb{L}_{p-}^{1,2}$) the class of processes in $\mathbb{L}^{1,2}$ such that (3.6) (resp. (3.7)) holds. We set $\mathbb{L}_p^{1,2} = \mathbb{L}_{p+}^{1,2} \cap \mathbb{L}_{p-}^{1,2}$. For $X \in \mathbb{L}_p^{1,2}$ we write

$$(\nabla X)_t = (D^+ X)_t + (D^- X)_t. \tag{3.8}$$

Let $X \in \mathbb{L}^{1,2}$. Suppose that the mapping $(s,t) \hookrightarrow D_s X_t$ is continuous from a neighborhood of the diagonal $V_\varepsilon = \{|s - t| < \varepsilon\}$ into $L^p(\Omega)$. Then $X \in \mathbb{L}_p^{1,2}$ and

$$(D^+ X)_t = (D^- X)_t = D_t X_t.$$

The following proposition provides an example of a process in the class $\mathbb{L}_2^{1,2}$.

**Proposition 3.1.1** *Consider a process of the form*

$$X_t = X_0 + \int_0^t u_s dW_s + \int_0^t v_s ds, \qquad (3.9)$$

*where $X_0 \in \mathbb{D}^{1,2}$ and $u \in \mathbb{D}^{2,2}(H)$, and $v \in \mathbb{L}^{1,2}$. Then, $X$ belongs to the class $\mathbb{L}_2^{1,2}$ and*

$$(D^+X)_t = u_t + D_t X_0 + \int_0^t D_t v_r dr + \int_0^t D_t u_r dW_r, \qquad (3.10)$$

$$(D^-X)_t = D_t X_0 + \int_0^t D_t v_r dr + \int_0^t D_t u_r dW_r. \qquad (3.11)$$

*Proof:*    First notice that $X$ belongs to $\mathbb{L}^{1,2}$, and for any $s$ we have

$$D_s X_t = u_s \mathbf{1}_{[0,t]}(s) + D_s X_0 + \int_0^t D_s v_r dr + \int_0^t D_s u_r dW_r.$$

Thus,

$$\int_0^1 \sup_{(s-\frac{1}{n})\vee 0 < t \leq s} E(|D_s X_t - (D^-X)_s|^2) ds$$

$$\leq \frac{2}{n} \int_0^1 \int_{(s-\frac{1}{n})\vee 0}^s E(|D_s v_r|^2) dr ds + 2 \int_0^1 \int_{(s-\frac{1}{n})\vee 0}^s E(|D_s u_r|^2) dr ds$$

$$+ 2 \int_0^1 \int_0^1 \int_{(s-\frac{1}{n})\vee 0}^s E(|D_\theta D_s u_r|^2) dr ds d\theta,$$

and this converges to zero as $n$ tends to infinity. In a similar way we show that $(D^+X)_t$ exists and is given by (3.10).    $\square$

In a similar way, $\mathbb{L}_{p-}^{1,2,f}$ is the class of processes $X$ in $\mathbb{L}^{1,2,f}$ such that there exists an element $D^-X \in L^p([0,1] \times \Omega)$ for which (3.7) holds. Suppose that $X_t$ is given by (3.9), where $X_0 \in \mathbb{D}^{1,2}$, $u \in \mathbb{L}^F$ and $v \in \mathbb{L}^{1,2,f}$, then, $X$ belongs to the class $\mathbb{L}_{2-}^{1,2,f}$ and $(D^-X)_t$ is given by (3.11).

Then we have the following result, which gives sufficient conditions for the existence of the Stratonovich integral and provides the relation between the Skorohod and the Stratonovich integrals.

**Theorem 3.1.1** *Let $u \in \mathbb{L}_{1,\mathrm{loc}}^{1,2}$. Then $u$ is Stratonovich integrable and*

$$\int_0^1 u_t \circ dW_t = \int_0^1 u_t dW_t + \frac{1}{2} \int_0^1 (\nabla u)_t dt. \qquad (3.12)$$

*Proof:*    By the usual localization argument we can assume that $u \in \mathbb{L}_1^{1,2}$. Then, from Eq. (3.4) and the above approximation results on the Skorohod integral, it suffices to show that

$$\sum_{i=0}^{n-1} \frac{1}{t_{i+1} - t_i} \int_{t_i}^{t_{i+1}} \int_{t_i}^{t_{i+1}} D_t u_s ds dt \to \frac{1}{2} \int_0^1 (\nabla u)_s ds,$$

in probability, as $|\pi| \to 0$. We will show that the expectation

$$E \left( \left| \sum_{i=0}^{n-1} \frac{1}{t_{i+1} - t_i} \int_{t_i}^{t_{i+1}} dt \int_t^{t_{i+1}} (D_t u_s) ds - \frac{1}{2} \int_0^1 (D^+ u)_t dt \right| \right)$$

converges to zero as $|\pi| \to 0$. A similar result can be proved for the operator $D^-$, and the desired convergence would follow. We majorize the above expectation by the sum of the following two terms:

$$E \left( \left| \sum_{i=0}^{n-1} \frac{1}{t_{i+1} - t_i} \int_{t_i}^{t_{i+1}} dt \int_t^{t_{i+1}} \left( D_t u_s - (D^+ u)_t \right) ds \right| \right)$$

$$+ E \left( \left| \int_{t_i}^{t_{i+1}} \sum_{i=0}^{n-1} \frac{t_{i+1} - t}{t_{i+1} - t_i} (D^+ u)_t dt - \frac{1}{2} \int_0^1 (D^+ u)_t dt \right| \right)$$

$$\leq \int_0^1 \sup_{t \leq s \leq (t+|\pi|) \wedge 1} E(|D_t u_s - (D^+ u)_t|) dt$$

$$+ E \left( \left| \int_0^1 (D^+ u)_t \left( \sum_{i=0}^{n-1} \frac{t_{i+1} - t}{t_{i+1} - t_i} \mathbf{1}_{(t_i, t_{i+1}]}(t) - \frac{1}{2} \right) dt \right| \right).$$

The first term in the above expression tends to zero by the definition of the class $\mathbb{L}_1^{1,2}$. For the second term we will use the convergence of the functions $\sum_{i=0}^{n-1} \frac{t_{i+1} - t}{t_{i+1} - t_i} \mathbf{1}_{(t_i, t_{i+1}]}(t)$ to the constant $\frac{1}{2}$ in the weak topology of $L^2([0,1])$. This weak convergence implies that

$$\int_0^1 (D^+ u)_t \left( \sum_{i=0}^{n-1} \frac{t_{i+1} - t}{t_{i+1} - t_i} \mathbf{1}_{(t_i, t_{i+1}]}(t) - \frac{1}{2} \right) dt$$

converges a.s. to zero as $|\pi| \to 0$. Finally, the convergence in $L^1(\Omega)$ follows by dominated convergence, using the definition of the space $\mathbb{L}_1^{1,2}$.    □

**Remarks:**

**1.** If the mapping $(s,t) \hookrightarrow D_s u_t$ is continuous from $V_\varepsilon = \{|s - t| < \varepsilon\}$ into $L^1(\Omega)$, then the second summand in formula (3.12) reduces to $\int_0^1 D_t u_t dt$.

**2.** Suppose that $X$ is a continuous semimartingale of the form (1.23). Then the Stratonovich integral of $X$ exists on any interval $[0,t]$ and coincides with the limit in probability of the sums (1.24). That is, we have

$$\int_0^t X_s \circ dW_s = \int_0^t X_s dW_s + \frac{1}{2} \langle X, W \rangle_t,$$

where $\langle X, W \rangle$ denotes the joint quadratic variation of the semimartingale $X$ and the Brownian motion. Suppose in addition that $X \in \mathbb{L}_1^{1,2}$. In that case, we have $(D^- X)_t = 0$ (because $X$ is adapted), and consequently,

$$\int_0^1 (D^+ X)_t \, dt = \langle X, W \rangle_1 = \lim_{|\pi| \downarrow 0} \sum_{i=0}^{n-1} (X(t_{i+1}) - X(t_i))(W(t_{i+1}) - W(t_i)),$$

where $\pi$ denotes a partition of $[0, 1]$. In general, for processes $u \in \mathbb{L}_1^{1,2}$ that are continuous in $L^2(\Omega)$, the joint quadratic variation of the process $u$ and the Brownian motion coincides with the integral

$$\int_0^1 ((D^+ u)_t - (D^- u)_t) dt$$

(see Exercise 3.1.2). Thus, the joint quadratic variation does not coincide in general with the difference between the Skorohod and Stratonovich integrals.

The approach we have described admits diverse extensions. For instance, we can use approximations of the following type:

$$\sum_{i=0}^{n-1} ((1-\alpha)u(t_i) + \alpha u(t_{i+1}))(W(t_{i+1}) - W(t_i)), \tag{3.13}$$

where $\alpha$ is a fixed number between 0 and 1. Assuming that $u \in \mathbb{L}_1^{1,2}$ and $E(u_t)$ is continuous in $t$, we can show (see Exercise 3.1.3) that expression (3.13) converges in $L^1(\Omega)$ to the following quantity:

$$\delta(u) + \alpha \int_0^1 (D^+ u)_t \, dt + (1 - \alpha) \int_0^1 (D^- u)_t \, dt. \tag{3.14}$$

For $\alpha = \frac{1}{2}$ we obtain the Stratonovich integral, and for $\alpha = 0$ expression (3.14) is the generalization of the Itô integral studied in [14, 30, 298].

## 3.1.2  The approach based on the $L^2$ development of the process

Suppose that $\{W(A), A \in \mathcal{B}, \mu(A) < \infty\}$ is an $L^2(\Omega)$-valued Gaussian measure associated with a measure space $(T, \mathcal{B}, \mu)$. We fix a complete orthonormal system $\{e_i, i \geq 1\}$ in the Hilbert space $H = L^2(T)$. We can compute the random Fourier coefficients of the paths of the process $u$ in that basis:

$$u(t) = \sum_{i=1}^{\infty} \langle u, e_i \rangle_H e_i(t).$$

Then one can define the stochastic integral of $u$ (wihch we will denote by $\int_0^1 u_t * dW_t$, following Rosinski [293]) as the sum of the series

$$\int_T u_t * dW_t = \sum_{i=1}^{\infty} \langle u, e_i \rangle_H W(e_i), \tag{3.15}$$

provided that it converges in probability (or in $L^2(\Omega)$) and the sum does not depend on the complete orthonormal system we have chosen. We will call it the $L^2$-integral.

We remark that if $u \in \mathbb{L}^{1,2}$, then using (1.48) we have for any $i$

$$\langle u, e_i \rangle_H \delta(e_i) = \delta(e_i \langle u, e_i \rangle_H) + \int_T \int_T D_s u_t e_i(s) e_i(t) \mu(ds) \mu(dt).$$

Consequently, if $u \in \mathbb{L}^{1,2}$ and the kernel $D_s u_t$ has a summable trace for all $\omega$ a.s., then the integral $\int_T u_t * dW_t$ exists, and we have the following relation (cf. Nualart and Zakai [263, Proposition 6.1]):

$$\int_T u_t * dW_t = \int_T u_t dW_t + \mathrm{T}(Du). \tag{3.16}$$

The following result (cf. Nualart and Zakai [267]) establishes the relationship between the Stratonovich integral and the $L^2$-integral when $T = [0, 1]$ and $\mu$ is the Lebesgue measure.

**Theorem 3.1.2** *Let $u$ be a measurable process such that $\int_0^1 u_t^2 dt < \infty$ a.s. Then if the integral $\int_0^1 u_t * dW_t$ exists, $u$ is Stratonovich integrable, and both integrals coincide.*

*Proof:* It suffices to show that for any increasing sequence of partitions $\{\pi(n), n \geq 1\}$ whose norm tends to zero and $\pi(n+1)$ is obtained by refining $\pi(n)$ at one point, the sequence $S^{\pi(n)}$ converges to $\int_0^1 u_t * dW_t$ as $n$ tends to infinity. The idea of the proof is to produce a particular complete orthonormal system for which $S^{\pi(n)}$ will be the partial sum of series (3.15). Without any loss of generality we may assume that $\pi(1) = \{0, 1\}$. Set $e_1 = 1$. For $n \geq 1$ define $e_{n+1}$ as follows. By our assumption $\pi(n+1)$ refines $\pi(n)$ in one point only. Assume that the point of refinement $\tau$ belongs to some interval $(s_n, t_n)$ of the partition $\pi(n)$. Then we set

$$e_{n+1}(t) = \left( \frac{t_n - \tau}{(\tau - s_n)(t_n - s_n)} \right)^{\frac{1}{2}}$$

if $t \in (s_n, \tau]$,

$$e_{n+1}(t) = -\left(\frac{\tau - s_n}{(t_n - \tau)(t_n - s_n)}\right)^{\frac{1}{2}}$$

if $t \in (\tau, t_n]$, and $e_{n+1}(t) = 0$ for $t \in (s_n, t_n]^c$. In this form we construct a complete orthonormal system in $L^2([0,1])$, which can be considered as a modified Haar system. Finally, we will show by induction that the partial sums of series (3.15) coincide with the approximations of the Stratonovich integral corresponding to the sequence of partitions $\pi(n)$. First notice that $S^{\pi(n+1)}$ differs from $S^{\pi(n)}$ only because of changes taking place in the interval $(s_n, t_n]$. Therefore,

$$S^{\pi(n+1)} - S^{\pi(n)} = -\left(\frac{1}{t_n - s_n}\int_{s_n}^{t_n} u_t dt\right)(W(t_n) - W(s_n))$$

$$+\left(\frac{1}{\tau - s_n}\int_{s_n}^{\tau} u_t dt\right)(W(\tau) - W(s_n))$$

$$+\left(\frac{1}{t_n - \tau}\int_{\tau}^{t_n} u_t dt\right)(W(t_n) - W(\tau)). \tag{3.17}$$

On the other hand,

$$\left(\int_0^1 e_{n+1}(t)u_t dt\right)\int_0^1 e_{n+1}(t)dW_t$$

$$= \frac{1}{t_n - s_n}\left(\left[\frac{t_n - \tau}{\tau - s_n}\right]^{\frac{1}{2}}\int_{s_n}^{\tau} u_t dt - \left[\frac{\tau - s_n}{t_n - \tau}\right]^{\frac{1}{2}}\int_{\tau}^{t_n} u_t dt\right)$$

$$\times\left(\left[\frac{t_n - \tau}{\tau - s_n}\right]^{\frac{1}{2}}(W(\tau) - W(s_n)) - \left[\frac{\tau - s_n}{t_n - \tau}\right]^{\frac{1}{2}}(W(t_n) - W(\tau))\right),$$

which is equal to

$$\frac{1}{t_n - s_n}\left\{\left(\frac{t_n - \tau}{\tau - s_n}\int_{s_n}^{\tau} u_t dt\right)(W(\tau) - W(s_n))\right.$$

$$-(\int_{s_n}^{\tau} u_t dt)(W(t_n) - W(\tau))) - \left(\int_{\tau}^{t_n} u_t dt\right)(W(\tau) - W(s_n))$$

$$\left.+\frac{\tau - s_n}{t_n - \tau}\left(\int_{\tau}^{t_n} u_t dt\right)(W(t_n) - (W(\tau))\right\}. \tag{3.18}$$

Comparing (3.17) with (3.18) obtains the desired result.    □

Using the previous theorem, we can show the existence of the Stratonovich integral under conditions that are different from those of Theorem 3.1.1.

**Proposition 3.1.2** *Let $u$ be a process in $\mathbb{L}_{\mathrm{loc}}^{1,2}$. Suppose that for all $\omega$ a.s. the integral operator from $H$ into $H$ associated with the kernel $Du(\omega)$ is a* *nuclear (or a trace class) operator. Then u is Stratonovich integrable, and we have*

$$\int_0^1 u_t \circ dW_t = \delta(u) + T(Du). \tag{3.19}$$

*Proof:* We have seen that for any complete orthonormal system $\{e_i, i \geq 1\}$ in $H$ the series

$$\sum_{i=1}^{\infty} \left( \int_0^1 u_t e_i(t) dt \right) \int_0^1 e_i(s) dW_s$$

converges in probability to $\delta(u) + T(Du)$. Then Proposition 3.1.2 follows from Theorem 3.1.2. □

## Exercises

**3.1.1** Let $u \in L^2([0,1] \times \Omega)$. Show that the families $u^\pi$ and $\hat{u}^\pi$ defined in Eqs. (3.2) and (3.3) satisfy condition (3.1) with $V = L^2([0,1] \times \Omega)$. If $u \in \mathbb{L}^{1,2}$, then (3.1) holds with $V = \mathbb{L}^{1,2}$.

**3.1.2** Let $u \in \mathbb{L}_1^{1,2}$ be a process continuous in $L^2(\Omega)$. Show that

$$\sum_{i=0}^{n-1} (u(t_{i+1}) - u(t_i))(W(t_{i+1}) - W(t_i))$$

converges in $L^1(\Omega)$ to $\int_0^1 ((D^+u)_t - (D^-u)_t) dt$ as $|\pi|$ tends to zero (see Nualart and Pardoux [249, Theorem 7.6]).

**3.1.3** Show the convergence of (3.13) to (3.14) in $L^1(\Omega)$ as $|\pi|$ tends to zero. The proof is similar to that of Theorem 3.1.1.

**3.1.4** Show that the process $u_t = W_{1-t}$ is not $L^2$-integrable but that it is Stratonovich integrable and

$$\int_0^1 W_{1-t} \circ dW_t = W_{\frac{1}{2}}^2 + 2 \int_0^1 W_{1-t} dW_t.$$

*Hint:* Consider the sequences of $\{\phi_n\}$ and $\{\psi_n\}$ of orthonormal functions in $L^2([0,1])$ given by $\phi_n(t) = \sqrt{2} \cos(2\pi nt)$ and $\psi_n(t) = \sqrt{2} \sin(2\pi n(1-t))$, $n \geq 1$, $t \in [0,1]$ (see Rosinski [293]).

**3.1.5** Let $\psi$ be a nonnegative $C^\infty$ function on $[-1,1]$ whose integral is 1 and such that $\psi(x) = \psi(-x)$. Consider the approximation of the identity $\psi_\epsilon(x) = \frac{1}{\epsilon}\psi(\frac{x}{\epsilon})$. Fix a process $u \in \mathbb{L}^{1,2}$, and define

$$I_\epsilon(u) = \int_0^1 u_t(\psi_\epsilon' * W)_t dt.$$

Show that $I_\epsilon(u)$ converges in $L^1(\Omega)$ to the Stratonovich integral of $u$. The convergence holds in probability if $u \in \mathbb{L}_{loc}^{1,2}$.

**3.1.6** Let $u = \{u_t, t \in [0,1]\}$ be a Stratonovich integrable process and let $F$ be a random variable. Show that $Fu_t$ is Stratonovich integrable and that

$$\int_0^1 Fu_t \circ dW_t = F \int_0^1 u_t \circ dW_t.$$

## 3.2  Stochastic calculus for anticipating integrals

In this section we will study the properties of the indefinite Skorohod integral as a stochastic process. In particular we discuss the regularity properties of its paths (continuity and quadratic variation) and obtain a version of the Itô formula for the indefinite Skorohod integral.

### 3.2.1  Skorohod integral processes

Fix a Brownian motion $W = \{W(t), t \in [0,1]\}$ defined on the canonical probability space $(\Omega, \mathcal{F}, P)$. Suppose that $u = \{u(t), 0 \le t \le 1\}$ is a Skorohod integrable process. In general, a process of the form $u\mathbf{1}_{(s,t]}$ may not be Skorohod integrable (see Exercise 3.2.1). Let us denote by $\mathbb{L}^s$ the set of processes $u$ such that $u\mathbf{1}_{[0,t]}$ is Skorohod integrable for any $t \in [0,1]$. Notice that the space $\mathbb{L}^{1,2}$ is included into $\mathbb{L}^s$. Suppose that $u$ belongs to $\mathbb{L}^s$, and define

$$X(t) = \delta(u\mathbf{1}_{[0,t]}) = \int_0^t u_s dW_s. \tag{3.20}$$

The process $X$ is not adapted, and its increments satisfy the following orthogonality property:

**Lemma 3.2.1** *For any process $u \in \mathbb{L}^s$ we have*

$$E\left(\int_s^t u_r dW_r \big| \mathcal{F}_{[s,t]^c}\right) = 0 \tag{3.21}$$

*for all $s < t$, where, as usual, $\mathcal{F}_{[s,t]^c}$ denotes the $\sigma$-field generated by the increments of the Brownian motion in the complement of the interval $[s,t]$.*

*Proof:*   To show (3.21) it suffices to take an arbitrary $\mathcal{F}_{[s,t]^c}$-measurable random variable $F$ belonging to the space $\mathbb{D}^{1,2}$, and check that

$$E\left(F \int_s^t u_r dW_r\right) = E\left(\int_0^1 u_r D_r F\mathbf{1}_{[s,t]}(r)dr\right) = 0,$$

which holds due to the duality relation (1.42) and Corollary 1.2.1.   $\square$

Let us denote by $\mathcal{M}$ the class of processes of the form $X(t) = \int_0^t u_s dW_s$, where $u \in \mathbb{L}^s$. One can show (see [237]) that if a process $X = \{X_t, t \in [0,1]\}$

satisfies $X(0) = 0$, $E(X(t)^2) < \infty$, for all $t$, and

$$\sup_{\pi=\{0=t_0<t_1<\cdots<t_n=1\}} \sum_{j=0}^{n-1} E[(X(t_{j+1}) - X(t_j))^2] < \infty,$$

then $X$ belongs to $\mathcal{M}$ if and only if condition (3.21) is satisfied. The necessity of (3.21) has been proved in Lemma 3.2.1.

A process $X$ in $\mathcal{M}$ is continuous in $L^2$. However, there exist processes $u$ in $\mathbb{L}^s$ such that the indefinite integral $\int_0^t u_s dW_s$ does not have a continuous version (see Exercise 3.2.3). This can be explained by the fact that the process $X$ is not a martingale, and we do not dispose of maximal inequalities to show the existence of a continuous version.

### 3.2.2 Continuity and quadratic variation of the Skorohod integral

If $u$ belongs to $\mathbb{L}^{1,2}$, then, under some integrability assumptions, we will show the existence of a continuous version for the indefinite Skorohod integral of $u$ by means of Kolmogorov's continuity criterion. To do this we need the following $L^p(\Omega)$ estimates for the Skorohod integral that are deduced by duality from Meyer's inequalities (see Proposition 1.5.8):

**Proposition 3.2.1** Let $u$ be a stochastic process in $\mathbb{L}^{1,2}$, and let $p > 1$. Then we have

$$\|\delta(u)\|_p \leq C_p \left( \left( \int_0^1 (E(u_t))^2 dt \right)^{\frac{1}{2}} + \|\left( \int_0^1 \int_0^1 (D_s u_t)^2 ds dt \right)^{\frac{1}{2}}\|_p \right). \quad (3.22)$$

The following proposition provides a sufficient condition for the existence of a continuous version of the indefinite Skorohod integral.

**Proposition 3.2.2** Let $u$ be a process in the class $\mathbb{L}^{1,2}$. Suppose that for some $p > 2$ we have $E \int_0^1 (\int_0^1 (D_s u_t)^2 ds)^{\frac{p}{2}} dt < \infty$. Then the integral process $\{\int_0^t u_s dW_s,\ 0 \leq t \leq 1\}$ has a continuous version.

*Proof:* We may assume that $E(u_t) = 0$ for any $t$, because the Gaussian process $\int_0^t E(u_s) dW_s$ always has a continuous version. Set $X_t = \int_0^t u_s dW_s$. Applying Proposition 3.2.1 and Hölder's inequality, we obtain for $s < t$

$$E(|X_t - X_s|^p) \leq C_p E(|\int_s^t \int_0^1 (D_\theta u_r)^2 d\theta dr|^{\frac{p}{2}})$$

$$\leq C_p |t - s|^{\frac{p}{2}-1} E\left( \int_s^t \left| \int_0^1 (D_\theta u_r)^2 d\theta \right|^{\frac{p}{2}} dr \right).$$

Set $A_r = \left| \int_0^1 (D_\theta u_r)^2 d\theta \right|^{\frac{p}{2}}$. Fix an exponent $2 < \alpha < 1 + \frac{p}{2}$, and assume that $p$ is close to 2. Applying Fubini's theorem we can write

$$E \left( \int_0^1 \int_0^1 \frac{|X_t - X_s|^p}{|t - s|^\alpha} dsdt \right)$$

$$\leq 2C_p E \left( \int_{\{s<t\}} |t - s|^{\frac{p}{2} - 1 - \alpha} \int_s^t A_r \, dr dsdt \right)$$

$$= \frac{2C_p}{\alpha - \frac{p}{2}} \int_{\{s<r\}} \left( |r - s|^{\frac{p}{2} - \alpha} - |1 - s|^{\frac{p}{2} - \alpha} \right) E(A_r) dr ds$$

$$= \frac{2C_p}{\left( \alpha - \frac{p}{2} \right) \left( \frac{p}{2} + 1 - \alpha \right)} \int_0^1 \left( r^{\frac{p}{2} + 1 - \alpha} + 1 - |1 - r|^{\frac{p}{2} + 1 - \alpha} \right) E(A_r) dr < \infty.$$

Hence, the random variable defined by

$$\Gamma = \int_0^1 \int_0^1 \frac{|X_t - X_s|^p}{|t - s|^\alpha} dsdt$$

satisfies $E(\Gamma) < \infty$ and by the results of Section A.3 we obtain

$$|X_t - X_s| \leq c_{p,\alpha} \Gamma^{\frac{1}{p}} |t - s|^{\frac{\alpha - 2}{p}}$$

for some constant $c_{p,\alpha}$.    □

For every $p > 1$ and any positive integer $k$ we will denote by $\mathbb{L}^{k,p}$ the space $L^p([0,1]; \mathbb{D}^{k,p}) \subset \mathbb{D}^{k,p}(L^2[0,1])$. Notice that for $p = 2$ and $k = 1$ this definition is consistent with the previous notation for the space $\mathbb{L}^{1,2}$.

The above proposition implies that for a process $u$ in $\mathbb{L}^{1,p}_{\text{loc}}$, with $p > 2$, the Skorohod integral $\int_0^t u_s dW_s$ has a continuous version. Furthermore, if $u$ in $\mathbb{L}^{1,p}$, $p > 2$, we have

$$E \left( \sup_{t \in [0,1]} \left| \int_0^t u_s dW_s \right|^p \right) < \infty.$$

It is possible to show the existence of a continuous version under different hypotheses (see Exercises 3.2.4, and 3.2.5.).

The next result will show the existence of a nonzero quadratic variation for the indefinite Skorohod integral (see [249]).

**Theorem 3.2.1** *Suppose that $u$ is a process of the space $\mathbb{L}^{1,2}_{\text{loc}}$. Then*

$$\sum_{i=0}^{n-1} \left( \int_{t_i}^{t_{i+1}} u_s dW_s \right)^2 \to \int_0^1 u_s^2 ds, \tag{3.23}$$

*in probability, as $|\pi| \to 0$, where $\pi$ runs over all finite partitions $\{0 = t_0 < t_1 < \cdots < t_n = 1\}$ of $[0,1]$. Moreover, the convergence is in $L^1(\Omega)$ if $u$ belongs to $\mathbb{L}^{1,2}$.*

*Proof:*     We will describe the details of the proof only for the case $u \in \mathbb{L}^{1,2}$. The general case would be deduced by an easy argument of localization.

For any process $u$ in $\mathbb{L}^{1,2}$ and for any partition $\pi = \{0 = t_0 < t_1 < \cdots < t_n = 1\}$ we define

$$V^\pi(u) = \sum_{i=0}^{n-1} \left( \int_{t_i}^{t_{i+1}} u_s dW_s \right)^2.$$

Suppose that $u$ and $v$ are two processes in $\mathbb{L}^{1,2}$. Then we have

$$E\left( |V^\pi(u) - V^\pi(v)| \right) \leq \left( E \sum_{i=0}^{n-1} \left( \int_{t_i}^{t_{i+1}} (u_s - v_s) dW_s \right)^2 \right)^{\frac{1}{2}}$$

$$\times \left( E \sum_{i=0}^{n-1} \left( \int_{t_i}^{t_{i+1}} (u_s + v_s) dW_s \right)^2 \right)^{\frac{1}{2}}$$

$$\leq \|u - v\|_{\mathbb{L}^{1,2}} \|u + v\|_{\mathbb{L}^{1,2}}. \tag{3.24}$$

It follows from this estimate that it suffices to show the result for a class of processes $u$ that is dense in $\mathbb{L}^{1,2}$. So we can assume that

$$u_t = \sum_{j=0}^{m-1} F_j \mathbf{1}_{(s_j, s_{j+1}]},$$

where $F_j$ is a smooth random variable for each $j$, and $0 = s_0 < \cdots < s_m = 1$. We can assume that the partition $\pi$ contains the points $\{s_0, \ldots, s_m\}$. In this case we have

$$V^\pi(u) = \sum_{j=0}^{m-1} \sum_{\{i : s_j \leq t_i < s_{j+1}\}} \left( F_j(W(t_{i+1}) - W(t_i)) - \int_{t_i}^{t_{i+1}} D_s F_j ds \right)^2$$

$$= \sum_{j=0}^{m-1} \left[ \sum_{\{i : s_j \leq t_i < s_{j+1}\}} F_j^2 (W(t_{i+1}) - W(t_i))^2 \right.$$

$$\left. -2(W(t_{i+1}) - W(t_i)) \int_{t_i}^{t_{i+1}} D_s F_j ds + \left( \int_{t_i}^{t_{i+1}} D_s F_j ds \right)^2 \right].$$

With the properties of the quadratic variation of the Brownian motion, this converges in $L^1(\Omega)$ to

$$\sum_{j=0}^{m-1} F_j^2 (s_{j+1} - s_j) = \int_0^1 u_s^2 ds,$$

as $|\pi|$ tends to zero. $\qquad \square$

As a consequence of the previous result, if $u \in \mathbb{L}_{loc}^{1,2}$ is a process such that the Skorohod integral $\int_0^t u_s dW_s$ has a continuous version with bounded variation paths, then $u = 0$.

Theorem 3.2.1 also holds if we assume that $u$ belongs to $\mathbb{L}_{loc}^F$.

### 3.2.3   Itô's formula for the Skorohod and Stratonovich integrals

In this section we will show the change-of-variables formula for the indefinite Skorohod integral. We start with the following version of this formula. Denote by $\mathbb{L}_{(4)}^{2,2}$ the space of processes $u \in \mathbb{L}^{2,2}$ such that $E(\|Du\|_{L^2([0,1]^2)}^4) < \infty$.

**Theorem 3.2.2**  *Consider a process of the form*

$$X_t = X_0 + \int_0^t u_s dW_s + \int_0^t v_s ds, \tag{3.25}$$

*where $X_0 \in \mathbb{D}_{loc}^{1,2}$, $u \in \mathbb{L}_{(4),loc}^{2,2}$ and $v \in \mathbb{L}_{loc}^{1,2}$. Suppose that the process $X$ has continuous paths. Let $F : \mathbb{R} \to \mathbb{R}$ be a twice continuously differentiable function. Then $F'(X_t)u_t$ belongs to $\mathbb{L}_{loc}^{1,2}$ and we have*

$$F(X_t) = F(X_0) + \int_0^t F'(X_s)dX_s + \frac{1}{2}\int_0^t F''(X_s)u_s^2$$

$$+ \int_0^t F''(X_s)(D^- X)_s u_s ds. \tag{3.26}$$

*Proof:*   Suppose that $(\Omega^{n,1}, X_0^n)$, $(\Omega^{n,2}, u^n)$ and $(\Omega^{n,3}, v^n)$ are localizing sequences for $X_0$, $u$ and $v$, respectively. For each positive integer $k$ let $\psi_k$ be a smooth function such that $0 \le \psi_k \le 1$, $\psi_k(x) = 0$ if $|x| \ge k + 1$, and $\psi_k(x) = 1$ if $|x| \le k$. Define

$$u_t^{n,k} = u_t^n \psi_k \left( \int_0^1 (u_s^n)^2 ds \right). \tag{3.27}$$

Set $X_t^{n,k} = X_0^n + \int_0^t u_s^{n,k} dW_s + \int_0^t v_s^n ds$ and consider the familuy of sets

$$G^{n,k} = \Omega^{n,1} \cap \Omega^{n,2} \cap \Omega^{n,3} \cap \left\{ \sup_{0 \le t \le 1} |X_t| \le k \right\} \cap \left\{ \int_0^1 (u_s^n)^2 ds \le k \right\}.$$

Define $F^k = F\psi_k$. Then, by a localization argument, it suffices to show the result for the processes $X_0^n$, $u^{n,k}$, and $v^n$ and for the function $F^k$. In this way we can assume that $X_0 \in \mathbb{D}^{1,2}$, $u \in \mathbb{L}_{(4)}^{2,2}$, $v \in \mathbb{L}^{1,2}$, $\int_0^1 u_s^2 ds \le k$, and that the functions $F$, $F'$ and $F''$ are bounded.

Set $t_i^n = \frac{it}{2^n}$, $0 \leq i \leq 2^n$. As usual, the basic argument in proving a change-of-variables formula is Taylor development. Going up to the second order, we get

$$F(X_t) = F(X_0) + \sum_{i=0}^{2^n-1} F'(X(t_i^n))(X(t_{i+1}^n) - X(t_i^n))$$

$$+ \sum_{i=0}^{2^n-1} \frac{1}{2} F''(\overline{X}_i)(X(t_{i+1}^n) - X(t_i^n))^2,$$

where $\overline{X}_i$ denotes a random intermediate point between $X(t_i^n)$ and $X(t_{i+1}^n)$. Now the proof will be decomposed in several steps.

*Step 1.* Let us show that

$$\sum_{i=0}^{2^n-1} F''(\overline{X}_i)(X(t_{i+1}^n) - X(t_i^n))^2 \rightarrow \int_0^t F''(X_s)u_s^2 ds, \qquad (3.28)$$

in $L^1(\Omega)$, as $n$ tends to infinity.

The increment $(X(t_{i+1}^n) - X(t_i^n))^2$ can be decomposed into

$$\left( \int_{t_i^n}^{t_{i+1}^n} u_s dW_s \right)^2 + \left( \int_{t_i^n}^{t_{i+1}^n} v_s ds \right)^2 + 2 \left( \int_{t_i^n}^{t_{i+1}^n} u_s dW_s \right) \left( \int_{t_i^n}^{t_{i+1}^n} v_s ds \right).$$

The contribution of the last two terms to the limit (3.28) is zero. In fact, we have

$$E \left| \sum_{i=0}^{2^n-1} F''(\overline{X}_i) \left( \int_{t_i^n}^{t_{i+1}^n} v_s ds \right)^2 \right| \leq \|F''\|_\infty t 2^{-n} \int_0^t E(v_s^2) ds,$$

and

$$E \left| \sum_{i=0}^{2^n-1} F''(\overline{X}_i) \left( \int_{t_i^n}^{t_{i+1}^n} u_s dW_s \right) \left( \int_{t_i^n}^{t_{i+1}^n} v_s ds \right) \right|$$

$$\leq \|F''\|_\infty \left( t 2^{-n} \int_0^t E(v_s^2) ds \right)^{\frac{1}{2}} \|u\|_{\mathbb{L}^{1,2}}.$$

Therefore, it suffices to show that

$$\sum_{i=0}^{2^n-1} F''(\overline{X}_i) \left( \int_{t_i^n}^{t_{i+1}^n} u_s dW_s \right)^2 \rightarrow \int_0^t F''(X_s)u_s^2 ds$$

in $L^1(\Omega)$, as $n$ tends to infinity. Suppose that $n \geq m$, and for any $i = 1, \ldots, n$ let us denote by $t_i^{(m)}$ the point of the $m$th partition   that is closer to $t_i^n$ from the left. Then we have

$$
\left| \sum_{i=0}^{2^n-1} F''(\overline{X}_i) \left( \int_{t_i^n}^{t_{i+1}^n} u_s dW_s \right)^2 - \int_0^t F''(X_s) u_s^2 ds \right|
$$

$$
\leq \left| \sum_{i=0}^{2^n-1} [F''(\overline{X}_i) - F''(X(t_i^{(m)}))] \left( \int_{t_i^n}^{t_{i+1}^n} u_s dW_s \right)^2 \right|
$$

$$
+ \left| \sum_{j=0}^{2^m-1} F''(X(t_j^m)) \sum_{i: t_i^n \in [t_j^m, t_{j+1}^m)} \left[ \left( \int_{t_i^n}^{t_{i+1}^n} u_s dW_s \right)^2 - \int_{t_i^n}^{t_{i+1}^n} u_s^2 ds \right] \right|
$$

$$
+ \left| \sum_{j=0}^{2^m-1} F''(X(t_j^m)) \int_{t_j^m}^{t_{j+1}^m} u_s^2 ds - \int_0^t F''(X_s) u_s^2 ds \right|
$$

$$
= b_1 + b_2 + b_3.
$$

The expectation of the term $b_3$ can be bounded by

$$
kE \left( \sup_{|s-r| \leq t2^{-m}} |F''(X_s) - F''(X_r)| \right),
$$

which converges to zero as $m$ tends to infinity by the continuity of the process $X_t$. In the same way the expectation of $b_1$ is bounded by

$$
E \left( \sup_{|s-r| \leq t2^{-m}} |F''(X_s) - F''(X_r)| \sum_{i=0}^{2^n-1} \left( \int_{t_i^n}^{t_{i+1}^n} u_s dW_s \right)^2 \right). \tag{3.29}
$$

Letting first $n$ tends to infinity and applying Theorem 3.2.1, (3.29) converges to

$$
E \left( \sup_{|s-r| \leq t2^{-m}} |F''(X_s) - F''(X_r)| \int_0^1 u_s^2 ds \right),
$$

which tends to zero as $m$ tends to infinity. Finally, the term $b_2$ converges to zero in $L^1(\Omega)$ as $n$ tends to infinity, for any fixed $m$, due to Theorem 3.2.1.

*Step 2.* Clearly,

$$
\sum_{i=0}^{2^n-1} F'(X(t_i^n)) \left( \int_{t_i^n}^{t_{i+1}^n} v_s ds \right) \rightarrow \int_0^t F'(X_s) v_s ds \tag{3.30}
$$

in $L^1(\Omega)$ as $n$ tends to infinity.

*Step 3.* From Proposition 1.3.5 we deduce

$$
F'(X(t_i^n)) \int_{t_i^n}^{t_{i+1}^n} u_s dW_s = \int_{t_i^n}^{t_{i+1}^n} F'(X(t_i^n)) u_s dW_s + \int_{t_i^n}^{t_{i+1}^n} D_s[F'(X(t_i^n))] u_s ds.
$$

Therefore, we obtain

$$\sum_{i=0}^{2^n-1} F'(X(t_i^n)) \int_{t_i^n}^{t_{i+1}^n} u_s dW_s = \sum_{i=0}^{2^n-1} \int_{t_i^n}^{t_{i+1}^n} F'(X(t_i^n)) u_s dW_s$$

$$+ \sum_{i=0}^{2^n-1} \int_{t_i^n}^{t_{i+1}^n} F''(X(t_i^n)) D_s X(t_i^n) u_s ds. \tag{3.31}$$

Let us first show that

$$\sum_{i=0}^{2^n-1} \int_{t_i^n}^{t_{i+1}^n} F''(X(t_i^n)) D_s X(t_i^n) u_s ds \rightarrow \int_0^t F''(X_s)(D^- X)_s u_s ds \tag{3.32}$$

in $L^1(\Omega)$ as $n$ tends to infinity. We have

$$E \left| \sum_{i=0}^{2^n-1} \int_{t_i^n}^{t_{i+1}^n} F''(X(t_i^n)) D_s X(t_i^n) u_s ds - \int_0^t F''(X_s)(D^- X)_s u_s ds \right|$$

$$\leq E \left| \sum_{i=0}^{2^n-1} F''(X(t_i^n)) \int_{t_i^n}^{t_{i+1}^n} \left[ D_s X(t_i^n) - (D^- X)_s \right] u_s ds \right|$$

$$+ E \left| \sum_{i=0}^{2^n-1} \int_{t_i^n}^{t_{i+1}^n} \left[ F''(X(t_i^n)) - F''(X_s) \right] (D^- X)_s u_s ds \right|$$

$$\leq \|F''\|_\infty \left\{ E \left( \int_0^t u_s^2 ds \right) \right\}^{1/2}$$

$$\times \left\{ \int_0^t \sup_{s-t2^{-n} \leq r \leq s} E \left( |D_s X_r - (D^- X)_s|^2 \right) ds \right\}^{1/2}$$

$$+ E \left( \sup_{|s-r| \leq t2^{-n}} |F''(X_s) - F''(X_r)| \int_0^t |(D^- X)_s u_s| ds \right)$$

$$: \quad = d_1^n + d_2^n.$$

The term $d_2^n$ tends to zero as $n$ tends to infinity because $E \int_0^t |(D^- X)_s u_s| \, ds < \infty$. The term $d_1^n$ tends to zero as $n$ tends to infinity because $X$ belongs to $\mathbb{L}_2^{1,2}$ by Proposition 3.1.1.

As a consequence of the convergences (3.28), (3.30) and (3.32) we have proved that the sequence

$$A_n := \sum_{i=0}^{2^n-1} \int_{t_i^n}^{t_{i+1}^n} F'(X(t_i^n)) u_s dW_s$$

converges in $L^1(\Omega)$ as $n$ tends to infinity to

$$
\begin{aligned}
\Phi_t \quad : \quad &= F(X_t) - F(X_0) - \int_0^t F'(X_s)v_s ds - \frac{1}{2}\int_0^t F''(X_s)u_s^2 \\
&\quad - \int_0^t F''(X_s)(D^- X)_s u_s ds.
\end{aligned}
\tag{3.33}
$$

*Step 4.* The process $u_t F'(X_t)$ belongs to $\mathbb{L}^{1,2}$ because $u \in \mathbb{L}^{2,2}$, $v \in \mathbb{L}^{1,2}$, $E(\|Du\|_{L^2([0,1]^2)}^4) < \infty$, and the processes $F'(X_t)$, $F''(X_t)$ and $\int_0^1 u_s^2 ds$ are uniformly bounded. In fact, we have

$$
\begin{aligned}
D_s\left[u_t F'(X_t)\right] \quad = \quad &u_t F_t''(X_t)\left(u_s \mathbf{1}_{\{s \le t\}} + D_s X_0 \right. \\
&\left. + \int_0^t D_s u_r dW_r + \int_0^t D_s v_r dr\right) + F'(X_t)D_s u_t,
\end{aligned}
$$

and all terms in the right-hand side of the above expression are square integrable. For the third term we use the duality relationship of the Skorohod integral:

$$
\begin{aligned}
&\int_0^1 \int_0^1 E\left(\left(u_t \int_0^t D_s u_r dW_r\right)^2\right)dsdt \\
&= E\left\{\int_0^1 \int_0^1 \int_0^1 D_s u_r \left[2u_t D_r u_t \left(\int_0^t D_s u_r dW_r\right) + u_t^2 D_s u_r\right.\right. \\
&\quad \left.\left. +u_t^2\left(\int_0^t D_r D_s u_\theta dW_\theta\right)\right]drdsdt\right\} \\
&\le \quad cE\left(\|Du\|_{L^2([0,1]^2)}^4 + \|D^2 u\|_{L^2([0,1]^3)}\right).
\end{aligned}
$$

*Step 5.* Using the duality relationship it is clear that for any smooth random variable $G \in \mathcal{S}$ we have

$$
\lim_{n\to\infty} E\left(G \sum_{i=0}^{2^n-1} \int_{t_i^n}^{t_{i+1}^n} F'(X(t_i^n))u_s dW_s\right) = E\left(G\int_0^t F'(X_s)u_s dW_s\right).
$$

On the other hand, we have seen that that $A_n$ converges in $L^1(\Omega)$ to (3.33). Hence, (3.26) holds. □

**Remarks:**

**1.** If the process $X$ is adapted then $D^- X = 0$, and we obtain the classical Itô's formula.

**2.** Also using the operator $\nabla$ introduced in (3.8) we can write

$$
F(X_t) = F(X_0) + \int_0^t F'(X_s)dX_s + \frac{1}{2}\int_0^t F''(X_s)(\nabla X)_s u_s ds.
$$

**3.** A sufficient condition for $X$ to have continuous paths is

$$E\left(\int_0^1 \|Du_t\|_H^p \, dt\right) < \infty$$

for some $p > 2$ (see Proposition 3.2.2).

**4.** One can show Theorem 3.2.2 under different types of hypotheses. More precisely, one can impose some conditions on $X$ and modify the assumptions on $X_0$, $u$ and $v$. For instance, one can assume either

(a) $u \in \mathbb{L}^{1,2} \cap L^\infty([0,1] \times \Omega)$, $v \in L^2([0,1] \times \Omega)$, $X \in \mathbb{L}_2^{1,2}$, and $X$ has a version which is continuous and $X_t \in \mathbb{D}^{1,2}$ for all $t$, or

(b) $u \in \mathbb{L}^{1,2} \cap L^4(\Omega; L^2([0,1]))$, $v \in L^2([0,1] \times \Omega)$, $X \in \mathbb{L}_2^{1,2}$, and $X$ has a version which is continuous and $X_t \in \mathbb{D}^{1,2}$ for all $t$, and $\int_0^1 \int_0^1 (D_s X_t)^2 ds dt + \int_0^1 (D^- X)_t^2 dt \in L^4(\Omega)$.

In fact, if $X$ has continuous paths, by means of a localization argument it suffices to show the result for each function $F_k$, which has compact support. On the other hand, the properties $u \in \mathbb{L}^{1,2}$, $v \in L^2([0,1] \times \Omega)$ and $X \in \mathbb{L}_2^{1,2}$ imply the convergences (3.28), (3.30) and (3.32). The boundedness or integrability assumptions on $u$, $DX$ and $D^- X$ are used in order to ensure that $E(\Phi_t^2) < \infty$, and $u_t F'(X_t) \in \mathbb{L}^{1,2}$.

**5.** If we assume the conditions $X_0 \in \mathbb{D}_{\text{loc}}^{1,2}$, $u \in \mathbb{L}_{\text{loc}}^{2,2}$ and $v \in \mathbb{L}_{\text{loc}}^{1,2}$, and $X$ has continuous paths, then we can conclude that $u_s F'(X_s) \mathbf{1}_{[0,t]}(s) \in (\text{Dom}\delta)_{\text{loc}}$ and Itô's formula (3.26) holds. In fact, steps 1, 2 and 3 of the proof are still valid. Finally, the sequence of processes

$$v^n := \sum_{i=0}^{2^n-1} F'(X(t_i^n))u_s \mathbf{1}_{(t_i^n, t_{i+1}^n]}(s)$$

converges in $L^2([0,1] \times \Omega)$ to $F'(X_s)u_s \mathbf{1}_{[0,t]}(s)$ as $n$ tends to infinity, and by Step 3, $\delta(v^n)$ converges in $L^1(\Omega)$ to $\Phi_t$. Then, by Proposition 1.3.6, $u_s F'(X_s)\mathbf{1}_{[0,t]}(s)$ belongs to the domain of the divergence and (3.26) holds.

In [10] the following version of Itô's formula is proved:

**Theorem 3.2.3** *Suppose that $X_t = X_0 + \int_0^t u_s dW_s + \int_0^t v_s ds$, where $X_0 \in \mathbb{D}_{\text{loc}}^{1,2}$, $u \in \left(\mathbb{L}^F \cap L^\infty(\Omega; L^2([0,1]))\right)_{\text{loc}}$, $v \in \mathbb{L}_{\text{loc}}^{1,2,f}$ and the process $X$ has continuous paths. Let $F : \mathbb{R} \to \mathbb{R}$ be a twice continuously differentiable function. Then $F'(X_s)u_s\mathbf{1}_{[0,t]}(s)$ belongs to $(\text{Dom}\delta)_{\text{loc}}$ and we have*

$$
\begin{aligned}
F(X_t) &= F(X_0) + \int_0^t F'(X_s)dX_s + \frac{1}{2}\int_0^t F''(X_s)u_s^2 \\
&\quad + \int_0^t F''(X_s)(D^- X)_s u_s ds.
\end{aligned}
\tag{3.34}
$$

*Proof:*     By a localization argument we can assume that the processes $F(X_t)$, $F'(X_t)$, $F''(X_t)$ are uniformly bounded and $\int_0^1 u_s^2 ds \leq k$. Then we proceed as in the steps 1, 2 and 3 of the proof of Theorem 3.2.2, and we conclude using Proposition 1.3.6.

Notice that for the proof of the convergence (3.28) we have to apply Theorem 3.2.1 for $u \in \mathbb{L}^F$. Also, (3.31) holds thanks to Proposition 1.3.5 applied to the set $A = [t_i^n, t_{i+1}^n]$.     □

**Remarks:**

**1.** A sufficient condition for the indefinite Skorohod integral $\int_0^t u_s dW_s$ of a process $u \in \mathbb{L}^F$ to have a continuous version is $E\left(\int_0^1 |u_t|^p dt\right) < \infty$ for some $p > 2$ (see Exercise 3.2.12).

**2.** The fact that $\int_0^1 u_t^2 dt$ is bounded is only used to insure that the right-hand side of (3.26) is square integrable, and it can be replaced by the conditions $\int_0^1 u_t^2 dt$, $\int_0^1 (D^- X)_t^2 dt \in L^2(\Omega)$.

**3.** If $X_0$ is a constant and $u$ and $v$ are adapted processes such that $\int_0^1 u_t^2 dt < \infty$ and $\int_0^1 v_t^2 dt < \infty$ a.s., then these processes satisfy the hypotheses of Theorem 3.2.3 because $u \in \left(\mathbb{L}^F \cap L^\infty(\Omega; L^2([0,1]))\right)_{\text{loc}}$ by Proposition 1.3.18, $v \in \mathbb{L}_{\text{loc}}^{1,2,f}$ (this property can be proved by a localization procedure similat to that used in the proof of Proposition 1.3.18), and the process $X_t = X_0 + \int_0^t u_s dW_s + \int_0^t v_s ds$ has continuous paths because it is a continuous semimartingale. Furthermore, in this case $D^- X$ vanishes and we obtain the classical Itô's formula.

**4.** In Theorem 3.2.3 we can replace the condition $v \in \mathbb{L}_{\text{loc}}^{1,2,f}$ by the fact that we can localize $v$ by processes such that $\int_0^1 |v_t^n| dt \in L^\infty(\Omega)$, and $\left\{V_t^n = \int_0^t v_s^n ds, t \in [0,1]\right\} \in \mathbb{L}_{2-}^{1,2,f}$. In this way the change-of-variable formula established in Theorem 3.2.3 is a generalization of Itô's formula.

The following result is a multidimensional and local version of the change-of-variables formula for the Skorohod integral.

**Theorem 3.2.4** *Let* $W = \{W_t, t \in [0,1]\}$ *be an* $N$-*dimensional Brownian motion. Suppose that* $X_0^i \in \mathbb{D}^{1,2}$, $u^{ij} \in \mathbb{L}_{(4)}^{2,2}$, *and* $v^i \in \mathbb{L}^{1,2}$, $1 \leq i \leq M$, $1 \leq j \leq N$ *are processes such that*

$$X_t^i = X_0^i + \sum_{j=1}^N \int_0^t u_s^{ij} dW_s^j + \int_0^t v_s^i ds, \ 0 \leq t \leq 1.$$

*Assume that $X_t^i$ has continuous paths. Let $F : \mathbb{R}^M \to \mathbb{R}$ be a twice contin-uously differentiable function. Then we have*

$$
\begin{aligned}
F(X_t) \;=\;& F(X_0) + \sum_{i=1}^{M} \int_0^t (\partial_i F)(X_s) dX_s^i \\
&+ \frac{1}{2} \sum_{i,j=1}^{M} \sum_{k=1}^{N} \int_0^t (\partial_i \partial_j F)(X_s) u_s^{ik} u_s^{jk} ds \\
&+ \sum_{i,j=1}^{M} \sum_{k=1}^{N} \int_0^t (\partial_i \partial_j F)(D^{k,-} X^j)_s u_s^{ik} ds,
\end{aligned}
$$

*where $D^k$ denotes the derivative with respect to the $k$th compoment of the Wiener process.*

Itô's formula for the Skorohod integral allows us to deduce a change-of-variables formula for the Stratonovich integral. Let us first introduce the following classes of processes:

The set $\mathbb{L}_S^{2,4}$ is the class of processes $u \in \mathbb{L}_1^{1,2} \cap \mathbb{L}_{(4)}^{2,2}$ continuous in $L^2(\Omega)$ and such that $\nabla u \in \mathbb{L}^{1,2}$.

**Theorem 3.2.5** *Let $F$ be a real-valued, twice continuously differentiable function. Consider a process of the form $X_t = X_0 + \int_0^t u_s \circ dW_s + \int_0^t v_s ds$, where $X_0 \in \mathbb{D}_{\mathrm{loc}}^{1,2}$, $u \in \mathbb{L}_{S,\mathrm{loc}}^{2,4}$ and $v \in \mathbb{L}_{\mathrm{loc}}^{1,2}$. Suppose that $X$ has continuous paths. Then we have*

$$
F(X_t) = F(X_0) + \int_0^t F'(X_s) v_s ds + \int_0^t [F'(X_s) u_s] \circ dW_s.
$$

*Proof:*   As in the proof of the change-of-variable formula for the Skorohod integral we can assume that the processes $F(X_t)$, $F'(X_t)$, $F''(X_t)$ and $\int_0^1 u_s^2 ds$ are uniformly bounded, $X_0 \in \mathbb{D}^{1,2}$, $u \in \mathbb{L}_S^{2,4}$ and $v \in \mathbb{L}^{1,2}$. We know, by Theorem 3.1.1 that the process $X_t$ has the following decomposition:

$$
X_t = X_0 + \int_0^t u_s dW_s + \int_0^t v_s ds + \int_0^t \frac{1}{2}(\nabla u)_s ds.
$$

This process verifies the assumptions of Theorem 3.2.2. Consequently, we can apply Itô's formula to $X$ and obtain

$$
\begin{aligned}
F(X_t) \;=\;& F(X_0) + \int_0^t F'(X_s) v_s ds + \frac{1}{2} \int_0^t F'(X_s)(\nabla u)_s ds \\
&+ \int_0^t F'(X_s) u_s dW_s + \frac{1}{2} \int_0^t F''(X_s)(\nabla X)_s u_s ds.
\end{aligned}
$$

The process $F'(X_t)u_t$ belongs to $\mathbb{L}_1^{1,2}$. In fact, notice first that as in the proof of Theorem 3.2.2 the boundedness of $F'(X_t)$, $F''(X_t)$ and $\int_0^1 u_s^2 ds$ and the fact that $u \in \mathbb{L}_{(4)}^{2,2}$, $v, \nabla u \in \mathbb{L}^{1,2}$ and $X_0 \in \mathbb{D}^{1,2}$ imply that this process belongs to $\mathbb{L}^{1,2}$ and

$$D_s\left[F'(X_t)u_t\right] = F'(X_t)D_s u_t + F''(X_t)D_s X_t u_t.$$

On the other hand, using that $u \in \mathbb{L}_1^{1,2}$, $u$ is continuous in $L^2(\Omega)$, and $X \in \mathbb{L}_2^{1,2}$ we deduce that $F'(X_t)u_t$ belongs to $\mathbb{L}_1^{1,2}$ and that

$$(\nabla\left(F'(X)u\right))_t = F'(X_t)(\nabla u)_t + F''(X_t)u_t(\nabla X)_t.$$

Hence, applying Theorem 3.1.1, we can write

$$
\begin{aligned}
\int_0^t [F'(X_s)u_s] \circ dW_s &= \int_0^t F'(X_s)u_s dW_s + \frac{1}{2}\int_0^t (\nabla\left(F'(X)u\right))_s ds \\
&= \int_0^t F'(X_s)u_s dW_s + \frac{1}{2}\int_0^t F'(X_s)(\nabla u)_s ds \\
&\quad + \frac{1}{2}\int_0^t F''(X_s)u_s(\nabla X)_s ds.
\end{aligned}
$$

Finally, notice that

$$(\nabla X)_t = 2(D^- X)_t + u_t.$$

This completes the proof of the theorem. $\qquad\square$

In the next theorem we state a multidimensional version of the change-of-variable formula for the Stratonovich integral.

**Theorem 3.2.6** *Let* $W = \{W_t, t \in [0,1]\}$ *be an* $N$-*dimensional Brownian motion. Suppose that* $X_0^i \in \mathbb{D}^{1,2}$, $u^{ij} \in \mathbb{L}_S^{2,4}$, *and* $v^i \in \mathbb{L}^{1,2}$, $1 \le i \le M$, $1 \le j \le N$ *are processes such that*

$$X_t^i = X_0^i + \sum_{j=1}^N \int_0^t u_s^{ij} \circ dW_s^j + \int_0^t v_s^i ds, \ 0 \le t \le 1.$$

*Assume that* $X_t^i$ *has continuous paths. Let* $F : \mathbb{R}^M \to \mathbb{R}$ *be a twice continuously differentiable function. Then we have*

$$F(X_t) = F(X_0) + \sum_{i=1}^M \sum_{j=1}^N \int_0^t (\partial_i F)(X_s)u_s^{ij} \circ dW_s^j + \sum_{i=1}^M \int_0^t (\partial_i F)(X_s)v_s^i ds.$$

Similar results can be obtained for the *forward and backward stochastic integrals*. In these cases we require some addtional continuity conditions. Let us consider the case of the forward integral. This integral is defined as the limit in probability of the forward Riemann suns:

**Definition 3.2.1** *Let* $u = \{u_t, t \in [0,1]\}$ *be a stochastic process. The forward stochastic integral* $\int_0^1 u_t d^- W_t$ *is defined as the limit in probability as* $|\pi| \to 0$ *of the Riemann sums*

$$\sum_{i=1}^{n-1} u_{t_i} (W_{t_{i+1}} - W_{t_i}).$$

The following proposition provides some sufficient conditions for the existence of this limit.

**Proposition 3.2.3** *Let* $u = \{u_t, t \in [0,1]\}$ *be a stochastic process which is continuous in the norm of the space* $\mathbb{D}^{1,2}$. *Suppose that* $u \in \mathbb{L}_{1-}^{1,2}$. *Then* $u$ *is forward integrable and*

$$\delta(u) = \int_0^1 u_t d^- W_t + \int_0^1 \left( D^- u \right)_s ds.$$

*Proof:*    We can write, using Eq. (3.4)

$$\delta \left( \sum_{i=1}^{n-1} u_{t_i} \mathbf{1}_{(t_i, t_{i+1}]} \right) = \sum_{i=1}^{n-1} u_{t_i} (W_{t_{i+1}} - W_{t_i}) - \sum_{i=1}^{n-1} \int_{t_i}^{t_{i+1}} D_s u_{t_i} ds. \quad (3.35)$$

The processes

$$u_t^{\pi-} = \sum_{i=1}^{n-1} u_{t_i} \mathbf{1}_{(t_i, t_{i+1}]}(t)$$

converge in the norm of the space $\mathbb{L}^{1,2}$ to the process $u$, due to the continuity of $t \to u_t$ in $\mathbb{D}^{1,2}$. Hence, left-hand side of Equation (3.35), which equals to $\delta(u^{\pi-})$, converges in $L^2(\Omega)$ to $\delta(u)$. On the other hand,

$$E \left( \left| \sum_{i=1}^{n-1} \int_{t_i}^{t_{i+1}} D_s u_{t_i} ds - \int_0^1 \left( D^- u \right)_s ds \right| \right)$$

$$\leq \sum_{i=1}^{n-1} \int_{t_i}^{t_{i+1}} E \left( \left| D_s u_{t_i} - \left( D^- u \right)_s \right| \right) ds$$

$$\leq \int_0^1 \sup_{(s-|\pi|) \vee 0 \leq t < s} E(|D_s X_t - (D^- X)_s|) ds,$$

which tends to zero as $|\pi$ tends to zero, by the definition of the space $\mathbb{L}_{1-}^{1,2}$.    $\square$

We can establish an Itô's formula for the forward integral as in the case of the Stratonovich integral. Notice that the forward integral follows the rules of the Itô stochastic calculus. Define the set $\mathbb{L}^{2,4}$ as the class of processes $u \in \mathbb{L}_{1-}^{1,2} \cap \mathbb{L}_{(4)}^{2,2}$ continuous in $\mathbb{D}^{1,2}$ and such that $D^- u \in \mathbb{L}^{1,2}$.

**Theorem 3.2.7** *Let $F$ be a real-valued, twice continuously differentiable function. Consider a process of the form $X_t = X_0 + \int_0^t u_s d^- W_s + \int_0^t v_s ds$, where $X_0 \in \mathbb{D}_{\text{loc}}^{1,2}$, $u \in \mathbb{L}_{-,\text{loc}}^{2,4}$ and $v \in \mathbb{L}_{\text{loc}}^{1,2}$. Suppose that $X$ has continuous paths. Then we have*

$$F(X_t) = F(X_0) + \int_0^t F'(X_s)v_s ds + \int_0^t [F'(X_s)u_s]d^- W_s + \frac{1}{2}\int_0^t F''(X_s)u_s^2 ds.$$
$$(3.36)$$

*Proof:*   As in the proof of the change-of-variable formula for the Skorohod integral we can assume that the processes $F(X_t)$, $F'(X_t)$, $F''(X_t)$ and $\int_0^1 u_s^2 ds$ are uniformly bounded, $X_0 \in \mathbb{D}^{1,2}$, $u \in \mathbb{L}_-^{2,4}$ and $v \in \mathbb{L}^{1,2}$. We know, by Proposition 3.2.3 that the process $X_t$ has the following decomposition:

$$X_t = X_0 + \int_0^t u_s dW_s + \int_0^t v_s ds + \int_0^t (D^- u)_s ds.$$

This process verifies the assumptions of Theorem 3.2.2. Consequently, we can apply Itô's formula to $X$ and obtain

$$\begin{aligned} F(X_t) &= F(X_0) + \int_0^t F'(X_s)v_s ds + \int_0^t F'(X_s)(D^- u)_s ds \\ &\quad + \int_0^t F'(X_s)u_s dW_s + \frac{1}{2}\int_0^t F''(X_s)(\nabla X)_s u_s ds. \end{aligned}$$

The process $F'(X_t)u_t$ belongs to $\mathbb{L}_{1-}^{1,2}$. In fact, notice first that as in the proof of Theorem 3.2.2 the boundedness of $F'(X_t)$, $F''(X_t)$ and $\int_0^1 u_s^2 ds$ and the fact that $u \in \mathbb{L}_{(4)}^{2,2}$, $v, D^- u \in \mathbb{L}^{1,2}$ and $X_0 \in \mathbb{D}^{1,2}$ imply that this process belongs to $\mathbb{L}^{1,2}$ and

$$D_s[F'(X_t)u_t] = F'(X_t)D_s u_t + F''(X_t)D_s X_t u_t.$$

On the other hand, using that $u \in \mathbb{L}_{1-}^{1,2}$, $u$ is continuous in $L^2(\Omega)$, $X$ has continuous paths, and $X \in \mathbb{L}_2^{1,2}$ we deduce that $F'(X_t)u_t$ belongs to $\mathbb{L}_{1-}^{1,2}$ and that

$$\left(D^-\left(F'(X)u\right)\right)_t = F'(X_t)(D^- u)_t + F''(X_t)u_t(D^- X)_t.$$

Hence, applying Proposition 3.2.3, we can write

$$\begin{aligned} \int_0^t [F'(X_s)u_s]d^- W_s &= \int_0^t F'(X_s)u_s dW_s + \int_0^t \left(D^-\left(F'(X)u\right)\right)_s ds \\ &= \int_0^t F'(X_s)u_s dW_s + \int_0^t F'(X_s)(D^- u)_s ds \\ &\quad + \int_0^t F''(X_s)u_s(D^- X)_s ds. \end{aligned}$$

Finally, notice that

$$(\nabla X)_t = 2(D^- X)_t + u_t.$$

This completes the proof of the theorem.                                      □

### 3.2.4   Substitution formulas

The aim of this section is to study the following problem. Suppose that $u = \{u_t(x), 0 \leq t \leq 1\}$ is a stochastic process parametrized by $x \in \mathbb{R}^m$, which is square integrable and adapted for each $x \in \mathbb{R}^m$. For each $x$ we can define the Itô integral

$$\int_0^1 u_t(x)dW_t.$$

Assume now that the resulting random field is a.s. continuous in $x$, and let $F$ be an $m$-dimensional random variable. Then we can evaluate the stochastic integral at $x = F$, that is, we can define the random variable

$$\int_0^1 u_t(x)dW_t|_{x=F}. \tag{3.37}$$

A natural question is under which conditions is the nonadapted process $\{u_t(F), 0 \leq t \leq 1\}$ Skorohod integrable, and what is the relationship between the Skorohod integral of this process and the random variable defined by (3.37). We will show that this problem can be handled if the random field $u_t(x)$ is continuously differentiable in $x$ and verifies some integrability conditions, and, on the other hand, the random variable $F$ belongs locally to $\mathbb{D}^{1,4}$. Notice, however, that no kind of smoothness in the sense of the Malliavin calculus will be required on the process $u_t(x)$. A similar question can be asked for the Stratonovich integral.

   To handle these problems we will make use of the following technical result.

**Lemma 3.2.2** *Suppose that* $\{Y_n(x), x \in \mathbb{R}^m\}$, $n \geq 1$ *is a sequence of random fields such that* $Y_n(x)$ *converges in probability to* $Y(x)$ *as* $n$ *tends to infinity, for each fixed* $x \in \mathbb{R}^m$. *Suppose that*

$$E(|Y_n(x) - Y_n(y)|^p) \leq c_K|x - y|^\alpha, \tag{3.38}$$

*for all* $|x|, |y| \leq K$, $n \geq 1$, $K > 0$ *and for some constants* $p > 0$ *and* $\alpha > m$. *Then, for any* $m$-*dimensional random variable* $F$ *one has*

$$\lim_{n \to \infty} Y_n(F) = Y(F)$$

*in probability. Moreover, the convergence is in* $L^p(\Omega)$ *if* $F$ *is bounded.*

*Proof:*    Fix $K > 0$. Replacing $F$ by $F_K := F\mathbf{1}_{\{|F| \leq K\}}$ we can assume that $F$ is bounded by $K$. Fix $\varepsilon > 0$ and consider a random variable $F_\varepsilon$ which takes finitely many values and such that $|F_\varepsilon| \leq K$ and $\|F - F_\varepsilon\|_\infty \leq \varepsilon$. We can write

$$|Y_n(F) - Y(F)| \leq |Y_n(F) - Y_n(F_\varepsilon)| + |Y_n(F_\varepsilon) - Y(F_\varepsilon)| + |Y(F_\varepsilon) - Y(F)|.$$

Take $0 < m' < \alpha - m$. By (3.38) and (A.10) there exist a constant $C_1$ and random variables $\Gamma_n$ such that

$$\begin{aligned}
|Y_n(x) - Y_n(y)|^p &\leq |x - y|^{m'}\Gamma_n, \\
E(\Gamma_n) &\leq C_1.
\end{aligned}$$

Moreover, Eq. (3.38) is also satisfied by the random field $Y(x)$ and we can also find a constant $C_2$ and a random variable $\Gamma$ such that

$$\begin{aligned}
|Y(x) - Y(y)|^p &\leq |x - y|^{m'}\Gamma, \\
E(\Gamma) &\leq C_2.
\end{aligned}$$

Hence,

$$E\left(|Y_n(F) - Y(F)|^p\right) \leq c_p\left((C_1 + C_2)\varepsilon^{m'} + E\left(|Y_n(F_\varepsilon) - Y(F_\varepsilon)|^p\right)\right),$$

and the desired convergence follows by taking first the limit as $n$ tends to infinity and then the limit as $\varepsilon$ tends to zero.    □

Consider a random field $u = \{u_t(x), 0 \leq t \leq 1, x \in \mathbb{R}^m\}$ satisfying the following conditions:

**(h1)** For each $x \in \mathbb{R}^m$ and $t \in [0,1]$, $u_t(x)$ is $\mathcal{F}_t$-measurable

**(h2)** There exist constants $p \geq 2$ and $\alpha > m$ such that

$$E(|u_t(x) - u_t(y)|^p) \leq C_{t,K}|x - y|^\alpha,$$

for all $|x|, |y| \leq K$, $K > 0$, where $\int_0^1 C_{t,K} dt < \infty$. Moreover $\int_0^1 E(|u_t(0)|^2) dt < \infty$.

Notice that under the above conditions for each $t \in [0,1]$ the random field $\{u_t(x), x \in \mathbb{R}^m\}$ possesses a continuous version, and the Itô integral $\int_0^1 u_t(x) dW_t$ possesses a continuous version in $(t,x)$. In fact, for all $|x|, |y| \leq K$, $K > 0$, we have

$$\begin{aligned}
&E\left(\sup_{t \in [0,1]} \left|\int_0^t (u_s(x) - u_s(y))\, dW_s\right|^p\right) \\
&\leq c_p E\left(\left|\int_0^t |u_s(x) - u_s(y)|^2\, ds\right|^{\frac{p}{2}}\right) \\
&\leq c_p \left(\int_0^1 C_{t,K} dt\right)|x - y|^\alpha.
\end{aligned}$$

The following theorem provides the relationship between the evaluated integral $\int_0^1 u_t(x)dW_t\big|_{x=F}$ and the Skorohod integral $\delta(u(F))$. We need the following hypothesis which is stronger than (h2):

**(h3)** For each $(t, \omega)$ the mapping $x \mapsto u_t(x)$ is continuously differentiable, and for each $K > 0$

$$\int_0^1 E\left( \sup_{|x| \leq K} |\nabla u_t(x)|^q \right) dt < \infty,$$

where $q \geq 4$ and $q > m$. Moreover $\int_0^1 E(|u_t(0)|^2)dt < \infty$.

**Theorem 3.2.8** *Suppose that $u = \{u_t(x), 0 \leq t \leq 1, x \in \mathbb{R}^m\}$ is a random field satisfying conditions* (h1) *and* (h3). *Let $F : \Omega \to \mathbb{R}^m$ be a bounded random variable such that $F^i \in \mathbb{D}^{1,4}$ for $1 \leq i \leq m$. Then the composition $u(F) = \{u_t(F), 0 \leq t \leq 1\}$ belongs to the domain of $\delta$ and*

$$\delta(u(F)) = \int_0^1 u_t(x)dW_t\bigg|_{x=F} - \sum_{j=1}^m \int_0^1 \partial_j u_t(F) D_t F^j dt. \qquad (3.39)$$

*Proof:* Consider the approximation of the process $u$ given by

$$u_t^n(x) = \sum_{i=1}^{n-1} n \left( \int_{\frac{i-1}{n}}^{\frac{i}{n}} u_s(x)ds \right) 1_{(\frac{i}{n}, \frac{i+1}{n}]}(t). \qquad (3.40)$$

Notice that $u(F)$ belongs to $L^2([0, 1] \times \Omega)$ and the sequence of processes $u^n(F)$ converges to $u(F)$ in $L^2([0, 1] \times \Omega)$ as $n$ tends to infinity. Indeed, if $F$ is bounded by $K$ we have

$$
\begin{aligned}
E\left( \int_0^1 u_t(F)^2 dt \right) &\leq \int_0^1 E\left( \sup_{|x| \leq K} |u_t(x)|^2 dt \right) \\
&\leq 2\int_0^1 E(|u_t(0)|^2)dt \\
&\quad +2K^2 \int_0^1 E\left( \sup_{|x| \leq K} |\nabla u_t(x)|^2 dt \right) \\
&< \infty.
\end{aligned}
$$

The convergence of $u^n(F)$ to $u(F)$ in $L^2([0, 1] \times \Omega)$ is obtained by first approximating $u(F)$ by an element of $C([0, 1]; L^2(\Omega))$.

From Proposition 1.3.4 and Exercise 3.2.10 we deduce that $u^n(F)$ belongs to Dom$\delta$ and

$$
\delta(u^n(F)) = \sum_{i=1}^{n-1} n \left( \int_{\frac{i-1}{n}}^{\frac{i}{n}} u_s(F)ds \right) \left( W_{\frac{i+1}{n}} - W_{\frac{i}{n}} \right) - \sum_{i=1}^{n-1} \sum_{j=1}^{m} n
$$
$$
\times \left( \int_{\frac{i-1}{n}}^{\frac{i}{n}} \partial_j u_s(F)ds \right) \left( \int_{\frac{i}{n}}^{\frac{i+1}{n}} D_r F^j dr \right). \tag{3.41}
$$

The Itô stochastic integrals $\int_0^1 u_t^n(x)dW_t$ satisfy

$$
E \left( \left| \int_0^1 u_t^n(x)dW_t - \int_0^1 u_t^n(y)dW_t \right|^q \right) \leq c_q |x - y|^q.
$$

Hence, by Lemma 3.2.2 the sequence $\int_0^1 u_t^n(x)dW_t \Big|_{x=F}$ converges in $L^q(\Omega)$ to the random variable $\int_0^1 u_t(x)dW_t \Big|_{x=F}$. On the other hand, the second summand in the right-hand side of (3.41) converges to

$$
\sum_{j=1}^{m} \int_0^1 \partial_j u_t(F) D_t F^j dt
$$

in $L^2(\Omega)$, as it follows from the estimate

$$
\left| \int_0^1 \partial_j u_t(F) D_t F^j dt - \int_0^1 \partial_j u_t^n(F) D_t F^j dt \right|
$$
$$
\leq \|DF^j\|_H \left( \int_0^1 |\partial_j u_t(F) - \partial_j u_t^n(F)|^2 dt \right)^{\frac{1}{2}}.
$$

The operator $\delta$ being closed the result follows.    □

The preceding theorem can be localized as follows:

**Theorem 3.2.9** *Suppose that* $u = \{u_t(x), 0 \leq t \leq 1, x \in \mathbb{R}^m\}$ *is a random field satisfying conditions* (h1) *and* (h3). *Let* $F : \Omega \to \mathbb{R}^m$ *be a random variable such that* $F^i \in \mathbb{D}_{loc}^{1,4}$ *for* $1 \leq i \leq m$. *Then the composition* $u(F) = \{u_t(F), 0 \leq t \leq 1\}$ *belongs to* $(\text{Dom}\delta)_{loc}$ *and*

$$
\delta(u(F)) = \int_0^1 u_t(x)dW_t \Big|_{x=F} - \sum_{j=1}^{m} \int_0^1 \partial_j u_t(F) D_t F^j dt. \tag{3.42}
$$

We recall that the operator $\delta$ is not known to be local in Dom $\delta$, and for this reason the value of $\delta(u(F))$ in the Theorem 3.2.9 might depend on the particular localizing sequence.

The Stratonovich integral also satisfies a commutativity relationship but in this case we do not need a complementary term. This fact is coherent with the the general behavior of the Stratonovich integral as an ordinary integral.

Let us introduce the following condition which is stronger than (h2):

**(h4)** There exists constants $p \geq 2$ and $\alpha > m$ such that

$$E(|u_t(x) - u_t(y)|^p) \leq c_K |x - y|^\alpha,$$
$$E(|u_t(x) - u_t(y) - u_s(x) + u_s(y)|^p) \leq c_K |x - y|^\alpha |t - s|^{\frac{p}{2}},$$

for all $|x|, |y| \leq K$, $K > 0$, and $s, t \in [0, 1]$. Moreover $\int_0^1 E(|u_t(0)|^2)$ $dt < \infty$, and $t \mapsto u_t(x)$ is continuous in $L^2(\Omega)$ for each $x$.

**Theorem 3.2.10** *Let* $u = \{u_t(x), 0 \leq t \leq 1, x \in \mathbb{R}^m\}$ *be a random field satisfying hypothesis* (h1) *and* (h4). *Suppose that for each* $x \in \mathbb{R}^m$ *the Stratonovich integral* $\int_0^1 u_t(x) \circ dW_t$ *exists. Consider an arbitrary random variable* $F$. *Then* $u(F)$ *is Stratonovich integrabl, and we have*

$$\int_0^1 u_t(F) \circ dW_t = \int_0^1 u_t(x) \circ dW_t \bigg|_{x=F}. \tag{3.43}$$

*Proof:*    Fix a partition $\pi = \{0 = t_0 < t_1 < \cdots < t_n = 1\}$. We can write

$$
\begin{aligned}
S^\pi &= \sum_{i=0}^{n-1} \frac{1}{t_{i+1} - t_i} \left( \int_{t_i}^{t_{i+1}} u_s(F) ds \right) (W(t_{i+1}) - W(t_i)) \\
&= \sum_{i=0}^{n-1} u_{t_i}(F)(W(t_{i+1}) - W(t_i)) \\
&\quad + \sum_{i=0}^{n-1} \frac{1}{t_{i+1} - t_i} \left( \int_{t_i}^{t_{i+1}} [u_s(F) - u_{t_i}(F)] ds \right) (W(t_{i+1}) - W(t_i)) \\
&= a_n + b_n.
\end{aligned}
$$

The continuity of $t \to u_t(x)$ in $L^2(\Omega)$ implies that for each $x \in \mathbb{R}^m$ the Riemann sums

$$\sum_{i=0}^{n-1} u_{t_i}(x)(W(t_{i+1}) - W(t_i))$$

converge in $L^2(\Omega)$, as $|\pi|$ tends to zero, to the Itô integral $\int_0^1 u_t(x) dW_t$. Moreover, we have for $|x|, |y| \leq K$,

$$E\left( \left| \sum_{i=0}^{n-1} [u_{t_i}(x) - u_{t_i}(y)] (W(t_{i+1}) - W(t_i)) \right|^p \right) \leq c_p c_K |x - y|^\alpha.$$

Hence, by Lemma 3.2.2 we deduce that $a_n$ converges in probability to $\int_0^1 u_t(x) \circ dW_t \big|_{x=F}$. Now we study the convergence of $b_n$. For each fixed $x \in \mathbb{R}^m$, the sums

$$R^\pi(x) := \sum_{i=0}^{n-1} \frac{1}{t_{i+1} - t_i} \left( \int_{t_i}^{t_{i+1}} [u_s(x) - u_{t_i}(x)] ds \right) (W(t_{i+1}) - W(t_i))$$

converge in probability, as $|\pi|$ tends to zero, to the difference

$$V(x) := \int_0^1 u_t(x) \circ dW_t - \int_0^1 u_t(x) dW_t .$$

So, it suffices to show that $R^\pi(F)$ converges in probability, as $|\pi|$ tends to zero, to $V(F)$. This convergence follows from Lemma 3.2.2 and the following estimates, where $0 < \varepsilon < 1$ verifies $\varepsilon\alpha > m$, and $|x|, |y| \leq K$

$$\|R^\pi(x) - R^\pi(x)\|_{p\varepsilon}$$

$$\leq \sum_{i=0}^{n-1} \frac{1}{t_{i+1} - t_i}$$

$$\times \int_{t_i}^{t_{i+1}} \|[u_s(x) - u_{t_i}(x) - u_s(y) + u_{t_i}(y)](W(t_{i+1}) - W(t_i))\|_{p\varepsilon} ds$$

$$\leq c_{p,\varepsilon} \sum_{i=0}^{n-1} |t_{i+1} - t_i|^{\frac{1}{2}} \sup_{s \in [t_i, t_{i+1}]} [E(|u_s(x) - u_{t_i}(x) - u_s(y) + u_{t_i}(y)|^p)]^{\frac{1}{p}}$$

$$\leq c_{p,\varepsilon} c_K |x - y|^{\frac{\alpha}{p}} .$$

$\square$

It is also possible to establish substitution formulas similar to those appearing in Theorems 3.2.9 and 3.2.10 for random fields $u_t(x)$ which are not adapted. In this case additional regularity assumptions are required. More precisely we need regularity in $x$ of $D_s u_t(x)$ for Theorem 3.2.9 and of $(\nabla u(x))_t$ for Theorem 3.2.10.

One can  also compute the differentials of processes of the form $F_t(X_t)$ where $\{X_t, t \in [0,1]\}$ and $\{F_t(x), t \in [0,1], x \in \mathbb{R}^m\}$ are generalized continuous semimartingales, i.e., they have a bounded variation component and an indefinite Skorohod (or Stratonovich) integral  component. This type of change-of-variables formula are known as Itô-Ventzell formulas. We show below a formula of this type that will be used to solve anticipating stochastic differential equations. We first state the following differentiation rule (see also Exercise 1.3.6).

**Lemma 3.2.3** *Let $F = (F^1, \ldots, F^m)$ be a random vector whose components belong to $\mathbb{D}^{1,p}$, $p > 1$ and suppose that $|F| \leq K$. Consider a measurable random field $u = \{u(x), x \in \mathbb{R}^m\}$ with continuously differentiable*

*paths, such that for any $x \in \mathbb{R}^m$, $u(x) \in \mathbb{D}^{1,r}$, $r > 1$, and the derivative*
*$Du(x)$ has a continuous version as an $H$-valued process. Suppose we have*

$$E\left(\sup_{|x|\leq K} \left[|u(x)|^r + \|Du(x)\|_H^r\right]\right) < \infty, \tag{3.44}$$

$$E\left(\sup_{|x|\leq K} |\nabla u(x)|^q\right) < \infty, \tag{3.45}$$

*where $\frac{1}{p} + \frac{1}{q} = \frac{1}{r}$. Then the composition $u(F)$ belongs to $\mathbb{D}^{1,r}$, and we have*

$$D\left(u(F)\right) = \sum_{i=1}^{m} \partial_i u(F) DF^i + (Du)(F). \tag{3.46}$$

*Proof:*    Consider an approximation of the identity $\psi_n$, and define

$$u_n(x) = \int_{\mathbb{R}^m} u(y)\psi_n(x-y)dy.$$

The sequence of random variables $u_n(F)$ converges almost surely and in in $L^r(\Omega)$ to $u(F)$ because of Condition (3.44). On the other hand, $u_n(F) \in \mathbb{D}^{1,r}$ and

$$
\begin{aligned}
D\left(u_n(F)\right) &= D\left(\int_{\mathbb{R}^m} u(y)\psi_n(F-y)dy\right) \\
&= \int_{\mathbb{R}^m} Du(y)\psi_n(F-y)dy + \sum_{i=1}^{m} \int_{\mathbb{R}^m} u(y)\partial_i\psi_n(F-y)DF^i dy \\
&= \int_{\mathbb{R}^m} Du(y)\psi_n(F-y)dy + \sum_{i=1}^{m} DF^i \int_{\mathbb{R}^m} \partial_i u(y)\psi_n(F-y)dy.
\end{aligned}
$$

Again by conditions (3.44) and (3.45), this expression converges in $L^r(\Omega; H)$ to the right-hand side of (3.46). $\qquad\square$

Let $W$ be an $N$-dimensional Wiener process. Consider two stochastic processes of the form:

$$X_t^i = X_0^i + \sum_{j=1}^{N} \int_0^t u_s^{ij} \circ dW_s^j + \int_0^t v_s^i ds, \; 1 \leq i \leq m$$

and

$$F_t(x) = F_0(x) + \sum_{j=1}^{N} \int_0^t H_s^j(x) \circ dW_s^j + \int_0^t G_s(x)ds, \; x \in \mathbb{R}^m.$$

We introduce the following set of hypotheses:

(A1) For all $1 \leq j \leq N$, $1 \leq i \leq m$, $X_0^i \in \mathbb{D}^{1,2}$, $u^{ij} \in \mathbb{L}_S^{2,4}$, $v^i \in \mathbb{L}^{1,2}$, and the process $X$ is continuous and bounded by a constant $K$. We assume also that $u$ is bounded.

(A2) $x \to F_t(x)$ is twice differentiable for all $t$; the processes $F_t(x)$, $\nabla F_t(x)$ and $\nabla^2 F_t(x)$ are continuous in $(t,x)$; $F(x) \in \mathbb{L}^{1,2}$, and there exist two versions of $D_s F_t(x)$ which are differentiable in $x$ and such that the functions $D_s F_t(x)$ and $\nabla(D_s F_t)(x)$ are continuous in the regions $\{s \leq t \leq 1, x \in \mathbb{R}^m\}$ and $\{t \leq s \leq 1, x \in \mathbb{R}^m\}$, respectively.

(A3) $x \to H_t^j(x)$ is differentiable for all $t$; the processes $H_t(x)$ and $\nabla H_t(x)$ are continuous in $(t,x)$; $H^j(x) \in \mathbb{L}_S^{2,4}$, and there exist two versions of $D_s H_t(x)$ which are differentiable in $x$ and such that the functions $D_s H_t(x)$ and $\nabla(D_s H_t)(x)$ are continuous in the regions $\{s \leq t \leq 1, x \in \mathbb{R}^m\}$ and $\{t \leq s \leq 1, x \in \mathbb{R}^m\}$, respectively.

(A4) The function $x \to G_t(x)$ is continuous for all $t$, and $G(x) \in \mathbb{L}^{1,2}$.

(A5) The following estimates hold:

$$E\left( \sup_{|x| \leq K, t \in [0,1]} \left( |F_t(x)|^4 + |\nabla F_t(x)|^4 + |\nabla^2 F_t(x)|^4 \right) \right) < \infty,$$

$$E\left( \sup_{|x| \leq K, t \in [0,1]} \left( |H_t(x)|^4 + |\nabla H_t(x)|^4 + |G_t(x)|^4 \right) \right) < \infty,$$

$$\int_0^1 E\left( \sup_{|x| \leq K, t \in [0,1]} \left( |D_s F_t(x)|^4 + |\nabla(D_s F_t)(x)|^4 \right) \right) ds < \infty,$$

$$\int_0^1 E\left( \sup_{|x| \leq K, t \in [0,1]} \left( |D_s H_t(x)|^4 + |\nabla(D_s H_t)(x)|^4 \right) \right) ds < \infty.$$

**Theorem 3.2.11** *Assume that the processes $X_t$ and $F_t(x)$ satisfy the above hypotheses. Then $\left\langle \nabla F_t(X_t), u_t^j \right\rangle$ and $H_t^j(X_t)$ are elements of $\mathbb{L}_1^{1,2}$ for all $j = 1, \ldots, N$, and*

$$F_t(X_t) = F_0(X_0) + \sum_{j=1}^N \int_0^t \left\langle \nabla F_s(X_s), u_s^j \right\rangle \circ dW_s^j + \int_0^t \left\langle \nabla F_s(X_s), v_s \right\rangle ds$$

$$+ \sum_{j=1}^N \int_0^t H_s^j(X_s) \circ dW_s^j + \int_0^t G_s(X_s) ds.$$

*Proof:*     To simplify then notation we will assume $N = m = 1$. The proof will be done in several steps.

*Step 1:*     Consider an approximation of the identity $\psi_n$ on $\mathbb{R}$. For each $x \in \mathbb{R}$ we can apply the multidimensional change-of-variables formula for the Stratonovich integral (Theorem 3.2.6) to the process $F_t(x)\psi_n(X_t - x)$ and we obtain

$$
\begin{aligned}
F_t(x)\psi_n(X_t - x) \;=\; & F_0(x)\psi_n(X_0 - x) + \int_0^t H_s(x)\psi_n(X_s - x) \circ dW_s \\
& + \int_0^t G_s(x)\psi_n(X_s - x)ds \\
& + \int_0^t F_s(x)\psi_n'(X_s - x)u_s \circ dW_s \\
& + \int_0^t F_s(x)\psi_n'(X_s - x)v_s ds.
\end{aligned}
$$

If we express the above Stratonovich integrals in terms of the corresponding Skorohod integrals we get

$$
\begin{aligned}
F_t(x)\psi_n(X_t - x) \;=\; & F_0(x)\psi_n(X_0 - x) + \int_0^t H_s(x)\psi_n(X_s - x)dW_s \\
& + \int_0^t \left[ G_s(x) + \frac{1}{2}(\nabla H)_s(x) \right] \psi_n(X_s - x)ds \\
& + \frac{1}{2} \int_0^t H_s(x)\psi_n'(X_s - x)\,(\nabla X)_s\, ds \\
& + \int_0^t F_s(x)\psi_n'(X_s - x)u_s dW_s \\
& + \frac{1}{2} \int_0^t (\nabla F)_s\,(x)\psi_n'(X_s - x)u_s ds \\
& + \int_0^t F_s(x)\psi_n'(X_s - x)\left( \frac{1}{2}(\nabla u)_s + v_s \right) ds \\
& + \frac{1}{2} \int_0^t F_s(x)\psi_n''(X_s - x)(\nabla X)_s u_s ds.
\end{aligned}
$$

Grouping the Skorohod integral terms and the Lebesgue integral terms together we can write

$$
F_t(x)\psi_n(X_t - x) = F_0(x)\psi_n(X_0 - x) + \int_0^t \alpha_n(s, x)dW_s + \int_0^t \beta_n(s, x)ds,
$$

where

$$
\alpha_n(t, x) = \psi_n(X_t - x)H_t(x) + \psi_n'(X_t - x)u_t F_t(x)
$$

and

$$\begin{aligned}
\beta_n(t,x) &= \psi_n(X_t - x)\left(G_t(x) + \frac{1}{2}(\nabla H)_t(x)\right) \\
&\quad + \psi'_n(X_t - x)\left(\frac{1}{2}H_t(x)\,(\nabla X)_t\right) \\
&\quad + F_t(x)\left(\frac{1}{2}(\nabla u)_t + v_t\right) + \frac{1}{2}u_t\,(\nabla F)_t\,(x) \\
&\quad + \frac{1}{2}\psi''_n(X_t - x)F_t(x)\,(\nabla X)_t\,u_t.
\end{aligned}$$

*Step 2:* We claim that the processes $H_t(X_t)$ and $F'_t(X_t)u_t$ belong to $\mathbb{L}_1^{1,2}$. In fact, first notice that Hypotheses (A2) and (A3) imply that for each $x$, $H_t(x)$ and $F'_t(x)u_t$ belong to $\mathbb{L}^{1,2}$, and $D_s(F'_t(x)) = (D_s F_t)'(x)$. Then we can apply Lemma 3.2.3 (suitably extended to stochastic processes) to $H_t(x)$ and $X_t$ and also to $F'_t(x)u_t$ and $X_t$ to deduce that $H_t(X_t)$ and $F'_t(X_t)u_t$ belong to $\mathbb{L}^{1,2}$, taking into account the following estimates

$$\int_0^1 E(H_s^2(X_s))ds \le \int_0^1 E\left(\sup_{|x|\le K} H_s^2(x)\right)ds < \infty,$$
$$\int_0^1 E\left(\sup_{|x|\le K}\|DH_s(x)\|_H^2\right)ds < \infty,$$
$$\int_0^1 E\left(\sup_{|x|\le K}|H'_s(x)|^4\right)ds < \infty,$$

and

$$\int_0^1 E((F'_s(X_s)u_s)^2)ds \le \|u\|_\infty^2 \int_0^1 E\left(\sup_{|x|\le K}|F'_s(x)|^2\right)ds < \infty,$$
$$\int_0^1 E(|F'_s(X_s)|^2\,\|Du_s\|_H^2)ds$$
$$\le \left(\int_0^1 E\left(\sup_{|x|\le K}|F'_s(x)|^4\right)ds\int_0^1 E\left(\|Du_s\|_H^4\right)ds\right)^{\frac{1}{2}} < \infty,$$
$$\int_0^1 E(|F''_s(X_s)|^2\,\|Du_s\|_H^2\,u_s^2)ds \le \|u\|_\infty^2$$
$$\times \left(\int_0^1 E\left(\sup_{|x|\le K}|F''_s(x)|^4\right)ds\int_0^1 E\left(\|Du_s\|_H^4\right)ds\right)^{\frac{1}{2}} < \infty,$$
$$\int_0^1 E(\|(DF_s)'\,(X_s)\|_H^2\,u_s^2)ds$$
$$\le \|u\|_\infty^2 \int_0^1 E\left(\sup_{|x|\le K}\|(DF_s)'\,(x)\|_H^2\right)ds < \infty.$$

The derivatives of these processes are given by

$$D_s[H_t(X_t)] = H'_t(X_t)D_s X_t + (D_s H_t)(X_t)$$

and

$$D_s[F'_t(X_t)u_t] = F''_t(X_t)D_s X_t u_t + (D_s F_t)'(X_t)u_t + F'_t(X_t)D_s u_t.$$

Then, from our hypotheses it follows that the processes $H_t(X_t)$ and $F_t'(X_t)u_t$ belong to $\mathbb{L}_1^{1,2}$, and

$$(\nabla (H.(X.)))_t = H_t'(X_t) \, (\nabla X)_t + (\nabla H)_t(X_t)$$

and

$$(\nabla (F_.'(X.)u.))_t = F_t''(X_t) \, (\nabla X)_t \, u_t + (\nabla F)_t'(X_t)u_t + F_t'(X_t) \, (\nabla u)_t \, .$$

*Step 3:*   We claim that

$$\int_{\mathbb{R}} \left( F_t(x)\psi_n(X_t - x) - F_0(x)\psi_n(X_0 - x) - \int_0^t \beta_n(s,x)ds \right) dx \quad (3.47)$$

converges in $L^1(\Omega)$ to the process

$$
\begin{aligned}
\Phi_t \;=\; & F_t(X_t) - F_0(X_0) - \int_0^t \Big[ G_s(X_s) + \frac{1}{2}(\nabla H)_s(X_s) \\
& + \frac{1}{2}H_s'(X_s)\,(\nabla X)_s + \frac{1}{2}(\nabla F)_s'(X_s)u_s + F_s'(X_s)\,(\nabla u)_s \\
& + F_s''(X_s)\,(\nabla X)_s \, u_s + F_s'(X_s)v_s \Big] ds.
\end{aligned}
$$

In fact, the convergence for each $(s, \omega)$ follows from the continuity assumption on $x$ of the processes $G_s(x)$, $(\nabla H)_s(x)$, $(\nabla F)_s'(x)$, and the fact that $H_s(x)$ is of class $C^1$, and $F_s'(x)$ is of class $C^2$. The $L^1(\Omega)$ convergence follows by the dominated convergence theorem because we have

$$E\left( \int_0^1 \sup_n \int_{\mathbb{R}} |\beta_n(s,x)| \, dx ds \right) < \infty.$$

From Step 2 we deduce

$$
\begin{aligned}
\Phi_t \;=\; & F_t(X_t) - F_0(X_0) - \int_0^t [G_s(X_s) + F_s'(X_s)v_s]\, ds \\
& - \int_0^t [H_s(X_s) + F_s'(X_s)u_s] \circ dW_s \\
& + \int_0^t [H_s(X_s) + F_s'(X_s)u_s]\, dW_s.
\end{aligned}
$$

*Step 4:*   Finally, we have

$$\int_{\mathbb{R}} \alpha_n(s,x)dx = \int_{\mathbb{R}} H_s(x)\psi_n(X_s - x)dx + \int_{\mathbb{R}} F_s(x)\psi_n'(X_s - x)u_s dx.$$

The integrals $\int_{\mathbb{R}} \alpha_n(s,x)dx$ converge as $n$ tends to infinity in the norm of $\mathbb{L}^{1,2}$ to $H_s(X_s) + F_s'(X_s)u_s$. Hence, the Skorohod integral

$$\int_0^t \left( \int_{\mathbb{R}} \alpha_n(s,x)dx \right) dW_s \quad (3.48)$$

converges in $L^2(\Omega)$ to

$$\int_0^t [H_s(X_s) + F_s'(X_s)u_s]\, dW_s.$$

By Fubini's theorem, (3.48) coincides with (3.47). Consequently, we get

$$\Phi_t = \int_0^t [H_s(X_s) + F_s'(X_s)u_s]\, dW_s,$$

which completes the proof.     □

## Exercises

**3.2.1** Consider the process $u$ defined by

$$u(t) = \mathbf{1}_{\{W(1)>0\}}(\mathbf{1}_{[0,\frac{1}{2}]}(t) - \mathbf{1}_{(\frac{1}{2},1]}(t)).$$

Show that this process is Skorohod integrable but the process $u\mathbf{1}_{[0,\frac{1}{2}]}$ is not.

**3.2.2** Let $u \in \mathbb{L}^{1,2}$, and fix $0 \le s < t \le 1$. Show that

$$E\left(\left(\int_s^t u_r dW_r\right)^2 |\mathcal{F}_{[s,t]^c}\right)$$
$$= E\left(\int_s^t u_r^2 dr + \int_s^t\int_s^t D_v u_r D_r u_v dr dv |\mathcal{F}_{[s,t]^c}\right).$$

**3.2.3** Consider the process $u(t) = \text{sign}(W_1 - t)\exp(W_t - \frac{t}{2})$, $0 \le t \le 1$. Show that this process is Skorohod integrable on any interval, and

$$\int_0^t u_s dW_s = \text{sign}(W_1 - t)\exp(W_t - \frac{t}{2}) - \text{sign}W_1.$$

*Hint:* Fix a smooth random variable $G$ and compute $E(\int_0^1 D_t G u_t dt)$ using Girsanov's theorem.

**3.2.4** Let $u \in \mathbb{L}^{2,2}$ be a process satisfying

$$\sup_{s,t\in[0,1]}\left[|E(D_s u_t)| + E\int_0^1 |D_s D_r u_t|^2 dr\right] < \infty.$$

Assume in addition that $E\int_0^1 |u_t|^p dt < \infty$ for some $p > 4$. Show that the Skorohod integral $\{\int_0^t u_s dW_s, t \in [0,1]\}$ possesses a continuous version.

**3.2.5** Let $u \in L^2([0,1] \times \Omega)$ be a stochastic process such that for any $t \in [0,1]$ the random variable $u_t$ belongs to the finite sum of chaos $\oplus_{n=0}^N \mathcal{H}_n$.

Show that the Skorohod integral $\{\int_0^t u_s dW_s, t \in [0,1]\}$ possesses a continuous version.

*Hint:* Use the fact that all the $p$ norms are equivalent on a finite chaos and apply a deterministic change of time (see Imkeller [148]).

**3.2.6** Let $u$ be a process in the space $\mathbb{L}^{2,4} = L^4([0,1]; \mathbb{D}^{2,4})$. Using the Gaussian formula (A.1) and Proposition 3.2.1, prove the following inequality:

$$E\left(\int_0^1 \left(\int_0^1 D_r u_\theta dW_\theta\right)^2 dr\right)^2 \leq C\left[E\left(\int_0^1 \int_0^1 (D_r u_\theta)^2 dr d\theta\right)^2\right.$$
$$\left. + E\left(\int_0^1 \left(\int_0^1 \int_0^1 (D_\alpha D_r u_\theta)^2 d\alpha dr\right)^2 d\theta\right)\right].$$

**3.2.7** Consider a random field $\{u_t(x), 0 \leq t \leq 1, x \in G\}$, where $G$ is a bounded open subset of $\mathbb{R}^m$, such that $u \in L^2(G \times [0,1] \times \Omega)$. Suppose that for each $x \in \mathbb{R}^m$, $u(x) \in \text{Dom } \delta$ and $E(\int_G |\delta(u(x))|^2 dx) < \infty$. Show that the process $\{\int_G u_t(x) dx, t \in [0,1]\}$ is Skorohod integrable and

$$\delta\left(\int_G u_t(x) dx\right) = \int_G \delta(u_\cdot(x)) dx.$$

**3.2.8** Let $X = \{X_t, t \in [0,1]\}$ and $Y = \{Y_t, t \in [0,1]\}$ be continuous processes in the space $\mathbb{L}^{1,2}_{\text{loc}}$ such that $X$ is adapted and $Y$ is backward adapted (that is, $Y_t$ is $\mathcal{F}_{[t,1]}$-measurable for any $t$). Consider a $C^1$ function $\Phi : \mathbb{R}^2 \to \mathbb{R}$. Show that the process $\{\Phi(X_t, Y_t), t \in [0,1]\}$ belongs to $\mathbb{L}^{1,2}_{\text{loc}}$ and that the sums

$$\sum_{i=0}^{n-1} \Phi(X(t_i), Y(t_{i+1}))(W(t_{i+1}) - W(t_i))$$

converge in probability to the Skorohod integral $\int_0^1 \Phi(X_t, Y_t) dW_t$ (see Pardoux and Protter [281]).

**3.2.9** Let $f, g : \mathbb{R} \to \mathbb{R}$ be bounded functions of bounded variation. Consider the stochastic process $X(t) = f(W(t))g(W(1) - W(t))$. Show that $X$ is Skorohod integrable, and compute $E(\delta(X)^2)$.

**3.2.10** Suppose that $W = \{W(h), h \in H = L^2(T, \mathcal{B}, \mu)\}$ is a centered isonormal Gaussian process on the complete probability space $(\Omega, \mathcal{F}, P)$. Assume $\mathcal{F} = \sigma(W)$. Let $G$ be a bounded domain of $\mathbb{R}^m$ with Lipschitz boundary, and let $A \in \mathcal{B}$, $\mu(A) < \infty$. Let $\{u(x), x \in G\}$ be an $\mathcal{F}_{A^c}$-measurable random field with continuously differentiable paths such that

$$E\left(\sup_{x \in G} |\nabla u(x)|^4\right) < \infty \quad \text{and} \quad E\left(\sup_{x \in G} |u(x)|^2\right) < \infty.$$

Set $h = 1_A$ and let $F \in \mathbb{D}^{1,4}$. Show that

(a) $u(F) \in \mathbb{D}^{2,h}$ and $D^h(u(F)) = \sum_{i=1}^m \partial_i u D^h F^i$.

(b) $u(F) 1_A$ belongs to the domain of $\delta$ and

$$\delta(u(F) 1_A) = u(F) W(A) - \sum_{i=1}^m \partial_i u(F) D^h F^i.$$

*Hint:* Approximate $u(F)$ by convolution with an approximation of the identity.

**3.2.11** Let $p \in (2, 4)$, and $\alpha = \frac{2p}{4-p}$. Consider a process $u = \{u_t, t \in [0,1]\}$ in $\mathbb{L}^F \cap L^\alpha([0,1] \times \Omega)$. Using the Itô's formula for the Skorohod integral show the following estimate

$$E \left| \int_s^t u_r dW_r \right|^p \leq C_p (t-s)^{\frac{p}{2}-1} \left( E \int_s^t |u_r|^\alpha dr + \right.$$

$$+ E \int_s^t \int_\theta^t (D_r u_\theta)^2 \, dr d\theta$$

$$\left. + E \int_s^t \int_\theta^t \int_\theta^t (D_\sigma D_r u_\theta)^2 \, dr d\sigma d\theta \right).$$

**3.2.12** Let $u = \{u_t, t \in [0,1]\}$ be a process in $\mathbb{L}^F$ such that $E\left(\int_0^1 |u_t|^p dt\right) < \infty$ for some $p > 2$. Show that the process $X_t = \int_0^t u_s dW_s$ has a continuous version whose trajectories are Hölder continuous of order less that $\frac{p-2}{4p}$.

**3.2.13** Show that the process $u^{n,k}$ defined in (3.27) belongs to $\mathbb{L}_{(4)}^{2,2}$.

## 3.3  Anticipating stochastic differential equations

The anticipating stochastic calculus developed in Section 3.2 can be applied to the study of stochastic differential equations where the solutions are nonadapted processes. Such kind of equations appear, for instance, when the initial condition is not independent of the Wiener process, or when we impose a condition relating the values of the process at the initial and final times. In this section we discuss several simple examples of anticipating stochastic differential equations of Skorohod and Stratonovich types.

### 3.3.1  Stochastic differential equations in the Sratonovich sense

In this section we will present examples of stochastic differential equations formulated using the Stratonovich integral. These equations can be solved taking into account that this integral satisfies the usual differentiation rules.

(A) *Stochastic differential equations with random initial condition*

Suppose that $W = \{W(t), t \in [0,1]\}$ is a $d$-dimensional Brownian motion on the canonical probability space $(\Omega, \mathcal{F}, P)$. Let $A_i : \mathbb{R}^m \to \mathbb{R}^m$, $0 \leq i \leq d$, be continuous functions. Consider the stochastic differential equation

$$X_t = X_0 + \sum_{i=1}^{d} \int_0^t A_i(X_s) \circ dW_s^i + \int_0^t A_0(X_s)ds, \qquad t \in [0,1], \quad (3.49)$$

with an anticipating initial condition $X_0 \in L^0(\Omega; \mathbb{R}^m)$. The stochastic integral is interpreted in the Stratonovich sense. There is a straightforward method to construct a solution to this equation, assuming that the coefficients are smooth enough. The idea is to compose the stochastic flow associated with the coefficients with the random initial condition $X_0$.

Suppose that the coefficients $A_i$, $1 \leq i \leq d$, are continuously differentiable with bounded derivatives. Define $B = A_0 + \frac{1}{2} \sum_{k=1}^{m} \sum_{i=1}^{d} A_i^k \partial_k A_i$, and suppose moreover that $B$ is Lipschitz.

Let $\{\varphi_t(x), t \in [0,1]\}$ denote the stochastic flow associated with the coefficients of Eq. (3.49), that is, the solution to the following stochastic differential equation with initial value $x \in \mathbb{R}^m$:

$$
\begin{aligned}
\varphi_t(x) &= x + \sum_{i=1}^{d} \int_0^t A_i(\varphi_s(x)) \circ dW_s^i + \int_0^t A_0(\varphi_s(x))ds \\
&= x + \sum_{i=1}^{d} \int_0^t A_i(\varphi_s(x))dW_s^i + \int_0^t B(\varphi_s(x))ds. \quad (3.50)
\end{aligned}
$$

We know (see, for instance Kunita [173]) that there exists a version of $\varphi_t(x)$ such that $(t,x) \hookrightarrow \varphi_t(x)$ is continuous, and we have

$$E\left(|\varphi_t(x) - \varphi_s(y)|^p\right) \leq C_{p,K}(|t - s|^{\frac{p}{2}} + |x - y|^p)$$

for all $s, t \in [0,1]$, $|x|, |y| \leq K$, $p \geq 2$, $K > 0$. Then we can establish the following result.

**Theorem 3.3.1** *Assume that $A_i$, $1 \leq i \leq d$, are of class $C^3$, and $A_i$, $1 \leq i \leq d$, and $B$ have bounded partial derivatives of first order. Then for any random vector $X_0$, the process $X = \{\varphi_t(X_0), t \in [0,1]\}$ satisfies the anticipating stochastic differential equation (3.49).*

*Proof:* Under the assumptions of the theorem we know that for any $x \in \mathbb{R}^m$ Eq. (3.50) has a unique solution $\varphi_t(x)$. Set $X = \{\varphi_t(X_0), t \in [0,1]\}$. By a localization argument we can assume that the coefficients $A_i$, $1 \leq i \leq d$, and $B$ have compact support. Then, it suffices to show that for any $i = 1, \ldots, d$ the process $u_i(t,x) = A_i(\varphi_t(x))$ satisfies the hypotheses of

Theorem 3.2.10. Condition (h1) is obvious and one easily check condition (h4) by means of Itô's formula. This completes the proof.     □

The uniqueness of a solution for Eq. (3.49) is more involved. We will show that there is a unique solution in the class of processes $\mathbb{L}_{2,\text{loc}}^{1,4}(\mathbb{R}^m)$, assuming some additional regulatiry conditions on the initial condition $X_0$.

**Lemma 3.3.1** *Suppose that $X_0 \in \mathbb{D}_{\text{loc}}^{1,p}(\mathbb{R}^m)$ for some $p > q \geq 2$, and assume that the coefficients $A_i$, $1 \leq i \leq d$, and $B$ are of class $C^2$ with bounded partial derivatives of first order. Then, we claim that $\{\varphi_t(X_0), t \in [0,1]\}$ belongs to $\mathbb{L}_{2,\text{loc}}^{1,q}(\mathbb{R}^m)$.*

*Proof:*     Let $(\Omega^n, X_0^n)$ be a localizing sequence for $X_0$ in $\mathbb{D}^{1,p}(\mathbb{R}^m)$. Set

$$G^{n,k,M} = \Omega^n \cap \{|X_0^n| \leq k\} \cap \{\sup_{|x|\leq K, t\in[0,1]} |\varphi_t(x)| \leq M\}.$$

On the set $G^{n,k,M}$ the process $\varphi_t(X_0)$ coincides with $\varphi_t^M(X_0^n \beta_k(X_0^n))$, where $\beta_k$ is a smooth function such that $0 \leq \beta_k \leq 1$, $\beta_k(x) = 1$ if $|x| \leq k$, and $\beta_k(x) = 0$ if $|x| \geq k+1$, and $\varphi_t^M(x)$ is the stochastic flow associated with the coefficients $\beta_M A_i$, $1 \leq i \leq d$, and $\beta_M B$. Hence, we can assume that the coefficients $A_i$, $1 \leq i \leq d$, and $B$ are bounded and have bounded derivatives up to the second order, $X_0 \in \mathbb{D}^{1,p}(\mathbb{R}^m)$, and $X_0$ is bounded by $k$. The following estimates hold

$$E\left(\sup_{|x|\leq K, t\in[0,1]} [|\varphi_t(x)|^r + \|D\varphi_t(x)\|_H^r]\right) < \infty, \qquad (3.51)$$

$$E\left(\sup_{|x|\leq K, t\in[0,1]} |\nabla\varphi_t(x)|^q\right) < \infty, \qquad (3.52)$$

for any $r \geq 2$. The estimates follow from our assumptions on the coefficients $A_i$, $1 \leq i \leq d$, and $B$, taking into account that

$$D_s^j[\varphi_t(x)] = A_j(\varphi_s(x)) + \sum_{k=1}^{m}\sum_{l=1}^{d}\int_s^t \partial_k A_l(\varphi_r(x)D_s^j[\varphi_r^k(x)] dW_r^l$$

$$+ \sum_{k=1}^{m}\int_s^t \partial_k B(\varphi_r(x)D_s^j[\varphi_r^k(x)] dr, \qquad (3.53)$$

for $1 \leq j \leq d$ and $0 \leq s \leq t \leq 1$. Hence, from Lemma 3.2.3 we obtain that $\{\varphi_t(X_0), t \in [0,1]\}$ belongs to $\mathbb{L}^{1,q}(\mathbb{R}^m)$ and

$$D_s(\varphi_t(X_0)) = \nabla\varphi_t(X_0)D_sX_0 + (D\varphi_t)(X_0).$$

Finally, from Eq. (3.53) for the derivative of $\varphi_t(X_0)$ it is easy to check that $\{\varphi_t(X_0), t \in [0,1]\}$ belongs to $\mathbb{L}_2^{1,q}(\mathbb{R}^m)$, and

$$(D(\varphi(X_0)))_t^{+,-} = \nabla\varphi_t(X_0)D_tX_0 + (D\varphi)_t^{+,-}(X_0).$$

□

**Theorem 3.3.2** *Assume that $A_i$, $1 \leq i \leq d$, and $B$ are of class $C^4$, with bounded partial derivatives of first order and $X_0$ belongs to $\mathbb{D}^{1,p}_{\mathrm{loc}}(\mathbb{R}^m)$ for some $p > 4$. Then the process $X = \{\varphi_t(X_0), t \in [0,1]\}$ is the unique solution to Eq. (3.49) in $\mathbb{L}^{1,4}_{2,\mathrm{loc}}(\mathbb{R}^m)$ which is continuous.*

*Proof:* By Lemma 3.3.1 we already know that $X$ belongs to $\mathbb{L}^{1,4}_{2,\mathrm{loc}}(\mathbb{R}^m)$. Let $Y$ be another continuous solution in this space. Suppose that $(\Omega^n, X_0^n)$ is a localizing sequence for $X_0$ in $\mathbb{D}^{1,p}_{\mathrm{loc}}(\mathbb{R}^m)$, and $(\Omega^{n,1}, Y^n)$ is a localizing sequence for $Y$ in $\mathbb{L}^{1,4}_{2,\mathrm{loc}}(\mathbb{R}^m)$. Set

$$\Omega^{n,k,M} = \Omega^n \cap \Omega^{n,1} \cap \{|X_0^n| \leq k\} \cap \{\sup_{|x| \leq K, t \in [0,1]} |\varphi_t(x)| \leq M\}$$

$$\cap \{\sup_{t \in [0,1]} |Y_t| \leq M\}.$$

The processes $Y_t^n \beta_M(Y_t^n)$, $(\beta_M A_i)(Y_t^n)$, $0 \leq i \leq d$, satisfy hypothesis (A1). On the set $\Omega^{n,k,M}$ we have

$$Y_t = X_0 + \sum_{i=1}^{d} \int_0^t A_i(Y_s) \circ dW_s^i + \int_0^t A_0(Y_s)ds, \qquad t \in [0,1].$$

Let us denote by $\varphi_t^{-1}(x)$ the inverse of the mapping $x \hookrightarrow \varphi_t(x)$. By the classical Itô's formula we can write

$$\varphi_t^{-1}(x) = x - \sum_{i=1}^{d} \int_0^t (\varphi_s'(\varphi_s^{-1}(x)))^{-1} A_i(x) \circ dW_s^i$$

$$- \int_0^t (\varphi_s'(\varphi_s^{-1}(x)))^{-1} A_0(x)ds.$$

We claim that the processes $F_t(x) = \varphi_t^{-1}(x)$, $H_t^i(x) = (\varphi_s'(\varphi_s^{-1}(x)))^{-1} A_i(x)$, and $G_t(x) = (\varphi_s'(\varphi_s^{-1}(x)))^{-1} A_0(x)$ with values in the space of $m \times m$ matrices satisfy hypotheses (A2) to (A5). This follows from the properties of the stochastic flow assicated with the coefficients $A_i$.

Consequently, we can apply Theorem 3.2.11 and we obtain, on the set $\Omega^{n,k,M}$

$$\varphi_t^{-1}(Y_t) = X_0 + \sum_{i=1}^{d} \int_0^t \left(\varphi_s^{-1}\right)'(Y_s) A_i(Y_s) \circ dW_s^i$$

$$+ \int_0^t \left(\varphi_s^{-1}\right)'(Y_s) A_0(Y_s)ds$$

$$- \sum_{i=1}^{d} \int_0^t (\varphi_s'(\varphi_s^{-1}(Y_s)))^{-1} A_i(Y_s) \circ dW_s^i$$

$$- \int_0^t (\varphi_s'(\varphi_s^{-1}(Y_s)))^{-1} A_0(Y_s)ds.$$

Notice that

$$\left(\varphi_s^{-1}\right)'(x) = -(\varphi_s'(\varphi_s^{-1}(x)))^{-1}.$$

Hence, we get $\varphi_t^{-1}(Y_t) = X_0$, that is, $Y_t = \varphi_t(X_0)$ which completes the proof of the uniqueness.    □

In [272] Ocone and Pardoux consider equations with a random drift and with time-dependent coefficients. In such a case one has to remove the drift before composing with the random initial condition. The solution is given by $X_t = \psi_t(Y_t)$, where $\{\psi_t(x), t \in [0, 1]\}$ denotes the flow associated with Eq. (3.49) with $b \equiv 0$, i.e.,

$$\psi_t(x) = x + \sum_{i=1}^{d} \int_0^t A_i(\psi_s(x)) \circ dW_s^i,$$

and $Y_t$ solves the following ordinary differential equation parametrized by $\omega$:

$$dY_t = (\psi_t^* A_0)(Y_t)dt, \qquad Y_0 = X_0,$$

where $(\psi_t^* A_0)(x) = \left[\psi_t'(x)\right]^{-1} A_0(\psi_t(x))$.

Then, a formal calculation based on Theorem 3.2.5 shows that the process $\{\psi_t(Y_t), t \geq 0\}$ solves Eq. (3.49). In fact, we have

$$\begin{aligned}
d\left[\psi_t(Y_t)\right] &= \sum_{i=1}^{d} A_i(\psi_t(Y_t)) \circ dW_t^i + \psi_t'(Y_t)Y_t'dt \\
&= \sum_{i=1}^{d} A_i(\psi_t(Y_t)) \circ dW_t^i + A_0(\psi_t(Y_t))dt.
\end{aligned}$$

(B) *One-dimensional stochastic differential equations with random coefficients*

Consider the one-dimensional equation

$$X_t = X_0 + \int_0^t \sigma(X_s) \circ dW_s + \int_0^t b(X_s)ds, \qquad (3.54)$$

where the coefficients $\sigma$ and $b$ and initial condition $X_0$ are random.

Suppose that the coefficients and the intial condition are deterministic and assume that $\sigma$ is of class $C^2$ with bounded first and second derivatives, and that $b$ is Lipschtiz. Then, there is a unique solution $X_t$ to Eq. (3.54) giben by (see Doss [84] and Exercise 2.2.2) $X_t = u(W_t, Y_t)$, where $u(x, y)$ is the solution of the ordinary differential equation

$$\frac{\partial u}{\partial x} = \sigma(u), \qquad u(0, y) = y,$$

and $Y_t$ is the solution to the ordinary differential equation

$$Y_t = X_0 + \int_0^t f(W_s, Y_s)ds, \qquad (3.55)$$

where $f(x, y) = b(u(x, y)) \exp(-\int_0^x \sigma'(u(z, y)dz)$.

The next result shows that in the random case, the process $X_t = u(W_t, Y_t)$ solves Eq. (3.54).

**Theorem 3.3.3** *Suppose that the coefficients and the intial condition of Eq. (3.54) are random and for each $\omega \in \Omega$, $\sigma(\omega)$ is of class $C^2$ with bounded first and second derivatives, and that $b(\omega)$ is Lipschtiz. Let $X_t = u(W_t, Y_t)$, where $Y_t$ is given by (3.55). Then, $\sigma(X_t)$ is Stratonovich integrable on any interval $[0, t]$ and (3.54) holds.*

*Proof:*    Consider a partition $\pi = \{0 = t_0 < t_1 < \cdots < t_n = 1\}$ of the interval $[0, 1]$. Denote by $W_t^\pi$ the polygonal approximation of the Brownian motion defined in (3.5). Set $X_t^\pi = u(W_t^\pi, Y_t)$. This process satisfies

$$X_t^\pi = X_0 + \int_0^t \sigma(X_s^\pi)\dot{W}_s ds + \int_0^t \exp\left(\int_{W_s}^{W_s^\pi} \sigma'(u(y, Y_s))dy\right) b(X_s)ds$$
$$= X_0 + A_t^\pi + B_t^\pi.$$

The term $A_t^\pi$ can be written as

$$A_t^\pi = \int_0^t [\sigma(X_s^\pi) - \sigma(X_s)]\dot{W}_s ds + \int_0^t \sigma(X_s)\dot{W}_s ds = A_t^{1,\pi} + A_t^{2,\pi}.$$

Clearly $A_t^\pi = X_t^\pi - X_0 - B_t^\pi$ converges in probability to $X_t - X_0 - \int_0^t b(X_s)ds$ as $|\pi|$ tends to zero. On the other hand, $A_t^{2,\pi}$ can be expressed as Riemann sum that approximates the Stratonovich integral $\int_0^t \sigma(X_s) \circ dW_s$. As a consequence, it suffices to show that $A_t^{1,\pi}$ converges in probability to zero as $|\pi|$ tends to zero. We have

$$\sigma(X_s^\pi) - \sigma(X_s) = \sigma\sigma'(X_s)(W_s^\pi - W_s)$$
$$+ \frac{1}{2}\left(\sigma(\sigma')^2 + \sigma^2\sigma''\right)(\xi)(W_s^\pi - W_s)^2,$$

where $\xi$ is a random intermediate point between $W_s^\pi$ and $W_s$. The last summand of the above expression does not contribute tot the limit of $A_t^{1,\pi}$. In order to handle the first summand we set $Z_s = \sigma\sigma'(X_s)$. To simplify we assume $t = 1$. Consider another partition of the interval $[0, 1]$ of the form

$\{s_j = \frac{j}{m}, 0 \le j \le m\}$. We can write

$$
\left| \int_0^1 Z_s \left( W_s^\pi - W_s \right) \dot{W}_s ds \right| \le \left| \sum_{j=1}^{m-1} Z_{s_j} \int_{s_j}^{s_{j+1}} \left( W_s^\pi - W_s \right) \dot{W}_s ds \right|
$$

$$
+ \left| \sum_{j=1}^{m-1} \int_{s_j}^{s_{j+1}} \left( Z_s - Z_{s_j} \right) \left( W_s^\pi - W_s \right) \dot{W}_s ds \right|
$$

$$
= C^{\pi,m} + D^{\pi,m}.
$$

For any fixed $m$ the term $C^{\pi,m}$ converges in probability to zero as $|\pi|$ tends to zero. In fact, if the points $s_j$ and $s_{j+1}$ belong to the partition $\pi$, we have

$$
\int_{s_j}^{s_{j+1}} \left( W_s^\pi - W_s \right) \dot{W}_s ds
$$

$$
= \sum_{i:t_i \in [s_j, s_{j+1})} \frac{W_{t_{i+1}} - W_{t_i}}{t_{i+1} - t_i} \int_{t_i}^{t_{i+1}} \left( W_s^\pi - W_s \right) ds
$$

$$
= - \sum_{i:t_i \in [s_j, s_{j+1})} \frac{1}{2(t_{i+1} - t_i)}
$$

$$
\times \int_{t_i}^{t_{i+1}} \left[ \left( W_{t_{i+1}} - W_s \right)^2 - \left( W_s - W_{t_i} \right)^2 \right] ds
$$

and this converges to zero. Finally, the term $D^{\pi,m}$ can be estimated as follows

$$
|D^{\pi,m}| \le \sup_{|t-s| \le \frac{1}{m}} |Z_s - Z_t| \sum_{j=1}^{m-1} \int_{s_j}^{s_{j+1}} \left| \left( W_s^\pi - W_s \right) \dot{W}_s \right| ds
$$

$$
\le \sup_{|t-s| \le \frac{1}{m}} |Z_s - Z_t| \sum_{j=0}^{m-1} \sum_{i:t_i \in [s_j, s_{j+1})} \frac{1}{2(t_{i+1} - t_i)}
$$

$$
\times \int_{t_i}^{t_{i+1}} \left[ \left( W_{t_{i+1}} - W_s \right)^2 + \left( W_s - W_{t_i} \right)^2 \right] ds.
$$

As $|\pi|$ tends to zero the right-hand side of the above inequality converges in probability to $\frac{1}{2}$. Hence, we obtain

$$
\limsup_{|\pi| \to 0} |D^{\pi,m}| \le \frac{1}{2} \sup_{|t-s| \le \frac{1}{m}} |Z_s - Z_t|,
$$

and this expression tends to zero in probability as $m$ tends to infinity, due to the continuity of the process $Z_t$.    □

### 3.3.2 Stochastic differential equations with boundary conditions

Consider the following Stratonovich differential equation on the time interval $[0, 1]$, where instead of giving the value $X_0$, we impose a linear relationship between the values of $X_0$ and $X_1$:

$$\begin{cases} dX_t = \sum_{i=1}^{d} A_i(X_t) \circ dW_t^i + A_0(X_t)dt, \\ h(X_0, X_1) = 0. \end{cases} \tag{3.56}$$

We are interested in proving the existence and uniqueness of a solution for this type of equations. We will discuss two particular cases:

(a) The case where the coefficients $A_i$, $0 \le i \le d$ and the function $h$ are affine (see Ocone and Pardoux [273]).

(b) The one-dimensional cased (see Donati-Martin [80]).

(A) *Linear stochastic differential equations with boundary conditions*

Consider the following stochastic boundary value problem for $t \in [0, 1]$

$$\begin{cases} X_t = X_0 + \sum_{i=1}^{d} \int_0^t A_i X_s \circ dW_s^i + \int_0^t A_0 X_s ds, \\ H_0 X_0 + H_1 X_1 = h. \end{cases} \tag{3.57}$$

We assume that $A_i$, $i = 0, \ldots, d$, $H_0$, and $H_1$ are $m \times m$ deterministic matrices and $h \in \mathbb{R}^m$. We will also assume that the $m \times 2m$ matrix $(H_0 : H_1)$ has rank $m$.

Concerning the boundary condition, two particular cases are interesting:

*Two-point boundary-value problem:*    Let $l \in \mathbb{N}$ be such that $0 < l < m$. Suppose that $H_0 = \begin{pmatrix} H_0' \\ 0 \end{pmatrix}$, $H_1 = \begin{pmatrix} 0 \\ H_1'' \end{pmatrix}$, where $H_0'$ is an $l \times m$ matrix and $H_1''$ is an $(m - l) \times m$ matrix. Condition $\text{rank}(H_0 : H_1) = m$ implies that $H_0'$ has rank $l$ and that $H_1''$ has rank $m - l$. If we write $h = \begin{pmatrix} h_0 \\ h_1 \end{pmatrix}$, where $h_0 \in \mathbb{R}^l$ and $h_1 \in \mathbb{R}^{m-l}$, then the boundary condition becomes

$$H_0' X_0 = h_0, \quad H_1' X_1 = h_1.$$

*Periodic solution:*    Suppose $H_0 = -H_1 = I$ and $h = 0$. Then the boundary condition becomes

$$X_0 = X_1.$$

With (3.57) we can associate an $m \times m$ adapted and continuous matrix-valued process $\Phi$ solution of the Stratonovich stochastic differential equation

$$\begin{cases} d\Phi_t = \sum_{i=1}^{d} A_i \Phi_t \circ dW_t^i + B\Phi_t dt, \\ \Phi_0 = I. \end{cases} \qquad (3.58)$$

Using the Itô formula for the Stratonovich integral, one can obtain an explicit expression for the solution to Eq. (3.57). By a solution we mean a continuous and adapted stochastic process $X$ such that $A_i(X_t)$ is Stratonovich integrable with respect to $W^i$ on any time interval $[0, t]$ and such that (3.57) holds.

**Theorem 3.3.4** *Suppose that the random matrix $H_0 + H_1\Phi_1$ is a.s. invertible. Then there exists a solution to the stochastic differential equation (3.57) which is unique among those continuous processes whose components belong to the space $\mathbb{L}_{S,\text{loc}}^{2,4}$.*

*Proof:*    Define

$$X_t = \Phi_t X_0, \qquad (3.59)$$

where $X_0$ is given by

$$X_0 = [H_0 + H_1\Phi_1]^{-1} h. \qquad (3.60)$$

Then it follows from this expression that $X^i$ belongs to $\mathbb{L}_{S,\text{loc}}^{2,4}$, for all $i = 1, \ldots, m$, and due to the change-of-variables formula for the Stratonovich integral (Theorem 3.2.6), this process satisfies Eq. (3.57).

In order to show the uniqueness, we proceed as follows. Let $Y \in \mathbb{L}_{S,\text{loc}}^{2,4}(\mathbb{R}^m)$ be a solution to (3.57). Then we have

$$\Phi_t^{-1} = I - \sum_{i=1}^{d} \int_0^t \Phi_s^{-1} A_i \circ dW_s^i - \int_0^t \Phi_s^{-1} A_0 ds.$$

By the change-of-variables formula for the Stratonovich integral (Theorem 3.2.6) we see that $\Phi_t^{-1}Y_t$, namely, $Y_t = \Phi_t Y_0$. Therefore, $Y$ satisfies (3.59). But (3.60) follows from (3.59) and the boundary condition $H_0 Y_0 + H_1 Y_1 = h$. Consequently, $Y$ satisfies (3.59) and (3.60), and it must be equal to $X$. $\square$

Notice that, in general, the solution to (3.57) will not belong to $\mathbb{L}_{S,\text{loc}}^{2,4}(\mathbb{R}^m)$. One can also treat non-homogeneous equations of the form

$$\begin{cases} X_t = X_0 + \sum_{i=1}^{d} \int_0^t A_i X_s \circ dW_s^i + \int_0^t A_0 X_s ds + V_t, \\ H_0 X_0 + H_1 X_1 = h, \end{cases} \qquad (3.61)$$

where $V_t$ is a continuous semimartingale. In that case,

$$X_t = \Phi_t [H_0 + H_1\Phi_1]^{-1} \left[ h - H_1\Phi_1 \int_0^1 \Phi_s^{-1} \circ dV_s \right] + \Phi_t \int_0^t \Phi_s^{-1} \circ dV_s$$

is a solution to Eq. (3.61). The uniqueness in the class $\mathbb{L}_{S,\text{loc}}^{2,4}(\mathbb{R}^m)$ can be established provided the process $V_t$ also belongs to this space. $\square$

(B) *One-dimensional stochastic differential equations with boundary conditions*

Consider the one-dimensional stochastic boundary value problem

$$\begin{cases} X_t = X_0 + \int_0^t \sigma(X_s) \circ dW_s + \int_0^t b(X_s)ds, \\ a_0 X_0 + a_1 X_1 = a_2, \end{cases} \tag{3.62}$$

Applying the techniques of the anticipating stochastic calculus we can show the following result.

**Theorem 3.3.5** *Suppose that the functions $\sigma$ and $b_1 := b + \frac{1}{2}\sigma\sigma'$ are of class $C^2$ with bounded derivatives and $a_0 a_1 > 0$. Then there exists a solution to Eq. (3.62). Furthermore, if the functions $\sigma$ and $b_1$ are of class $C^4$ with bounded derivatives then the solution is unique in the class $\mathbb{L}_{S,\text{loc}}^{2,4}$.*

*Proof:*  Let $\varphi_t(x)$ be the stochastic flow associated with the coefficients $\sigma$ and $b_1$. By Theoren 3.3.1 for any random variable $X_0$ the process $\varphi_t(X_0)$ satisfies

$$X_t = X_0 + \int_0^t \sigma(X_s) \circ dW_s + \int_0^t b(X_s)ds.$$

Hence, in order to show the existence of a solution it suffices to prove that there is a unique random variable $X_0$ such that

$$\varphi_1(X_0) = \frac{a_2 - a_0 X_0}{a_1}. \tag{3.63}$$

The mapping $g(x) = \varphi_1(x)$ is strictly increasing and this implies the existence of a unique solution to Eq. (3.63).

Taking into account Theorem 3.3.2 to show the uniqueness it suffices to check that the unique solution $X_0$ to Eq. (3.63) belongs to $\mathbb{D}_{\text{loc}}^{1,p}$ for some $p > 4$. By the results of Doss (see [84] and Exercise 2.2.2) one can represent the flow $\varphi_t(x)$ as a Fréchet differentiable function of the Brownian motion $W$. Using this fact and the implicit function theorem one deduces that $X_0 \in \mathbb{D}_{\text{loc}}^{1,p}$ for all $p \geq 2$.                    □

### 3.3.3  *Stochastic differential equations in the Skorohod sense*

Let $W = \{W_t, t \in [0,1]\}$ be a one-dimensional Brownian motion defined on the canonical probability space $(\Omega, \mathcal{F}, P)$. Consider the stochastic differential equation

$$X_t = X_0 + \int_0^t \sigma(s, X_s)dW_s + \int_0^t b(s, X_s)ds, \tag{3.64}$$

$0 \leq t \leq 1$, where $X_0$ is $\mathcal{F}_1$-measurable and $\sigma$ and $b$ are deterministic functions. First notice that the usual Picard iteration procedure cannot be applied in that situation. In fact the estimation of the $L^2$-norm of the Skorohod integral requires a bound for the derivative of $X$, and this derivative can be estimated only in terms of the second derivative. So we are faced with a nonclosed procedure. In some sense, Eq. (3.64) is an infinite dimensional hyperbolic partial differential equation. In fact, this equation can be formally written as

$$X_t = X_0 + \int_0^t \sigma(s, X_s) \circ dW_s + \frac{1}{2} \int_0^t \sigma'(s, X_s) \left[ (D^+ X)_s + (D^- X)_s \right] ds$$
$$+ \int_0^t b(s, X_s) ds.$$

Nevertheless, if the diffusion coefficient is linear it is possible to show that there is a unique global solution using the techniques developed by Buckdahn in [48, 49], based on the classical Girsanov transformation. In order to illustrate this approcah let consider the following particular case:

$$X_t = X_0 + \sigma \int_0^t X_s dW_s. \tag{3.65}$$

When $X_0$ is deterministic, the solution to this equation is the martingale

$$X_t = X_0 e^{\sigma W_t - \frac{1}{2}\sigma^2 t}.$$

If $X_0 = \text{sign} W_1$, then a solution to Eq. (3.65) is (see Exercise 3.2.3)

$$X_t = \text{sign} \left( W_1 - \sigma t \right) e^{\sigma W_t - \frac{1}{2}\sigma^2 t}.$$

More generally, by Theorem 3.3.6 below, if $X_0 \in L^p(\Omega)$ for some $p > 2$, then

$$X_t = X_0(A_t) e^{\sigma W_t - \frac{1}{2}\sigma^2 t}$$

is a solution to Eq. (3.65), where $A_t(\omega)_s = \omega_s - \sigma(t \wedge s)$. In terms of the Wick product (see [53]) one can write

$$X_t = X_0 \Diamond e^{\sigma W_t - \frac{1}{2}\sigma^2 t}.$$

Let us now turn to the case of a general linear diffusion coefficient and consider the equation

$$X_t = X_0 + \int_0^t \sigma_s X_s dW_s + \int_0^t b(s, X_s) ds, \qquad 0 \le t \le 1, \tag{3.66}$$

where $\sigma \in L^2([0, 1])$, $X_0$ is a random variable and $b$ is a random function satisfying the following condition:

**(H.1)**    $b : [0,1] \times \mathbb{R} \times \Omega \to \mathbb{R}$   is a measurable function such that there exist an integrable function $\gamma_t$ on $[0,1]$, $\gamma_t \geq 0$, a constant $L > 0$, and a set $N_1 \in \mathcal{F}$ of probability one, verifying

$$|b(t,x,\omega) - b(t,y,\omega)| \leq \gamma_t |x - y|, \qquad \int_0^1 \gamma_t dt \leq L,$$

$$|b(t,0,\omega)| \leq L,$$

for all $x, y \in \mathbb{R}$, $t \in [0,1]$ and $\omega \in N_1$.

Let us introduce some notation. Consider the family of transformations $T_t, A_t : \Omega \to \Omega$, $t \in [0,1]$, given by

$$T_t(\omega)_s = \omega_s + \int_0^{t \wedge s} \sigma_u du,$$

$$A_t(\omega)_s = \omega_s - \int_0^{t \wedge s} \sigma_u du.$$

Note that $T_t A_t = A_t T_t = $ Identity. Define

$$\varepsilon_t = \exp\left(\int_0^t \sigma_s dW_s - \frac{1}{2} \int_0^t \sigma_s^2 ds\right).$$

Then, by Girsanov's theorem (see Proposition 4.1.2) $E\left[F(A_t)\varepsilon_t\right] = E[F]$ for any random variable $F \in L^1(\Omega)$. For each $x \in \mathbb{R}$ and $\omega \in \Omega$ we denote by $Z_t(\omega, x)$ the solution of the integral equation

$$Z_t(\omega, x) = x + \int_0^t \varepsilon_s^{-1}\left(T_t(\omega)\right) b\left(s, \varepsilon_s\left(T_t(\omega)\right) Z_s(\omega, x), T_s(\omega)\right) ds. \quad (3.67)$$

Notice that for $s \leq t$ we have $\varepsilon_s(T_t) = \exp\left(\int_0^s \sigma_u dW_u + \frac{1}{2}\int_0^s \sigma_u^2 du\right) = \varepsilon_s(T_s)$. Henceforth we will omit the dependence on $\omega$ in order to simplify the notation.

**Theorem 3.3.6** *Fix an initial condition $X_0 \in L^p(\Omega)$ for some $p > 2$, and define*

$$X_t = \varepsilon_t Z_t\left(A_t, X_0(A_t)\right). \quad (3.68)$$

*Then the process $X = \{X_t, 0 \leq t \leq 1\}$ satisfies $\mathbf{1}_{[0,t]}\sigma X \in \mathrm{Dom}\,\delta$ for all $t \in [0,1]$, $X \in L^2([0,1] \times \Omega)$, and $X$ is the unique solution of Eq. (3.66) verifying these conditions.*

*Proof:*
*Existence:*    Let us prove first that the process $X$ given by (3.68) satisfies the desired conditions. By Gronwall's lemma and using hypothesis (H.1), we have

$$|X_t| \leq \varepsilon_t e^{tL}\left(|X_0(A_t)| + L \int_0^t \varepsilon_s^{-1}(T_s) ds\right),$$

which implies $\sup_{t \in [0,1]} E(|X_t|^q) < \infty$, for all $2 \le q < p$, as it follows easily from Girsanov's theorem and Hölder's inequality. Indeed, we have

$$
\begin{aligned}
E(|X_t|^q) &\le c_q E \left( \varepsilon_t^q e^{qtL} \left( |X_0(A_t)|^q + L^q \int_0^t \varepsilon_s^{-q}(T_s) ds \right) \right) \\
&\le c_q e^{qL} \left\{ E \left( \varepsilon_t^{q-1}(T_t) |X_0|^q + L^q \varepsilon_t^{q-1}(T_t) \int_0^t \varepsilon_s^{-q}(T_s^2) ds \right) \right\} \\
&\le C \left\{ E(|X_0|^p)^{\frac{q}{p}} + 1 \right\}.
\end{aligned}
$$

Now fix $t \in [0,1]$ and let us prove that $\mathbf{1}_{[0,t]} \sigma X \in \mathrm{Dom}\, \delta$ and that (3.66) holds. Let $G \in \mathcal{S}$ be a smooth random variable. Using (3.68) and Girsanov's theorem, we obtain

$$
\begin{aligned}
E \left( \int_0^t \sigma_s X_s D_s G ds \right) &= E \left( \int_0^t \sigma_s \varepsilon_s Z_s \left( A_s, X_0(A_s) \right) D_s G ds \right) \\
&= E \left( \int_0^t \sigma_s Z_s(X_0) (D_s G)(T_s) ds \right). \quad (3.69)
\end{aligned}
$$

Notice that $\frac{d}{ds} G(T_s) = \sigma_s (D_s G)(T_s)$. Therefore, integrating by parts in (3.69) and again applying Girsanov's theorem yield

$$
\begin{aligned}
E \left( \int_0^t Z_s(X_0) \frac{d}{ds} G(T_s) ds \right) =\ & E \Big[ Z_t(X_0) G(T_t) - Z_0(X_0) G \\
& - \int_0^t \varepsilon_s^{-1}(T_t) b\left(s, \varepsilon_s(T_s) Z_s(X_0), T_s\right) G(T_s) ds \Big] \\
=\ & E \left( \varepsilon_t Z_t \left( A_t, X_0(A_t) \right) G \right) - E \left( Z_0(X_0) G \right) \\
& - \int_0^t E \left( b\left(s, \varepsilon_s Z_s \left( A_s, X_0(A_s) \right) \right) G \right) ds \\
=\ & E \left( X_t G \right) - E \left( X_0 G \right) - \int_0^t E \left( b(s, X_s) G \right) ds.
\end{aligned}
$$

Because the random variable $X_t - X_0 - \int_0^t b(s, X_s) ds$ is square integrable, we deduce that $\mathbf{1}_{[0,t]} \sigma X$ belongs to the domain of $\delta$ and that (3.66) holds.

*Uniqueness:*  Let $Y$ be a solution to Eq. (3.66) such that $Y$ belongs to $L^2([0,1] \times \Omega)$ and $\mathbf{1}_{[0,t]} \sigma Y \in \mathrm{Dom}\, \delta$ for all $t \in [0,1]$. Fix $t \in [0,1]$ and let $G$ be a smooth random variable. Multiplying both members of (3.66) by $G(A_t)$ and taking expectations yield

$$
\begin{aligned}
E(Y_t G(A_t)) =\ & E(Y_0 G(A_t)) + E \left( \int_0^t b(s, Y_s) G(A_t) ds \right) \\
& + E \left( \int_0^t \sigma_s Y_s D_s \left( G(A_t) \right) \right) ds. \quad (3.70)
\end{aligned}
$$

Notice that $\frac{d}{ds}G(A_s) = -\sigma(D_sG)(A_s)$. Therefore, integrating by parts obtains

$$
E\left(Y_tG(A_t)\right) = E\left(Y_0G\right) - \int_0^t E\left(Y_0\sigma_s(D_sG)(A_s)\right)ds
$$
$$
+E\left(\int_0^t b(s,Y_s)G(A_s)ds\right) - E\left(\int_0^t\int_0^r b(s,Y_s)\sigma_r(D_rG)(A_r)dsdr\right)
$$
$$
+E\left(\int_0^t \sigma_sY_s(D_sG)(A_s)ds\right)
$$
$$
-E\left(\int_0^t\int_0^r \sigma_sY_s(D_rD_sG)(A_r)\sigma_r dsdr\right). \tag{3.71}
$$

If we apply Eq. (3.71) to the smooth random variable $\sigma_r(D_rG)(A_r)$ for each fixed $r \in [0,t]$, the negative terms in the above expression cancel out with the term $E\left(\int_0^t \sigma_sY_s(D_sG)(A_s)ds\right)$, and we obtain

$$
E\left(Y_tG(A_t)\right) = E\left(Y_0G\right) + E\left(\int_0^t b(s,Y_s)G(A_s)\,ds\right).
$$

By Girsanov's theorem this implies

$$
E\left(Y_t(T_t)\varepsilon_t^{-1}(T_t)G\right) = E\left(Y_0G\right) + E\left(\int_0^t b\left(s,Y_s(T_s),T_s\right)\varepsilon_s^{-1}(T_s)G\right)ds.
$$

Therefore, we have

$$
Y_t(T_t)\varepsilon_t^{-1}(T_t) = Y_0 + \int_0^t b\left(s,Y_s(T_s),T_s\right)\varepsilon^{-1}(T_s)ds, \tag{3.72}
$$

and from (3.72) we get $Y_t(T_t)\varepsilon_t^{-1}(T_t) = Z_t(Y_0)$ a.s. That is,

$$
Y_t = \varepsilon_t Z_t\left(A_t, Y_0(A_t)\right) = X_t
$$

a.s., which completes the proof of the uniqueness.    □

When the diffusion coefficient is not linear one can show that there exists a solution up to a random time.

## Exercises

**3.3.1** Let $f$ be a continuously differentiable function with bounded derivative. Solve the linear Skorohod stochastic differential equation

$$
\begin{aligned}
dX_t &= X_t dW_t, \quad t \in [0,1]\\
X_0 &= f(W_1).
\end{aligned}
$$

**3.3.2** Consider the stochastic boundary-value problem

$$
\begin{aligned}
dX_t^1 &= X_t^2 \circ dW_t, \\
dX_t^2 &= 0, \\
X_0^1 &= 0, \quad X_1^1 = 1.
\end{aligned}
$$

Find the unique solution of this system, and show that $E(|X_t|^2) = \infty$ for all $t \in [0, 1]$.

**3.3.3** Find an explicit solution for the stochastic boundary-value problem

$$
\begin{aligned}
dX_t^1 &= (X_t^1 + X_t^2) \circ dW_t^1, \\
dX_t^2 &= X_t^2 \circ dW_t^2, \\
X_0^1 + X_0^2 &= 1, \quad X_1^2 = 1.
\end{aligned}
$$

**Notes and comments**

**[3.1]**    The Skorohod integral is an extension of the Itô integral introduced in Section 1.3 as the adjoint of the derivative operator. In Section 3.1, following [249], we show that it can be obtained as the limit of two types of modified Riemann sums, including the conditional expectation operator or subtracting a complementary term that converges to the trace of the derivative.

The forward stochastic integral, defined as

$$
\delta(u) + \int_0^1 D_t^- u_t dt,
$$

is also an extension of the Itô integral which has been studied by different authors. Berger and Mizel [30] introduced this integral in order to solve stochastic Volterra equations. In [14], Asch and Potthoff prove that it satisfies a change-of-variables formula analogous to that of the Itô calculus. An approach using a convolution of the Brownian path with a rectangular function can be found in Russo and Vallois [298]. In [176] Kuo and Russek study the anticipating stochastic integrals in the framework of the white noise calculus.

The definition of the stochastic integral using an orthonormal basis of $L^2([0, 1])$ is due to Paley and Wiener in the case of deterministic integrands. For random integrands this analytic approach has been studied by Balkan [16], Ogawa [274, 275, 276], Kuo and Russek [176] and Rosinski [293], among others.

**[3.2]**    The stochastic calculus for the Skorohod and Stratonovich integrals was developed by Nualart and Pardoux [249]. In particular, the local property introduced there has allowed us to extend the change-of-variables formula and to deal with processes that are only locally integrable or possess locally integrable derivatives. Another extensive work on the stochastic calculus for the Skorohod integral in $L^2$ is Sekiguchi and Shiota [305].

Other versions of the change-of-variables formula for the Skorohod integral can be found in Sevljakov [306], Hitsuda [136], and Üstünel [330].

**[3.3]**     For a survey of this kind of applications, we refer the reader to Pardoux [278].

Stochastic differential equations in the Skorohod sense were first studied by Shiota [311] using Wiener chaos expansions. A different method was used by Üstünel [331]. The approach described in Section 3.3, based on the Girsanov transformation, is due to Buckdahn and allows us to solve a wide class of quasilinear stochastic differential equations in the Skorohod sense with a constant or adapted diffusion coefficient. When the diffusion coefficient is random, one can use the same ideas by applying the anticipating version of Girsanov's theorem (see [50]). In [49] Buckdahn considers Skorohod stochastic differential equations of the form (3.64), where the diffusion coefficient $\sigma$ is not necessarily linear. In this case the situation is much more complicated, and an existence theorem can be proved only in some random neighborhood of zero.

Stochastic differential equations in the sense of Stratonovich have been studied by Ocone and Pardoux in [272]. In this paper they prove the existence of a unique solution to Eq. (3.49) assuming that the coefficient $A_0(s,x)$ is random. In [273] Ocone and Pardoux treat stochastic differential equations of the Stratonovich type with boundary conditions of the form (3.56), assuming that the coefficients $A_i$ and the function $h$ are affine, and they also investigate the Markov properties of the solution.

# 4

# Transformations
# of the Wiener measure

In this chapter we discuss different extensions of the classical Girsanov theorem to the case of a transformation of the Brownian motion induced by a nonadapted process. This generalized version of Girsanov's theorem will be applied to study the Markov property of solutions to stochastic differential equations with boundary conditions.

## 4.1 Anticipating Girsanov theorems

In this section we will work in the context of an abstract Wiener space. That is, we will assume that the underlying probability space $(\Omega, \mathcal{F}, P)$ is such that $\Omega$ is a separable Banach space, $P$ is a Gaussian measure on $\Omega$ with full support, and $\mathcal{F}$ is the completion of the Borel $\sigma$-field of $\Omega$ with respect to $P$. Moreover, there is a separable Hilbert space $H$ that is continuously and densely embedded in $\Omega$, with injection $i : H \to \Omega$, and such that

$$\int_\Omega e^{i\langle x,y \rangle} P(dx) = \exp(-\frac{1}{2}\|y\|_H^2)$$

for any $y \in \Omega^* \subset H$ (here we identify $H$ with its dual).

The triple $(\Omega, H, P)$ is called an *abstract Wiener space*. Note that each element $y \in \Omega^*$ defines a Gaussian random variable. If we denote this variable by $W(y)$, then the mapping $y \to W(y)$, from $\Omega^*$ into $L^2(\Omega)$, is continuous with respect to the norm of $H$, and it can be extended to $H$. In that way $H$ is isometric to a Gaussian subspace in $L^2(\Omega)$ that generates $\mathcal{F}$

and that is denoted by $\mathcal{H}_1$ (i.e., $\mathcal{H}_1$ is the first Wiener chaos). The image of $H$ by the injection $i$ is denoted by $H^1 \subset \Omega$.

The classical Wiener space is a particular case of an abstract Wiener space if we take $\Omega = C_0([0,1])$, $H = L^2([0,1])$, $i(h)(t) = \int_0^t h(s)ds$, and $P$ is the Wiener measure. The space $H^1$ is here the Cameron-Martin space.

Consider a measurable mapping $u : \Omega \to H$, and define the transformation $T : \Omega \to \Omega$ by

$$T(\omega) = \omega + i(u(\omega)). \tag{4.1}$$

In the white noise case, $H$ can be represented as $H = L^2(T, \mathcal{B}, \mu)$, and $H$-valued random variables are stochastic processes parametrized by $T$. Along this chapter we will use this terminology.

We want to discuss the following problems:

**(A)** When is the image measure $P \circ T^{-1}$ absolutely continuous with respect to $P$ ?

**(B)** Find a new probability $Q$ absolutely continuous with respect to $P$ such that $Q \circ T^{-1} = P$.

Furthermore, we are also interested in finding expressions for the Radon-Nikodym densities $\frac{d[P \circ T^{-1}]}{dP}$ and $\frac{dQ}{dP}$ in terms of $u$.

### 4.1.1   The adapted case

Consider the case of the Wiener space, that is, $\Omega = C_0([0,1])$. Then $u = \{u_t, 0 \le t \le 1\}$ is a random process such that $\int_0^1 u_t^2 dt < \infty$ a.s., and the associated transformation is given by

$$T(\omega)_t = \omega_t + \int_0^t u_s(\omega)ds. \tag{4.2}$$

Suppose that the process $u$ is adapted, and define

$$\xi_t = \exp\left(-\int_0^t u_s dW_s - \frac{1}{2}\int_0^t u_s^2 ds\right), \quad t \in [0,1]. \tag{4.3}$$

The following two propositions provide a complete answer to questions (A) and (B).

**Proposition 4.1.1** *The probability $P \circ T^{-1}$ is absolutely continuous with respect to $P$.*

**Proposition 4.1.2** *(Girsanov theorem) There exists a probability $Q$ absolutely continuous with respect to $P$ such that $Q \circ T^{-1} = P$ (that is, $W_t + \int_0^t u_s ds$ has the law of a Brownian motion under $Q$) if and only if $E(\xi_1) = 1$ and in this case $\frac{dQ}{dP} = \xi_1$.*

*Proof of Proposition 4.1.2:*   Suppose first that $E(\xi_1) = 1$. Consider the increasing family of stopping times defined by

$$\tau_k = \inf\{t : \int_0^t u_s^2 ds \geq k\}.$$

By Itô's formula $\{\xi_{t \wedge \tau_k}, t \in [0,1]\}$ is a positive continuous martingale. Hence $\{\xi_t, t \in [0,1]\}$ is a continuous local martingale, and is a martingale because $E(\xi_1) = 1$.

Fix $0 \leq s < t \leq 1$ and $A \in \mathcal{F}_s$. Applying Itô's formula yields

$$E\left(e^{i\lambda(W_t - W_s + \int_s^t u_r dr)} \mathbf{1}_A \xi_1\right)$$

$$= E\left(E\left(e^{i\lambda(W_t - W_s + \int_s^t u_r dr) - \int_s^t u_r dW_r - \frac{1}{2}\int_s^t u_r^2 dr} | \mathcal{F}_s\right) \mathbf{1}_A \xi_s\right)$$

$$= E(\mathbf{1}_A \xi_1) - \int_s^t \frac{\lambda^2}{2} E\left(e^{i\lambda(W_u - W_s + \int_s^u u_\theta d\theta)} \mathbf{1}_A \xi_1\right) du.$$

Hence,

$$E\left(e^{i\lambda(W_t - W_s + \int_s^t u_r dr)} \mathbf{1}_A \xi_1\right) = E(\mathbf{1}_A \xi_1) e^{-\frac{\lambda^2}{2}(t-s)},$$

and we obtain

$$E(F(T)\xi_1) = E(F)$$

for any functional $F : \Omega \to \mathbb{C}$ of the form

$$F = \exp\left(i \sum_{j=0}^{m-1} \lambda_j (W(t_{j+1}) - W(t_j))\right),$$

where $\lambda_j \in \mathbb{R}$ and $0 = t_0 < t_1 < \cdots < t_m = 1$. Therefore, the probability $Q$ given by $\frac{dQ}{dP} = \xi_1$ satisfies $Q \circ T^{-1} = P$.

Conversely, assume $Q \circ T^{-1} = P$ and $\frac{dQ}{dP} = \eta$. For any integer $k \geq 1$ we define the transformation

$$T_k(\omega)_t = \omega_t + \int_0^{t \wedge \tau_k} u_s ds.$$

In view of $E(\xi_{\tau_k}) = 1$ we can apply the arguments of the first part of the proof to the transformation $T_k$, deducing

$$E(F(T_k)\xi_{\tau_k}) = E(F)$$

for any nonnegative and bounded random variable $F$. If $F$ is $\mathcal{F}_{\tau_k}$-measurable, then $F(T_k) = F(T)$. On the other hand, we know that

$$E(F(T)\eta) = E(F).$$

Hence, $\xi_{\tau_k} = E(\eta|T^{-1}(\mathcal{F}_{\tau_k}))$, and letting $k$ tend to infinity obtains $\xi_1 = \eta$.     $\square$

*Proof of Proposition 4.1.1:*     Let $B$ be a Borel subset of $\Omega$ with $P(B) = 0$. Consider the stopping times $\tau_k$ and the transformations $T_k$ introduced in the proof of Proposition 4.1.2. We know that $P \circ T_k^{-1}$ and $P$ are mutually absolutely continuous. Hence,

$$
\begin{aligned}
P(T^{-1}(B)) &= P(T^{-1}(B) \cap \{\tau_k = 1\}) + P(T^{-1}(B) \cap \{\tau_k < 1\}) \\
&= P(T_k^{-1}(B) \cap \{\tau_k = 1\}) + P(T^{-1}(B) \cap \{\tau_k < 1\}) \\
&\leq P\{\tau_k < 1\} = P\left\{ \int_0^1 u_t^2 dt > k \right\},
\end{aligned}
$$

which converges to zero as $k$ tends to infinity. This completes the proof. $\square$

### 4.1.2  General results on absolute continuity of transformations

In this subsection we will assume that $(\Omega, \mathcal{F}, P)$ is an arbitrary complete probability space and $T : \Omega \longrightarrow \Omega$ is a measurable transformation. First we will see that questions (A) and (B) are equivalent if the measures $P \circ T^{-1}$ and $P$ (or $Q$ and $P$) are equivalent (i.e., they are mutually absolutely continuous).

**Lemma 4.1.1** *The following statements are equivalent:*

*(i)  The probabilities $P \circ T^{-1}$ and $P$ are equivalent.*

*(ii)  There exists a probability $Q$ equivalent to $P$ such that $Q \circ T^{-1} = P$.*

Under the above assumptions, and setting $X = \frac{d[P \circ T^{-1}]}{dP}$ and $Y = \frac{dQ}{dP}$, we have $E(Y \mid T) = \frac{1}{X(T)}$, $P$ a.s. If we assume, moreover, that $\{T^{-1}(A), A \in \mathcal{F}\} = \mathcal{F}$ a.s. (which is equivalent to the existence of a left inverse $T_l$ such that $T_l^{-1} \circ T = I$ a.s.), we have

$$
\frac{dQ}{dP} = \left( \frac{d[P \circ T^{-1}]}{dP} \circ T \right)^{-1}. \tag{4.4}
$$

*Proof:*     Let us first show that (i) implies (ii). Set $X = \frac{d[P \circ T^{-1}]}{dP}$. We know that $P\{X = 0\} = 0$ and $P\{X(T) = 0\} = (P \circ T^{-1})\{X = 0\} = 0$. Define the measure $Q$ by $dQ = \frac{1}{X(T)} dP$. Then for any $B \in \mathcal{F}$ we have

$$
Q(T^{-1}(B)) = \int_{T^{-1}(B)} \frac{1}{X(T)} dP = \int_B \frac{1}{X} d[P \circ T^{-1}] = P(B).
$$

Clearly, $P << Q$ because $\frac{dQ}{dP} \neq 0$.

Conversely, assume that (ii) holds. Then (i) follows from the implications

$$P(B) = 0 \quad \Longleftrightarrow \quad Q(T^{-1}(B)) = 0 \quad \Longleftrightarrow \quad P(T^{-1}(B)) = 0.$$

In order to show the equality $E(Y \mid T) = \frac{1}{X(T)}$ a.s., we write for any $B \in \mathcal{F}$

$$\int_{T^{-1}(B)} E(Y|T)dP = \int_{T^{-1}(B)} Y dP = Q(T^{-1}(B))$$

$$= \int_{T^{-1}(B)} \frac{1}{X(T)} dP.$$

The last statement follows from $E(Y \mid T) = Y$. $\qquad\qquad\qquad\square$

**Remarks:** If we only assume the absolute continuity $P \circ T^{-1} << P$ (or $Q << P$), the above equivalence is no longer true. We can only affirm that $Q \circ T^{-1} = P$ and $P << Q$ imply $P \circ T^{-1} << P$ . The following examples illustrate this point.

**Examples:** Let $\Omega = C_0([0,1])$ and $P$ be the Wiener measure. Consider the following two examples.

**(1)** Suppose $u_t = f(W_1)$, where $f : \mathbb{R} \to \mathbb{R}$ is a measurable function. The probability $P \circ T^{-1}$ is absolutely continuous with respect to $P$ if and only if the distribution of the random variable $Y + f(Y)$ is absolutely continuous, where $Y$ has the law $N(0,1)$ (see Exercise 4.1.1). If we take $f(x) = x^2$, then $P \circ T^{-1} << P$, but it is not possible to find a probability $Q$ on $\Omega$ such that under $Q$ the process $\{W_t + tW_1^2, t \in [0,1]\}$ is a Brownian motion.

**(2)** Let

$$u_t = \frac{2W_t}{(1-t)^2} \mathbf{1}_{[0,S]}(t),$$

where $S = \inf\{t : W_t^2 = 1 - t\}$. Since $P(0 < S < 1) = 1$, we have $\int_0^1 u_t^2 dt = 4 \int_0^S \frac{W_t^2}{(1-t)^4} dt < \infty$ a.s. However, $E(\xi_1) < 1$ (see Exercise 4.1.2), where $\xi_1$ is given by (4.3), and, from Proposition 4.1.2, it is not possible to find a probability $Q$ absolutely continuous with respect to $P$ such that under $Q$ the process

$$W_t + \int_0^t u_s ds = W_t + 2 \int_0^{S \wedge t} \frac{W_s}{(1-s)^2} ds$$

is a Brownian motion (see [200, p. 224]).

In the following sections we will look for a random variable $\eta$ such that

$$E(\eta \mathbf{1}_{T^{-1}(B)}) \le P(B) \qquad\qquad (4.5)$$

for all $B \in \mathcal{F}$. If $\eta > 0$ a.s., this implies that $P \circ T^{-1} << P$. On the other hand, if the equality in (4.5) holds for all $B \in \mathcal{F}$ (which is equivalent to imposing that $E(\eta) = 1$), then $P$ is equivalent to $P \circ T^{-1}$, and the probability $Q$ given by $\frac{dQ}{dP} = \eta$ verifies $Q \circ T^{-1} = P$.

## 4.1.3  Continuously differentiable variables in the direction of $H^1$

Let $(\Omega, H, P)$ be an abstract Wiener space. We will need the following notion of differentiability.

**Definition 4.1.1** *We will say that a random variable $F$ is (a.s.) $H$-continuously differentiable if for (almost) all $\omega \in \Omega$ the mapping $h \to F(\omega + i(h))$ is continuously differentiable in $H$. The set of a.s. $H$-continuously differentiable functions will be denoted by $C_H^1$.*

We will prove that $C_H^1 \subset \mathbb{D}_{\text{loc}}^{1,2}$. In order to show this inclusion we have to introduce some preliminary tools. For any set $A \in \mathcal{F}$ define

$$\rho_A(\omega) = \inf\{\|h\|_H : \omega + i(h) \in A\},$$

where the infimum is taken to be infinite if the set is empty. Clearly, $\rho_A(\omega) = 0$ if $\omega \in A$, and $\rho_A(\omega) \le \epsilon$ if $\omega$ belongs to the $\epsilon$-neighborhood of $A$ constructed with the distance of $H^1$. Also, $\rho_A(\omega) = \infty$ if $\omega \notin A + H^1$. Other properties of this distance are the following:

(i) $A \subset B \implies \rho_A(\omega) \ge \rho_B(\omega)$;

(ii) $|\rho_A(\omega + i(h)) - \rho_A(\omega)| \le \|h\|_H$;

(iii) $A_n \uparrow A \implies \rho_{A_n}(\omega) \downarrow \rho_A(\omega)$;

(iv) If $G$ is $\sigma$-compact and $\phi \in C_0^\infty(\mathbb{R})$, then $\phi(\rho_G)$ belongs to $\mathbb{D}^{1,p}$ for all $p \ge 2$ (with the convention $\phi(\infty) = 0$), and $\|D[\phi(\rho_G)]\|_H \le \|\phi'\|_\infty$.

*Proof of these properties:*   Properties (i) and (ii) are immediate consequences of the definition of $\rho_A$. In order to show (iii) it suffices to see that $\inf_n \rho_{A_n}(\omega) \le \rho_A(\omega)$ for all $\omega \in \Omega$. If $\omega \notin A + H^1$, this inequality is clear. Suppose $\omega \in A + H^1$. By definition, for any $\epsilon > 0$ there exists $h \in H$ such that $\omega + i(h) \in A$ and $\|h\|_H \le \rho_A(\omega) + \epsilon$. We can find an index $n_0$ such that $\omega + i(h) \in A_n$ for all $n \ge n_0$. Therefore, $\rho_{A_n}(\omega) \le \rho_A(\omega) + \epsilon$ for all $n \ge n_0$, which implies the desired result.

Let us prove (iv). Set $G = \cup_n K_n$, where $\{K_n\}$ is an increasing sequence of compact sets. By (iii) we have $\rho_{K_n} \downarrow \rho_G$, hence $\phi(\rho_{K_n}) \to \phi(\rho_G)$. For each $n$, $\rho_{K_n}$ is a measurable function because it can be written as

$$\rho_{K_n}(\omega) = \begin{cases} \infty & \text{on } (K_n + H^1)^c \\ d_{H^1}((\omega - K_n) \cap H^1, 0) & \text{on } \quad K_n + H^1, \end{cases}$$

where $d_{H^1}$ is the metric of $H^1$ and $\omega \to (\omega - K_n) \cap H^1$ is a measurable function from $K_n + H^1$ into the closed subsets of $H^1$. Consequently, $\phi(\rho_G)$ is a bounded and measurable function such that

$$|\phi(\rho_G(\omega + i(h))) - \phi(\rho_G(\omega))| \le \|\phi'\|_\infty \|h\|_H, \tag{4.6}$$

and it suffices to apply Exercise 1.2.9 and Proposition 1.5.5.    $\square$

**Proposition 4.1.3** *We have $C^1_H \subset \mathbb{D}^{1,2}_{\mathrm{loc}}$.*

*Proof:*    Suppose that $F \in C^1_H$. Define the following sequence of sets:

$$A_n = \Big\{ \omega \in \Omega : \quad \sup_{\|h\|_H \le \frac{1}{n}} |F(\omega + i(h))| \le n,$$

$$\sup_{\|h\|_H \le \frac{1}{n}} \|DF(\omega + i(h))\|_H \le n \Big\}.$$

The fact that $F$ is a.s. $H$-continuously differentiable implies that almost surely $\bigcup_{n=1}^{\infty} A_n = \Omega$. For each $n$ we can find a $\sigma$-compact set $G_n \subset A_n$ such that $P(G_n) = P(A_n)$.

Let $\phi \in C^{\infty}_0(\mathbb{R})$ be a nonnegative function such that $|\phi(t)| \le 1$ and $|\phi'(t)| \le 4$ for all $t$, $\phi(t) = 1$ for $|t| \le \frac{1}{3}$, and $\phi(t) = 0$ for $|t| \ge \frac{2}{3}$. Define

$$F_n = \phi(n \rho_{G_n}) F.$$

We have

(a)  $F_n = F$ on the set $G_n$.

(b)  $F_n$ is bounded by $n$. Indeed,

$$|F_n| \le \mathbf{1}_{\{\rho_{G_n} < \frac{2}{3n}\}} |F| \le n,$$

because $\rho_{G_n} < \frac{2}{3n}$ implies that there exists $h \in H$ with $\omega + i(h) \in G_n \subset A_n$ and $\|h\|_H < \frac{2}{3n} \le \frac{1}{n}$.

(c)  For any $k \in H$ with $\|k\|_H \le \frac{1}{3n}$ we have, using (4.6),

$$|F_n(\omega + i(k)) - F_n(\omega)| \le$$
$$|\phi(n\rho_{G_n}(\omega + i(k))) F(\omega + i(k)) - \phi(n\rho_{G_n}(\omega)) F(\omega + i(k))|$$
$$+ |\phi(n\rho_{G_n}(\omega)) F(\omega + i(k)) - \phi(n\rho_{G_n}(\omega)) F(\omega)|$$
$$\le 4n\|k\|_H \mathbf{1}_{\{\rho_{G_n} < \frac{2}{3n} \ \ \mathrm{or} \ \ \rho_{G_n - i(k)} < \frac{2}{3n}\}} |F(\omega + i(k))|$$
$$+ \mathbf{1}_{\{\rho_{G_n} < \frac{2}{3n}\}} |F(\omega + i(k)) - F(\omega)|$$
$$\le 8n^2 \|k\|_H + \mathbf{1}_{\{\rho_{G_n} < \frac{2}{3n}\}} \|k\|_H \int_0^1 \|DF(\omega + ti(k))\|_H dt$$
$$\le (8n^2 + n)\|k\|_H.$$

So, by Exercise 1.2.9 we obtain $F_n \in \mathbb{D}^{1,2}$, and the proof is complete. □

**Lemma 4.1.2** *Let $F$ be a random variable in $C^1_H$ such that $E(F^2) < \infty$ and $E(\|DF\|^2_H) < \infty$. Then $F \in \mathbb{D}^{1,2}$.*

*Proof:*     For any $h \in H$ and any smooth random variable $G \in \mathcal{S}_b$ we have

$$E\left(\langle DF, h \rangle_H G\right)$$
$$= \lim_{\epsilon \to 0} \frac{1}{\epsilon} E\left((F(\omega + \epsilon i(h)) - F(\omega)) G\right)$$
$$= \lim_{\epsilon \to 0} \frac{1}{\epsilon} E\left(FG(\omega - \epsilon i(h)) e^{\epsilon W(h) - \frac{\epsilon^2}{2}\|h\|_H^2} - FG\right)$$
$$= E\left(F\left(GW(h) - \langle DG, h \rangle_H\right)\right).$$

The result then follows from Lemma 1.3.1.     □

For any $h \in H$ the random variable $W(h)$ is a.s. $H$-continuously differentiable and $D(W(h)) = h$ (see Exercise 4.1.3).

### 4.1.4 Transformations induced by elementary processes

Denote by $\{e_i, \ i \geq 1\}$ a complete orthonormal system in $H$. We will assume that $e_i \in \Omega^*$; in that way $W(e_i)$ is a continuous linear functional on $\Omega$. Consider a smooth elementary process $u \in \mathcal{S}_H$ (see Section 1.3.1) of the form

$$u = \sum_{j=1}^{N} \psi_j(W(e_1), \ldots, W(e_N)) e_j, \tag{4.7}$$

where $\psi \in C_b^\infty(\mathbb{R}^N; \mathbb{R}^N)$. The transformation $T : \Omega \to \Omega$ induced by $u$ is

$$T(\omega) = \omega + \sum_{j=1}^{N} \psi_j(W(e_1), \ldots, W(e_N)) i(e_j).$$

Define the random variable

$$\eta(u) = |\det(I + \Delta)| \exp\left(-\sum_{j=1}^{N} \langle u, e_j \rangle_H W(e_j) - \frac{1}{2} \sum_{j=1}^{N} \langle u, e_j \rangle_H^2\right), \tag{4.8}$$

where

$$\Delta = \psi'(W^{(N)})$$

and

$$W^{(N)} = (W(e_1), \ldots, W(e_N)).$$

We claim that $\eta$ can also be written as

$$\eta(u) = |\det_2(I + Du)| \exp(-\delta(u) - \frac{1}{2}\|u\|_H^2), \tag{4.9}$$

where $\det_2(I + Du)$ denotes the Carleman-Fredholm determinant of the square integrable kernel $I + Du$ (see the appendix, section A4). In order to deduce (4.9), let us first remark that for an elementary process $u$ of the form (4.7) the Skorohod integral $\delta(u)$ can be evaluated as follows:

$$
\begin{aligned}
\delta(u) &= \sum_{j=1}^{N} \psi_j(W^{(N)}) W(e_j) - \sum_{j=1}^{N} \sum_{i=1}^{N} \frac{\partial \psi_j}{\partial x_i}(W^{(N)}) \langle e_i, e_j \rangle_H \\
&= \sum_{j=1}^{N} \langle u, e_j \rangle_H W(e_j) - \mathrm{T}\,\Delta.
\end{aligned}
\tag{4.10}
$$

On the other hand, the derivative $Du$ is equal to

$$
Du = \sum_{j=1}^{N} \sum_{i=1}^{N} \Delta_{ij} e_j \otimes e_i, \quad \Delta_{ij} = \frac{\partial \psi_j}{\partial x_i}(W^{(N)}),
\tag{4.11}
$$

and the Carleman-Fredholm determinant of the kernel $I + Du$ is

$$
\det(I + \Delta) \exp(-\mathrm{T}\,\Delta),
$$

which completes the proof of (4.9).

The following proposition contains the basic results on the problems (A) and (B) for the transformation $T$ induced by $u$, when $u \in \mathcal{S}_H$.

**Proposition 4.1.4** *Suppose that $u$ is a smooth elementary process of the form (4.7). It holds that*

(i) *If $T$ is one to one, we have*

$$
E[\eta(u) F(T)] \leq E[F]
\tag{4.12}
$$

*for any positive and bounded random variable $F$.*

(ii) *If $\det_2(I + Du) \neq 0$ a.s. (which is equivalent to saying that the matrix $I + \psi'(x)$ is invertible a.e.), then*

$$
P \circ T^{-1} \leq P.
\tag{4.13}
$$

(iii) *If $T$ is bijective, then*

$$
E[\eta(u) F(T)] = E[F],
\tag{4.14}
$$

*for any positive and bounded random variable $F$.*

*Proof:*    Let us first remark that $T$ is one to one (onto) if and only if the mapping $\varphi(x) = x + \psi(x)$ is one to one (onto) on $\mathbb{R}^N$. We start by showing (i). It suffices to assume that the random variable $F$ has the form

$$
F = f(W(e_1), \ldots, W(e_N), G)
\tag{4.15}
$$

where $G$ denotes the vector $(W(e_{N+1}), \ldots, W(e_M))$, $M > N$, and $f$ belongs to the space $C_b^\infty(\mathbb{R}^M)$. The composition $F(T)$ is given by

$$F(T) = f(W(e_1) + \langle u, e_1 \rangle_H, \ldots, W(e_N) + \langle u, e_N \rangle_H, G).$$

Note that $\langle u, e_j \rangle_H = \psi_j(W^{(N)})$. Applying the change-of-variables formula and using the one to one character of $\varphi$, we obtain

$$E[\eta F(T)] = E[|\det(I + \Delta)| \exp\left(-\sum_{j=1}^N \langle u, e_j \rangle_H W(e_j) - \frac{1}{2}\sum_{j=1}^N \langle u, e_j \rangle_H^2\right)$$

$$\times f(W(e_1) + \langle u, e_1 \rangle_H, \ldots, W(e_N) + \langle u, e_N \rangle_H, G)]$$

$$= E\int_{\mathbb{R}^N} \left|\det\left(I + \psi'(x)\right)\right| \exp\left(-\frac{1}{2}\sum_{j=1}^N (x_j + \psi_j(x))^2\right)$$

$$\times f(x_1 + \psi_1(x), \ldots, x_N + \psi_N(x), G)(2\pi)^{-\frac{N}{2}} dx_1 \cdots dx_N$$

$$= E\int_{\varphi(\mathbb{R}^N)} \exp\left(-\frac{1}{2}\sum_{j=1}^N y_j^2\right) f(y_1, \ldots, y_N, G)(2\pi)^{-\frac{N}{2}} dy_1 \cdots dy_N$$

$$\leq E\int_{\mathbb{R}^N} \exp\left(-\frac{1}{2}\sum_{j=1}^N y_j^2\right) f(y_1, \ldots, y_N, G)(2\pi)^{-\frac{N}{2}} dy_1 \cdots dy_N$$

$$= E(F). \tag{4.16}$$

This concludes the proof of (i). Property (ii) follows from the fact that $\varphi$ is locally one to one. Indeed, we can find a countable family of open sets $G_n \subset \mathbb{R}^N$ such that their union is $\{\det(I + \psi') \neq 0\}$ and $\varphi$ is one to one on each $G_n$. This implies by (i) that

$$E[\eta(u)\mathbf{1}_{\{W^{(N)} \in G_n\}} F(T)] \leq E[F]$$

for any nonnegative and bounded random variable $F$, which yields the desired absolute continuity. Finally, property (iii) is proved in the same way as (i). $\qquad\square$

**Remarks:** Condition $\det_2(I + Du) \neq 0$ a.s. implies a positive answer to problem (A). On the other hand, the bijectivity of $T$ allows us to deduce a Girsanov-type theorem. Notice that (4.12) implies (4.13) and (4.12), and the additional hypothesis $E[\eta(u)] = 1$ implies (4.14).

In the next section we will extend these results to processes that are not necessarily elementary.

### 4.1.5 Anticipating Girsanov theorems

Let us first consider the case of a process whose derivative is strictly less than one in the norm of $H \otimes H$. We need the following approximation result.

**Lemma 4.1.3** *Let $u \in \mathbb{L}^{1,2}$ be such that $\|Du\|_{H \otimes H} < 1$ a.s. Then there exists a sequence of smooth elementary processes $u_n \in \mathcal{S}_H$ such that a.s. $u_n \to u$, $\eta(u_n) \to \eta(u)$, and $\|Du_n\|_{H \otimes H} < 1$ for all $n$.*

*Proof:*   It suffices to find a sequence $u_n$ such that $u_n$ converges to $u$ in the norm of $\mathbb{L}^{1,2}$, and $\|Du_n\|_{H \otimes H} \leq 1$. In fact, the almost sure convergence of $u_n$ and $\eta(u_n)$ is then deduced by choosing a suitable subsequence (note that $Du$ and $\delta(u)$ are continuous functions of $u$ with respect to the norm of $\mathbb{L}^{1,2}$, and $\eta(u)$ is a continuous function of $u$, $Du$, and $\delta(u)$). Moreover, to get the strict inequality $\|Du\|_{H \otimes H} < 1$, we replace $u_n$ by $(1 - \frac{1}{n})u_n$.

The desired sequence $u_n$ is obtained as follows. Fix a complete orthonormal system $\{e_i, i \geq 1\} \subset \Omega^*$ in $H$, and denote by $\mathcal{F}_n$ the $\sigma$-field generated by $W^{(n)} = (W(e_1), \ldots, W(e_n))$. Define

$$P_n u = \sum_{i=1}^{n} E[\langle u, e_i \rangle_H | \mathcal{F}_n] e_i.$$

Then $P_n u$ converges to $u$ in the norm of $\mathbb{L}^{1,2}$, and

$$\|D[P_n u]\|_{H \otimes H} \leq \|E[Du | \mathcal{F}_n]\|_{H \otimes H} \leq 1.$$

By Exercise 1.2.8 we know that $E[\langle u, e_i \rangle_H | \mathcal{F}_n] = f_i(W^{(n)})$, where the function $f_i$ belongs to the Sobolev space $\mathbb{W}^{1,2}(\mathbb{R}^n, N(0, I_n))$.

Replacing $f_i$ by $\Psi_N \circ f_i$, where $\Psi_N \in C_0^{\infty}(\mathbb{R})$ is a function equal to $x$ if $|x| \leq N$ and $|\Psi'_N(x)| \leq 1$, we can assume that $f_i$ and its partial derivatives are bounded. Finally, it suffices to smooth $f_i$ by means of the convolution with an approximation of the identity. In fact, we have

$$\sum_{i,j=1}^{n} (\varphi_\epsilon * \partial_j f_i)(x)^2 \leq \sum_{i,j=1}^{n} [\varphi_\epsilon * (\partial_j f_i)^2](x) \leq \sum_{i,j=1}^{n} (\partial_j f_i)^2(x) \leq 1$$

a.e., which allows us to complete the proof.  □

The following result is due to Buckdahn and Enchev (cf. [48] and [91]). See also Theorem 6.1 in Kusuoka's paper [178].

**Proposition 4.1.5** *Suppose that $u \in \mathbb{L}^{1,2}$ and $\|Du\|_{H \otimes H} < 1$ a.s. Then the transformation $T(\omega) = \omega + i(u)$ verifies (4.12), and $P \circ T^{-1} << P$.*

*Proof:*   Let $u_n$ be the sequence provided by Lemma 4.1.3. The transformation $T$ associated with each process $u_n$ is one to one, because $\varphi_n$ is a contraction (we assume $\|Du_n\|_{H \otimes H} \leq 1 - \frac{1}{n}$). Then, by (4.12), for each $n$ we have

$$E[\eta(u_n)F(T_n)] \leq E[F],$$

for any nonnegative, continuous, and bounded function $F$. By Fatou's lemma this inequality holds for the limit process $u$, and (4.12) is true. Finally, the absolute continuity follows from the fact that $\eta(u) > 0$ a.s. because $\|Du\|_{H \otimes H} < 1$ a.s.    □

If in Proposition 4.1.5 we assume in addition that $E[\eta(u)] = 1$, then (4.14) holds. A sufficient condition for $E[\eta(u)] = 1$ has been given by Buckdahn in [48] (see also Exercise 4.1.4).

In order to remove the hypothesis that the derivative is strictly bounded by one, we will show that if $u$ is $H$-continuously differentiable then the corresponding transformation can be locally approximated by the composition of a linear transformation and a transformation induced by a process whose derivative is less than one. The $H$ differentiability property seems to be fundamental in order to obtain this decomposition. The following two lemmas will be useful in completing the localization procedure.

**Lemma 4.1.4** *Suppose that there exists a sequence of measurable sets $B_n$ and an element $u \in \mathbb{L}_{\text{loc}}^{1,2}$ such that $\cup_n B_n = \Omega$ a.s., and*

$$E[\eta(u)\mathbf{1}_{B_n} F(T)] \leq E[F] \qquad (4.17)$$

*for any $n \geq 1$, and for any nonnegative, measurable and bounded function $F$. Then*

*(i) if $\det_2(I + Du) \neq 0$ a.s., we have $P \circ T^{-1} << P$;*

*(ii) if there exists a left inverse $T_l$ such that $T_l \circ T = Id$ a.s., then*

$$E[\eta(u)F(T)] \leq E[F]$$

*for any nonnegative, measurable, and bounded function $F$.*

*Proof:*    Part (i) is obvious. In order to show (ii) we can assume that the sets $B_n$ are pairwise disjoint. We can write $B_n = T^{-1}(T_l^{-1}(B_n))$, and we have

$$E[\eta(u)F(T)] = \sum_n E[\eta(u)\mathbf{1}_{B_n} F(T)] \leq \sum_n E[\mathbf{1}_{T_l^{-1}(B_n)} F(T)] \leq E[F].$$

□

We will denote by $C_H^1(H)$ the class of a.s. $H$-continuously differentiable processes (or $H$-valued random variables).

**Lemma 4.1.5** *Let $u_1 \in C_H^1(H)$, $u_2, u_3 \in \mathbb{L}^{1,2}$, and denote by $T_1$, $T_2$, and $T_3$ the corresponding transformations. Suppose that*

*(i) $P \circ T_2^{-1} << P$;*

*(ii) $T_3 = T_1 \circ T_2$    (or $u_3 = u_2 + u_1(T_2)$).*

*Then we have*

(a) $I + Du_3 = [I + (Du_1)(T_2)](I + Du_2)$;

(b) $\eta(u_3) = \eta(u_1)(T_2)\eta(u_2)$.

**Remarks:**   We recall that the composition $(Du_1)(Du_2)$ verifies

$$(Du_1)(Du_2) = \sum_{i,j,k=1}^{\infty} D_{e_k}(\langle u_1, e_i \rangle_H) D_{e_j}(\langle u_2, e_k \rangle_H) e_i \otimes e_j.$$

*Proof of Lemma 4.1.5:*   From Lemma 4.1.2 we deduce

$$Du_3 = Du_2 + (Du_1)(T_2) + (Du_1)(T_2)(Du_2),$$

which implies (a). In order to prove (b) we use the following property of the Carleman-Fredholm determinant:

$$\det_2[(I + K_1)(I + K_2)] = \det_2(I + K_1)\det_2(I + K_2)\exp(-\mathrm{T}(K_1 K_2)).$$

Hence,

$$
\begin{aligned}
\eta(u_3) \;=\; & \det_2\;(I + (Du_1)(T_2))\det_2(I + Du_2) \\
& \times \exp(-\mathrm{T}[(Du_1)(T_2)(Du_2)]) \\
& \times \exp\left[-\delta(u_2) - \delta(u_1(T_2)) - \frac{1}{2}\|u_2\|_H^2 \right. \\
& \left. - \frac{1}{2}\|u_1(T_2)\|_H^2 - \langle u_2, u_1(T_2) \rangle_H \right].
\end{aligned}
$$

Finally, we use the property

$$
\begin{aligned}
\delta(u_1(T_2)) \;=\; & \delta(f(W(h) + \langle h, u_2 \rangle_H)g) \\
=\; & f(W(h) + \langle h, u_2 \rangle_H)W(g) - f'(W(h) + \langle h, u_2 \rangle_H)(\langle h, g \rangle_H \\
& - \langle Du_2, h \otimes g \rangle_H) \\
=\; & \delta(u_1)(T_2) - \langle u_1(T_2), u_2 \rangle_H - \mathrm{T}[(Du_1)(T_2)(Du_2)],
\end{aligned}
$$

which allows us to complete the proof.                                      □

We are now able to show the following result.

**Proposition 4.1.6** *Let $u \in C_H^1(H)$, and assume that $\det_2(I + Du) \neq 0$ (which is equivalent to saying that $I + Du$ is invertible a.s.). Then there exists a sequence of measurable subsets $G_n \subset \Omega$ such that $\cup_n G_n = \Omega$ a.s., and three sequences of processes $u_{n,1}$, $u_{n,2}$, and $u_{n,3}$ such that*

(a) $u_{n,1} = v_n$ *is deterministic;*

(b) $u_{n,2}$ *belongs to $\mathbb{L}^{1,2}$ and $\|Du_{n,2}\|_{H \otimes H} \leq c < 1$;*

(c) $u_{n,3}$ is a smooth elementary process given by a linear map $\psi_n$ such that $\det(I + \psi_n) \neq 0$;

(d) $T = T_{n,1} \circ T_{n,2} \circ T_{n,3}$ on $G_n$, where $T$ and $T_{n,i}$ $(i = 1, 2, 3)$ are the transformations associated with the processes $u$ and $u_{n,i}$, respectively. In other words, $u = v_n + u_{n,2}(T_{n,3}) + u_{n,3}$, on $G_n$.

*Proof:*    Let $\{e_i, i \geq 1\} \subset \Omega^*$ be a complete orthonormal system in $H$. Fix a countable and dense subset $H_0$ of $H$. Consider the set $\mathcal{K} \subset H \otimes H$ formed by the kernels of the form

$$K = \sum_{i,j=1}^{n} \lambda_{ij} e_i \otimes e_j,$$

where $n \geq 1$, $\lambda_{ij} \in \mathbb{Q}$, and $\det(I + \lambda) \neq 0$. The set $\mathcal{K}$ is dense in the set $\{K \in H \otimes H : \det_2(I + K) \neq 0\}$, and for any $K \in \mathcal{K}$ the operator $I + K$ has a bounded inverse. We will denote by $\tilde{K}$ the smooth elementary process associated with $K$, namely,

$$\tilde{K} = \sum_{i,j=1}^{n} \lambda_{ij} W(e_j) e_i.$$

So, we have $D_{e_i} \tilde{K} = K(e_i)$, $i \geq 1$. Fix a positive number $0 < a < \frac{c}{9}$, where $0 < c < 1$ is a fixed number. Set

$$\gamma(K) = \|(I + K)^{-1}\|_{\mathcal{L}(H,H)}^{-1}.$$

Consider the set of triples $\nu = (K, v, n)$, with $n \geq 1$, $K \in \mathcal{K}$, and $v \in H_0$, such that $\gamma(K) < \frac{n}{3a}$. This countable set will be denoted by $\mathcal{I}$. For each $\nu \in \mathcal{I}$ we define the following set:

$$C_\nu = \Big\{\omega \in \Omega : \sup_{\|h\|_H \leq \frac{1}{n}} \|Du(\omega + i(h)) - K\|_{H \otimes H} \leq a\gamma(K),$$

$$\|u(\omega) - \tilde{K}(\omega) - v\|_H \leq \frac{a\gamma(K)}{n}\Big\}.$$

Then the union of the sets $C_\nu$ when $\nu \in \mathcal{I}$ is the whole space $\Omega$ a.s., because $\det_2(I + Du) \neq 0$ and $u$ is a.s. $H$-continuously differentiable. We can find $\sigma$-compact sets $G_\nu \subset C_\nu$ such that $P(G_\nu) = P(C_\nu)$. The sets $G_\nu$ constitute the family we are looking for, and we are going to check that these sets satisfy properties (a) through (d) for a suitable sequence of processes. Let $\phi \in C_0^\infty(\mathbb{R})$ be a nonnegative function such that $|\phi(t)| \leq 1$ and $|\phi'(t)| \leq 4$ for all $t$, $\phi(t) = 1$ for $|t| \leq \frac{1}{3}$, and $\phi(t) = 0$ for $|t| \geq \frac{2}{3}$. Define

$$u_{\nu,1} = v,$$
$$u_{\nu,3} = \tilde{K},$$
$$u_{\nu,2} = \phi\left[n\rho_{G_\nu}(T_{\nu,3}^{-1})\right]\left[u(T_{\nu,3}^{-1}) - u_{\nu,3}(T_{\nu,3}^{-1}) - v\right].$$

Clearly, these $u_{\nu,1}$ and $u_{\nu,3}$ satisfy conditions (a) and (c), respectively. Then it suffices to check properties (b) and (d). The process $u_{\nu,2}$ is the product of two factors: The first one is a random variable that is bounded and belongs to $\mathbb{D}^{1,p}$ for all $p \geq 2$, due to property (iv) of the function $\rho$. The second factor is $H$-continuously differentiable.

Let us now show that the derivative of $u_{\nu,2}$ is bounded by $c < 1$ in the norm of $H \otimes H$. Define

$$v_\nu = u_{\nu,2}(T_{\nu,3}) = \phi(n\rho_{G_\nu})[u - \tilde{K} - v]. \tag{4.18}$$

Then $u_{\nu,2} = v_\nu(T_{\nu,3}^{-1})$, and we have

$$
\begin{aligned}
\|Du_{\nu,2}\|_{H\otimes H} &= \left\|[I + K]^{-1}(Dv_\nu)(T_{\nu,3}^{-1})\right\|_{H\otimes H} \\
&\leq \gamma(K)^{-1}\|(Dv_\nu)(T_{\nu,3}^{-1})\|_{H\otimes H}.
\end{aligned}
$$

So it suffices to check that $\|Dv_\nu\|_{H\otimes H} \leq c\gamma(K)$. We have

$$
\begin{aligned}
\|Dv_\nu\|_{H\otimes H} &\leq \mathbf{1}_{\{\rho_{G_\nu}<\frac{1}{n}\}}(\|Du - K\|_{H\otimes H} \\
&\quad +4n\|u - u_{\nu,3} - v\|_H).
\end{aligned}
$$

Suppose that $\rho_{G_\nu}(\omega) < \frac{1}{n}$. Then there exists an element $h \in H$ such that $\omega + i(h) \in G_\nu$ and $\|h\|_H < \frac{1}{n}$. Consequently, from the definition of the set $C_\nu$ we obtain for this element $\omega$

$$\|(Du)(\omega) - K\|_{H\otimes H} \leq a\gamma(K),$$

and

$$
\begin{aligned}
\|u(\omega) - u_{\nu,3}(\omega) - v\|_H &\leq \|u(\omega + i(h)) - u_{\nu,3}(\omega + i(h)) - v\|_H \\
&\quad +\|u(\omega + i(h)) - u(\omega) - \langle K, h\rangle\|_H \\
&\leq \frac{a\gamma(K)}{n} + \|\int_0^1 \langle(Du)(\omega + ti(h)) - K, h\rangle_H\, dt\|_H \\
&\leq \frac{2a\gamma(K)}{n}.
\end{aligned}
$$

Hence, we have $\|Dv_\nu\|_{H\otimes H} \leq 9a\gamma(K)$. Also this argument implies that $u_{\nu,2}$ is bounded. As a consequence (see Exercise 4.1.5), $u_{\nu,2}$ belongs to $\mathbb{L}^{1,2}$.

Finally, we will check property (d). Notice first that $\omega \in G_\nu$ implies $\rho_{G_\nu}(\omega) = 0$. Therefore, from (4.18) we obtain

$$u(\omega) = v + u_{\nu,2}(T_{\nu,3}(\omega)) + u_{\nu,3}(\omega).$$

$\square$

Using the above proposition we can prove the following absolute continuity result.

**Theorem 4.1.1** *Let $u$ be a process that is a.s. $H$-continuously differentiable. Suppose that $I + Du$ is invertible a.s. (or $\det_2(I + Du) \neq 0$ a.s.). Denote by $T$ the transformation defined by $T(\omega) = \omega + i(u(\omega))$. Then $P \circ T^{-1} << P$.*

*Proof:*   Consider the decomposition $T = T_{n,1} \circ T_{n,2} \circ T_{n,3}$ and the sequence of sets $G_n$ introduced in Proposition 4.1.6. The transformations $T_{n,i}$, $i = 1, 2, 3$, verify (4.12). For $i = 1, 3$ this follows from part (i) of Proposition 4.1.4, and for $i = 2$ it follows from Proposition 4.1.5. From Lemma 4.1.5 and using the local property of $\eta(u)$, we have

$$\eta(u) = \eta(v)(T_{n,2} \circ T_{n,3})\eta(u_{n,2})(T_{n,3})\eta(u_{n,3}) \quad \text{on} \quad G_n.$$

Consequently, for any nonnegative and bounded random variable $F$ we obtain

$$\begin{aligned}
E\left[\mathbf{1}_{G_n} F(T)\eta(u)\right] &= \\
E\left[\mathbf{1}_{G_n} F(T_{n,1} \circ T_{n,2} \circ T_{n,3})\eta(v)(T_{n,2} \circ T_{n,3})\eta(u_{n,2})(T_{n,3})\eta(u_{n,3})\right] \\
&\leq E\left[F(T_{n,1} \circ T_{n,2})\eta(v)(T_{n,2})\eta(u_{n,2})\right] \\
&\leq E\left[F(T_{n,1})\eta(v)\right] \leq E[F],
\end{aligned}$$

and the result follows from Lemma 4.1.4.   □

The main result of this section is the following version of Girsanov's theorem due to Kusuoka ([178], Theorem 6.4).

**Theorem 4.1.2** *Let $u$ be a process that is $H$-continuously differentiable, and denote by $T$ the transformation defined by $T(\omega) = \omega + i(u(\omega))$. Suppose that*

*(i) $T$ is bijective;*

*(ii) $I + Du$ is invertible a.s. (or $\det_2(I + Du) \neq 0$ a.s.).*

*Then there exists a probability $Q$ equivalent to $P$, such that $Q \circ T^{-1} = P$, given by*

$$\frac{dQ}{dP} = |\det_2(I + Du)| \exp(-\delta(u) - \frac{1}{2} \|u\|_H^2). \tag{4.19}$$

*Proof:*   From Theorem 4.1.1 we know that Eq. (4.17) of Lemma 4.1.4 holds with the sequence of sets $G_n$ given by Proposition 4.1.6. Then part (ii) of Lemma 4.1.4 yields

$$E[\eta(u)F(T)] \leq E[F], \tag{4.20}$$

for any nonnegative and bounded random variable $F$. It is not difficult to see, using the implicit function theorem, that the inverse transformation $T^{-1}$ is also associated with an a.s. $H$-continuously differentiable process.

Let us denote this process by $v$. From Lemma 4.1.5 applied to the composition $I = T^{-1} \circ T$, we get $1 = \eta(v)(T)\eta(u)$. Therefore, $P \circ T << P$, and applying (4.20) to $T^{-1}$ we get

$$E(F) = E[\eta(v)\eta(u)(T^{-1})F] \le E[\eta(u)F(T)],$$

so the equality in (4.20) holds and the proof is complete.     □

## Exercises

**4.1.1** Show that the law of the continuous process $W_t + tf(W_1)$ is absolutely continuous with respect to the Wiener measure on $C_0([0,1])$ if and only if $Y + f(Y)$ has a density when $Y$ has $N(0,1)$ law. If $f$ is locally Lipschitz, a sufficient condition for this is $f'(x) \ne -1$ a.e.

**4.1.2** Using Itô's formula, show that

$$E\left(\exp\left(-\int_0^S \frac{2W_t}{(1-t)^2} dW_t - 2\int_0^S \frac{W_t^2}{(1-t)^4} dt\right)\right) \le e^{-1},$$

where $S = inf\{t : W_t^2 = 1 - t\}$.

**4.1.3**    Show that for any $h \in H$ the random variable $W(h)$ is a.s. $H$ continuously differentiable and $D(W(h)) = h$.

**4.1.4** Let $u \in \mathbb{L}^{1,2}$ be a process satisfying the following properties:

(i) $\| Du \|_{H \otimes H} < c < 1$;

(ii) $E\left(\exp\left(\frac{q}{2}\int_0^1 u_t^2 dt\right)\right) < \infty$ for some $q > 1$.

Show that $E(\eta) = 1$ (see Enchev [91]).

**4.1.5** Let $F \in C_H^1$, and $G \in \mathbb{D}^{1,p}$ for all $p > 1$. Assume that $E(F^2 G^2) < \infty$ and

$$E(\|DF\|_H^2 G^2 + \|DG\|_H^2 F^2) < \infty.$$

Show that $FG$ belongs to $\mathbb{D}^{1,2}$.

## 4.2    Markov random fields

We start by introducing the notion of conditional independence.

**Definition 4.2.1** *Let $\mathcal{F}_1$, $\mathcal{F}_2$, and $\mathcal{F}_3$ be three sub-$\sigma$-algebras in a probability space $(\Omega, \mathcal{F}, P)$. We will say that $\mathcal{F}_1$ and $\mathcal{F}_2$ are conditionally independent given $\mathcal{F}_3$ if*

$$P(A_1 \cap A_2 | \mathcal{F}_3) = P(A_1 | \mathcal{F}_3) P(A_2 | \mathcal{F}_3)$$

*for all $A_1 \in \mathcal{F}_1$ and $A_2 \in \mathcal{F}_2$. In this case we will write $\mathcal{F}_1 \underset{\mathcal{F}_3}{\perp\!\!\!\perp} \mathcal{F}_2$.*

The conditional independence $\mathcal{F}_1 \underset{\mathcal{F}_3}{\perp\!\!\!\perp} \mathcal{F}_2$ is equivalent to the property

$$P(A_1|\mathcal{F}_3) = P(A_1|\mathcal{F}_2 \vee \mathcal{F}_3)$$

for all $A_1 \in \mathcal{F}_1$. We refer to Rozanov [296] for a detailed analysis of this notion and its main properties.

The conditional independence allows us to introduce several definitions of Markov's property. Let $\{X(t), t \in [0,1]\}$ be a continuous stochastic process.

(a) We will say that $X$ is a *Markov process* if for all $t \in [0,1]$ we have

$$\sigma\{X_r, r \in [0,t]\} \underset{\sigma\{X_t\}}{\perp\!\!\!\perp} \sigma\{X_r, r \in [t,1]\}.$$

(b) We will say that $X$ is a *Markov random field* if for all $0 \le s < t \le 1$ we have

$$\sigma\{X_r, r \in [s,t]\} \underset{\sigma\{X_s,X_t\}}{\perp\!\!\!\perp} \sigma\{X_r, r \in [0,1] - (s,t)\}.$$

Property (a) is stronger than (b) (see Exercise 4.2.3). The converse is not true (see Exercise 4.2.4).

In the next section we will apply Theorem 4.1.2 to study the Markov field property of solutions to stochastic differential equations with boundary conditions.

## 4.2.1  Markov field property for stochastic differential equations with boundary conditions

Let $\{W_t, t \in [0,1]\}$ be a standard Brownian motion defined on the canonical probability space $(\Omega, \mathcal{F}, P)$. Consider the stochastic differential equation

$$\begin{cases} X_t = X_0 - \int_0^t f(X_s)ds + W_t \\ X_0 = g(X_1 - X_0), \end{cases} \tag{4.21}$$

where $f, g : \mathbb{R} \to \mathbb{R}$ are two continuous functions.

Observe that the periodic condition $X_0 = X_1$ is not included in this formulation. In order to handle this and other interesting cases, one should consider more general boundary conditions of the form

$$X_0 = g(e^{-\lambda}X_1 - X_0),$$

with $\lambda \in \mathbb{R}$. The periodic case would correspond to $\lambda \ne 0$ and $g(x) = (e^{-\lambda} - 1)^{-1}x$. In order to simplify the exposition we will assume henceforth that $\lambda = 0$.

When $f \equiv 0$, the solution of (4.21) is

$$Y_t = W_t + g(W_1). \tag{4.22}$$

Denote by $\Sigma$ the set of continuous functions $x : [0,1] \to \mathbb{R}$ such that $x_0 = g(x_1 - x_0)$. The mapping $\omega \to Y(\omega)$ is a bijection from $\Omega$ into $\Sigma$. Consider the process $Y = \{Y_t, t \in [0,1]\}$ given by (4.22). Define the transformation $T : \Omega \to \Omega$ by

$$T(\omega)_t = \omega_t + \int_0^t f(Y_s(\omega))ds. \tag{4.23}$$

**Lemma 4.2.1** *The transformation $T$ is a bijection of $\Omega$ if and only if Eq. (4.21) has a unique solution for each $\omega \in \Omega$; in this case this solution is given by $X = Y(T^{-1}(\omega))$.*

*Proof:*    If $T(\eta) = \omega$, then the function $X_t = Y_t(\eta)$ solves Eq. (4.21) for $W_t = \omega_t$. Indeed:

$$X_t = X_0 + \eta_t = X_0 + \omega_t - \int_0^t f(Y_s(\eta))ds = X_0 + W_t - \int_0^t f(X_s)ds.$$

Conversely, given a solution $X$ to Eq. (4.21), we have $T(Y^{-1}(X)) = W$. Indeed, if we set $Y^{-1}(X) = \eta$, then

$$
\begin{aligned}
T(\eta)_t &= \eta_t + \int_0^t f(Y_s(\eta))ds = \eta_t + \int_0^t f(X_s)ds \\
&= \eta_t + W_t - X_t + X_0 = W_t.
\end{aligned}
$$

$\square$

There are sufficient conditions for $T$ to be a bijection (see Exercise 4.2.10). Henceforth we will impose the following assumptions:

**(H.1)** There exists a unique solution to Eq. (4.21) for each $\omega \in \Omega$.

**(H.2)** $f$ and $g$ are of class $C^1$.

Now we turn to the discussion of the Markov field property. First notice that the process $Y$ is a Markov random field (see Exercise 4.2.3). Suppose that $Q$ is a probability on $\Omega$ such that $P = Q \circ T^{-1}$. Then $\{T(\omega)_t, 0 \le t \le 1\}$ will be a Wiener process under $Q$, and, consequently, the law of the process $X$ under the probability $P$ coincides with the law of $Y$ under $Q$. In this way we will translate the problem of the Markov property of $X$ into the problem of the Markov property of the process $Y$ under a new probability $Q$. This problem can be handled, provided $Q$ is absolutely continuous with respect to the Wiener measure $P$ and we can compute an explicit expression for its Radon-Nikodym derivative. To do this we will make use of Theorem 4.1.2, applied to the process

$$u_t = f(Y_t). \tag{4.24}$$

Notice that $T$ is bijective by assumption (H.1) and that $u$ is $H$-continuously differentiable by (H.2). Moreover,

$$D_s u_t = f'(Y_t)[g'(W_1) + \mathbf{1}_{\{s \le t\}}].$$

(4.25)

The the Carleman-Fredholm determinant of kernel (4.25) is computed in the next lemma.

**Lemma 4.2.2** *Set* $\alpha_t = f'(Y_t)$. *Then*

$$\det{}_2(I + Du) = \left(1 + g'(W_1)\left(1 - e^{-\int_0^1 \alpha_t dt}\right)\right) e^{-g'(W_1)\int_0^1 \alpha_t dt}.$$

*Proof:*   From (A.12) applied to the kernel $Du$, we obtain

$$\det{}_2(I + Du) = 1 + \sum_{n=2}^{\infty} \frac{\gamma_n}{n!},$$

(4.26)

where

$$\begin{aligned}
\gamma_n &= \int_{[0,1]^n} \det(\mathbf{1}_{\{i \ne j\}} D_{t_i} u_{t_j}) dt_1 \ldots dt_n \\
&= n! \int_{\{t_1 < t_2 < \cdots < t_n\}} \det(\mathbf{1}_{\{i \ne j\}} D_{t_i} u_{t_j}) dt_1 \ldots dt_n \\
&= \left(\int_0^1 \alpha(t) dt\right)^n \det B_n,
\end{aligned}$$

and the matrix $B_n$ is given by

$$B = \begin{bmatrix}
0 & g'(W_1) + 1 & g'(W_1) + 1 & \cdots & g'(W_1) + 1 \\
g'(W_1) & 0 & g'(W_1) + 1 & \cdots & g'(W_1) + 1 \\
g'(W_1) & g'(W_1) & 0 & \cdots & g'(W_1) + 1 \\
\cdots & \cdots & \cdots & \cdots & \cdots \\
g'(W_1) & g'(W_1) & g'(W_1) & \cdots & 0
\end{bmatrix}.$$

Simple computations show that for all $n \ge 1$

$$\det B_n = (-1)^n g'(W_1)^n (g'(W_1) + 1) + (-1)^{n+1} g'(W_1)(g'(W_1) + 1)^n.$$

Hence,

$$\begin{aligned}
\det{}_2(I + Du) &= 1 + \sum_{n=1}^{\infty} \frac{1}{n!} \left(\int_0^1 \alpha(t) dt\right)^n \\
&\quad \times \left[(-1)^n g'(W_1)^n (g'(W_1) + 1) + (-1)^{n+1} g'(W_1)(g'(W_1) + 1)^n\right] \\
&= (g'(W_1) + 1) e^{-g'(W_1)\int_0^1 \alpha_t dt} - g'(W_1) e^{-(g'(W_1)+1)\int_0^1 \alpha_t dt} \\
&= \left(1 + g'(W_1)\left(1 - e^{-\int_0^1 \alpha_t dt}\right)\right) e^{-g'(W_1)\int_0^1 \alpha_t dt}.
\end{aligned}$$

$\square$

Therefore, the following condition implies that $\det_2(I + Du) \neq 0$ a.s.:

**(H.3)**  $1 + g'(y)\left(1 - e^{-f'(x+g(y))}\right) \neq 0$, for almost all $x, y$ in $\mathbb{R}$.

Suppose that the functions $f$ and $g$ satisfy conditions (H.1) through (H.3). Then the process $u$ given by (4.24) satisfies the conditions of Theorem 4.1.2, and we obtain

$$\eta(u) = \frac{dQ}{dP} = \left|1 + g'(W_1)\left[1 - \exp\left(-\int_0^1 f'(Y_t)dt\right)\right]\right| \qquad (4.27)$$
$$\times \exp\left(-g'(W_1)\int_0^1 f'(Y_t)dt - \int_0^1 f(Y_t)dW_t\right.$$
$$\left.-\frac{1}{2}\int_0^1 f(Y_t)^2 dt\right).$$

We will denote by $\Phi$ the term

$$\Phi = 1 + g'(W_1)\left[1 - \exp\left(-\int_0^1 f'(Y_t)dt\right)\right],$$

and let $L$ be the exponential factor in (4.27). Using the relationship between the Skorohod and Stratonovich integrals, we can write

$$\int_0^1 f(Y_t)dW_t = \int_0^1 f(Y_t) \circ dW_t - \frac{1}{2}\int_0^1 f'(Y_t)dt - g'(W_1)\int_0^1 f'(Y_t)dt.$$

Consequently, the term $L$ can be written as

$$L = \exp\left(-\int_0^1 f(Y_t) \circ dW_t + \frac{1}{2}\int_0^1 f'(Y_t)dt - \frac{1}{2}\int_0^1 f(Y_t)^2 dt\right). \qquad (4.28)$$

In this form we get

$$\eta(u) = |\Phi|L.$$

The main result about the Markov field property of the process $X$ is the following:

**Theorem 4.2.1** *Suppose that the functions $f$ and $g$ are of class $C^2$ and $f'$ has linear growth. Suppose furthermore that the equation*

$$\begin{cases} X_t = X_0 - \int_0^t f(X_s)ds + W_t \\ X_0 = g(X_1 - X_0) \end{cases} \qquad (4.29)$$

*has a unique solution for each $W \in \Omega$ and that (H.3) holds. Then the process $X$ verifies the Markov field property if and only if one of the following conditions holds:*

*(a) $f(x) = ax + b$, for some constants $a, b \in \mathbb{R}$,*

(b) $g' = 0$,

(c) $g' = -1$.

## Remarks:

**1.** If condition (b) or (c) is satisfied, we have an initial or final fixed value. In this case, assuming only that $f$ is Lipschitz, it is well known that there is a unique solution that is a Markov proces (not only a Markov random field).

**2.** Suppose that (a) holds, and assume that the implicit equation $x = g((e^{-a} - 1)x + y)$ has a unique continuous solution $x = \varphi(y)$. Then Eq. (4.29) admits a unique solution that is a Markov random field (see Exercise 4.2.6).

*Proof:*     Taking into account the above remarks, it suffices to show that if $X$ is a Markov random field then one of the above conditions is satisfied. Let $Q$ be the probability measure on $C_0([0,1])$ given by (4.27). The law of the process $X$ under $P$ is the same as the law of $Y$ under $Q$. Therefore, $Y$ is a Markov field under $Q$.

For any $t \in (0,1)$, we define the $\sigma$-algebras

$$\mathcal{F}_t^i = \sigma\left\{Y_u, \, 0 \le u \le t\right\} = \sigma\{W_u, \, 0 \le u \le t, g(W_1)\},$$
$$\mathcal{F}_t^e = \sigma\left\{Y_u, \, t \le u \le 1, \, Y_0\right\} = \sigma\{W_u, \, t \le u \le 1\}, \quad \text{and}$$
$$\mathcal{F}_t^0 = \sigma\{Y_0, Y_t\} = \sigma\{W_t, g(W_1)\}.$$

The random variable $L$ defined in (4.28) can be written as $L = L_t^i L_t^e$, where

$$L_t^i = \exp\left(-\int_0^t f(Y_s) \circ dW_s + \frac{1}{2}\int_0^t f'(Y_s)ds - \frac{1}{2}\int_0^t f(Y_s)^2 ds\right)$$

and

$$L_t^e = \exp\left(-\int_t^1 f(Y_s) \circ dW_s + \frac{1}{2}\int_t^1 f'(Y_s)ds - \frac{1}{2}\int_t^1 f(Y_s))^2 ds\right).$$

Notice that $L_t^i$ is $\mathcal{F}_t^i$-measurable and $L_t^e$ is $\mathcal{F}_t^e$-measurable. For any nonnegative random variable $\xi$, $\mathcal{F}_t^i$-measurable, we define (see Exercise 4.2.11)

$$\wedge_\xi = E_Q(\xi|\mathcal{F}_t^e) = \frac{E(\xi\eta(u) \mid \mathcal{F}_t^e)}{E(\eta(u) \mid \mathcal{F}_t^e)} = \frac{E(\xi|\Phi|L_t^i \mid \mathcal{F}_t^e)}{E(|\Phi|L_t^i \mid \mathcal{F}_t^e)}.$$

The denominator in the above expression is finite a.s. because $\eta(u)$ is integrable with respect to $P$. The fact that $Y$ is a Markov field under $Q$ implies that the $\sigma$-fields $\mathcal{F}_t^i$ and $\mathcal{F}_t^e$ are conditionally independent given $\mathcal{F}_t^0$. As a consequence, $\wedge_\xi$ is $\mathcal{F}_t^0$-measurable. Choosing $\xi = (L_t^i)^{-1}$ and $\xi = \chi(L_t^i)^{-1}$, where $\chi$ is a nonnegative, bounded, and $\mathcal{F}_t^i$-measurable random variable, we obtain that

$$\frac{E(|\Phi| \mid \mathcal{F}_t^e)}{E(|\Phi|L_t^i \mid \mathcal{F}_t^e)} \quad \text{and} \quad \frac{E(\chi|\Phi| \mid \mathcal{F}_t^e)}{E(|\Phi|L_t^i \mid \mathcal{F}_t^e)}$$

are $\mathcal{F}_t^0$-measurable. Consequently,

$$G_\chi = \frac{E(\chi|\Phi| \mid \mathcal{F}_t^e)}{E(|\Phi| \mid \mathcal{F}_t^e)}$$

is also $\mathcal{F}_t^0$-measurable.

The next step will be to translate this measurability property into an analytical condition using Lemma 1.3.3. First notice that if $\chi$ is a smooth random variable that is bounded and has a bounded derivative, then $G_\chi$ belongs to $\mathbb{D}_{\mathrm{loc}}^{1,2}$ because $f'$ has linear growth. Applying Lemma 1.3.3 to the random variable $G_\chi$ and to the $\sigma$-field $\sigma\{W_t, W_1\}$ yields

$$\frac{d}{ds}D_s[G_\chi] = 0$$

a.e. on $[0,1]$. Notice that $\frac{d}{ds}D_s\chi = 0$ a.e. on $[t,1]$, because $\chi$ is $\mathcal{F}_t^i$-measurable (again by Lemma 1.3.3). Therefore, for almost all $s \in [t,1]$, we get

$$E\left[\chi\frac{d}{ds}D_s|\Phi| \mid \mathcal{F}_t^e\right]E\left[|\Phi| \mid \mathcal{F}_t^e\right] = E\left[\chi|\Phi| \mid \mathcal{F}_t^e\right]E\left[\frac{d}{ds}D_s|\Phi| \mid \mathcal{F}_t^e\right].$$

The above equality holds true if $\chi$ is $\mathcal{F}_t^e$-measurable. So, by a monotone class argument, it holds for any bounded and nonnegative random variable $\chi$, and we get that

$$\frac{1}{\Phi}\frac{d}{ds}D_s\Phi = \frac{-g'(W_1)Zf''(W_s + g(W_1))}{1 + g'(W_1)(1 - Z)}$$

is $\mathcal{F}_t^e$-measurable for almost all $s \in [t,1]$ (actually for all $s \in [t,1]$ by continuity), where

$$Z = \exp\left(-\int_0^1 f'(W_r + g(W_1))dr\right).$$

Suppose now that condition (a) does not hold, that is, there exists a point $y \in \mathbb{R}$ such that $f''(y) \neq 0$. By continuity we have $f''(x) \neq 0$ for all $x$ in some interval $(y - \epsilon, y + \epsilon)$. Given $t < s < 1$, define

$$A_s = \{f''(W_s + g(W_1)) \in (y - \epsilon, y + \epsilon)\}.$$

Then $P(A_s) > 0$, and

$$1_{A_s}\frac{g'(W_1)Z}{1 + g'(W_1)(1 - Z)}$$

is $\mathcal{F}_t^e$-measurable. Again applying Lemma 1.3.3, we obtain that

$$\frac{d}{dr} D_r \left[ \frac{g'(W_1)Z}{1 + g'(W_1)(1 - Z)} \right] = 0$$

for almost all $r \in [0, t]$ and $\omega \in A_s$. This implies

$$g'(W_1)(1 + g'(W_1))f''(W_r + g(W_1)) = 0$$

a.e. on $[0, t] \times A_s$. Now, if

$$B = A_s \cap \{f''(W_r + g(W_1)) \in (y - \epsilon, y + \epsilon)\},$$

we have that $P(B) \neq 0$ and

$$g'(W_1)(1 + g'(W_1)) = 0,$$

a.s. on $B$. Then if (b) and (c) do not hold, we can find an interval $I$ such that if $W_1 \in I$ then $g'(W_1)(1 + g'(W_1)) \neq 0$. The set $B \cap \{W_1 \in I\}$ has nonzero probability, and this implies a contradiction.  □

Consider the stochastic differential equation (4.21) in dimension $d > 1$. One can ask under which conditions the solution is a Markov random field. This problem is more difficult, and a complete solution is not available. First we want to remark that, unlike in the one-dimensional case, the solution can be a Markov process even though $f$ is nonlinear. In fact, suppose that the boundary conditions are of the form

$$
\begin{aligned}
X_0^{i_k} &= a_k; & 1 \leq k \leq l, \\
X_1^{j_k} &= b_k; & 1 \leq k \leq d - l,
\end{aligned}
$$

where $\{i_1, \ldots, i_l\} \cup \{j_1, \ldots, j_{d-l}\}$ is a partition of $\{1, \ldots, d\}$. Assume in addition that $f$ is triangular, that means, $f^k(x)$ is a function of $x^1, \ldots, x^k$ for all $k$. In this case, if for each $k$, $f^k$ satisfies a Lipschitz and linear growth condition on the variable $x^k$, one can show that there exists a unique solution of the equation $dX_t + f(X_t) = dW_t$ with the above boundary conditions, and the solution is a Markov process. The Markov field property for triangular functions $f$ and triangular boundary conditions has been studied by Ferrante [97]. Other results in the general case obtained by a change of probability argument are the following:

**(1)**   In dimension one, and assuming a linear boundary condition of the type $F_0 X_0 + F_1 X_1 = h_0$, Donati-Martin (cf. [80]) has obtained the existence and uniqueness of a solution for the equation

$$dX_t = \sigma(X_t) \circ dW_t + b(X_t)$$

when the coefficients $b$ and $\sigma$ are of class $C^4$ with bounded derivatives, and $F_0 F_1 \neq 0$. On the other hand, if $\sigma$ is linear ($\sigma(x) = \alpha x$), $h_0 \neq 0$, and assuming that $b$ is of class $C^2$, then one can show that the solution $X$ is a Markov random field only if the drift is of the form $b(x) = Ax + Bx \log |x|$, where $|B| < 1$. See also [5] for a discussion of this example using the approach developed in Section 4.2.3.

**(2)**   In the $d$-dimensional case one can show the following result, which is similar to Theorem 2.1 (cf. Ferrante and Nualart [98]):

**Theorem 4.2.2** *Suppose $f$ is infinitely differentiable, $g$ is of class $C^2$, and* $\det(I - \phi(1)g'(W_1) + g'(W_1)) \neq 0$ *a.s., where $\phi(t)$ is the solution of the linear equation $d\phi(t) = f'(Y_t)\phi(t)dt$, $\phi(0) = I$. We also assume that the equation*

$$
\begin{cases}
X_t = X_0 - \int_0^t f(X_s)ds + W_t \\
X_0 = g(X_1 - X_0)
\end{cases}
$$

*has a unique solution for each $W \in C_0([0,1];\mathbb{R}^d)$, and that the following condition holds:*

**(H.4)**   span $\langle \partial_{i_1} \cdots \partial_{i_m} f'(x); i_1, \ldots, i_m \in \{1, \ldots, d\}, m \geq 1 \rangle = \mathbb{R}^{d \times d}$,
            *for all $x \in \mathbb{R}^d$.*

   *Then we have that $g'(x)$ is zero or $-I_d$, that is, the boundary condition is of the form $X_0 = a$ or $X_1 = b$.*

**(3)**   It is also possible to have a dichotomy similar to the one-dimensional case in higher dimensions (see Exercise 4.2.12).

### 4.2.2   Markov field property for solutions to stochastic partial differential equations

In this section we will review some results on the germ Markov field (GMF) property for solutions to stochastic partial differential equations driven by a white noise which have been obtained by means of the technique of change of probability.

   Let $D$ be a bounded domain in $\mathbb{R}^k$ with smooth boundary, and consider a continuous stochastic process $X = \{X_z, z \in D\}$. We will say that $X$ is a germ Markov field (GMF) if for any $\epsilon > 0$ and any open subset $A \subset D$, the $\sigma$-fields $\sigma\{X_z, z \in A\}$ and $\sigma\{X_z, z \in D - A^c\}$ are conditionally independent given the $\sigma$-field $\sigma\{X_z, z \in (\partial A)_\epsilon\}$, where $(\partial A)_\epsilon$ denotes the $\epsilon$-neighborhood of the boundary of $A$.

   We will first discuss in some detail the case of an elliptic stochastic partial differential equation with additive white noise.

**(A)**  *Elliptic stochastic partial differential equations*

Let $D$ be a bounded domain in $\mathbb{R}^k$ with smooth boundary, and assume $k = 1, 2, 3$. Let $\lambda^k$ denote the Lebesgue measure on $D$, and set $H = L^2(D, \mathcal{B}(D), \lambda^k)$. Consider an isonormal Gaussian process $W = \{W(h), h \in H\}$ associated with $H$. That is, if we set $W(A) = W(\mathbf{1}_A)$, then $W = \{W(A), A \in \mathcal{B}(D)\}$ is a zero-mean Gaussian process with covariance

$$E(W(A)W(B)) = \lambda^k(A \cap B).$$

We want to study the equation

$$\begin{cases} -\Delta U(x) + f(U(x)) = \dot{W}(x), & x \in D, \\ U|_{\partial D} = 0. \end{cases} \tag{4.30}$$

Let us first introduce the notion of the solution to (4.30) in the sense of distributions.

**Definition 4.2.2** *We will say that a continuous process $U = \{U(x), x \in \overline{D}\}$ that vanishes in the boundary of $D$ is a solution to (4.30) if*

$$-\langle U, \Delta\varphi\rangle_H + \langle f(U), \varphi\rangle_H = \int_D \varphi(x)W(dx)$$

*for all $\varphi \in C^\infty(D)$ with compact support.*

We will denote by $G(x, y)$ the Green function associated with the Laplace operator $\Delta$ with Dirichlet boundary conditions on $D$. That is, for any $\varphi \in L^2(D)$, the elliptic linear equation

$$\begin{cases} -\Delta\psi(x) = \varphi(x), & x \in D, \\ \psi|_{\partial D} = 0 \end{cases} \tag{4.31}$$

possesses a unique solution in the Sobolev space $H_0^1(D)$, which can be written as

$$\psi(x) = \int_D G(x, y)\varphi(y)dy.$$

We recall that $H_0^1(D)$ denotes the completion of $C_0^\infty(D)$ for the Sobolev norm $\|\cdot\|_{1,2}$. We will use the notation $\psi = G\varphi$. We recall that $G$ is a symmetric function such that $G(x, \cdot)$ is harmonic on $D - \{x\}$.

One can easily show (see [54]) that $U$ is a solution to the elliptic equation (4.30) if and only if it satisfies the integral equation

$$U(x) + \int_D G(x, y)f(U(y))dy = \int_D G(x, y)W(dy). \tag{4.32}$$

Note that the right-hand side of (4.32) is a well-defined stochastic integral because the Green function is square integrable. More precisely, we have

$$\sup_{x \in \overline{D}} \int_D G^2(x, y)dy < \infty. \tag{4.33}$$

In dimension $k > 3$ this property is no longer true, and for this reason the analysis stops at dimension three.

We will denote by $U_0$ the solution of (4.30) for $f = 0$, that is,

$$U_0(x) = \int_D G(x, y)W(dy). \tag{4.34}$$

Using Kolmogorov's criterion one can show (see Exercise 4.2.13) that the process $\{U_0(x), x \in D\}$ has Lipschitz paths if $k = 1$, Hölder continuous paths of order $1 - \epsilon$ if $k = 2$, and Hölder continuous paths of order $\frac{3}{8} - \epsilon$ if $k = 3$, for any $\epsilon > 0$.

The following result was established by Buckdahn and Pardoux in [54].

**Theorem 4.2.3** *Let $D$ be a bounded domain of $\mathbb{R}^k$, $k = 1, 2, 3$, with a smooth boundary. Let $f$ be a continuous and nondecreasing function. Then Eq. (4.32) possesses a unique continuous solution.*

A basic ingredient in the proof of this theorem is the following inequality:

**Lemma 4.2.3** *There exists a constant $a > 0$ such that for any $\varphi \in L^2(D)$,*

$$\langle G\varphi, \varphi \rangle_H \geq a\|G\varphi\|_H^2. \tag{4.35}$$

*Proof:*    Set $\psi = G\varphi$. Then $\psi$ solves Eq. (4.31). Multiplying this equation by $\psi$ and integrating by parts, we obtain

$$\sum_{i=1}^k \left\| \frac{\partial \psi}{\partial x_i} \right\|_H^2 = \langle \varphi, \psi \rangle_H.$$

From Poincaré's inequality (cf. [120, p. 157]) there exists a constant $a > 0$ such that for any $\psi \in H_0^1(D)$,

$$\sum_{i=1}^k \left\| \frac{\partial \psi}{\partial x_i} \right\|_H^2 \geq a\|\psi\|_H^2.$$

The result follows.    □

We are going to reformulate the above existence and uniqueness theorem in an alternative way. Consider the Banach space

$$B = \{\omega \in C(\overline{D}), \omega \mid_{\partial D} = 0\},$$

equipped with the supremum norm, and the transformation $T : B \to B$ given by

$$T(\omega)(x) = \omega(x) + \int_D G(x, y)f(\omega(y))dy. \tag{4.36}$$

Note that $\{U(x), x \in D\}$ is a solution to (4.32) if and only if

$$T(U(x)) = U_0(x).$$

Then Theorem 4.2.3 is a consequence of the following result.

**Lemma 4.2.4** *Let $f$ be a continuous and nondecreasing function. Then the transformation $T$ given by (4.36) is bijective.*

*Proof:*     Let us first show that $T$ is one to one. Let $u, v \in B$ such that $T(u) = T(v)$. Then

$$u - v + G[f(u) - f(v)] = 0. \qquad (4.37)$$

Multiplying this equation by $f(u) - f(v)$, we obtain

$$\langle u - v, f(u) - f(v) \rangle_H + \langle G[f(u) - f(v)], f(u) - f(v) \rangle_H = 0.$$

Using the fact that $f$ is nondecreasing, and Lemma 4.2.3, it follows that

$$a\|G[f(u) - f(v)]\|_H^2 \le 0.$$

By (4.37) this is equivalent to $a\|u - v\|_H^2 \le 0$, so $u = v$ and $T$ is one to one.

In order to show that $T$ is onto, we will assume that $f$ is bounded. The general case would be obtained by a truncation argument. Let $v \in B$, and let $\{v_n, n \in \mathbb{N}\}$ be a sequence of functions in $C^2(D)$, with compact support in $D$, such that $\|v - v_n\|_\infty$ tends to zero as $n$ tends to infinity. Set $h_n = -\Delta v_n$. It follows from Lions ([199], Theorem 2.1, p. 171) that the elliptic partial differential equation

$$\begin{cases} -\Delta u_n + f(u_n) = h_n \\ u_n \mid_{\partial D} = 0 \end{cases}$$

admits a unique solution $u_n \in H_0^1(D)$. Then,

$$u_n + G[f(u_n)] = G h_n = v_n, \qquad (4.38)$$

that is, $T(u_n) = v_n$. We now prove that $u_n$ is a Cauchy sequence in $L^2(D)$. Multiplying the equation

$$u_n - u_m + G[f(u_n) - f(u_m)] = v_n - v_m$$

by $f(u_n) - f(u_m)$, and using Lemma 4.2.3 and the monotonicity property of $f$, we get

$$a\|G[f(u_n) - f(u_m)]\|_H^2 \le \langle v_n - v_m, f(u_n) - f(u_m) \rangle_H,$$

which implies, using the above equation,

$$a\|u_n - u_m\|_H \le \langle v_n - v_m, f(u_n) - f(u_m) + 2a(u_n - u_m) \rangle_H.$$

Since $\{v_n\}$ is a Cauchy sequence in $L^2(D)$ and $f$ is bounded, $\{u_n\}$ is a Cauchy sequence in $L^2(D)$. Define $u = \lim u_n$. Then $f(u_n)$ converges to $f(u)$ in $L^2(D)$. Taking the limit in (4.38), we obtain

$$u + G[f(u)] = v.$$

Thus $u \in B$ ($f$ is bounded) and $T(u) = v$.     □

Let us now discuss the germ Markov field property of the process $U(x)$. First we will show that the Gaussian process $U_0(x)$ verifies the germ Markov field property. To do this we shall use a criterion expressed in terms of the reproducing kernel Hilbert space (RKHS) $\mathcal{H}$ associated to $U_0$. Let $\mathcal{H}_1 \subset L^2(\Omega)$ be the Gaussian space (i.e., the first chaos) generated by $W$. An element $v \in B$ belongs to the RKHS $\mathcal{H}$ iff there exists a random variable $X \in \mathcal{H}_1$ such that

$$v(x) = E[XU_0(x)],$$

for all $x \in D$, i.e., iff there exists $\phi \in L^2(D)$ such that $v = G\phi$. In other words, $\mathcal{H} = \{v \in B : \Delta v \in L^2(D)\}$, and $\langle v_1, v_2 \rangle_{\mathcal{H}} = \langle \Delta v_1, \Delta v_2 \rangle_H$.

We now have the following result (see Pitt [285] and Künsch [177]).

**Proposition 4.2.1** *A continuous Gaussian field $U = \{U(x), x \in D\}$ possesses the germ Markov field property iff its RKHS $\mathcal{H} \subset B$ is local in the sense that it satisfies the following two properties:*

(i) *Whenever $u$, $v$ in $\mathcal{H}$ have disjoint supports, $\langle u, v \rangle_{\mathcal{H}} = 0$.*

(ii) *If $v \in \mathcal{H}$ is of the form $v = v_1 + v_2$ with $v_1, v_2 \in B$ with disjoint supports, then $v_1, v_2 \in \mathcal{H}$.*

The RKHS associated to the process $U_0$ verifies conditions (i) and (ii), and this implies the germ Markov field property of $U_0$. Concerning the process $U$, one can prove the following result.

**Theorem 4.2.4** *Assume that $f$ is a $C^2$ function such that $f' > 0$ and $f'$ has linear growth. Then the solution $\{U(x), x \in D\}$ of the elliptic equation (4.30) has the germ Markov property if and only if $f'' = 0$.*

This theorem has been proved by Donati-Martin and Nualart in [82]. In dimension one, Eq. (4.30) is a second-order stochastic differential equation studied by Nualart and Pardoux in [251]. In that case the germ $\sigma$-field corresponding to the boundary points $\{s, t\}$ is generated by the variables $\{X_s, \dot{X}_s, X_t, \dot{X}_t\}$, and the theorem holds even if the function $f$ depends on $X_t$ and $\dot{X}_t$ (assuming in that case more regularity on $f$). The main difference between one and several parameters is that in dimension one, one can explicitly compute the Carleman-Fredholm determinant of $Du$. Similar to the work done in [81] and [82] we will give a proof for the case $k = 2$ or $k = 3$.

*Proof of Theorem 4.2.4:*    The proof follows the same lines as the proof of Theorem 2.1. We will indicate the main steps of the argument.

*Step 1:*    We will work on the abstract Wiener space $(B, H, \mu)$, where $\mu$ is the law of $U_0$, and the continuous injection $i : H \to B$ is defined as follows:

$$i(h)(x) = \int_D G(x, y)h(y)dy.$$

From Lemma 4.2.3 we deduce that $i$ is one to one, and from Eq. (4.33) we see that $i$ is continuous. The image $i(H)$ is densely included in $B$. We identify $H$ and $H^*$, and in this way $B^*$ can be viewed as a dense subset of $H$, the inclusion map being given by

$$\alpha \to \int_D G(y, \cdot)\alpha(dy) = G^*\alpha.$$

Finally, for any $\alpha \in B^*$ we have

$$
\begin{aligned}
\int_B e^{i\langle \alpha, \omega \rangle} \mu(d\omega) &= E\left[\exp(i \int_D U_0(x)\alpha(dx))\right] \\
&= E\left[\exp(i \int_D \int_D G(x, y)dW_y\alpha(dx))\right] \\
&= \exp\left(-\frac{1}{2}\int_D \left(\int_D G(x, y)\alpha(dx)\right)^2 dy\right) \\
&= \exp\left(-\frac{1}{2}\|G^*\alpha\|_H^2\right),
\end{aligned}
$$

which implies that $(B, H, \mu)$ is an abstract Wiener space. Note that $i(H)$ coincides with the RKHS $\mathcal{H}$ introduced before, and that $U_0(x) = w(x)$ is now the canonical process in the space $(B, \mathcal{B}(B), \mu)$.

We are interested in the germ Markov field property of the process $U(x) = T^{-1}(U_0)(x)$. Let $\nu$ be the probability on $B$ defined by $\mu = \nu \circ T^{-1}$. That is, $\nu$ is the law of $U$.

*Step 2:*     Let us show that the transformation $T$ verifies the hypotheses of Theorem 4.1.2. We already know from Lemma 4.2.4 that $T$ is bijective. Notice that we can write

$$T(\omega) = \omega + i(f(\omega)),$$

so we have to show that:

(i) the mapping $\omega \to i(f(\omega))$ from $B$ to $H$ is $H$-continuously differentiable;

(ii) the mapping $I_H + Du(\omega) : H \to H$ is invertible for all $\omega \in B$, where $Du(\omega)$ is the Hilbert-Schmidt operator given by the kernel

$$Du(\omega)(x, y) = f'(\omega(x))G(x, y).$$

Property (i) is obvious and to prove (ii), from the Fredholm alternative, it suffices to check that $-1$ is not an eigenvalue of $Du(\omega)$. Let $h \in H$ be an element such that

$$h(x) + f'(\omega(x)) \int_D G(x, y)h(y)dy = 0.$$

Multiplying this equality by $\frac{h(x)}{f'(\omega(x))}$ and integrating over $D$, we obtain

$$\int_D \frac{h^2(x)}{f'(\omega(x))}dx + \langle h, Gh\rangle_H = 0.$$

From Lemma 4.2.3, $\langle h, Gh\rangle_H \geq a\|Gh\|_H^2$, thus $\|Gh\|_H = 0$ and $h = 0$. Therefore, by Theorem 4.1.2 we obtain

$$\frac{d\nu}{d\mu} = |\det_2(I + Du)| \exp(-\delta(u) - \tfrac{1}{2}\|u\|_H^2). \tag{4.39}$$

Set $L = \exp(-\delta(u) - \tfrac{1}{2}\|u\|^2)$.

*Step 3:*    For a fixed domain $A$ with smooth boundary $\Gamma$ and such that $\overline{A} \subset D$, we denote

$$\mathcal{F}^i = \sigma\{U_0(x), x \in A\}, \quad \mathcal{F}^e = \sigma\{U_0(x), x \in D - \overline{A}\},$$

and

$$\mathcal{F}^0 = \cap_{\epsilon>0}\sigma\{U_0(x), x \in (\partial A)_\epsilon\}.$$

Consider the factorization $L = L^i L^e$, where

$$L^i = \exp\left(-\delta(u\mathbf{1}_A) - \frac{1}{2}\|u\mathbf{1}_A\|_H^2\right)$$

and

$$L^e = \exp\left(-\delta(u\mathbf{1}_{D-\overline{A}}) - \frac{1}{2}\|u\mathbf{1}_{D-\overline{A}}\|_H^2\right).$$

We claim that $J^i$ is $\mathcal{F}^i$-measurable and $J^e$ is $\mathcal{F}^e$-measurable. This follows from the fact that the Skorohod integrals

$$\delta(u\mathbf{1}_A) = \int_A f(U_0(x))W(dx)$$

and

$$\delta(u\mathbf{1}_{D-\overline{A}}) = \int_{D-\overline{A}} f(U_0(x))W(dx)$$

are $\mathcal{F}^i$-measurable and $\mathcal{F}^e$-measurable, respectively (see [81]).

*Step 4:*    From Step 3 it follows that if $f'' = 0$, the Radon-Nikodym density given by (4.39) can be expressed as the product of two factors, one being $\mathcal{F}^i$-measurable, and the second one being $\mathcal{F}^e$-measurable. This factorization implies the germ Markov field property of $X$ under $\mu$.

*Step 5:*    Suppose conversely that $U$ possesses the germ Markov property under $\mu$. By the same arguments as in the proof of Theorem 2.1 we can show that for any nonnegative random variable $\xi$ that is $\mathcal{F}^i$-measurable, the quotient

$$G_\xi = \frac{E[\xi\Phi \mid \mathcal{F}^e]}{E[\Phi \mid \mathcal{F}^e]}$$

is $\mathcal{F}^0$-measurable, where $\Phi = \det_2(I + f'(U_0(x))G(x,y))$. Observe that $\Phi \geq 0$ because the eigenvalues of the kernel $f'(U_0(x))G(x,y)$ are positive.

*Step 6:*     The next step will be to translate the above measurability property into an analytical condition. Fix $\epsilon > 0$ such that $A_\epsilon^- = A - \overline{\Gamma}_\epsilon$ and $A_\epsilon^+ = (D - \overline{A}) - \overline{\Gamma}_\epsilon$ are nonempty sets. We have that $G_\xi$ is $\sigma\{U_0(x), x \in \Gamma_\epsilon\}$-measurable. If we assume that $\xi$ is a smooth random variable, then $G_\xi$ is in $\mathbb{D}^{1,2}_{\text{loc}}$, and by Lemma 1.3.3 we obtain that

$$DG_\xi \in \langle G(x,\cdot), x \in \Gamma_\epsilon \rangle_H.$$

This implies that for any function $\phi \in C_0^\infty(D - \Gamma_\epsilon)$ we have $\langle \phi, \Delta DG_\xi \rangle_H = 0$ a.s. Suppose that $\phi \in C_0^\infty(A_\epsilon^+)$. In that case we have in addition that $\langle \phi, \Delta D\xi \rangle_H = 0$, because $\xi$ is $\mathcal{F}^i$-measurable. Consequently, for such a function $\phi$ we get

$$E\left[\xi\langle\phi, \Delta D\Phi\rangle_H \mid \mathcal{F}^e\right] E[\Phi \mid \mathcal{F}^e] = E\left[\xi\Phi \mid \mathcal{F}^e\right] E[\langle\phi, \Delta D\phi\rangle_H \mid \mathcal{F}^e].$$

The above equality holds true for any bounded and nonnegative random variable $\xi$, therefore, we obtain that

$$\frac{1}{\Phi}\langle\phi, \Delta D\Phi\rangle_H \tag{4.40}$$

is $\mathcal{F}^e$-measurable.

*Step 7:*     The derivative of the random variable $\Phi$ can be computed using the expression for the derivative of the Carleman-Fredholm determinant. We have

$$D_z\Phi = \Phi\mathrm{T}\left(((I + f'(U_0(x))G(x,y))^{-1} - I)D_z[f'(U_0(x))G(x,y)]\right).$$

So, from (4.40) we get that

$$\mathrm{T}\left(((I + f'(U_0(x))G(x,y))^{-1} - I)\langle\phi, \Delta D.[f'(U_0(x))G(x,y)]\rangle_H\right)$$

is $\mathcal{F}^e$-measurable. Note that

$$D_z[f'(U_0(x))G(x,y)] = f''(U_0(x))G(x,z)G(x,y)$$

and

$$\langle\Delta\phi, G(x,\cdot)\rangle_H = \phi.$$

Thus, we have that

$$\mathrm{T}\left(((I + f'(U_0(x))G(x,y))^{-1} - I)(\phi(x)f''(U_0(x))G(x,y))\right)$$

is $\mathcal{F}^e$-measurable, and we conclude that for any $x \in A_\epsilon^+$,

$$f''(U_0(x)) \int_D K(y,x)G(x,y)dy$$

is $\mathcal{F}^e$-measurable, where $K(x,y) = ((I+Du)^{-1} - I)(x,y)$. Suppose now that there exists a point $b \in \mathbb{R}$ such that $f''(b) \neq 0$. Then $f''$ will be nonzero in some interval $J$. Set

$$A = \{\omega \in B : f''(U_0(x)) \in J\}.$$

The set $A$ has nonzero probability, it belongs to $\mathcal{F}^e$, and

$$\mathbf{1}_A \int_D K(y,x)G(x,y)dy$$

is $\mathcal{F}^e$-measurable. Applying again Lemma 1.3.3 and using the same arguments as above, we obtain that on the set $A$

$$\left(G(I+Du)^{-1}[f''(U_0(x_1))\psi(x_1)G(x_1,x_2)](I+Du)^{-1}\right)(x,x) = 0$$

for any function $\psi \in C_0^\infty(A_\epsilon^-)$. So we get

$$\mathbf{1}_{\{f''(U_0(x))\in J\}}\mathbf{1}_{\{f''(U_0(x_1))\in J\}}G(I+Du)^{-1}(x,x_1)G(I+Du)^{-1}(x_1,x) = 0$$

for all $x, x_1$ such that $x \in A_\epsilon^+$ and $x_1 \in A_\epsilon^-$. Notice that the operator $G(I+Du)^{-1}$ has a singularity in the diagonal of the same type as $G$. So from the above equality we get

$$\mathbf{1}_{\{f''(U_0(x))\in J\}} = 0,$$

which is not possible.                                                              $\square$

**(B)**  *Parabolic stochastic partial differential equations*

Consider the following equation studied in Section 2.4.2:

$$\begin{cases} \frac{\partial u}{\partial t} - \frac{\partial^2 u}{\partial x^2} = f(u(t,x)) + \frac{\partial^2 W}{\partial t \partial x}, & (t,x) \in [0,T] \times [0,1], \\ u(t,0) = u(t,1) = 0, & 0 \leq t \leq T. \end{cases}$$

We will impose two different types of boundary conditions:

**(B.1)**   $u(0,x) = u_0(x),$

**(B.2)**   $u(0,x) = u(1,x), \quad 0 \leq x \leq 1.$

In case (B.1) we are given a initial condition $u_0 \in C([0,1])$ such that $u_0(0) = u_0(1) = 0$, and in case (B.2) we impose a periodic boundary condition in time. Under some hypotheses on the function $f$ there exists a unique continuous solution of the corresponding integral equation:

For (B.1) a sufficient condition is that $f$ is Lipschitz (see Theorem 2.4.3).

For (B.2) (see [253]) we require that there exists a constant $0 < c < 2$ such that

$$(z - y)(f(z) - f(y)) \leq c(z - y)^2$$

for all $y, z \in \mathbb{R}$.

From the point of view of the Markov property of the solution, the behavior of these equations is completely different. In case (B.1) the GMF property always holds. On the other hand, assuming that $f(z)$ is of class $C_b^2$ and that the boundary condition (B.2) holds, then the solution $u$ has the GMF property if and only if $f'' = 0$. These results have been proved under more general assumptions on the function $f$ in [253].

### 4.2.3  *Conditional independence and factorization properties*

In this section we prove a general characterization of the conditional independence and apply it to give an alternative proof of Theorem 4.2.1. More precisely, we discuss the following general problem: Consider two independent sub-$\sigma$-fields $\mathcal{F}_1$, $\mathcal{F}_2$ of a probability space, and let $X$ and $Y$ be two random variables determined by a system of the form

$$\begin{cases} X = g_1(Y, \omega) \\ Y = g_2(X, \omega) \end{cases}$$

where $g_i(y, \cdot)$ is $\mathcal{F}_i$-measurable ($i = 1, 2$). Under what conditions on $g_1$ and $g_2$ are $\mathcal{F}_1$ and $\mathcal{F}_2$ conditionally independent given $X$ and $Y$? We will see that this problem arises in a natural way when treating stochastic equations with boundary conditions.

Let $(\Omega, \mathcal{F}, P)$ be a complete probability space and let $\mathcal{F}_1$ and $\mathcal{F}_2$ be two independent sub-$\sigma$-fields of $\mathcal{F}$. Consider two functions $g_1, g_2 : \mathbb{R} \times \Omega \to \mathbb{R}$ such that $g_i$ is $\mathcal{B}(\mathbb{R}) \otimes \mathcal{F}_i$-measurable, for $i = 1, 2$, and that they verify the following conditions for some $\varepsilon_0 > 0$:

**H1**  For every $x \in \mathbb{R}$ and $y \in \mathbb{R}$ the random variables $g_1(y, \cdot)$ and $g_2(x, \cdot)$ possess absolutely continuous laws and the function

$$\delta(x, y) = \sup_{0 < \varepsilon < \varepsilon_0} \frac{1}{\varepsilon^2} P\{|x - g_1(y)| < \varepsilon, |y - g_2(x)| < \varepsilon\}$$

is locally integrable in $\mathbb{R}^2$.

**H2**  For almost all $\omega \in \Omega$ and for any $|\xi| < \varepsilon_0$, $|\eta| < \varepsilon_0$ the system

$$\begin{cases} x - g_1(y, \omega) = \xi \\ y - g_2(x, \omega) = \eta \end{cases} \tag{4.41}$$

has a unique solution $(x, y) \in \mathbb{R}^2$.

**H3** For almost all $\omega \in \Omega$ the functions $y \to g_1(y)$ and $x \to g_2(x)$ are continuously differentiable and there exists a nonnegative random variable $H$ such that $E(H) < \infty$ and

$$\sup_{\substack{|y-g_2(x)|<\varepsilon_0 \\ |x-g_1(y)|<\varepsilon_0}} |1 - g_1'(y)g_2'(x)|^{-1} \le H \qquad \text{a.s.}$$

Hypothesis H2 implies the existence of two random variables $X$ and $Y$ determined by the system

$$\begin{cases} X(\omega) = g_1(Y(\omega), \omega) \\ Y(\omega) = g_2(X(\omega), \omega). \end{cases} \qquad (4.42)$$

**Theorem 4.2.5** *Let $g_1$ and $g_2$ be two functions satisfying hypotheses H1 through H3. Then the following statements are equivalent:*

*(i) $\mathcal{F}_1$ and $\mathcal{F}_2$ are conditionally independent given the random variables $X, Y$.*

*(ii) There exist two functions $F_i : \mathbb{R}^2 \times \Omega \to \mathbb{R}$, $i = 1, 2$, which are $\mathcal{B}(\mathbb{R}^2) \otimes \mathcal{F}_i$ -measurable for $i = 1, 2$, such that*

$$|1 - g_1'(Y)g_2'(X)| = F_1(X, Y)F_2(X, Y) \qquad \text{a.s.}$$

*Proof:*    Let $G_1$ and $G_2$ be two bounded nonnegative random variables such that $G_i$ is $\mathcal{F}_i$-measurable for $i = 1, 2$. Suppose that $f : \mathbb{R}^2 \longrightarrow \mathbb{R}$ is a nonnegative continuous and bounded function. For any $x \in \mathbb{R}$ we will denote by $f_i(x, \cdot)$ the density of the law of $g_i(x)$, for $i = 1, 2$. For each $\varepsilon > 0$, define $\varphi^\varepsilon(z) = \frac{1}{2\varepsilon} \mathbf{1}_{[-\varepsilon, \varepsilon]}(z)$. Set

$$J(x, y) = |1 - g_1'(y)g_2'(x)|^{-1} \, .$$

We will first show the equality

$$E[G_1 G_2 J(X, Y) f(X, Y)] \qquad (4.43)$$

$$= \int_{\mathbb{R}^2} E[G_1 | g_1(y) = x] f_1(y, x) E[G_2 | g_2(x) = y] f_2(x, y) f(x, y) dx dy.$$

Actually, we will see that both members arise when we compute the limit of

$$\int_{\mathbb{R}^2} E[G_1 G_2 \varphi^\varepsilon(x - g_1(y))\varphi^\varepsilon(y - g_2(x))] f(x, y) dx dy \qquad (4.44)$$

as $\varepsilon$ tends to zero in two different ways.

For any $\omega \in \Omega$ we introduce the mapping $\Phi_\omega : \mathbb{R}^2 \to \mathbb{R}^2$ defined by

$$\Phi_\omega(x, y) = (x - g_1(y, \omega), y - g_2(x, \omega)) = (\bar{x}, \bar{y}).$$

Notice that $\Phi_\omega(X(\omega), Y(\omega)) = (0,0)$. Denote by $D_{\varepsilon_0}(\omega)$ the set

$$D_{\varepsilon_0}(\omega) = \{(x,y) \in \mathbb{R}^2 : |x - g_1(y,\omega)| < \varepsilon_0, |y - g_2(x,\omega)| < \varepsilon_0\}.$$

Hypotheses H2 and H3 imply that for almost all $\omega$ the mapping $\Phi_\omega$ is a $C^1$-diffeomorphism from $D_{\varepsilon_0}(\omega)$ onto $(-\varepsilon_0, \varepsilon_0)^2$. Therefore, making the change of variable $(\bar{x}, \bar{y}) = \Phi_\omega(x, y)$, we obtain for any $\varepsilon < \varepsilon_0$

$$\int_{\mathbb{R}^2} \varphi^\varepsilon(x - g_1(y))\varphi^\varepsilon(y - g_2(x))f(x,y)dxdy$$

$$= \int_{\mathbb{R}^2} \varphi^\varepsilon(\bar{x})\varphi^\varepsilon(\bar{y})J(\Phi_\omega^{-1}(\bar{x}, \bar{y}))f(\Phi_\omega^{-1}(\bar{x}, \bar{y}))d\bar{x}d\bar{y}.$$

By continuity this converges to $J(X,Y)f(X,Y)$ as $\varepsilon$ tends to zero. The convergence of the expectations follows by the dominated convergence theorem, because from hypothesis H3 we have

$$J(\Phi_\omega^{-1}(\bar{x}, \bar{y})) \leq \sup_{\substack{|y - g_2(x,\omega)| < \varepsilon_0 \\ |x - g_1(y,\omega)| < \varepsilon_0}} |1 - g_1'(y)g_2'(x)|^{-1} \leq H \in L^1(\Omega)$$

if $|\bar{x}| < \varepsilon_0$, and $|\bar{y}| < \varepsilon_0$.

Consequently, (4.44) converges to the left-hand side of (4.43) as $\varepsilon$ tends to zero. Let us now turn to the proof that the limit of (4.44) equals the right-hand side of (4.43). We can write

$$[G_1 G_2 \varphi^\varepsilon(x - g_1(y))\varphi^\varepsilon(y - g_2(x))]$$

$$= E[G_1 \varphi^\varepsilon(x - g_1(y))] E[G_2 \varphi^\varepsilon(y - g_2(x))]$$

$$= \left(\int_{\mathbb{R}} \varphi^\varepsilon(x - \alpha)E[G_1 |g_1(y) = \alpha] f_1(y, \alpha)d\alpha\right)$$

$$\times \left(\int_{\mathbb{R}} \varphi^\varepsilon(y - \beta)E[G_2 |g_2(x) = \beta] f_2(x, \beta)d\beta\right).$$

We are going to take the limit of both factors as $\varepsilon$ tends to zero. For the first one, the Lebesgue differentiation theorem tell us that for any $y \in \mathbb{R}$ there exists a set $N^y$ of zero Lebesgue measure such that for all $x \notin N^y$,

$$\lim_{\varepsilon \downarrow 0} \int_{\mathbb{R}} \varphi^\varepsilon(x - \alpha)E[G_1 |g_1(y) = \alpha] f_1(y, \alpha)d\alpha = E[G_1 |g_1(y) = x] f_1(y, x).$$

In the same way, for the second integral, for each fixed $x \in \mathbb{R}$, there will be a set $N^x$ of zero Lebesgue measure such that for all $y \notin N^x$,

$$\lim_{\varepsilon \downarrow 0} \int_{\mathbb{R}} \varphi^\varepsilon(y - \beta)E[G_2 |g_2(x) = \beta] f_2(x, \beta)d\beta = E[G_2 |g_2(x) = y] f_2(x, y).$$

We conclude that, except on the set

$$N = \{(x,y) : x \in N^y \text{ or } y \in N^x\}$$

we will have the convergence

$$\lim_{\varepsilon \downarrow 0} E\left[G_1 G_2 \; \varphi^\varepsilon(x - g_1(y))\varphi^\varepsilon(y - g_2(x))\right]$$
$$= E\left[G_1 \left| g_1(y) = x\right.\right] f_1(y, x) E\left[G_2 \left| g_2(x) = y\right.\right] f_2(x, y).$$

Thus, this convergence holds almost everywhere. The preceding equality provides the pointwise convergence of the integrands appearing in expression (4.44). The corresponding convergence of the integral is derived through the dominated convergence theorem, using hypothesis H1.

Consequently, (4.43) holds for any continuous and bounded function $f$, and this equality easily extends to any measurable and bounded function $f$. Taking $f = \mathbf{1}_B$, where $B$ is a set of zero Lebesgue measure, and putting $G_1 = G_2 = 1$, we deduce from (4.43) that $P\{(X, Y) \in B\} = 0$ because $J(X, Y) > 0$ a.s. As a consequence, the law of $(X, Y)$ is absolutely continuous with a density given by

$$f_{XY}(x, y) = \frac{f_1(x, y) f_2(y, x)}{E\left[J(X, Y) \left| X = x, Y = y\right.\right]}.$$

Therefore, (4.43) implies that

$$E\left[G_1 G_2 J(X, Y) \left| X = x, Y = y\right.\right] f_{XY}(x, y)$$
$$= \quad E\left[G_1 \left| g_1(y) = x\right.\right] f_1(y, x) E\left[G_2 \left| g_2(x) = y\right.\right] f_2(x, y), \quad (4.45)$$

almost surely with respect to the law of $(X, Y)$. Putting $G_2 = 1$, we obtain

$$E\left[G_1 J(X, Y) \left| X = x, Y = y\right.\right] f_{XY}(x, y) \qquad\qquad (4.46)$$
$$= E\left[G_1 \left| g_1(y) = x\right.\right] f_1(y, x) f_2(x, y),$$

and with $G_1 \equiv 1$ we get

$$E\left[G_2 J(X, Y) \left| X = x, Y = y\right.\right] f_{XY}(x, y) \qquad\qquad (4.47)$$
$$= E\left[G_2 \left| g_2(x) = y\right.\right] f_1(y, x) f_2(x, y).$$

Substituting (4.46) and (4.47) into (4.45) yields

$$E\left[G_1 G_2 J(X, Y) \left| X = x, Y = y\right.\right] E\left[J(X, Y) \left| X = x, Y = y\right.\right] \quad (4.48)$$
$$= E\left[G_1 J(X, Y) \left| X = x, Y = y\right.\right] E\left[G_2 J(X, Y) \left| X = x, Y = y\right.\right].$$

Conditioning first by the bigger $\sigma$-fields $\sigma(X, Y) \vee \mathcal{F}_1$ and $\sigma(X, Y) \vee \mathcal{F}_2$ in the right-hand side of (4.48), we obtain

$$E\left[G_1 G_2 J(X, Y) \left| XY\right.\right] E\left[J(X, Y) \left| XY\right.\right] \qquad\qquad (4.49)$$
$$= E\left[G_1 E\left[J(X, Y) \left| X, Y, \mathcal{F}_1\right.\right] \left| X, Y\right.\right]$$
$$\times E\left[G_2 E\left[J(X, Y) \left| X, Y, \mathcal{F}_2\right.\right] \left| X, Y\right.\right].$$

Suppose first that $\mathcal{F}_1 \underset{X,Y}{\perp\!\!\!\perp} \mathcal{F}_2$. This allows us to write Eq. (4.49) as follows:

$$E\left[G_1 G_2 J(X,Y) E\left[J(X,Y)\,|\,X,Y\right]\,|\,X,Y\right]$$
$$= E\left[G_1 G_2 E\left[J(X,Y)\,|\,X,Y,\mathcal{F}_1\right] E\left[J(X,Y)\,|\,X,Y,\mathcal{F}_2\right]\,|\,X,Y\right].$$

Taking the expectation of both members of the above equality, we obtain

$$J(X,Y)^{-1} = \frac{E\left[J(X,Y)\,|\,X,Y\right]}{E\left[J(X,Y)\,|\,X,Y,\mathcal{F}_1\right] E\left[J(X,Y)\,|\,X,Y,\mathcal{F}_2\right]}.$$

This implies the desired factorization because any random variable that is $\sigma(X,Y) \vee \mathcal{F}_i$-measurable $(i = 1,2)$ can be written as $F(X(\omega), Y(\omega), \omega)$ for some $\mathcal{B}(\mathbb{R}^2) \otimes \mathcal{F}_i$-measurable function $F : \mathbb{R}^2 \times \Omega \to \mathbb{R}$.

Conversely, suppose that (ii) holds. Then we have from (4.49)

$$E\left[G_1 G_2\,|\,X,Y\right] = E\left[G_1 G_2 F_1(X,Y) F_2(X,Y)\ J(X,Y)\,|\,X,Y\right]$$

$$= \frac{E\left[G_1 F_1(X,Y) J(X,Y)\,|\,X,Y\right] E\left[G_2 F_2(X,Y)\ J(X,Y)\,|\,X,Y\right]}{E\left[J(X,Y)\,|\,X,Y\right]}.$$

Writing this equality for $G_1 \equiv 1$, $G_2 \equiv 1$, and for $G_1 \equiv G_2 \equiv 1$, we conclude that

$$E\left[G_1 G_2\,|\,X,Y\right] = E\left[G_1\,|\,X,Y\right] E\left[G_2\,|\,X,Y\right].$$

$\square$

**Remarks:**  Some of the conditions appearing in the preceding hypotheses can be weakened or modified, and the conclusion of Theorem 4.2.5 will continue to hold. In particular, in hypothesis H3 we can replace $H(\omega)$ by $H_1(\omega) H_2(\omega)$, with $H_i(\omega)$ $\mathcal{F}_i$-measurable for $i = 1,2$, and assume only $H_1(\omega) H_2(\omega) < \infty$ a.s. In H1 the local integrability of the function $\delta(x,y)$ holds if the densities $f_1(y,z)$ and $f_2(x,z)$ of $g_1(y)$ and $g_2(x)$ are locally bounded in $\mathbb{R}^2$.

If the variables $X$ and $Y$ are discrete, then the conditional independence $\mathcal{F}_1 \underset{X,Y}{\perp\!\!\!\perp} \mathcal{F}_2$ is always true (see Exercise 4.2.9).

The following two lemmas allow us to reformulate the factorization property appearing in the preceding theorem.

**Lemma 4.2.5** *Suppose that $(A_1, B_1)$ and $(A_2, B_2)$ are $\mathbb{R}^2$-valued independent random variables such that $A_1 A_2 = B_1 B_2$. Then either*

(i) $A_1 = 0$ a.s. or $A_2 = 0$ a.s.; or

(ii) there is a constant $k \neq 0$ such that $A_1 = k B_1$ a.s. and $A_2 = k^{-1} B_2$ a.s.

*Proof:*    Suppose that (i) does not hold. Then $P(A_1 \neq 0) \neq 0$ and $P(A_2 \neq 0) \neq 0$. Without loss of generality we can assume that the underlying probability space is a product space $(\Omega, \mathcal{F}, P) = (\Omega_1, \mathcal{F}_1, P_1) \times (\Omega_2, \mathcal{F}_2, P_2)$. Let $\omega_2$ be such that $A_2(\omega_2) \neq 0$. Then condition (ii) follows from the relationship

$$A_1(\omega_1) = \frac{B_2(\omega_2)}{A_2(\omega_2)} B_1(\omega_1).$$

□

**Lemma 4.2.6** *Consider two independent $\sigma$-fields $\mathcal{F}_1$, $\mathcal{F}_2$ and two random variables $G_1$, $G_2$ such that $G_i$ is $\mathcal{F}_i$-measurable for $i = 1, 2$. The following statements are equivalent:*

*(a) There exist two random variables $H_1$ and $H_2$ such that $H_i$ is $\mathcal{F}_i$-measurable, $i = 1, 2$, and*

$$1 - G_1 G_2 = H_1 H_2.$$

*(b) $G_1$ or $G_2$ is constant a.s.*

*Proof:*    The fact that (b) implies (a) is obvious. Let us show that (a) implies (b). As before we can assume that the underlying probability space is a product space $(\Omega_1 \times \Omega_2, \mathcal{F}_1 \otimes \mathcal{F}_2, P_1 \times P_2)$. Property (a) implies that

$$[\tilde{G}_1(\tilde{\omega}_1) - G_1(\omega_1)]G_2(\omega_2) = [H_1(\omega_1) - \tilde{H}_1(\tilde{\omega}_1)]H_2(\omega_2),$$

where $\tilde{G}_1$ and $\tilde{H}_1$ are independent copies of $G_1$ and $H_1$ on some space $(\tilde{\Omega}_1, \tilde{\mathcal{F}}_1, P_1)$. Lemma 4.2.5 applied to the above equality implies either

(PA)  $G_2 = 0$ a.s. or $\tilde{G}_1 - G_1 = 0$ a.s.; or

(PB)  $G_2 = kH_2$ for some constant $k \neq 0$.

Then (PA) leads to property (b) directly and (PB) implies that $1 = [H_1 + kG_1]H_2$. Again applying Lemma 4.2.5 to this identity yields that $H_2$ and thus $G_2$ are a.s. constant.    □

**Corollary 4.2.1** *Under the hypotheses of Theorem 4.2.5, assume in addition that $1 - g_1'(Y)g_2'(X)$ has constant sign. Then conditions (i) and (ii) are equivalent to the following statement:*

*(iii) One (or both) of the variables $g_1'(Y)$ and $g_2'(X)$ is almost surely constant with respect to the conditional law given $X, Y$.*

As an application of the above criterion of conditional independence we are going to provide an alternative proof of Theorem 4.2.1 under slightly different hypotheses. Let $W = \{W_t, t \in [0, 1]\}$ be a Brownian motion defined in the canonical probability space $(\Omega, \mathcal{F}, P)$.

**Theorem 4.2.6** *Let $f$ and $\psi$ be functions of class $C^1$ such that $|f'| \leq K$ and $\psi' \leq 0$. Consider the equation*

$$\begin{cases} X_t = X_0 - \int_0^t f(X_s)ds + W_t \\ X_0 = \psi(X_1). \end{cases} \tag{4.50}$$

*This equation has a unique solution $X = \{X_t, t \in [0,1]\}$ that is a Markov random field if and only if one of the following conditions holds:*

(a) $f(x) = ax + b$, *for some constants $a, b \in \mathbb{R}$;*

(b) $\psi' \equiv 0$.

**Remarks:**

**1.** The fact that a unique solution exists is easy (see Exercise 4.2.7).

**2.** The case $X_1$ constant (condition (c) of Theorem 4.2.1) is not included in the above formulation.

*Proof:* We will only show that if $X$ is a Markov random field then one of conditions (a) or (b) holds. We will assume that $X$ is a Markov random field and $\psi'(x_0) \neq 0$ for some $x_0 \in \mathbb{R}$. Fix $0 < s < t \leq 1$. The Markov field property implies

$$\sigma\{X_r, r \in [s,t]\} \underset{\sigma\{X_s, X_t\}}{\perp\!\!\!\perp} \sigma\{X_r, r \notin (s,t)\}. \tag{4.51}$$

From the definition of the conditional independence we deduce

$$\mathcal{F}^i_{s,t} \underset{\sigma\{X_s, X_t\}}{\perp\!\!\!\perp} \mathcal{F}^e_{s,t}, \tag{4.52}$$

where

$$\mathcal{F}^i_{s,t} = \sigma\{W_r - W_s, s \leq r \leq t\}, \quad \text{and}$$
$$\mathcal{F}^e_{s,t} = \sigma\{W_r, 0 \leq r \leq s; W_r - W_t, t \leq r \leq 1\}.$$

Indeed, (4.51) implies (4.52) because

$$\mathcal{F}^i_{s,t} \subset \sigma\{X_r, r \in [s,t]\}$$

and

$$\mathcal{F}^e_{s,t} \subset \sigma\{X_r, r \notin (s,t)\}.$$

Define

$$\varphi_{s,t}(y) = y - \int_s^t f(\varphi_{s,r}(y))dr + W_t - W_s,$$

for any $s \leq t$ and $y \in \mathbb{R}$. Consider the random functions $g_1, g_2 : \mathbb{R} \times \Omega \to \mathbb{R}$ given by

$$\begin{cases} g_1(y) = \varphi_{s,t}(y) \\ g_2(x) = \varphi_{0,s}(\psi(\varphi_{t,1}(x))). \end{cases} \tag{4.53}$$

We have

$$\begin{cases} X_t = g_1(X_s) \\ X_s = g_2(X_t), \end{cases}$$

and we are going to apply Theorem 4.2.5 to the functions $g_1$ and $g_2$. Let us first verify that these functions satisfy conditions H1 through H3 of this theorem.

*Proof of H1:*    We have that $g_1(y), g_2(x) \in \mathbb{D}^{1,p}$ for all $p \geq 2$, and, moreover,

$$D_r g_1(y) = e^{-\int_r^t f'(\varphi_{s,u}(y))du} \mathbf{1}_{[s,t]}(r),$$

and

$$D_r g_2(x) = e^{-\int_r^s f'(\varphi_{0,u}(\psi(\varphi_{t,1}(x))))du} \mathbf{1}_{[0,s]}(r)$$
$$+ \varphi'_{0,s}(\psi(\varphi_{t,1}(x)))\psi'(\varphi_{t,1}(x))e^{-\int_r^1 f'(\varphi_{t,u}(x))du} \mathbf{1}_{[t,1]}(r).$$

From these explicit expressions for the derivative of the functions $g_1$ and $g_2$ we can deduce the absolute continuity of the laws of these variables using the criteria established in Chapter 2. In fact, we have $\|D(g_i)\|_H > 0$, $i = 1, 2$, and we can apply Theorem 2.1.3. Furthermore, we have

$$\frac{1}{2\varepsilon} P\left(|x - g_1(y)| < \varepsilon\right) \leq e^K (t - s)^{-\frac{1}{2}} \tag{4.54}$$

and

$$\frac{1}{2\varepsilon} P\left(|y - g_2(x)| < \varepsilon\right) \leq e^K s^{-\frac{1}{2}}, \tag{4.55}$$

which imply the boundedness of the function $\delta(x, y)$ introduced in H1. In fact, let us check Eq. (4.54). Set $h = \mathbf{1}_{[s,t]}$ and $\psi_\varepsilon(z) = \frac{1}{2\varepsilon} \int_{-\infty}^z \mathbf{1}_{[x-\varepsilon, x+\varepsilon]}(r)dr$. We have $D_h(g_1(y)) \geq e^{-K}(t - s)$. The duality formula (1.42) implies

$$\begin{aligned} \frac{1}{2\varepsilon} P\left(|x - g_1(y)|\right) &= E\left(\frac{D_h[\psi_\varepsilon(g_1(y))]}{D_h(g_1(y))}\right) \\ &\leq \frac{E(D_h[\psi_\varepsilon(g_1(y))])}{e^{-K}(t - s)} = \frac{E((W_t - W_s)\psi_\varepsilon(g_1(y)))}{e^{-K}(t - s)} \\ &\leq e^K(t - s)^{-\frac{1}{2}}, \end{aligned}$$

and (4.54) holds.

*Proof of H2:*    We are going to show that for all $\omega \in \Omega$ the transformation

$$(x, y) \longmapsto (x - g_1(y, \omega), y - g_2(x, \omega))$$

is bijective from $\mathbb{R}^2$ to $\mathbb{R}^2$. Let $(\bar{x}, \bar{y}) \in \mathbb{R}^2$. Set $x = \bar{x} + \varphi_{s,t}(y)$. It suffices to show that the mapping

$$y \xrightarrow{\Gamma} \varphi_{0,s}\left(\psi\left(\varphi_{t,1}(\bar{x} + \varphi_{s,t}(y))\right)\right) + \bar{y}$$

has a unique fixed point, and this follows from

$$\frac{d\Gamma}{dy} = \varphi'_{0,s}\left(\psi\left(\varphi_{t,1}(\bar{x}+\varphi_{s,t}(y))\right)\right)\psi'\left(\varphi_{t,1}(\bar{x}+\varphi_{s,t}(y))\right)$$
$$\times\varphi'_{t,1}(\bar{x}+\varphi_{s,t}(y))\varphi'_{s,t}(y) \le 0.$$

*Proof of H3:*     We have

$$1 - g'_1(y)g'_2(x) = 1 - \varphi'_{s,t}(y)\varphi'_{0,s}\left(\psi\left(\varphi_{t,1}(x)\right)\right)\psi'\left(\varphi_{t,1}(x)\right)\varphi'_{t,1}(x) \ge 1.$$

Note that

$$g'_1(X_s) = \exp(-\int_s^t f'(X_r)dr)$$

and

$$g'_2(X_t) = \psi'(X_1)\exp(-\int_{[0,s]\cup[t,1]} f'(X_r)dr).$$

In view of Corollary 4.2.1, the conditional independence (4.52) implies that one of the variables $g'_1(X_s)$ and $g'_2(X_t)$ is constant a.s. with respect to the conditional probability, given $X_s$ and $X_t$. Namely, there exists a measurable function $h : \mathbb{R}^2 \to \mathbb{R}$ such that either

(1) $\exp(-\int_s^t f'(X_r)dr) = h(X_s, X_t),$     or

(2) $\psi'(X_1)\exp(-\int_{[0,s]\cup[t,1]} f'(X_r)dr) = h(X_s, X_t).$

We will show that (1) implies that $f'$ is constant. Case (2) would be treated in a similar way. Suppose that there exist two points $x_1, x_2 \in \mathbb{R}$ such that $f'(x_1) < \alpha < f'(x_2)$. Let $\mathcal{C}$ be the class of continuous functions $y : [0, 1] \to \mathbb{R}$ such that $y_0 = \psi(y_1)$. The stochastic process $X$ takes values in $\mathcal{C}$, which is a closed subset of $C([0, 1])$, and the topological support of the law of $X$ is $\mathcal{C}$. As a consequence, we can assume that (1) holds for any $X \in \mathcal{C}$. For any $0 < 2\delta < t - s$ we can find two trajectories $z_1, z_2, \in \mathcal{C}$, depending on $\delta$ such that:

(i) $z_i(s) = z_i(t) = \frac{1}{2}(x_1 + x_2)$.

(ii) The functions $z_1$ and $z_2$ coincide on $[0, 1] - (s, t)$, and they do not depend on $\delta$ on this set.

(iii) $z_i(r) = x_i$ for all $i = 1, 2$ and $r \in (s + \delta, t - \delta)$.

(iv) The functions $z_i$ are linear on the intervals $(s, s + \delta)$ and $(t - \delta, t)$.

We have

$$\lim_{\delta\downarrow 0}\exp(\int_s^t f'(z_i(r))dr) = e^{(t-s)f'(x_i)}.$$

We can thus find a $\delta$ small enough such that

$$\exp\left(\int_s^t f'(z_1(r))dr\right) < e^{(t-s)\alpha} < \exp\left(\int_s^t f'(z_2(r))dr\right).$$

By continuity we can find neighborhoods $V_i$ of $z_i$ in $\mathcal{C}$, $i = 1, 2$, such that

$$\exp\left(\int_s^t f'(x(r))dr\right) < e^{(t-s)\alpha} < \exp\left(\int_s^t f'(y(r))dr\right),$$

for all $x \in V_1$ and $y \in V_2$. Therefore, $h(X_s, X_t) > e^{-(t-s)\alpha}$ if $X \in V_1$, and $h(X_s, X_t) < e^{-(t-s)\alpha}$ if $X \in V_2$. This is contradictory because there is a $\gamma > 0$ such that when $x$ runs over $V_i$, $i = 1, 2$, the point $(x(s), x(t))$ takes all possible values on some rectangle $[\frac{1}{2}(x_1 + x_2) - \gamma, \frac{1}{2}(x_1 + x_2) + \gamma]^2$. $\square$

## Exercises

**4.2.1** Let $\mathcal{F}_1, \mathcal{F}_2, \mathcal{G}$ be three sub-$\sigma$-fields in a probability space such that $\mathcal{F}_1 \underset{\mathcal{G}}{\perp\!\!\!\perp} \mathcal{F}_2$. Show the following properties:

(a) $\mathcal{F}_1 \vee \mathcal{G} \underset{\mathcal{G}}{\perp\!\!\!\perp} \mathcal{F}_2 \vee \mathcal{G}$.

(b) $\mathcal{F}_1 \underset{\mathcal{G}_1}{\perp\!\!\!\perp} \mathcal{F}_2$   if   $\mathcal{G} \subset \mathcal{G}_1 \subset \mathcal{F}_1$.

(c) $\mathcal{F}_1 \underset{\mathcal{H}}{\perp\!\!\!\perp} \mathcal{F}_2$   if   $\mathcal{G} \subset \mathcal{F}_1 \cap \mathcal{F}_2$, and $\mathcal{H}$ is a $\sigma$-field containing $\mathcal{G}$ of the form $\mathcal{H} = \mathcal{H}_1 \vee \mathcal{H}_2$, where $\mathcal{H}_1 \subset \mathcal{F}_1$ and $\mathcal{H}_2 \subset \mathcal{F}_2$.

**4.2.2** Let $\{\mathcal{G}_n, n \geq 1\}$ be a sequence of $\sigma$-fields such that $\mathcal{F}_1 \underset{\mathcal{G}_n}{\perp\!\!\!\perp} \mathcal{F}_2$ for each $n$. Show that $\mathcal{F}_1 \underset{\mathcal{G}}{\perp\!\!\!\perp} \mathcal{F}_2$, where $\mathcal{G} = \vee_n \mathcal{G}_n$ if the sequence is increasing, and $\mathcal{G} = \cap_n \mathcal{G}_n$ if it is decreasing.

**4.2.3** Let $X = \{X_t, t \in [0, 1]\}$ be a continuous Markov process. Show that it satisfies the Markov field property.

**4.2.4** Let $W = \{W_t, t \in [0, 1]\}$ be a Brownian motion, and let $g : \mathbb{R} \to \mathbb{R}$ be a measurable function. Show that $X_t = W_t + g(W_1)$ is a Markov random field. Assume that $g(x) = ax + b$. Show that in this case $X$ is a Markov process if and only if $a = -1$ or $a = 0$.

**4.2.5** For any $0 < \epsilon < 1$ consider the function $f_\epsilon : \mathbb{R}_+ \to \mathbb{R}$ defined by

$$f_\epsilon(t) = \begin{cases} 1 - t & \text{if} & 0 \leq t < 1 - \epsilon \\ \epsilon & \text{if} & 1 - \epsilon \leq t < 2 + \epsilon \\ 2 + t & \text{if} & 2 + \epsilon \leq t. \end{cases}$$

Let $X^\epsilon = \{X^\epsilon_t, t \geq 0\}$ be a stochastic process such that $P\{X^\epsilon_t = f_\epsilon(t), \forall t \geq 0\} = \frac{1}{2}$ and $P\{X^\epsilon_t = -f_\epsilon(t), \forall t \geq 0\} = \frac{1}{2}$. Show that $X^\epsilon$ is a Markov process but $X = \lim_{\epsilon \downarrow 0} X^\epsilon$ does not have the Markov property.

**4.2.6** Consider the stochastic differential equation

$$\begin{cases} X_t = X_0 - a \int_0^t X_s ds - b + W_t \\ X_0 = g(X_1 - X_0), \end{cases}$$

where $a, b \in \mathbb{R}$. Suppose that the implicit equation $x = g((e^{-a} - 1)x + y)$ has a unique continuous solution $x = \varphi(y)$. Show that the above equation admits a unique explicit solution that is a Markov random field.

**4.2.7** Show that Eq. (4.50) has a unique continuous solution which is a Markov random field if $f(x) = ax + b$.

**4.2.8** Check the estimates (4.54) and (4.55) integrating by parts on the Wiener space.

**4.2.9** Let $A$, $B$ be two independent discrete random variables taking values in some countable set $S$. Consider two measurable functions $f, g : \mathbb{R} \times S \to \mathbb{R}$, and suppose that the system

$$\begin{cases} x = f(y, a) \\ y = g(x, b) \end{cases}$$

has a unique solution for each $(a, b) \in S$ such that $P\{A = a, B = b\} > 0$. Let $X$, $Y$ be the random variables determined by the equations

$$\begin{cases} X = f(Y, A) \\ Y = g(X, B). \end{cases}$$

Show that $A$ and $B$ are conditionally independent given $X, Y$.

**4.2.10** Suppose that $f, g$ are two real-valued functions satisfying the following conditions:

(i) $f$ is of class $C^1$, and there exist $K > 0$ and $\lambda \in \mathbb{R}$ such that $-\lambda \leq f'(x) \leq K$ for all $x$.

(ii) $g$ is of class $C^1$, and $e^{\lambda' }|g'(x)| \leq |1 + g'(x)|$ for all $x$ and for some $\lambda' > \lambda$.

Show that Eq. (4.21) has a unique solution for each $\omega \in C_0([0, 1])$ ([250] and [97]).

**4.2.11** Let $Q << P$ be two probabilities in a measurable space $(\Omega, \mathcal{F})$, and set $\eta = \frac{dQ}{dP}$. Show that for any nonnegative (or $Q$-integrable) random variable $\xi$ and for any sub-$\sigma$-algebra $\mathcal{G} \subset \mathcal{F}$ we have

$$E_Q(\xi|\mathcal{G}) = \frac{E_P(\xi\eta|\mathcal{G})}{E_P(\eta|\mathcal{G})}.$$

**4.2.12** Consider the equation in $\mathbb{R}^2$

$$\begin{cases} dX_t + f(X_t) = dW_t \\ X_1^1 = X_0^2 = 0, \end{cases}$$

where $f(x^1, x^2) = (x^1 - x^2, -f_2(x^1))$ and $f_2$ is a twice continuously differentiable function such that $0 \leq f_2'(x) \leq K$ for some positive constant $K$. Show that there exists a unique solution, which is a Markov random field if and only if $f_2'' \equiv 0$ (cf. [250]).

**4.2.13** Let $G(x, y)$ be the Green function of $-\Delta$ on a bounded domain $D$ of $\mathbb{R}^k$, $k = 2, 3$. Let $W = \{W(A), A \in \mathcal{B}(D)\}$ be a Brownian measure on $D$. Define the random field $U_0(x) = \int_D G(x, y)W(dy)$, $x \in D$. Show that the process $\{U_0(x), x \in D\}$ has Hölder continuous paths of order $1 - \epsilon$ if $k = 2$, and Hölder continuous paths of order $\frac{3}{8} - \epsilon$ if $k = 3$, for any $\epsilon > 0$.

  *Hint:* Write $G(x, y)$ as the sum of a smooth function plus a function with a singularity of the form $\log |x - y|$ if $k = 2$ and $|x - y|^{-1}$ if $k = 3$, and use Kolmogorov's continuity criterion.

## Notes and comments

[**4.1**]    Proposition 4.1.2 is a fundamental result on nonlinear transformations of the Wiener measure and was obtained by Girsanov in [121]. Absolute continuity of the Wiener measure under linear (resp. nonlinear) transformations was discussed by Cameron and Martin in [56] (resp. [57]). We refer to Liptser and Shiryayev [200] for a nice and complete presentation of the absolute continuity of the transformations of the Wiener measure under adapted shifts.

  The extension of Girsanov's theorem to nonlinear transformations was discussed by Ramer [290] and Kusuoka [178] in the context of an abstract Wiener space. The notion of $H$-continuously differentiable random variable and the material of Sections 4.1.3 and 4.1.5 have been taken from Kusuoka's paper [178].

  The case of a contraction (i.e., $\|Du\|_{H \otimes H} < 1$) has been studied by Buckdahn in [48]. In that case, and assuming some additional assumptions, one can show that there exists a transformation $A : \Omega \to \Omega$ verifying $A \circ T = T \circ A = Id$ a.s., and the random variable $\eta(u)$ has the following expression (in the case of the classical Wiener space):

$$\eta(u) = \frac{d[P \circ A^{-1}]}{dP} = \exp\left( -\delta(u) - \frac{1}{2}\|u\|_H^2 dt \right.$$
$$\left. - \int_0^1 \int_0^t D_s u_t(D_t(u_s(A_t)))(T_t)dsdt \right),$$

where $\{T_t, 0 \le t \le 1\}$ is the one-parameter family of transformations of $\Omega$ defined by

$$(T_t\omega)_s = \omega_s + \int_0^{s \wedge t} u_r(\omega)dr$$

and $\{A_t, 0 \le t \le 1\}$ is the corresponding family of inverse transformations.

In [338] Üstünel and Zakai proved Proposition 4.1.5 under the hypothesis $\|Du\|_{\mathcal{L}(H,H)} < 1$ a.s.

Theorem 4.1.2 has been generalized in different directions. On one hand, local versions of this theorem can be found in Kusuoka [179]. On the other hand, Üstünel and Zakai [337] discuss the case where the transformation $T$ is not bijective (a multiplicity function must be introduced in this case) and $u$ is locally $H$-continuously differentiable.

The case of a one-parameter family of transformations on the classical Wiener space $\{T_t, 0 \le t \le 1\}$ defined by the integral equations

$$(T_t\omega)(s) = \omega_s + \int_0^{t \wedge s} u_r(T_r\omega)dr$$

has been studied by Buckdahn in [49]. Assuming that $u \in \mathbb{L}^{1,2}$ is such that $\int_0^1 \|u_t\|_\infty^2 dt + \int_0^1 \|\|Du_t\|_H\|_\infty^2 dt < \infty$, Buckdahn has proved that for each $t \in [0,1]$ there exists a transformation $A_t : \Omega \to \Omega$ such that $T_t \circ A_t = A_t \circ T_t = Id$ a.s., $P \circ T_t^{-1} << P$, $P \circ A_t^{-1} << P$, and the density functions of $P \circ T_t^{-1}$ and $P \circ A_t^{-1}$ are given by

$$\begin{aligned}
M_t &= \frac{d[P \circ A_t^{-1}]}{dP} = \exp\left\{ -\int_0^t u_s(T_s)dW_s - \frac{1}{2}\int_0^t u_s(T_s)^2 ds \right. \\
&\qquad \left. -\int_0^t \int_0^s (D_r u_s)(T_s) D_s[u_r(T_r)]drds \right\},
\end{aligned}$$

$$\begin{aligned}
L_t &= \frac{d[P \circ T_t^{-1}]}{dP} = \exp\left\{ \int_0^t u_s(T_s A_t)dW_s - \frac{1}{2}\int_0^t u_s(T_s A_t)^2 ds \right. \\
&\qquad \left. -\int_0^t \int_0^s (D_s u_r)(T_r A_t) D_r[u_s(T_s A_t)]drds \right\}.
\end{aligned}$$

The process $\{L_t, 0 \le t \le 1\}$ satisfies the Skorohod linear stochastic differential equation $L_t = 1 + \int_0^t u_s L_s dW_s$. This provides a generalization of the results for this type of equations presented in Chapter 2.

Üstünel and Zakai (cf. [335]) have extended this result to processes $u$ such that:

$$E\int_0^1 \exp(\lambda u_r^2)dr < \infty \quad \text{and} \quad \int_0^1 \|\|Du_t\|_H\|_\infty^4 \, dt < \infty$$

for some $\lambda > 0$. They use a general expression of the Radon-Nikodym derivative associated with smooth flows of transformations of $\Omega$ (see Cruzeiro [71]).

In [93] Enchev and Stroock extend the above result to the case where the process $u$ is Lipschitz, that means,

$$\|Du\|_{H\otimes H} \le c \quad \text{a.s.}$$

We refer to the monograph by Üstünel and Zakai (cf. [339]) for a complete analysis of transformations on the Wiener space and their induced measures.

[**4.2**]    We refer to Rozanov [296] for a detailed analysis of the notion of conditional independence. The study of the Markov property for solutions to stochastic differential equations with boundary conditions by means of the change of probability technique was first done in [250]. Further applications to different type of equations can be found in [80], [97], [98], [246], [251], and [252]. The study of the germ Markov field property for solutions to stochastic partial differential equations driven by a white noise has been done in [81], [82], and [253]. The characterization of the conditional independence presented in Section 4.2.3 has been obtained in [5].

# 5

# Fractional Brownian motion

The fractional Brownian motion is a self-similar centered Gaussian process with stationary increments and variance equals $t^{2H}$, where $H$ is a parameter in the interval $(0,1)$. For $H = \frac{1}{2}$ this process is a classical Brownian motion. In this chapter we will present the application of the Malliavin Calculus to develop a stochastic calculus with respect to the fractional Brownian motion.

## 5.1 Definition, properties and construction of the fractional Brownian motion

A centered Gaussian process $B = \{B_t, t \geq 0\}$ is called *fractional Brownian motion* (fBm) of Hurst parameter $H \in (0,1)$ if it has the covariance function

$$R_H(t,s) = E(B_t B_s) = \frac{1}{2}\left(s^{2H} + t^{2H} - |t-s|^{2H}\right). \qquad (5.1)$$

Fractional Brownian motion has the following *self-similar* property: For any constant $a > 0$, the processes $\{a^{-H} B_{at}, t \geq 0\}$ and $\{B_t, t \geq 0\}$ have the same distribution. This property is an immediate consequence of the fact that the covariance function (5.1) is homogeneous of order $2H$.

From (5.1) we can deduce the following expression for the variance of the increment of the process in an interval $[s, t]$:

$$E\left(|B_t - B_s|^2\right) = |t - s|^{2H}. \tag{5.2}$$

This implies that fBm has *stationary increments*.

By Kolmogorov's continuity criterion and (5.2) we deduce that fBm has a version with continuous trajectories. Moreover, by Garsia-Rodemich-Rumsey Lemma (see Lemma A.3.1), we can deduce the following modulus of continuity for the trajectories of fBm: For all $\varepsilon > 0$ and $T > 0$, there exists a nonnegative random variable $G_{\varepsilon,T}$ such that $E\left(|G_{\varepsilon,T}|^p\right) < \infty$ for all $p \geq 1$, and

$$|B_t - B_s| \leq G_{\varepsilon,T}|t - s|^{H-\varepsilon},$$

for all $s, t \in [0, T]$. In other words, the parameter $H$ controls the regularity of the trajectories, which are Hölder continuous of order $H - \varepsilon$, for any $\varepsilon > 0$.

For $H = \frac{1}{2}$, the covariance can be written as $R_{\frac{1}{2}}(t, s) = t \wedge s$, and the process $B$ is a standard Brownian motion. Hence, in this case the increments of the process in disjoint intervals are independent. However, for $H \neq \frac{1}{2}$, the increments are not independent.

Set $X_n = B_n - B_{n-1}$, $n \geq 1$. Then $\{X_n, n \geq 1\}$ is a Gaussian stationary sequence with covariance function

$$\rho_H(n) = \frac{1}{2}\left((n+1)^{2H} + (n-1)^{2H} - 2n^{2H}\right).$$

This implies that two increments of the form $B_k - B_{k-1}$ and $B_{k+n} - B_{k+n-1}$ are positively correlated (i.e. $\rho_H(n) > 0$) if $H > \frac{1}{2}$ and they are negatively correlated (i.e. $\rho_H(n) < 0$) if $H < \frac{1}{2}$. In the first case the process presents an aggregation behaviour and this property can be used to describe cluster phenomena. In the second case it can be used to model sequences with intermittency.

In the case $H > \frac{1}{2}$ the stationary sequence $X_n$ exhibits *long range dependence*, that is,

$$\lim_{n \to \infty} \frac{\rho_H(n)}{H(2H-1)n^{2H-2}} = 1$$

and, as a consequence, $\sum_{n=1}^{\infty} \rho_H(n) = \infty$.

In the case $H < \frac{1}{2}$ we have

$$\sum_{n=1}^{\infty} |\rho_H(n)| < \infty.$$

## 5.1.1 Semimartingale property

We have seen that for $H \neq \frac{1}{2}$ fBm does not have independent increments. The following proposition asserts that it is not a semimartingale.

**Proposition 5.1.1** *The fBm is not a semimartingale for $H \neq \frac{1}{2}$.*

*Proof:*     For $p > 0$ set

$$Y_{n,p} = n^{pH-1} \sum_{j=1}^{n} \left| B_{j/n} - B_{(j-1)/n} \right|^p .$$

By the self-similar property of fBm, the sequence $\{Y_{n,p}, n \geq 1\}$ has the same distribution as $\{\widetilde{Y}_{n,p}, n \geq 1\}$, where

$$\widetilde{Y}_{n,p} = n^{-1} \sum_{j=1}^{n} \left| B_j - B_{j-1} \right|^p .$$

The stationary sequence $\{B_j - B_{j-1}, j \geq 1\}$ is mixing. Hence, by the Ergodic Theorem $\widetilde{Y}_{n,p}$ converges almost surely and in $L^1(\Omega)$ to $E\left(|B_1|^p\right)$ as $n$ tends to infinity. As a consequence, $Y_{n,p}$ converges in probability as $n$ tends to infinity to $E\left(|B_1|^p\right)$. Therefore,

$$V_{n,p} = \sum_{j=1}^{n} \left| B_{j/n} - B_{(j-1)/n} \right|^p$$

converges in probability to zero as $n$ tends to infinity if $pH > 1$, and to infinity if $pH < 1$. Consider the following two cases:

i) If $H < \frac{1}{2}$, we can choose $p > 2$ such that $pH < 1$, and we obtain that the $p$-variation of fBm (defined as the limit in probability $\lim_{n\to\infty} V_{n,p}$) is infinite. Hence, the quadratic variation ($p = 2$) is also infinite.

ii) If $H > \frac{1}{2}$, we can choose $p$ such that $\frac{1}{H} < p < 2$. Then the $p$-variation is zero, and, as a consequence, the quadratic variation is also zero. On the other hand, if we choose $p$ such that $1 < p < \frac{1}{H}$ we deduce that the total variation is infinite.

Therefore, we have proved that for $H \neq \frac{1}{2}$ the fractional Brownian motion cannot be a semimartingale.     $\square$

In [65] Cheridito has introduced the notion of *weak semimartingale* as a stochastic process $\{X_t, t \geq 0\}$ such that for each $T > 0$, the set of random variables

$$\left\{ \sum_{j=1}^{n} f_j (X_{t_j} - X_{t_{j-1}}), n \geq 1, 0 \leq t_0 < \cdots < t_n \leq T, \right.$$

$$\left. |f_j| \leq 1, f_j \text{ is } \mathcal{F}^X_{t_{j-1}}\text{-measurable} \right\}$$

is bounded in $L^0(\Omega)$, where for each $t \geq 0$, $\mathcal{F}_t^X$ is the $\sigma$-field generated by the random variables $\{X_s, 0 \leq s \leq t\}$. It is important to remark that this $\sigma$-field is not completed with the null sets. Then, in [65] it is proved that fBm is not a weak semimartingale if $H \neq \frac{1}{2}$.

Let us mention the following surprising result also proved in [65]. Suppose that $\{B_t, t \geq 0\}$ is a fBm with Hurst parameter $H \in (0, 1)$, and $\{W_t, t \geq 0\}$ is an ordinary Brownian motion. Assume they are independent. Set
$$M_t = B_t + W_t.$$
Then $\{M_t, t \geq 0\}$ is not a weak semimartingale if $H \in (0, \frac{1}{2}) \cup (\frac{1}{2}, \frac{3}{4}]$, and it is a semimartingale, equivalent in law to Brownian motion on any finite time interval $[0, T]$, if $H \in (\frac{3}{4}, 1)$.

### 5.1.2 Moving average representation

Mandelbrot and Van Ness obtained in [217] the following integral representation of fBm in terms of a Wiener process on the whole real line (see also Samorodnitsky and Taqqu [301]).

**Proposition 5.1.2** *Let $\{W(A), A \in \mathcal{B}(\mathbb{R}), \mu(A) < \infty\}$ be a white noise on $\mathbb{R}$. Then*
$$B_t = \frac{1}{C_1(H)} \int_{\mathbb{R}} \left[ ((t-s)^+)^{H-\frac{1}{2}} - ((-s)^+)^{H-\frac{1}{2}} \right] dW_s,$$
*is a fractional Brownian motion with Hurst parameter $H$, if*
$$C_1(H) = \left( \int_0^\infty \left( (1+s)^{H-\frac{1}{2}} - s^{H-\frac{1}{2}} \right)^2 ds + \frac{1}{2H} \right)^{\frac{1}{2}}.$$

*Proof:*     Set $f_t(s) = ((t-s)^+)^{H-\frac{1}{2}} - ((-s)^+)^{H-\frac{1}{2}}$, $s \in \mathbb{R}$, $t \geq 0$. Notice that $\int_{\mathbb{R}} f_t(s)^2 ds < \infty$. In fact, if $H \neq \frac{1}{2}$, as $s$ tends to $-\infty$, $f_t(s)$ behaves as $(-s)^{H-\frac{3}{2}}$ which is square integrable at infinity. For $t \geq 0$ set
$$X_t = \int_{\mathbb{R}} \left[ ((t-s)^+)^{H-\frac{1}{2}} - ((-s)^+)^{H-\frac{1}{2}} \right] dW_s.$$

We have
$$
\begin{aligned}
E(X_t^2) &= \int_{\mathbb{R}} \left[ ((t-s)^+)^{H-\frac{1}{2}} - ((-s)^+)^{H-\frac{1}{2}} \right]^2 ds \\
&= t^{2H} \int_{\mathbb{R}} \left[ ((1-u)^+)^{H-\frac{1}{2}} - ((-u)^+)^{H-\frac{1}{2}} \right]^2 du \\
&= t^{2H} \left( \int_{-\infty}^0 \left[ (1-u)^{H-\frac{1}{2}} - (-u)^{H-\frac{1}{2}} \right]^2 du + \int_0^1 (1-u)^{2H-1} du \right) \\
&= C_1(H)^2 t^{2H}. \qquad\qquad (5.3)
\end{aligned}
$$

Similarly, for any $s < t$ we obtain

$$
\begin{aligned}
E(|X_t - X_s|^2) &= \int_{\mathbb{R}} \left[ \left((t-u)^+\right)^{H-\frac{1}{2}} - \left((s-u)^+\right)^{H-\frac{1}{2}} \right]^2 du \\
&= \int_{\mathbb{R}} \left[ \left((t-s-u)^+\right)^{H-\frac{1}{2}} - \left((-u)^+\right)^{H-\frac{1}{2}} \right]^2 du \\
&= C_1(H)^2 |t-s|^{2H}.
\end{aligned}
\tag{5.4}
$$

From (5.3) and (5.4) we deduce that the centered Gaussian process $\{X_t, t \geq 0\}$ has the covariance $R_H$ of a fBm with Hurst parameter $H$.  □

Notice that the above integral representation implies that the function $R_H$ defined in (5.1) is a covariance function, that is, it is symmetric and nonnegative definite.

It is also possible to establish the following spectral representation of fBm (see Samorodnitsky and Taqqu [301]):

$$
B_t = \frac{1}{C_2(H)} \int_{\mathbb{R}} \frac{e^{its} - 1}{is} |s|^{\frac{1}{2} - H} d\widetilde{W}_s,
$$

where $\widetilde{W} = W^1 + iW^2$ is a complex Gaussian measure on $\mathbb{R}$ such that $W^1(A) = W^1(-A)$, $W^2(A) = -W^2(A)$, and $E(W^1(A)^2) = E(W^2(A)^2) = \frac{|A|}{2}$, and

$$
C_2(H) = \left( \frac{\pi}{H\Gamma(2H)\sin H\pi} \right)^{\frac{1}{2}}.
$$

### 5.1.3 Representation of fBm on an interval

Fix a time interval $[0, T]$. Consider a fBm $\{B_t, t \in [0, T]\}$ with Hurst parameter $H \in (0, 1)$. We denote by $\mathcal{E}$ the set of step functions on $[0, T]$. Let $\mathcal{H}$ be the Hilbert space defined as the closure of $\mathcal{E}$ with respect to the scalar product

$$
\langle \mathbf{1}_{[0,t]}, \mathbf{1}_{[0,s]} \rangle_{\mathcal{H}} = R_H(t, s).
$$

The mapping $\mathbf{1}_{[0,t]} \longrightarrow B_t$ can be extended to an isometry between $\mathcal{H}$ and the Gaussian space $\mathcal{H}_1$ associated with $B$. We will denote this isometry by $\varphi \longrightarrow B(\varphi)$. Then $\{B(\varphi), \varphi \in \mathcal{H}\}$ is an isonormal Gaussian process associated with the Hilbert space $\mathcal{H}$ in the sense of Definition 1.1.1.

In this subsection we will establish the representation of fBm as a Volterra process using some computations inspired in the works [10] (case $H > \frac{1}{2}$) and [240] (general case).

Case $H > \frac{1}{2}$

It is easy to see that the covariance of fBm can be written as

$$
R_H(t, s) = \alpha_H \int_0^t \int_0^s |r - u|^{2H-2} du \, dr,
\tag{5.5}
$$

where $\alpha_H = H(2H-1)$. Formula (5.5) implies that

$$\langle \varphi, \psi \rangle_{\mathcal{H}} = \alpha_H \int_0^T \int_0^T |r-u|^{2H-2} \varphi_r \psi_u \, du \, dr \qquad (5.6)$$

for any pair of step functions $\varphi$ and $\psi$ in $\mathcal{E}$.

We can write

$$|r-u|^{2H-2} = \frac{(ru)^{H-\frac{1}{2}}}{\beta(2-2H, H-\frac{1}{2})}$$
$$\times \int_0^{r \wedge u} v^{1-2H}(r-v)^{H-\frac{3}{2}}(u-v)^{H-\frac{3}{2}} \, dv, \qquad (5.7)$$

where $\beta$ denotes the Beta function. Let us show Equation (5.7). Suppose $r > u$. By means of the change of variables $z = \frac{r-v}{u-v}$ and $x = \frac{r}{uz}$, we obtain

$$\int_0^u v^{1-2H}(r-v)^{H-\frac{3}{2}}(u-v)^{H-\frac{3}{2}} \, dv$$
$$= (r-u)^{2H-2} \int_{\frac{r}{u}}^\infty (zu-r)^{1-2H} z^{H-\frac{3}{2}} \, dz$$
$$= (ru)^{\frac{1}{2}-H}(r-u)^{2H-2} \int_0^1 (1-x)^{1-2H} x^{H-\frac{3}{2}} \, dx$$
$$= \beta(2-2H, H-\frac{1}{2})(ru)^{\frac{1}{2}-H}(r-u)^{2H-2}.$$

Consider the square integrable kernel

$$K_H(t,s) = c_H s^{\frac{1}{2}-H} \int_s^t (u-s)^{H-\frac{3}{2}} u^{H-\frac{1}{2}} \, du, \qquad (5.8)$$

where $c_H = \left[ \frac{H(2H-1)}{\beta(2-2H, H-\frac{1}{2})} \right]^{1/2}$ and $t > s$.

Taking into account formulas (5.5) and (5.7) we deduce that this kernel verifies

$$\int_0^{t \wedge s} K_H(t,u) K_H(s,u) \, du = c_H^2 \int_0^{t \wedge s} \left( \int_u^t (y-u)^{H-\frac{3}{2}} y^{H-\frac{1}{2}} \, dy \right)$$
$$\times \left( \int_u^s (z-u)^{H-\frac{3}{2}} z^{H-\frac{1}{2}} \, dz \right) u^{1-2H} \, du$$
$$= c_H^2 \beta(2-2H, H-\frac{1}{2}) \int_0^t \int_0^s |y-z|^{2H-2} \, dz \, dy$$
$$= R_H(t,s). \qquad (5.9)$$

Formula (5.9) implies that the kernel $R_H$ is nonnegative definite and provides an explicit representation for its square root as an operator.

From (5.8) we get

$$\frac{\partial K_H}{\partial t}(t,s) = c_H \left(\frac{t}{s}\right)^{H-\frac{1}{2}} (t-s)^{H-\frac{3}{2}}. \tag{5.10}$$

Consider the linear operator $K_H^*$ from $\mathcal{E}$ to $L^2([0,T])$ defined by

$$(K_H^*\varphi)(s) = \int_s^T \varphi(t)\frac{\partial K_H}{\partial t}(t,s)dt. \tag{5.11}$$

Notice that

$$\left(K_H^*\mathbf{1}_{[0,t]}\right)(s) = K_H(t,s)\mathbf{1}_{[0,t]}(s). \tag{5.12}$$

The operator $K_H^*$ is an isometry between $\mathcal{E}$ and $L^2([0,T])$ that can be extended to the Hilbert space $\mathcal{H}$. In fact, for any $s,t \in [0,T]$ we have using (5.12) and (5.9)

$$\begin{aligned}
\left\langle K_H^*\mathbf{1}_{[0,t]}, K_H^*\mathbf{1}_{[0,s]}\right\rangle_{L^2([0,T])} &= \left\langle K_H(t,\cdot)\mathbf{1}_{[0,t]}, K_H(s,\cdot)\mathbf{1}_{[0,s]}\right\rangle_{L^2([0,T])} \\
&= \int_0^{t\wedge s} K_H(t,u)K_H(s,u)du \\
&= R_H(t,s) = \left\langle \mathbf{1}_{[0,t]}, \mathbf{1}_{[0,s]}\right\rangle_{\mathcal{H}}.
\end{aligned}$$

The operator $K_H^*$ can be expressed in terms of fractional integrals:

$$(K_H^*\varphi)(s) = c_H\Gamma(H-\frac{1}{2})s^{\frac{1}{2}-H}(I_{T-}^{H-\frac{1}{2}}u^{H-\frac{1}{2}}\varphi(u))(s). \tag{5.13}$$

This is an immediate consequence of formulas (5.10), (5.11) and (A.14).

For any $a \in [0,T]$, the indicator function $\mathbf{1}_{[0,a]}$ belongs to the image of $K_H^*$ and applying the rules of the fractional calculus yields (Exercise 5.1.6)

$$(K_H^*)^{-1}(\mathbf{1}_{[0,a]}) = \frac{1}{c_H\Gamma(H-\frac{1}{2})}s^{\frac{1}{2}-H}\left(D_{a-}^{H-\frac{1}{2}}u^{H-\frac{1}{2}}\right)(s)\mathbf{1}_{[0,a]}(s). \tag{5.14}$$

Consider the process $W = \{W_t, t \in [0,T]\}$ defined by

$$W_t = B((K_H^*)^{-1}(\mathbf{1}_{[0,t]})). \tag{5.15}$$

Then $W$ is a Wiener process, and the process $B$ has the integral representation

$$B_t = \int_0^t K_H(t,s)dW_s. \tag{5.16}$$

Indeed, for any $s,t \in [0,T]$ we have

$$\begin{aligned}
E(W_tW_s) &= E\left(B((K_H^*)^{-1}(\mathbf{1}_{[0,t]}))B((K_H^*)^{-1}(\mathbf{1}_{[0,s]}))\right) \\
&= \left\langle (K_H^*)^{-1}(\mathbf{1}_{[0,t]}), (K_H^*)^{-1}(\mathbf{1}_{[0,s]})\right\rangle_{\mathcal{H}} \\
&= \left\langle \mathbf{1}_{[0,t]}, \mathbf{1}_{[0,s]}\right\rangle_{L^2([0,T])} = s\wedge t.
\end{aligned}$$

Moreover, for any $\varphi \in \mathcal{H}$ we have

$$B(\varphi) = \int_0^T (K_H^* \varphi)(t) dW_t. \tag{5.17}$$

Notice that from (5.14), the Wiener process $W$ is adapted to the filtration generated by the fBm $B$ and (5.15) and (5.16) imply that both processes generate the same filtration. Furthermore, the Wiener process $W$ that provides the integral representation (5.16) is unique. Indeed, this follows from the fact that the image of the operator $K_H^*$ is $L^2([0,T])$, because this image contains the indicator functions.

The elements of the Hilbert space $\mathcal{H}$ may not be functions but distributions of negative order (see Pipiras and Taqqu [283], [284]). In fact, from (5.13) it follows that $\mathcal{H}$ coincides with the space of distributions $f$ such that $s^{\frac{1}{2}-H} I_{0+}^{H-\frac{1}{2}} (f(u) u^{H-\frac{1}{2}})(s)$ is a square integrable function.

We can find a linear space of functions contained in $\mathcal{H}$ in the following way. Let $|\mathcal{H}|$ be the linear space of measurable functions $\varphi$ on $[0,T]$ such that

$$\|\varphi\|_{|\mathcal{H}|}^2 = \alpha_H \int_0^T \int_0^T |\varphi_r| |\varphi_u| |r-u|^{2H-2} \, dr du < \infty. \tag{5.18}$$

It is not difficult to show that $|\mathcal{H}|$ is a Banach space with the norm $\|\cdot\|_{|\mathcal{H}|}$ and $\mathcal{E}$ is dense in $|\mathcal{H}|$. On the other hand, it has been shown in [284] that the space $|\mathcal{H}|$ equipped with the inner product $\langle \varphi, \psi \rangle_{\mathcal{H}}$ is not complete and it is isometric to a subspace of $\mathcal{H}$. The following estimate has been proved in [222].

**Lemma 5.1.1** *Let* $H > \frac{1}{2}$ *and* $\varphi \in L^{\frac{1}{H}}([0,T])$. *Then*

$$\|\varphi\|_{|\mathcal{H}|} \leq b_H \|\varphi\|_{L^{\frac{1}{H}}([0,T])}, \tag{5.19}$$

*for some constant* $b_H > 0$.

*Proof:* Using Hölder's inequality with exponent $q = \frac{1}{H}$ in (5.18) we get

$$\|\varphi\|_{|\mathcal{H}|}^2 \leq \alpha_H \left( \int_0^T |\varphi_r|^{\frac{1}{H}} dr \right)^H \left( \int_0^T \left( \int_0^T |\varphi_u| |r-u|^{2H-2} du \right)^{\frac{1}{1-H}} dr \right)^{1-H}.$$

The second factor in the above expression, up to a multiplicative constant, is equal to the $\frac{1}{1-H}$ norm of the left-sided fractional integral $I_{0+}^{2H-1} |\varphi|$. Finally is suffices to apply the Hardy-Littlewood inequality (see [317, Theorem 1, p.119])

$$\|I_{0+}^\alpha f\|_{L^q(0,\infty)} \leq c_{\alpha,q} \|f\|_{L^p(0,\infty)} \tag{5.20}$$

where $0 < \alpha < 1$, $1 < p < q < \infty$ satisfy $\frac{1}{q} = \frac{1}{p} - \alpha$, with the particular values $\alpha = 2H - 1$, $q = \frac{1}{1-H}$ and $p = \frac{1}{H}$. $\qquad\square$

As a consequence

$$L^2([0,T]) \subset L^{\frac{1}{H}}([0,T]) \subset |\mathcal{H}| \subset \mathcal{H}.$$

The inclusion $L^2([0,T]) \subset |\mathcal{H}|$ can be proved by a direct argument:

$$\int_0^T \int_0^T |\varphi_r| \, |\varphi_u| \, |r-u|^{2H-2} \, dr du \quad \le \quad \int_0^T \int_0^T |\varphi_u|^2 \, |r-u|^{2H-2} \, dr du$$

$$\le \quad \frac{T^{2H-1}}{H-\frac{1}{2}} \int_0^T |\varphi_u|^2 \, du.$$

This means that the Wiener-type integral $\int_0^T \varphi(t) dB_t$ (which is equal to $B(\varphi)$, by definition) can be defined for functions $\varphi \in |\mathcal{H}|$, and

$$\int_0^T \varphi(t) dB_t = \int_0^T (K_H^* \varphi)(t) dW_t. \tag{5.21}$$

Case $H < \frac{1}{2}$

To find a square integrable kernel that satisfies (5.9) is more difficult than in the case $H > \frac{1}{2}$. The following proposition provides the answer to this problem.

**Proposition 5.1.3** *Let* $H < \frac{1}{2}$. *The kernel*

$$K_H(t,s) \;=\; c_H \left[ \left(\frac{t}{s}\right)^{H-\frac{1}{2}} (t-s)^{H-\frac{1}{2}} \right. $$
$$\left. -(H-\frac{1}{2}) s^{\frac{1}{2}-H} \int_s^t u^{H-\frac{3}{2}} (u-s)^{H-\frac{1}{2}} du \right],$$

*where* $c_H = \sqrt{\frac{2H}{(1-2H)\beta(1-2H,H+1/2)}}$, *satisfies*

$$R_H(t,s) = \int_0^{t\wedge s} K_H(t,u) K_H(s,u) du. \tag{5.22}$$

In the references [78] and [284] Eq. (5.22) is proved using the analyticity of both members as functions of the parameter $H$. We will give here a direct proof using the ideas of [240]. Notice first that

$$\frac{\partial K_H}{\partial t}(t,s) = c_H (H-\frac{1}{2}) \left(\frac{t}{s}\right)^{H-\frac{1}{2}} (t-s)^{H-\frac{3}{2}}. \tag{5.23}$$

*Proof:* Consider first the diagonal case $s = t$. Set $\phi(s) = \int_0^s K_H(s,u)^2 du$. We have

$$
\begin{aligned}
\phi(s) \;=\; c_H^2 &\Bigg[ \int_0^s (\frac{s}{u})^{2H-1}(s-u)^{2H-1} du \\
&-(2H-1)\int_0^s s^{H-\frac{1}{2}} u^{1-2H}(s-u)^{H-\frac{1}{2}} \\
&\times \left( \int_u^s v^{H-\frac{3}{2}}(v-u)^{H-\frac{1}{2}} dv \right) du \\
&+(H-\frac{1}{2})^2 \int_0^s u^{1-2H} \left( \int_u^s v^{H-\frac{3}{2}}(v-u)^{H-\frac{1}{2}} dv \right)^2 du \Bigg].
\end{aligned}
$$

Making the change of variables $u = sx$ in the first integral and using Fubini's theorem yields

$$
\begin{aligned}
\phi(s) \;=\; c_H^2 &\Big[ s^{2H}\beta(2-2H,2H) \\
&-(2H-1)s^{H-\frac{1}{2}} \int_0^s v^{H-\frac{3}{2}} \\
&\times \left( \int_0^v u^{1-2H}(s-u)^{H-\frac{1}{2}}(v-u)^{H-\frac{1}{2}} du \right) dv \\
&+2(H-\frac{1}{2})^2 \int_0^s \int_0^v \int_0^w u^{1-2H}(v-u)^{H-\frac{1}{2}}(w-u)^{H-\frac{1}{2}} \\
&\times w^{H-\frac{3}{2}} v^{H-\frac{3}{2}} du\,dw\,dv \Big].
\end{aligned}
$$

Now we make the change of variable $u = vx$, $v = sy$ for the second term and $u = wx$, $w = vy$ for the third term and we obtain

$$
\begin{aligned}
\phi(s) \;=\; c_H^2 s^{2H} &\Big[ \beta(2-2H,2H) - (2H-1)(\frac{1}{4H}+\frac{1}{2}) \\
&\times \int_0^1 \int_0^1 x^{1-2H}(1-xy)^{H-\frac{1}{2}}(1-x)^{H-\frac{1}{2}} dx\,dy \Big] \\
=\; s^{2H}&.
\end{aligned}
$$

Suppose now that $s < t$. Differentiating Equation (5.22) with respect to $t$, we are aimed to show that

$$
H(t^{2H-1} - (t-s)^{2H-1}) = \int_0^s \frac{\partial K_H}{\partial t}(t,u)K_H(s,u)du. \tag{5.24}
$$

Set $\phi(t,s) = \int_0^s \frac{\partial K_H}{\partial t}(t,u) K_H(s,u) du$. Using (5.23) yields

$$
\begin{aligned}
\phi(t,s) &= c_H^2 (H - \frac{1}{2}) \int_0^s \left(\frac{t}{u}\right)^{H-\frac{1}{2}} (t-u)^{H-\frac{3}{2}} \left(\frac{s}{u}\right)^{H-\frac{1}{2}} (s-u)^{H-\frac{1}{2}} du \\
&\quad - c_H^2 (H - \frac{1}{2})^2 \int_0^s \left(\frac{t}{u}\right)^{H-\frac{1}{2}} (t-u)^{H-\frac{3}{2}} u^{\frac{1}{2}-H} \\
&\quad \times \left( \int_u^s v^{H-\frac{3}{2}} (v-u)^{H-\frac{1}{2}} dv \right) du.
\end{aligned}
$$

Making the change of variables $u = sx$ in the first integral and $u = vx$ in the second one we obtain

$$
\begin{aligned}
\phi(t,s) &= c_H^2 (H - \frac{1}{2}) (ts)^{H-\frac{1}{2}} \gamma(\frac{t}{s}) \\
&\quad - c_H^2 (H - \frac{1}{2})^2 t^{H-\frac{1}{2}} \int_0^s v^{H-\frac{3}{2}} \gamma(\frac{t}{v}) \, dv,
\end{aligned}
$$

where $\gamma(y) = \int_0^1 x^{1-2H} (y-x)^{H-\frac{3}{2}} (1-x)^{H-\frac{1}{2}} dx$ for $y > 1$. Then, (5.24) is equivalent to

$$
\begin{aligned}
&c_H^2 \left[ (H - \frac{1}{2}) s^{H-\frac{1}{2}} \gamma(\frac{t}{s}) - (H - \frac{1}{2})^2 \int_0^s v^{H-\frac{3}{2}} \gamma(\frac{t}{v}) \, dv \right] \\
&= H(t^{H-\frac{1}{2}} - t^{\frac{1}{2}-H} (t-s)^{2H-1}).
\end{aligned}
\tag{5.25}
$$

Differentiating the left-hand side of equation (5.25) with respect to $t$ yields

$$
\begin{aligned}
&c_H^2 (H - \frac{3}{2}) \left[ (H - \frac{1}{2}) s^{H-\frac{3}{2}} \delta(\frac{t}{s}) - (H - \frac{1}{2})^2 \int_0^s v^{H-\frac{5}{2}} \delta(\frac{t}{v}) \, dv \right] \\
&: \; = \mu(t,s),
\end{aligned}
\tag{5.26}
$$

where, for $y > 1$,

$$
\delta(y) = \int_0^1 x^{1-2H} (y-x)^{H-\frac{5}{2}} (1-x)^{H-\frac{1}{2}} dx.
$$

By means of the change of variables $z = \frac{y(1-x)}{y-x}$ we obtain

$$
\delta(y) = \beta(2 - 2H, H + \frac{1}{2}) y^{-H-\frac{1}{2}} (y-1)^{2H-2}.
\tag{5.27}
$$

Finally, substituting (5.27) into (5.26) yields

$$
\begin{aligned}
\mu(t,s) &= c_H^2 \beta(2 - 2H, H + \frac{1}{2})(H - \frac{3}{2})(H - \frac{1}{2}) \\
&\quad \times t^{-H-\frac{1}{2}} s(t-s)^{2H-2} + \frac{1}{2} t^{-H-\frac{1}{2}} ((t-s)^{2H-1} - t^{2H-1}) \\
&= H(1 - 2H) \\
&\quad \times \left( t^{-H-\frac{1}{2}} s(t-s)^{2H-2} + \frac{1}{2}(t-s)^{2H-1} t^{-H-\frac{1}{2}} - \frac{1}{2} t^{H-\frac{3}{2}} \right).
\end{aligned}
$$

This last expression coincides with the derivative with respect to $t$ of the right-hand side of (5.25). This completes the proof of the equality (5.22).$\square$

The kernel $K_H$ can also be expressed in terms of fractional derivatives:

$$K_H(t,s) = c_H \Gamma(H + \frac{1}{2}) s^{\frac{1}{2}-H} \left( D_{t-}^{\frac{1}{2}-H} u^{H-\frac{1}{2}} \right)(s). \tag{5.28}$$

Consider the linear operator $K_H^*$ from $\mathcal{E}$ to $L^2([0,T])$ defined by

$$(K_H^* \varphi)(s) = K_H(T,s)\varphi(s) + \int_s^T (\varphi(t) - \varphi(s)) \frac{\partial K_H}{\partial r}(t,s)dt. \tag{5.29}$$

Notice that

$$\left(K_H^* \mathbf{1}_{[0,t]}\right)(s) = K_H(t,s)\mathbf{1}_{[0,t]}(s). \tag{5.30}$$

From (5.22) and (5.30) we deduce as in the case $H > \frac{1}{2}$ that the operator $K_H^*$ is an isometry between $\mathcal{E}$ and $L^2([0,T])$ that can be extended to the Hilbert space $\mathcal{H}$.

The operator $K_H^*$ can be expressed in terms of fractional derivatives:

$$(K_H^* \varphi)(s) = d_H \, s^{\frac{1}{2}-H}(D_{T-}^{\frac{1}{2}-H} u^{H-\frac{1}{2}} \varphi(u))(s), \tag{5.31}$$

where $d_H = c_H \Gamma(H + \frac{1}{2})$. This is an immediate consequence of (5.29) and the equality

$$\left(D_{t-}^{\frac{1}{2}-H} u^{H-\frac{1}{2}}\right)(s)\mathbf{1}_{[0,t]}(s) = \left(D_{T-}^{\frac{1}{2}-H} u^{H-\frac{1}{2}} \mathbf{1}_{[0,t]}(u)\right)(s).$$

As a consequence,

$$C^\gamma([0,T]) \subset \mathcal{H} \subset L^2([0,T]),$$

if $\gamma > \frac{1}{2} - H$.

Using the alternative expression for the kernel $K_H$ given by

$$K_H(t,s) = c_H(t-s)^{H-\frac{1}{2}} + s^{H-\frac{1}{2}} F_1(\frac{t}{s}), \tag{5.32}$$

where

$$F_1(z) = c_H(\frac{1}{2} - H) \int_0^{z-1} \theta^{H-\frac{3}{2}}(1 - (\theta+1)^{H-\frac{1}{2}})d\theta,$$

one can show that $\mathcal{H} = I_{T-}^{\frac{1}{2}-H}(L^2)$ (see [78] and Proposition 6 of [9]). In fact, from (5.29) and (5.32) we obtain, for any function $\varphi$ in $I_{T-}^{\frac{1}{2}-H}(L^2)$

$$
\begin{aligned}
(K_H^* \varphi)(s) &= c_H(T-s)^{H-\frac{1}{2}}\varphi(s) \\
&\quad + c_H(H - \frac{1}{2}) \int_s^T (\varphi(r) - \varphi(s))(r-s)^{H-\frac{3}{2}} dr \\
&\quad + s^{H-\frac{3}{2}} \int_s^T \varphi(r) F_1'\left(\frac{T}{s}\right) dr \\
&= c_H \Gamma(\frac{1}{2} + H) D_{T-}^{\frac{1}{2}-H} \varphi(s) + \Lambda\varphi(s),
\end{aligned}
$$

where the operator

$$\Lambda\varphi(s) = c_H(\tfrac{1}{2} - H) \int_s^T \varphi(r)(r - s)^{H - \frac{3}{2}} \left(1 - \left(\frac{r}{s}\right)^{H - \frac{1}{2}}\right) dr$$

is bounded in $L^2$.

On the other hand, (5.31) implies that

$$\mathcal{H} = \{f : \exists \phi \in L^2(0, T) : f(s) = d_H^{-1} s^{\frac{1}{2} - H} (I_{T-}^{\frac{1}{2} - H} u^{H - \frac{1}{2}} \phi(u))(s)\},$$

with the inner product

$$\langle f, g \rangle_{\mathcal{H}} = \int_0^T \phi(s)\psi(s)ds,$$

if

$$f(s) = d_H^{-1} s^{\frac{1}{2} - H} (I_{T-}^{\frac{1}{2} - H} u^{H - \frac{1}{2}} \phi(u))(s)$$

and

$$g(s) = d_H^{-1} s^{\frac{1}{2} - H} (I_{T-}^{\frac{1}{2} - H} u^{H - \frac{1}{2}} \psi(u))(s).$$

Consider process $W = \{W_t, t \in [0, T]\}$ defined by

$$W_t = B((K_H^*)^{-1} (\mathbf{1}_{[0,t]})).$$

As in the case $H > \frac{1}{2}$, we can show that $W$ is a Wiener process, and the process $B$ has the integral representation

$$B_t = \int_0^t K_H(t, s)dW_s.$$

Therefore, in this case the Wiener-type integral $\int_0^T \varphi(t)dB_t$ can be defined for functions $\varphi \in I_{T-}^{\frac{1}{2} - H}(L^2)$, and (5.21) holds.

**Remark**

In [9] these results have been generalized to Gaussian Volterra processes of the form

$$X_t = \int_0^t K(t, s)dW_s,$$

where $\{W_t, t \geq 0\}$ is a Wiener process and $K(t, s)$ is a square integrable kernel. Two different types of kernels can be considered, which correspond to the cases $H < \frac{1}{2}$ and $H > \frac{1}{2}$:

i) *Singular case*: $K(\cdot, s)$ has bounded variation on any interval $(u, T]$, $u > s$, but $\int_s^T |K|(dt, s) = \infty$ for every $s$.

ii) *Regular case*: The kernel satisfies $\int_s^T |K|((s, T], s)^2 ds < \infty$ for each $s$.

Define the left and right-sided fractional derivative operators on the whole real line for $0 < \alpha < 1$ by

$$D^{\alpha}_{-} f(s) := \frac{\alpha}{\Gamma(1-\alpha)} \int_0^{\infty} \frac{f(s) - f(s+u)}{u^{1+\alpha}} du$$

and

$$D^{\alpha}_{+} f(s) := \frac{\alpha}{\Gamma(1-\alpha)} \int_0^{\infty} \frac{f(s) - f(s-u)}{u^{1+\alpha}} du,$$

$s \in \mathbb{R}$, respectively. Then, the scalar product in $\mathcal{H}$ has the following simple expression

$$\langle f, g \rangle_{\mathcal{H}} = e_H^2 \left\langle D^{\frac{1}{2}-H}_{-} f, D^{\frac{1}{2}-H}_{+} g \right\rangle_{L^2(\mathbb{R})}, \tag{5.33}$$

where $e_H = C_1(H)^{-1}\Gamma(H+\frac{1}{2})$, $f, g \in \mathcal{H}$, and by convention $f(s) = g(s) = 0$ if $s \notin [0, T]$.

## Exercises

**5.1.1** Show that $B = \{B_t, t \geq 0\}$ is a fBm with Hurst parameter $H$ if and only if it is a centered Gaussian process with stationary increments and variance $t^{2H}$.

**5.1.2** Using the self-similarity property of the fBm show that for all $p > 0$

$$E\left( \sup_{0 \leq t \leq T} |B_t|^p \right) = C_{p,H} T^{pH}.$$

**5.1.3** Let $B = \{B_t, t \geq 0\}$ be a fBm with Hurst parameter $H \in (0, \frac{1}{2}) \cup (\frac{1}{2}, 1)$. Show that the following process is a martingale

$$M_t = \int_0^t s^{\frac{1}{2}-H}(t-s)^{\frac{1}{2}-H} dB_s$$

with variance $c_{1,H} t^{2-2H}$ and compute the constant $c_{1,H}$.
    *Hint:* Use the representation (5.17).

**5.1.4** Show that the fBm admits the representation $B_t = c_{2,H} \int_0^t s^{H-\frac{1}{2}} dY_s$, where $Y_t = \int_0^t (t-s)^{H-\frac{1}{2}} s^{H-\frac{1}{2}} dW_s$, and $W$ is an ordinary Brownian motion.

**5.1.5** Suppose $H > \frac{1}{2}$. If $\tau$ is a stopping time with values in $[0, T]$, show that for all $p > 0$

$$E\left( \sup_{0 \leq t \leq \tau} |B_t|^p \right) \leq C_{p,H} E(\tau^{pH}).$$

    *Hint:* Use the representation established in Exercise 5.1.4.

**5.1.6** Show formula (5.14).

**5.1.7** Show that $|\mathcal{H}|$ is a Banach space with the norm $\|\cdot\|_{|\mathcal{H}|}$ and $\mathcal{E}$ is dense in $|\mathcal{H}|$.

**5.1.8** Show formula (5.33).

## 5.2  Stochastic calculus with respect to fBm

In this section we develop a stochastic calculus with respect to the fBm. There are essentially two different approaches to construct stochastic integrals with respect to the fBm:

(i) *Path-wise approach.* If $u = \{u_t, t \in [0,T]\}$ is a stochastic process with $\gamma$-Hölder continuous trajectories, where $\gamma > 1 - H$, then by the results of Young ([354]) the Riemann Stieltjes integral $\int_0^T u_t dB_t$ exists path-wise. This method is particularly useful in the case $H > \frac{1}{2}$, because it includes processes of the form $u_t = F(B_t)$, where $F$ is a continuously differentiable function.

(ii) *Malliavin calculus.* We have seen in Chapter 1 that in the case of an ordinary Brownian motion, the adapted processes in $L^2([0,T] \times \Omega)$ belong to the domain of the divergence operator, and on this set the divergence operator coincides with Itô's stochastic integral. Actually, the divergence operator coincides with an extension of Itô's stochastic integral introduced by Skorohod in [315]. In this context a natural question is to ask in which sense the divergence operator with respect to a fractional Brownian motion $B$ can be interpreted as a stochastic integral. Note that the divergence operator provides an isometry between the Hilbert Space $\mathcal{H}$ associated with the fBm $B$ and the Gaussian space $\mathcal{H}_1$, and gives rise to a notion of stochastic integral for classes of deterministic functions included in $\mathcal{H}$. If $H < \frac{1}{2}$, then $\mathcal{H} = I_{T_-}^{\frac{1}{2}-H}(L^2)$ is a class of functions that contains $C^\gamma([0,T])$ if $\gamma > \frac{1}{2} - H$. If $H = \frac{1}{2}$, then $\mathcal{H} = L^2([0,T])$, and if $H > \frac{1}{2}$, $\mathcal{H}$ contains the space $|\mathcal{H}|$ of functions. We will see that in the random case, (see Propositions 5.2.3 and 5.2.4) the divergence equals to a path-wise integral minus the trace of the derivative.

### 5.2.1  Malliavin Calculus with respect to the fBm

Let $B = \{B_t, t \in [0,T]\}$ be a fBm with Hurst parameter $H \in (0,1)$. The process $\{B(\varphi), \varphi \in \mathcal{H}\}$ is an isonormal Gaussian process associated with the Hilbert space $\mathcal{H}$ in the sense of Definition 1.1.1. We will denote by $D$ and $\delta$ the derivative and divergence operators associated with this process.

Recall that the operator $K_H^*$ is an isometry between $\mathcal{H}$ and a closed subspace of $L^2([0,T])$. Moreover, $W_t = B((K_H^*)^{-1}(\mathbf{1}_{[0,t]}))$ is a Wiener process such that

$$B_t = \int_0^t K_H(t,s) dW_s,$$

and for any $\varphi \in \mathcal{H}$ we have $B(\varphi) = W(K_H^* \varphi)$.

In this framework there is a *transfer principle* that connects the derivative and divergence operators of both processes $B$ and $W$. Its proof is left as an exercise (Exercise 5.2.1).

**Proposition 5.2.1** *For any* $F \in \mathbb{D}_W^{1,2} = \mathbb{D}^{1,2}$

$$K_H^* DF = D^W F,$$

*where* $D^W$ *denotes the derivative operator with respect to the process* $W$, *and* $\mathbb{D}_W^{1,2}$ *the corresponding Sobolev space.*

**Proposition 5.2.2** $\mathrm{Dom}\,\delta = (K_H^*)^{-1} (\mathrm{Dom}\,\delta_W)$, *and for any* $\mathcal{H}$*-valued random variable* $u$ *in* $\mathrm{Dom}\,\delta$ *we have* $\delta(u) = \delta_W(K_H^* u)$, *where* $\delta_W$ *denotes the divergence operator with respect to the process* $W$.

Suppose $H > \frac{1}{2}$. We denote by $|\mathcal{H}| \otimes |\mathcal{H}|$ the space of measurable functions $\varphi$ on $[0,T]^2$ such that

$$\|\varphi\|_{|\mathcal{H}|\otimes|\mathcal{H}|}^2 = \alpha_H^2 \int_{[0,T]^4} |\varphi_{r,\theta}| \, |\varphi_{u,\eta}| \, |r-u|^{2H-2} \, |\theta-\eta|^{2H-2} \, dr\,du\,d\theta\,d\eta < \infty.$$

Then, $|\mathcal{H}| \otimes |\mathcal{H}|$ is a Banach space with respect to the norm $\|\cdot\|_{|\mathcal{H}|\otimes|\mathcal{H}|}$. Furthermore, equipped with the inner product

$$\langle \varphi, \psi \rangle_{\mathcal{H}\otimes\mathcal{H}} = \alpha_H^2 \int_{[0,T]^4} \varphi_{r,\theta}\psi_{u,\eta} \, |r-u|^{2H-2} \, |\theta-\eta|^{2H-2} \, dr\,du\,d\theta\,d\eta$$

the space $|\mathcal{H}| \otimes |\mathcal{H}|$ is isometric to a subspace of $\mathcal{H} \otimes \mathcal{H}$. A slight extension of the inequality (5.19) yields

$$\|\varphi\|_{|\mathcal{H}|\otimes|\mathcal{H}|} \leq b_H \|\varphi\|_{L^{\frac{1}{H}}([0,T]^2)}. \tag{5.34}$$

For any $p > 1$ we denote by $\mathbb{D}^{1,p}(|\mathcal{H}|)$ the subspace of $\mathbb{D}^{1,p}(\mathcal{H})$ formed by the elements $u$ such that $u \in |\mathcal{H}|$ a.s., $Du \in |\mathcal{H}| \otimes |\mathcal{H}|$ a.s., and

$$E\left(\|u\|_{|\mathcal{H}|}^p\right) + E\left(\|Du\|_{|\mathcal{H}|\otimes|\mathcal{H}|}^p\right) < \infty.$$

## 5.2.2  Stochastic calculus with respect to fBm. Case $H > \frac{1}{2}$

We can introduce the Stratonovich integral as the limit of symmetric Riemann sums as we have done in Chapter 3 in the framework of the anticipating stochastic calculus for the Brownian motion. However, here we will consider only uniform partitions, because this simplify the proof of the results.

Consider a measurable process $u = \{u_t, t \in [0,T]\}$ such that $\int_0^T |u_t|\,dt < \infty$ a.s. Let us define the aproximating sequences of processes

$$(\pi^n u)_t = \sum_{i=0}^{n-1} \Delta_n^{-1}\left(\int_{t_i}^{t_{i+1}} u_s\,ds\right) \mathbf{1}_{(t_i,t_{i+1}]}(t), \tag{5.35}$$

where $t_i = i\Delta_n$, $i = 0, \ldots, n$, and $\Delta_n = \frac{T}{n}$. Set

$$S^n = \sum_{i=0}^{n-1} \Delta_n^{-1} \left( \int_{t_i}^{t_{i+1}} u_s ds \right) (B_{t_{i+1}} - B_{t_i}).$$

**Definition 5.2.1** *We say that a measurable process $u = \{u_t, 0 \le t \le T\}$ such that $\int_0^T |u_t| dt < \infty$ a.s. is Stratonovich integrable with respect to the fBm if the sequence $S^n$ converges in probability as $|\pi| \to 0$, and in this case the limit will be denoted by $\int_0^T u_t \circ dB_t$.*

The following proposition establishes the relationship between the Stratonovich integral and the divergence integral. It is the counterpart of Theorem 3.1.1 for the fBm.

**Proposition 5.2.3** *Let $u = \{u_t, t \in [0, T]\}$ be a stochastic process in the space $\mathbb{D}^{1,2}(|\mathcal{H}|)$. Suppose also that a.s.*

$$\int_0^T \int_0^T |D_s u_t| \, |t - s|^{2H-2} \, ds dt < \infty. \tag{5.36}$$

*Then $u$ is Stratonovich integrable and we have*

$$\int_0^T u_t \circ dB_t = \delta(u) + \alpha_H \int_0^T \int_0^T D_s u_t \, |t - s|^{2H-2} \, ds dt. \tag{5.37}$$

*Proof:* The proof will be done in two steps.

*Step 1.* Notice first that $\int_0^T |u_t| dt < \infty$ a.s. because $|\mathcal{H}| \subset L^1([0, T])$. We claim that

$$\|\pi^n u\|_{|\mathcal{H}|}^2 \le d_H \|u\|_{|\mathcal{H}|}^2, \tag{5.38}$$

for some positive constant $d_H$. In fact, we have that

$$\|\pi^n u\|_{|\mathcal{H}|}^2 = \alpha_H \int_0^T \int_0^T |(\pi^n u)_s| \, |(\pi^n u)_t| \, |s - t|^{2H-2} ds dt$$

$$\le \alpha_H \int_0^T \int_0^T |u_s| \, |u_t| \, \phi_n(s, t) ds dt,$$

where

$$\phi_n(s, t) = \Delta_n^{-2} \sum_{i,j=0}^{n-1} \mathbf{1}_{(t_i, t_{i+1}]}(s) \mathbf{1}_{(t_j, t_{j+1}]}(t) \int_{t_i}^{t_{i+1}} \int_{t_j}^{t_{j+1}} |\sigma - \theta|^{2H-2} d\sigma d\theta.$$

Hence, in order to show (5.38) it suffices to check that

$$\phi_n(s, t) |s - t|^{2-2H} \le d_H.$$

Notice that

$$\alpha_H \int_{t_i}^{t_{i+1}} \int_{t_j}^{t_{j+1}} |\sigma - \theta|^{2H-2} d\sigma d\theta = E((B_{t_{i+1}} - B_{t_i})(B_{t_{j+1}} - B_{t_j})).$$

Thus, for $s, t \in (t_i, t_{i+1}]$

$$\phi_n(s,t)|s - t|^{2-2H} = \Delta_n^{-2} \alpha_H^{-1} \Delta_n^{2H} |s - t|^{2-2H} \le \alpha_H^{-1},$$

and for $s \in (t_i, t_{i+1}]$, $t \in (t_j, t_{j+1}]$ with $i < j$

$$
\begin{aligned}
&\phi_n(s,t)|s - t|^{2-2H} \\
&= \frac{|s - t|^{2-2H}}{2\alpha_H \Delta_n^2} \left[ (t_j - t_{i+1})^{2H} + (t_j - t_{i+1} + 2\Delta_n)^{2H} \right. \\
&\qquad \left. - 2(t_j - t_{i+1} + \Delta_n)^{2H} \right] \\
&\le \frac{1}{2\alpha_H} \max_{k \ge 0} \left[ k^{2H} + (k+2)^{2H} - 2(k+1)^{2H} \right] (k+2)^{2-2H}.
\end{aligned}
$$

Therefore (5.38) holds with

$$d_H = \alpha_H^{-1} \max_{k \ge 0}(1, \frac{1}{2} \left[ k^{2H} + (k+2)^{2H} - 2(k+1)^{2H} \right] (k+2)^{2-2H}).$$

We can find a sequence of step processes $u_k$ such that $\|u_k - u\|_{|\mathcal{H}|}^2 \to 0$ as $k$ tends to infinity. Then

$$
\begin{aligned}
\|\pi^n u - u\|_{|\mathcal{H}|} &\le \|\pi^n u - \pi^n u_k\|_{|\mathcal{H}|} + \|\pi^n u_k - u_k\|_{|\mathcal{H}|} + \|u_k - u\|_{|\mathcal{H}|} \\
&\le (\sqrt{d_H} + 1) \|u_k - u\|_{|\mathcal{H}|} + \|\pi^n u_k - u_k\|_{|\mathcal{H}|},
\end{aligned}
$$

and letting first $n$ tend to infinity and then $k$ tend to infinity we get

$$\lim_{n \to \infty} \|\pi^n u - u\|_{|\mathcal{H}|} = 0$$

a.s., and by dominated convergence we obtain

$$\lim_{n \to \infty} E(\|\pi^n u - u\|_{|\mathcal{H}|}^2) = 0.$$

In a similar way we can show that

$$\|D\pi^n u\|_{|\mathcal{H}| \otimes |\mathcal{H}|}^2 \le f_H \|Du\|_{|\mathcal{H}| \otimes |\mathcal{H}|}^2,$$

for some constant $f_H > 0$, and as a consequence

$$\lim_{n \to \infty} E(\|D\pi^n u - Du\|_{|\mathcal{H}| \otimes |\mathcal{H}|}^2) = 0.$$

Therefore, $\pi^n u \to u$ in the norm of the space $\mathbb{D}^{1,2}(|\mathcal{H}|)$.

*Step 2.* Using Proposition 1.3.3 we can write

$$\sum_{i=0}^{n-1} \Delta_n^{-1} \left( \int_{t_i}^{t_{i+1}} u_s ds \right) (B_{t_{i+1}} - B_{t_i}) = \delta(\pi^n u) + T_n(u), \tag{5.39}$$

where

$$T_n(u) = \sum_{i=0}^{n-1} \Delta_n^{-1} \int_{t_i}^{t_{i+1}} \langle Du_s, \mathbf{1}_{[t_i, t_{i+1}]} \rangle_{\mathcal{H}} \, ds.$$

By Step 1 $\delta(\pi^n u)$ will converge in $L^2(\Omega)$ to $\delta(u)$ as $n$ tends to infinity. Then it suffices to show that $T_n(u)$ converges almost surely to

$$\alpha_H \int_0^T \int_0^T D_s u_t \, |t - s|^{2H-2} \, ds dt.$$

We can write

$$T_n(u) = \alpha_H \int_0^T \int_0^T D_s u_t \psi_n(s, t) ds dt,$$

where

$$\psi_n(s, t) = \sum_{i=0}^{n-1} \mathbf{1}_{[t_i, t_{i+1}]}(t) \Delta_n^{-1} \int_{t_i}^{t_{i+1}} |s - \sigma|^{2H-2} d\sigma.$$

By dominated convergence it suffices to show that

$$\psi_n(s, t) |s - t|^{2-2H} \le e_H$$

for some constant $e_H > 0$. If $s, t \in (t_i, t_{i+1}]$ then,

$$
\begin{aligned}
\psi_n(s, t) |s - t|^{2-2H} &\le \Delta_n^{-1} \Delta_n^{2-2H} (2H - 1)^{-1} \\
&\quad \times \left[ (t_{i+1} - s)^{2H-1} + (s - t_i)^{2H-1} \right] \\
&\le \frac{2}{2H - 1}.
\end{aligned}
$$

On the other hand, if $s \in (t_i, t_{i+1}]$, $t \in (t_j, t_{j+1}]$ with $i < j$ we have

$$
\begin{aligned}
\psi_n(s, t) |s - t|^{2-2H} &\le \frac{1}{2H - 1} \sup_{\substack{k \ge 0 \\ \lambda \in [0,1]}} (k + 2)^{2-2H} \\
&\quad \times [(k + 1 + \lambda)^{2H-1} - (k + \lambda)^{2H-1}] \\
&= \frac{g_H}{2H - 1}.
\end{aligned}
$$

Hence, (5.2.2) holds with $e_H = \frac{1}{2H-1} \max(2, g_H)$. This completes the proof of the proposition. $\qquad \square$

**Remark 1** A sufficient condition for (5.36) is

$$\int_0^T \left( \int_s^T |D_s u_t|^p \, dt \right)^{1/p} ds < \infty$$

for some $p > \frac{1}{2H-1}$.

**Remark 2** Let $u = \{u_t, t \in [0,T]\}$ be a stochastic process which is continuous in the norm of $\mathbb{D}^{1,2}$ and (5.36) holds. Then the Riemann sums

$$\sum_{i=0}^{n-1} u_{s_i}(B_{t_{i+1}} - B_{t_i}),$$

where $t_i \leq s_i \leq t_{i+1}$, converge in probability to the right-hand side of (5.37). In particulat, the forward and backward integrals of $u$ with respect to the fBm exists and they coincide with the Stratonovich integral.

(A) *The divergence integral*

Suppose that $u = \{u_t, t \in [0,T]\}$ is a stochastic process in the space $\mathbb{D}^{1,2}(|\mathcal{H}|)$. Then, for any $t \in [0,T]$ the process $u\mathbf{1}_{[0,t]}$ also belongs to $\mathbb{D}^{1,2}(|\mathcal{H}|)$ and we can define the indefnite divergence integral denoted by

$$\int_0^t u_s dB_s = \delta\left( u\mathbf{1}_{[0,t]} \right).$$

If (5.36) holds, then by Proposition 5.2.3 we have

$$\int_0^t u_s \circ dB_s = \int_0^t u_s dB_s + \alpha_H \int_0^t \int_0^T D_r u_s \, |s-r|^{2H-2} \, dr ds.$$

By Proposition 1.5.8, if $p > 1$, a process $u \in \mathbb{D}^{1,p}(|\mathcal{H}|)$ belongs to the domain of the divergence in $L^p(\Omega)$, and we have

$$E\left(|\delta(u)|^p\right) \leq C_{H,p}\left( \|E(u)\|_{|\mathcal{H}|}^p + E\left( \|Du\|_{|\mathcal{H}|\otimes|\mathcal{H}|}^p \right) \right).$$

As a consequence, applying (5.34) we obtain

$$E\left(|\delta(u)|^p\right) \leq C_{H,p}\left( \|E(u)\|_{L^{1/H}([0,T])}^p + E\left( \|Du\|_{L^{1/H}([0,T]^2)}^p \right) \right). \quad (5.40)$$

Let $pH > 1$. Denote by $\mathbb{L}_H^{1,p}$ the space of processes $u \in \mathbb{D}^{1,2}(|\mathcal{H}|)$ such that

$$\|u\|_{p,1} := \left[ \int_0^T E(|u_s|^p)ds + E\left( \int_0^T \left( \int_0^T |D_r u_s|^{\frac{1}{H}} \, dr \right)^{pH} ds \right) \right]^{\frac{1}{p}} < \infty.$$

Assume $pH > 1$ and suppose that $u \in \mathbb{L}_H^{1,p}$ and consider the indefinite divergence integral $X_t = \int_0^t u_s dB_s$. The following results have been established in [6]:

(i) *Maximal inequality* for the divergence integral:

$$E\left(\sup_{t\in[0,T]}|X_t|^p\right)\le C\,\|u\|_{p,1}^p,$$

where the constant $C>0$ depends on $p$, $H$ and $T$. This follows from the $L^p$ estimate (5.40) and a convolution argument.

(ii) *Continuity:* The process $X_t$ has a version with continuous trajectories and for all $\gamma<H-\frac{1}{p}$ there exists a random variable $C_\gamma$ such that

$$|X_t-X_s|\le C_\gamma\,|t-s|^\gamma.$$

As a consequence, for a process $u\in\cap_{p>1}\mathbb{L}_H^{1,p}$, the indefinite integral process $X=\left\{\int_0^t u_s dB_s,\ t\in[0,T]\right\}$ is $\gamma$-Hölder continuous for all $\gamma<H$.

(B) *Itô's formula for the divergence integral*

Suppose that $f,g:[0,T]\longrightarrow\mathbb{R}$ are Hölder continuous functions of orders $\alpha$ and $\beta$ with $\alpha+\beta>1$. Young [354] proved that the Riemann-Stieltjes integral $\int_0^T f_s dg_s$ exists. Moreover, if $h_t=\int_0^t f_s dg_s$ and $F$ is of class $C^2$ the following change of variables formula holds:

$$F(h_t)=F(0)+\int_0^t F'(h_s)f_s dg_s.$$

As a consequence, if $F$ is a function of class $C^2$, and $H>\frac{1}{2}$, the Stratonovich integral integral $\int_0^t F'(B_s)\circ dB_s$ introduced in Definition 5.2.1 is actually a path-wise Riemann-Stieltjes integral and for any function $F$ of class $C^2$ we have

$$F(B_t)=F(0)+\int_0^t F'(B_s)\circ dB_s. \tag{5.41}$$

Suppose that $F$ is a function of class $C^2(\mathbb{R})$ such that

$$\max\left\{|F(x)|,|F'(x)|,|F''(x)|\right\}\le ce^{\lambda x^2}, \tag{5.42}$$

where $c$ and $\lambda$ are positive constants such that $\lambda<\frac{1}{4T^{2H}}$. This condition implies

$$E\left(\sup_{0\le t\le T}|F(B_t)|^p\right)\le c^p E\left(e^{p\lambda\sup_{0\le t\le T}|B_t|^2}\right)<\infty$$

for all $p<\frac{T^{-2H}}{2\lambda}$. In particular, we can take $p=2$. The same property holds for $F'$ and $F''$.

Then, if $F$ satisfies the growth condition (5.42), the process $F'(B_t)$ belongs to the space $\mathbb{D}^{1,2}(|\mathcal{H}|)$ and (5.36) holds. As a consequence, from Proposition 5.2.3 we obtain

$$
\begin{aligned}
\int_0^t F'(B_s) \circ dB_s &= \int_0^t F'(B_s)dB_s + \alpha_H \int_0^t \int_0^s F''(B_s)(s-r)^{2H-2}drds \\
&= \int_0^t F'(B_s)dB_s + H \int_0^t F''(B_s)s^{2H-1}ds. \qquad (5.43)
\end{aligned}
$$

Therefore, putting together (5.41) and (5.43) we deduce the following Itô's formula for the divergence process

$$
F(B_t) = F(0) + \int_0^t F'(B_s)dB_s + H \int_0^t F''(B_s)s^{2H-1}ds. \qquad (5.44)
$$

We recall that the divergence operator has the local property and $\delta(u)$ is defined without ambiguity in $\mathbb{D}_{loc}^{1,2}(\mathcal{H})$.

We state the following general version of Itô's formula (see [11]).

**Theorem 5.2.1** *Let $F$ be a function of class $C^2(\mathbb{R})$. Assume that $u = \{u_t, t \in [0,T]\}$ is a process in the space $\mathbb{D}_{loc}^{2,2}(|\mathcal{H}|)$ such that the indefinite integral $X_t = \int_0^t u_s dB_s$ is a.s. continuous. Assume that $\|u\|_2$ belongs to $\mathcal{H}$. Then for each $t \in [0,T]$ the following formula holds*

$$
\begin{aligned}
F(X_t) = F(0) &+ \int_0^t F'(X_s)u_s dB_s \\
&+ \alpha_H \int_0^t F''(X_s)\, u_s \left( \int_0^T |s-\sigma|^{2H-2} \left( \int_0^s D_\sigma u_\theta dB_\theta \right) d\sigma \right) ds \\
&+ \alpha_H \int_0^t F''(X_s)u_s \left( \int_0^s u_\theta\, (s-\theta)^{2H-2}\, d\theta \right) ds. \qquad (5.45)
\end{aligned}
$$

**Remark 1** If the process $u$ is adapted, then the third summand in the right-hand side of (5.45) can be written as

$$
\alpha_H \int_0^t F''(X_s)\, u_s \left( \int_0^s \left( \int_0^\theta |s-\sigma|^{2H-2} D_\sigma u_\theta d\sigma \right) dB_\theta \right) ds.
$$

**Remark 2** $\frac{2H-1}{s^{2H-1}}(s-\theta)^{2H-2}\mathbf{1}_{[0,s]}(\theta)$ is an approximation of the identity as $H$ tends to $\frac{1}{2}$. Therefore, taking the limit as $H$ converges to $\frac{1}{2}$ in Equation (5.45) we recover the usual Itô's formula for the the Skorohod integral (see Theorem 3.2.2).

### 5.2.3  Stochastic integration with respect to fBm in the case $H < \frac{1}{2}$

The extension of the previous results to the case $H < \frac{1}{2}$ is not trivial and new difficulties appear. In order to illustrate these difficulties, let us first remark that the forward integral $\int_0^T B_t dB_t$ defined as the limit in $L^2$ of the Riemann sums

$$\sum_{i=0}^{n-1} B_{t_i}(B_{t_{i+1}} - B_{t_i}),$$

where $t_i = \frac{iT}{n}$, does not exists. In fact, a simple argument shows that the expectation of this sum diverges:

$$\sum_{i=1}^{n} E\left(B_{t_{i-1}}(B_{t_i} - B_{t_{i-1}})\right) = \frac{1}{2}\sum_{i=1}^{n}\left[t_i^{2H} - t_{i-1}^{2H} - (t_i - t_{i-1})^{2H}\right]$$

$$= \frac{1}{2}T^{2H}\left(1 - n^{1-2H}\right) \to -\infty,$$

as $n$ tends to infinity. Notice, however, that the expectation of symmetric Riemann sums is constant:

$$\frac{1}{2}\sum_{i=1}^{n} E\left((B_{t_i} + B_{t_{i-1}})(B_{t_i} - B_{t_{i-1}})\right) = \frac{1}{2}\sum_{i=1}^{n}\left[t_i^{2H} - t_{i-1}^{2H}\right] = \frac{T^{2H}}{2}.$$

We recall that for $H < \frac{1}{2}$ the operator $K_H^*$ given by (5.31) is an isometry between the Hilbert space $\mathcal{H}$ and $L^2([0,T])$. We have the estimate :

$$\left|\frac{\partial K}{\partial t}(t,s)\right| \leq c_H\left(\frac{1}{2} - H\right)(t - s)^{H-\frac{3}{2}}. \tag{5.46}$$

Also from (5.32) (see Exercise 5.2.3) it follow sthat

$$|K(t,s)| \leq C(t - s)^{H-\frac{1}{2}}. \tag{5.47}$$

Consider the following seminorm on the set $\mathcal{E}$ of step functions on $[0,T]$:

$$\|\varphi\|_K^2 = \int_0^T \varphi^2(s)(T - s)^{2H-1}ds$$

$$+ \int_0^T\left(\int_s^T |\varphi(t) - \varphi(s)|(t - s)^{H-\frac{3}{2}}dt\right)^2 ds.$$

We denote by $\mathcal{H}_K$ the completion of $\mathcal{E}$ with respect to this seminorm. The space $\mathcal{H}_K$ is the class of functions $\varphi$ on $[0,T]$ such that $\|\varphi\|_K^2 < \infty$, and it is continuously included in $\mathcal{H}$. If $u \in \mathbb{D}^{1,2}(\mathcal{H}_K)$, then $u \in \text{Dom}\,\delta$.

Note that if $u = \{u_t, t \in [0,T]\}$ is a process in $\mathbb{D}^{1,2}(\mathcal{H}_K)$, then there is a sequence $\{\varphi_n\}$ of bounded simple $\mathcal{H}_K$-valued processes of the form

$$\varphi_n = \sum_{j=0}^{n-1} F_j 1_{(t_j, t_{j+1}]}, \tag{5.48}$$

where $F_j$ is a smooth random variable of the form

$$F_j = f_j(B_{s_1^j}, ..., B_{s_{m(j)}^j}),$$

with $f_j \in C_b^\infty(\mathbb{R}^{m(j)})$, and $0 = t_0 < t_1 < ... < t_n = T$, such that

$$E\|u - \varphi_n\|_K^2 + E \int_0^T \|D_r u - D_r \varphi_n\|_K^2 \, dr \longrightarrow 0, \qquad \text{as} \qquad n \to \infty. \tag{5.49}$$

In the case $H < \frac{1}{2}$ it is more convenient to consider the symmetric integral introduced by Russo and Vallois in [298]. For a process $u = \{u_t, t \in [0,T]\}$ with integrable paths and $\varepsilon > 0$, we denote by $u_t^\varepsilon$ the integral $(2\varepsilon)^{-1} \int_{t-\varepsilon}^{t+\varepsilon} u_s ds$, where we use the convention $u_s = 0$ for $s \notin [0,T]$. Also we put $B_s = B_T$ for $s > T$ and $B_s = 0$ for $s < 0$.

**Definition 5.2.2** *The symmetric integral of a process $u$ with integrable paths with respect to the fBm is defined as the limit in probability of*

$$(2\varepsilon)^{-1} \int_0^T u_s \left( B_{s+\varepsilon} - B_{s-\varepsilon} \right) ds.$$

*as $\varepsilon \downarrow 0$ if it exists. We denote this limit by $\int_0^T u_r \circ dB_r$.*

The following result is the counterpart of Proposition 5.2.3 in the case $H < \frac{1}{2}$, for the symmetric integral.

**Proposition 5.2.4** *Let $u = \{u_t, t \in [0,T]\}$ be a stochastic process in the space $\mathbb{D}^{1,2}(\mathcal{H}_K)$. Suppose that the trace defined as the limit in probability*

$$\mathrm{Tr} Du := \lim_{\varepsilon \to 0} \frac{1}{2\varepsilon} \int_0^T \left\langle Du_s, 1_{[s-\varepsilon, s+\varepsilon] \cap [0,T]} \right\rangle_{\mathcal{H}} ds$$

*exists. Then the symmetric stochastic integral of $u$ with respect to fBm in the sense of Definition 5.2.1 exists and*

$$\int_0^T u_t \circ dB_t = \delta(u) + \mathrm{Tr} Du.$$

In order to prove this theorem, we need the following technical result.

**Lemma 5.2.1** *Let $u$ be a simple process of the form (5.48). Then $u^\varepsilon$ converges to $u$ in $\mathbb{D}^{1,2}(\mathcal{H}_K)$ as $\varepsilon \downarrow 0$.*

*Proof:* Let $u$ be given by the right-hand side of (5.48). Then $u$ is a bounded process. Hence, by the dominated convergence theorem

$$E \int_0^T (u_s - u_s^\varepsilon)^2 (T-s)^{2H-1} ds \longrightarrow 0 \qquad \text{as} \qquad \varepsilon \downarrow 0. \qquad (5.50)$$

Fix an index $i \in \{0, 1, ..., n-1\}$. Using that $u_t - u_s = 0$ for $s, t \in [t_i, t_{i+1}]$ we obtain

$$\int_{t_i}^{t_{i+1}} \left( \int_s^T |u_t^\varepsilon - u_s^\varepsilon - (u_t - u_s)| (t-s)^{H-\frac{3}{2}} dt \right)^2 ds$$

$$\leq 2 \int_{t_i}^{t_{i+1}} \left( \int_s^{t_{i+1}} |u_t^\varepsilon - u_s^\varepsilon| (t-s)^{H-\frac{3}{2}} dt \right)^2 ds$$

$$+ 2 \int_{t_i}^{t_{i+1}} \left( \int_{t_{i+1}}^T |u_t^\varepsilon - u_s^\varepsilon - (u_t - u_s)| (t-s)^{H-\frac{3}{2}} dt \right)^2 ds$$

$$= 2A_1(i, \varepsilon) + 2A_2(i, \varepsilon). \qquad (5.51)$$

The convergence of the expectation of the term $A_2(i, \varepsilon)$ to 0, as $\varepsilon \downarrow 0$, follows from the dominated convergence theorem, the fact that $u$ is a bounded process and that for a.a. $0 \leq s < t \leq T$,

$$|u_t^\varepsilon - u_s^\varepsilon - (u_t - u_s)| (t-s)^{H-\frac{3}{2}} \longrightarrow 0 \qquad \text{as} \qquad \varepsilon \downarrow 0.$$

Suppose that $\varepsilon < \frac{1}{4} \min_{0 \leq i \leq n-1} |t_{i+1} - t_i|$. Then $u_t^\varepsilon - u_s^\varepsilon = 0$ if $s$ and $t$ belong to $[t_i + 2\varepsilon, t_{i+1} - 2\varepsilon]$, we can make the following decomposition

$$E(A_1(i, \varepsilon))$$

$$\leq 8 \int_{t_i}^{t_i+2\varepsilon} \left( \int_s^{t_i+2\varepsilon} |u_t^\varepsilon - u_s^\varepsilon| (t-s)^{H-\frac{3}{2}} dt \right)^2 ds$$

$$+ 8 \int_{t_{i+1}-2\varepsilon}^{t_{i+1}} \left( \int_s^{t_{i+1}} |u_t^\varepsilon - u_s^\varepsilon| (t-s)^{H-\frac{3}{2}} dt \right)^2 ds$$

$$+ 8 \int_{t_i}^{t_i+2\varepsilon} \left( \int_{t_i+2\varepsilon}^{t_{i+1}} |u_t^\varepsilon - u_s^\varepsilon| (t-s)^{H-\frac{3}{2}} dt \right)^2 ds$$

$$+ 8 \int_{t_i}^{t_{i+1}-2\varepsilon} \left( \int_{t_{i+1}-2\varepsilon}^{t_{i+1}} |u_t^\varepsilon - u_s^\varepsilon| (t-s)^{H-\frac{3}{2}} dt \right)^2 ds.$$

The first and second integrals converge to zero, due to the estimate

$$|u_t^\varepsilon - u_s^\varepsilon| \le \frac{c}{\varepsilon}|t - s|.$$

On the other hand, the third and fourth term of the above expression converge to zero because $u_t^\varepsilon$ is bounded. Therefore we have proved that

$$E \|u - u^\varepsilon\|_K^2 \longrightarrow 0 \qquad \text{as} \qquad \varepsilon \to 0.$$

Finally, it is easy to see by the same arguments that we also have

$$E \int_0^T \|D_r u - D_r u^\varepsilon\|_K^2 \, dr \longrightarrow 0 \qquad \text{as} \qquad \varepsilon \to 0.$$

Thus the proof is complete. □

Now we are ready to prove Proposition 5.2.4.

*Proof of Proposition 5.2.4:* From the properties of the divergence operator, and applying Fubini's theorem we have

$$(2\varepsilon)^{-1} \int_0^T u_s \left(B_{s+\varepsilon} - B_{s-\varepsilon}\right) ds = (2\varepsilon)^{-1} \int_0^T \delta \left(u_s 1_{[s-\varepsilon, s+\varepsilon]}(\cdot)\right) ds$$
$$+ (2\varepsilon)^{-1} \int_0^T \left\langle Du_s, 1_{[s-\varepsilon, s+\varepsilon]} \right\rangle_{\mathcal{H}} ds$$
$$= (2\varepsilon)^{-1} \int_0^T \left( \int_{r-\varepsilon}^{r+\varepsilon} u_s ds \right) dB_r$$
$$+ (2\varepsilon)^{-1} \int_0^T \left\langle D_\cdot u_s, 1_{[s-\varepsilon, s+\varepsilon]}(\cdot) \right\rangle_{\mathcal{H}} ds$$
$$= \int_0^T u_r^\varepsilon dB_r + B^\varepsilon.$$

By our hypothesis we get that $B^\varepsilon$ converges to $TrDu$ in probability as $\varepsilon \downarrow 0$. In order to see that $\int_0^T u_r^\varepsilon dB_r$ converges to $\delta(u)$ in $L^2(\Omega)$ as $\varepsilon$ tends to zero, we will show that $u^\varepsilon$ converges to $u$ in the norm of $\mathbb{D}^{1,2}(\mathcal{H}_K)$. Fix $\delta > 0$. We have already noted that the definition of the space $\mathbb{D}^{1,2}(\mathcal{H}_K)$ implies that there is a bounded simple $\mathcal{H}_K$-valued processes $\varphi$ as in (5.48) such that

$$E \|u - \varphi\|_K^2 + E \int_0^T \|D_r u - D_r \varphi\|_K^2 \, dr \le \delta. \qquad (5.52)$$

Therefore, Lemma 5.2.1 implies that for $\varepsilon$ small enough,

$$
\begin{aligned}
& E \left\| u - u^\varepsilon \right\|_K^2 + E \int_0^T \left\| D_r \left( u - u^\varepsilon \right) \right\|_K^2 dr \\
\leq \quad & cE \left\| u - \varphi \right\|_K^2 + cE \int_0^T \left\| D_r \left( u - \varphi \right) \right\|_K^2 dr \\
& + cE \left\| \varphi - \varphi^\varepsilon \right\|_K^2 + cE \int_0^T \left\| D_r \left( \varphi - \varphi^\varepsilon \right) \right\|_K^2 dr \\
& + cE \left\| \varphi^\varepsilon - u^\varepsilon \right\|_K^2 + cE \int_0^T \left\| D_r \left( \varphi^\varepsilon - u^\varepsilon \right) \right\|_K^2 dr \\
\leq \quad & 2c\delta + cE \left\| \varphi^\varepsilon - u^\varepsilon \right\|_K^2 + cE \int_0^T \left\| D_r \left( \varphi^\varepsilon - u^\varepsilon \right) \right\|_K^2 dr. \quad (5.53)
\end{aligned}
$$

We have

$$
\begin{aligned}
& \int_0^T E \left( \varphi_s^\varepsilon - u_s^\varepsilon \right)^2 \left( T - s \right)^{2H-1} ds \\
\leq \quad & \int_0^T E \left( \frac{1}{2\varepsilon} \int_{s-\varepsilon}^{s+\varepsilon} \left( \varphi_r - u_r \right) dr \right)^2 \left( T - s \right)^{2H-1} ds \\
\leq \quad & \int_0^T E \left( \varphi_r - u_r \right)^2 \left( \frac{1}{2\varepsilon} \int_{(r-\varepsilon)\vee 0}^{(r+\varepsilon)\wedge T} \left( T - s \right)^{2H-1} ds \right) dr.
\end{aligned}
$$

From property (i) it follows that

$$
(2\varepsilon)^{-1} \int_{(r-\varepsilon)\vee 0}^{(r+\varepsilon)\wedge T} K(T,t)^2 dt \leq c \left[ (T-r)^{-2\alpha} + r^{-2\alpha} \right].
$$

Hence, by the dominated convergence theorem and condition (4.21) we obtain

$$
\begin{aligned}
& \limsup_{\varepsilon \downarrow 0} \int_0^T E \left( \varphi_s^\varepsilon - u_s^\varepsilon \right)^2 K(T,s)^2 ds \\
\leq \quad & \int_0^T E \left( \varphi_s - u_s \right)^2 K(T,s)^2 ds \leq \delta. \quad (5.54)
\end{aligned}
$$

On the other hand,

$$E \int_0^T \left( \int_s^T |\varphi_t^\varepsilon - u_t^\varepsilon - \varphi_s^\varepsilon + u_s^\varepsilon| (t-s)^{H-\frac{3}{2}} \, dt \right)^2 ds$$

$$\leq \frac{1}{4\varepsilon^2} E \int_0^T \left( \int_{-\varepsilon}^\varepsilon \int_s^T |(\varphi - u)_{t-\theta} - (\varphi - u)_{s-\theta}| (t-s)^{H-\frac{3}{2}} \, dt d\theta \right)^2 ds$$

$$= \frac{1}{4\varepsilon^2} E \int_0^T \left( \int_{s-\varepsilon}^{s+\varepsilon} \int_r^{T+r-s} |(\varphi - u)_t - (\varphi - u)_r| (t-r)^{H-\frac{3}{2}} \, dt dr \right)^2 ds$$

$$\leq \frac{1}{2\varepsilon} E \int_0^T \int_{s-\varepsilon}^{s+\varepsilon} \left( \int_r^{T+\varepsilon} |(\varphi - u)_t - (\varphi - u)_r| (t-r)^{H-\frac{3}{2}} \, dt \right)^2 dr ds$$

$$= \frac{1}{2\varepsilon} E \int_{-\varepsilon}^{T+\varepsilon} \int_{(r-\varepsilon)\vee 0}^{(r+\varepsilon)\wedge T} \left( \int_r^{T+\varepsilon} |\varphi_t - u_t - \varphi_r + u_r| (t-r)^{H-\frac{3}{2}} \, dt \right)^2 ds dr$$

$$\leq E \int_{-\varepsilon}^{T+\varepsilon} \left( \int_r^{T+\varepsilon} |\varphi_t - u_t - \varphi_r + u_r| (t-r)^{H-\frac{3}{2}} \, dt \right)^2 dr. \qquad (5.55)$$

By (5.54) and (5.55) we obtain

$$\limsup_{\varepsilon \downarrow 0} E \, \|\varphi^\varepsilon - u^\varepsilon\|_K^2 \leq 2\delta.$$

By a similar argument,

$$\limsup_{\varepsilon \downarrow 0} E \int_0^T \|D_r \, (\varphi^\varepsilon - u^\varepsilon)\|_K^2 \, dr \leq 2\delta.$$

Since $\delta$ is arbitrary, $u^\varepsilon$ converges to $u$ in the norm of $\mathbb{D}^{1,2} \, (\mathcal{H}_K)$ as $\varepsilon \downarrow 0$, and, as a consequence, $\int_0^T u_r^\varepsilon dB_r$ converges in $L^2 \, (\Omega)$ to $\delta \, (u)$. Thus the proof is complete. $\qquad \square$

Consider the particular case of the process $u_t = F(B_t)$, where $F$ is a continuously differentiable function satisfying the growth condition (5.42). If $H > \frac{1}{4}$, the process $F(B_t)$ the process belongs to $\mathbb{D}^{1,2}(\mathcal{H}_K)$. Moreover, $\mathrm{Tr} Du$ exists and

$$\mathrm{Tr} Du = H \int_0^T F'(B_t) t^{2H-1} dt.$$

As a consequence we obtain

$$\int_0^T F(B_t) \circ dB_t = \int_0^T F(B_t) dB_t + H \int_0^T F'(B_t) t^{2H-1} dt.$$

(C) *Itô's formulas for the divergence integral in the case $H < \frac{1}{2}$*

An Itô's formula similar to (5.44) was proved in [9] for general Gaussian processes of Volterra-type of the form $B_t = \int_0^t K(t,s)dW_s$, where $K(t,s)$ is a singular kernel. In particular, the process $B_t$ can be a fBm with Hurst parameter $\frac{1}{4} < H < \frac{1}{2}$. Moreover, in this paper, an Itô's formula for the indefinite divergence process $X_t = \int_0^t u_s dB_s$ similar to (5.45) was also proved.

On the other hand, in the case of the fractional Brownian motion with Hurst parameter $\frac{1}{4} < H < \frac{1}{2}$, an Itô's formula for the indefinite symmetric integral $X_t = \int_0^t u_s dB_s$ has been proved in [7] assuming again $\frac{1}{4} < H < \frac{1}{2}$.

Let us explain the reason for the restriction $\frac{1}{4} < H$. In order to define the divergence integral $\int_0^T F'(B_s)dB_s$, we need the process $F'(B_s)$ to belong to $L^2(\Omega; \mathcal{H})$. This is clearly true, provided $F$ satisfies the growth condition (5.42), because $F'(B_s)$ is Hölder continuous of order $H - \varepsilon > \frac{1}{2} - H$ if $\varepsilon < 2H - \frac{1}{2}$. If $H \leq \frac{1}{4}$, one can show (see [66]) that

$$P(B \in \mathcal{H}) = 0,$$

and the space $\mathbb{D}^{1,2}(\mathcal{H})$ is too small to contains processes of the form $F'(B_t)$.

Following the approach of [66] we are going to extend the domain of the divergence operator to processes whose trajectories are not necessarily in the space $\mathcal{H}$.

Using (5.31) and applying the integration by parts formula for the fractional calculus (A.17) we obtain for any $f, g \in \mathcal{H}$

$$
\begin{aligned}
\langle f, g \rangle_{\mathcal{H}} &= \langle K_H^* f, K_H^* g \rangle_{L^2([0,T])} \\
&= d_H^2 \left\langle s^{\frac{1}{2}-H} D_{T-}^{\frac{1}{2}-H} s^{H-\frac{1}{2}} f, s^{\frac{1}{2}-H} D_{T-}^{\frac{1}{2}-H} s^{H-\frac{1}{2}} g \right\rangle_{L^2([0,T])} \\
&= d_H^2 \left\langle f, s^{H-\frac{1}{2}} s^{\frac{1}{2}-H} D_{0+}^{\frac{1}{2}-H} s^{1-2H} D_{T-}^{\frac{1}{2}-H} s^{H-\frac{1}{2}} g \right\rangle_{L^2([0,T])}.
\end{aligned}
$$

This implies that the adjoint of the operator $K_H^*$ in $L^2([0,T])$ is

$$\left( K_H^{*,a} f \right)(s) = d_H s^{\frac{1}{2}-H} D_{0+}^{\frac{1}{2}-H} s^{1-2H} D_{T-}^{\frac{1}{2}-H} s^{H-\frac{1}{2}} f.$$

Set $\mathcal{H}_2 = \left( K_H^* \right)^{-1} \left( K_H^{*,a} \right)^{-1} (L^2([0,T]))$. Denote by $\mathcal{S}_{\mathcal{H}}$ the space of smooth and cylindrical random variables of the form

$$F = f(B(\phi_1), \ldots, B(\phi_n)), \tag{5.56}$$

where $n \geq 1$, $f \in C_b^\infty(\mathbb{R}^n)$ ($f$ and all its partial derivatives are bounded), and $\phi_i \in \mathcal{H}_2$.

**Definition 5.2.3** *Let $u = \{u_t, t \in [0,T]\}$ be a measurable process such that*

$$E\left( \int_0^T u_t^2 dt \right) < \infty.$$

We say that $u \in \mathrm{Dom}^* \delta$ if there exists a random variable $\delta(u) \in L^2(\Omega)$ such that for all $F \in \mathcal{S}_{\mathcal{H}}$ we have

$$\int_{\mathbb{R}} E(u_t K_H^{*,a} K_H^* D_t F) dt = E(\delta(u) F).$$

This extended domain of the divergence operator satisfies the following elementary properties:

1. $\mathrm{Dom}\delta \subset \mathrm{Dom}^* \delta$, and $\delta$ restricted to $\mathrm{Dom}\delta$ coincides with the divergence operator.

2. If $u \in \mathrm{Dom}^* \delta$ then $E(u)$ belongs to $\mathcal{H}$.

3. If $u$ is a deterministic process, then $u \in \mathrm{Dom}^* \delta$ if and only if $u \in \mathcal{H}$.

This extended domain of the divergence operator leads to the following version of Itô's formula for the divergence process, established by Cheridito and Nualart in [66].

**Theorem 5.2.2** *Suppose that $F$ is a function of class $C^2(\mathbb{R})$ satisfying the growth condition (5.42). Then for all $t \in [0, T]$, the process $\{F'(B_s)\mathbf{1}_{[0,t]}(s)\}$ belongs to $\mathrm{Dom}^* \delta$ and we have*

$$F(B_t) = F(0) + \int_0^t F'(B_s) dB_s + H \int_0^t F''(B_s) s^{2H-1} ds. \qquad (5.57)$$

*Proof:*   Notice that $F'(B_s)\mathbf{1}_{[0,t]}(s) \in L^2([0,T] \times \Omega)$ and

$$F(B_t) - F(0) - H \int_0^t F''(B_s) s^{2H-1} ds \in L^2(\Omega).$$

Hence, it suffices to show that for any $F \in \mathcal{S}_H$

$$\langle E(F'(B_s)\mathbf{1}_{[0,t]}(s)), D_s F \rangle_{\mathcal{H}}$$
$$= E\left(\left(F(B_t) - F(0) - H \int_0^t F''(B_s) s^{2H-1} ds\right) F\right). \qquad (5.58)$$

Take $F = H_n(B(\varphi))$, where $\varphi \in \mathcal{H}_2$ and $H_n$ is the $n$th-Hermite polynomial. We have

$$D_t F = H_{n-1}(B(\varphi))\varphi_t$$

Hence, (5.58) can be written as

$$E\left(H_{n-1}(B(\varphi))\langle F'(B_s)\mathbf{1}_{[0,t]}(s), \varphi_s \rangle_{\mathcal{H}}\right)$$
$$= E((F(B_t) - F(0) - H \int_0^t F''(B_s) s^{2H-1} ds) H_n(B(\varphi)))$$

Using (5.33) we obtain

$$e_H^2 \int_0^t E\left(F'(B_s)H_{n-1}(B(\varphi))\right)(D_+^{\frac{1}{2}-H}D_-^{\frac{1}{2}-H}\varphi)(s)ds$$

$$= E\left((F(B_t^H) - F(0) - H\int_0^t F''(B_s)s^{2H-1}ds)H_n(B(\varphi))\right). \quad (5.59)$$

In order to show (5.59) we will replace $F$ by

$$F_k(x) = k\int_{-1}^1 F(x-y)\varepsilon(ky)dy,$$

where $\varepsilon$ is a nonnegative smooth function supported by $[-1,1]$ such that $\int_{-1}^1 \varepsilon(y)dy = 1$.

We will make use of the following equalities:

$$E(F(B_t)H_n(B(\varphi))) = \frac{1}{n!}E(F(B_t)\delta^n(\varphi^{\otimes n}))$$

$$= \frac{1}{n!}E\langle D^n(F(B_t)), \varphi^{\otimes n}\rangle_{\mathcal{H}^{\otimes n}}$$

$$= \frac{1}{n!}E(F^{(n)}(B_t))\langle \mathbf{1}_{(0,t]}, \varphi\rangle_{\mathcal{H}}^n.$$

Let $p(\sigma,y) := (2\pi\sigma)^{-\frac{1}{2}}\exp\left(-\frac{y^2}{2\sigma}\right)$. Note that $\frac{\partial p}{\partial\sigma} = \frac{1}{2}\frac{\partial^2 p}{\partial y^2}$. For all $n \geq 0$ and $s \in (0,t]$,

$$\frac{d}{ds}E(F^{(n)}(B_s)) = \frac{d}{ds}\int_{\mathbb{R}} p(s^{2H},y)F^{(n)}(y)dy$$

$$= \int_{\mathbb{R}} \frac{\partial p}{\partial\sigma}(s^{2H},y)2Hs^{2H-1}F^{(n)}(y)dy$$

$$= Hs^{2H-1}\int_{\mathbb{R}} \frac{\partial^2 p}{\partial y^2}(s^{2H},y)F^{(n)}(y)dy$$

$$= Hs^{2H-1}\int_{\mathbb{R}} p(s^{2H},y)F^{(n+2)}(y)dy$$

$$= Hs^{2H-1}E(F^{(n+2)}(B_s)). \quad (5.60)$$

For $n = 0$ the left hand side of (5.59) is zero. On the other hand it follows from (5.60) that

$$E\left(F(B_t)\right) - F(0) - H\int_0^t E(F''(B_s))s^{2H-1}ds = 0$$

This shows that (5.59) is valid for $n = 0$.

Fix $n \geq 1$. Equation (5.60) implies that for all $s \in (0, t]$,

$$\frac{d}{ds} \left( E(F^{(n)}(B_s)) \langle \mathbf{1}_{(0,s]}, \varphi \rangle_{\mathcal{H}}^n \right) = H s^{2H-1} E(F^{(n+2)}(B_s)) \langle \mathbf{1}_{(0,s]}, \varphi \rangle_{\mathcal{H}}^n$$
$$+ e_H^2 n E(F^{(n)}(B_s))$$
$$\times \langle \mathbf{1}_{(0,s]}, \varphi \rangle_{\mathcal{H}}^{n-1} (D_+^{\frac{1}{2}-H} D_-^{\frac{1}{2}-H} \varphi)(s).$$

It follows that

$$E(F^{(n)}(B_t)) \langle \mathbf{1}_{(0,s]}, \varphi \rangle_{\mathcal{H}}^n = H \int_0^t E(F^{(n+2)}(B_s)) \langle \mathbf{1}_{(0,s]}, \varphi \rangle_{\mathcal{H}}^n s^{2H-1} ds$$
$$+ e_H^2 n \int_0^t E(F^{(n)}(B_s))$$
$$\times \langle \mathbf{1}_{(0,s]}, \varphi \rangle_{\mathcal{H}}^{n-1} (D_+^{\frac{1}{2}-H} D_-^{\frac{1}{2}-H} \varphi)(s), \quad (5.61)$$

(5.61) is equivalent to (5.59) because

$$E(F^{(n)}(B_t)) \langle \mathbf{1}_{(0,s]}, \varphi \rangle_{\mathcal{H}}^n = n! E(F(B_t) H_n(B(\varphi))),$$

$$E(F^{(n)}(B_s)) \langle \mathbf{1}_{(0,s]}, \varphi \rangle_{\mathcal{H}}^{n-1} = (n-1)! E(F'(B_s) H_{(n-1)}(B(\varphi))),$$

and

$$E(F^{(n+2)}(B_s)) \langle \mathbf{1}_{(0,s]}, \varphi \rangle_{\mathcal{H}}^n = n! E(F''(B_s) H_n(B(\varphi))).$$

This completes the proof (5.59) for the function $F_k$. Finally it suffices to let $k$ tend to infinity.    □

## (D) Local time and Tanaka's formula for fBm

Berman proved in [22] that that fractional Brownian motion $B = \{B_t, t \geq 0\}$ has a local time $l_t^a$ continuous in $(a, t) \in \mathbb{R} \times [0, \infty)$ which satisfies the occupation formula

$$\int_0^t g(B_s) ds = \int_{\mathbb{R}} g(a) l_t^a da \qquad (5.62)$$

for every continuous and bounded function $g$ on $\mathbb{R}$. Moreover, $l_t^a$ is increasing in the time variable. Set

$$L_t^a = 2H \int_0^t s^{2H-1} l^a (ds).$$

It follows from (5.62) that

$$2H \int_0^t g(B_s) s^{2H-1} ds = \int_{\mathbb{R}} g(a) L_t^a da.$$

This means that $a \to L_t^a$ is the density of the occupation measure

$$\mu(C) = 2H \int_0^t \mathbf{1}_C(B_s) s^{2H-1} ds,$$

where $C$ is a Borel subset of $\mathbb{R}$. Furthermore, the continuity property of $l_t^a$ implies that $L_t^a$ is continuous in $(a, t) \in \mathbb{R} \times [0, \infty)$.

As an extension of the Itô's formula (5.57), the following result has been proved in [66]:

**Theorem 5.2.3** *Let $0 < t < \infty$ and $a \in \mathbb{R}$. Then*

$$\mathbf{1}_{\{B_s > a\}} \mathbf{1}_{[0,t]}(s) \in \mathrm{Dom}^* \delta,$$

*and*

$$(B_t - a)^+ = (-a)^+ + \int_0^t \mathbf{1}_{\{B_s > a\}} dB_s + \frac{1}{2} L_t^a. \qquad (5.63)$$

This result can be considered as a version of Tanaka's formula for the fBm. In [69] it is proved that for $H > \frac{1}{3}$, the process $\mathbf{1}_{\{B_s > a\}} \mathbf{1}_{[0,t]}(s)$ belongs to $\mathrm{Dom}\delta$ and (5.63) holds.

The local time $l_t^a$ has Hölder continuous paths of order $\delta < 1 - H$ in time, and of order $\gamma < \frac{1-H}{2H}$ in the space variable, provided $H \geq \frac{1}{3}$ (see Table 2 in [117]). Moreover, $l_t^a$ is absolutely continuous in $a$ if $H < \frac{1}{3}$, it is continuously differentiable if $H < \frac{1}{5}$, and its smoothness in the space variable increases when $H$ decreases.

In a recent paper, Eddahbi, Lacayo, Solé, Tudor and Vives [88] have proved that $l_t^a \in \mathbb{D}^{\alpha,2}$ for all $\alpha < \frac{1-H}{2H}$. That means, the regularity of the local time $l_t^a$ in the sense of Malliavin calculus is the same order as its Hölder continuity in the space variable. This result follows from the Wiener chaos expansion (see [69]):

$$l_t^a = \sum_{n=0}^{\infty} \int_0^t s^{-nH} p(s^{2H}, a) H_n(as^{-H}) I_n \left( \mathbf{1}_{[o,s]}^{\otimes n} \right) ds.$$

In fact, the series

$$\sum_{n=0}^{\infty} (1+n)^\alpha E \left[ \left( \int_0^t s^{-nH} p(s^{2H}, a) H_n(as^{-H}) I_n \left( \mathbf{1}_{[o,s]}^{\otimes n} \right) ds \right)^2 \right]$$

$$= \sum_{n=0}^{\infty} (1+n)^\alpha n! \int_0^t \int_0^t (sr)^{-nH} p(s^{2H}, a) p(r^{2H}, a) H_n(as^{-H}) H_n(ar^{-H})$$

$$\times R_H(r, s)^n dr ds$$

is equivalent to

$$\sum_{n=1}^{\infty} n^{-\frac{1}{2}+\alpha} \int_0^t \int_0^t R_H(u,v)(uv)^{-nH-1} du dv$$

$$= \sum_{n=0}^{\infty} n^{-\frac{1}{2}+\alpha} \int_0^1 R_H(1,z)z^{-nH-1} dz.$$

Then, the result follows from the estimate

$$\left| \int_0^1 R_H(1,z)z^{-nH-1} dz \right| \le Cn^{-\frac{1}{2H}}.$$

## Exercises

**5.2.1** Show the tranfer principle stated in Propositions 5.2.1 and 5.2.2.

**5.2.2** Show the inequality (5.34).

**5.2.3** Show the estimate (5.47).

**5.2.4** Deduce the Wiener chaos expansion of the local time $l_t^a$.

## 5.3 Stochastic differential equations driven by a fBm

In this section we will establish the existence and uniqueness of a solution for stochastic differential equations driven by a fractional Brownian motion with Hurst parameter $H > \frac{1}{2}$, following an approach based on the fractional calculus. We first introduce a notion of Stieltjes integral based on the fractional integration by parts formula (A.17).

### 5.3.1 Generalized Stieltjes integrals

Given a function $g : [0,T] \to \mathbb{R}$, set $g_{T-}(s) = g(s) - \lim_{\varepsilon \downarrow 0}(T - \varepsilon)$ provided this limit exists. Take $p, q \ge 1$ such that $\frac{1}{p} + \frac{1}{q} \le 1$ and $0 < \alpha < 1$. Suppose that $f$ and $g$ are functions on $[0,T]$ such that $g(T-)$ exists, $f \in I_{0+}^{\alpha}(L^p)$ and $g_{T-} \in I_{T-}^{1-\alpha}(L^q)$. Then the *generalized Stieltjes integral* of $f$ with respect to $g$ is defined by (see [357])

$$\int_0^T f_s dg_s = \int_0^T D_{0+}^{\alpha} f_{a+}(s) D_{T-}^{1-\alpha} g_{T-}(s) ds. \qquad (5.64)$$

In [357] it is proved that this integral coincides with the Riemann-Stieltjes integral if $f$ and $g$ are Hölder continuous of orders $\alpha$ and $\beta$ with $\alpha + \beta > 1$.

Fix $0 < \alpha < \frac{1}{2}$. Denote by $W_0^{\alpha,\infty}(0,T)$ the space of measurable functions $f : [0,T] \to \mathbb{R}$ such that

$$\|f\|_{\alpha,\infty} := \sup_{t\in[0,T]} \left( |f(t)| + \int_0^t \frac{|f(t) - f(s)|}{(t-s)^{\alpha+1}} ds \right) < \infty.$$

We have, for all $0 < \varepsilon < \alpha$

$$C^{\alpha+\varepsilon}(0,T) \subset W_0^{\alpha,\infty}(0,T) \subset C^{\alpha-\varepsilon}(0,T).$$

Denote by $W_T^{1-\alpha,\infty}(0,T)$ the space of measurable functions $g : [0,T] \to \mathbb{R}$ such that

$$\|g\|_{1-\alpha,\infty,T} := \sup_{0<s<t<T} \left( \frac{|g(t) - g(s)|}{(t-s)^{1-\alpha}} + \int_s^t \frac{|g(y) - g(s)|}{(y-s)^{2-\alpha}} dy \right) < \infty.$$

We have, for all $0 < \varepsilon < \alpha$

$$C^{1-\alpha+\varepsilon}(0,T) \subset W_T^{1-\alpha,\infty}(0,T) \subset C^{1-\alpha}(0,T).$$

For $g \in W_T^{1-\alpha,\infty}(0,T)$ define

$$\Lambda_\alpha(g) := \frac{1}{\Gamma(1-\alpha)} \sup_{0<s<t<T} \left| \left(D_{t-}^{1-\alpha} g_{t-}\right)(s) \right|$$

$$\leq \frac{1}{\Gamma(1-\alpha)\Gamma(\alpha)} \|g\|_{1-\alpha,\infty,T}.$$

Finally, denote by $W_0^{\alpha,1}(0,T)$ the space of measurable functions $f$ on $[0,T]$ such that

$$\|f\|_{\alpha,1} := \int_0^T \frac{|f(s)|}{s^\alpha} ds + \int_0^T \int_0^s \frac{|f(s) - f(y)|}{(s-y)^{\alpha+1}} dy\, ds < \infty.$$

**Lemma 5.3.1** *Fix* $0 < \alpha < \frac{1}{2}$ *and consider two functions* $g \in W_T^{1-\alpha,\infty}(0,T)$ *and* $f \in W_0^{\alpha,1}(0,T)$. *Then the generalized Stieltjes integral*

$$\int_0^t f_s dg_s := \int_0^T f_s \mathbf{1}_{[0,t]}(s) dg_s$$

*exists for all* $t \in [0,T]$ *and for any* $s < t$ *we have*

$$\int_0^t f_u dg_u - \int_0^s f_u dg_u = \int_s^t D_{s+}^\alpha (f)(r) \left(D_{t-}^{1-\alpha} g_{t-}\right)(r) dr.$$

This lemma follows easily from the definition of the fractional derivative. We have

$$\left| \int_0^t f_s dg_s \right| \leq \Lambda_\alpha(g) \, \|f\|_{\alpha,1} \, .$$

Indeed,

$$\left| \int_0^t f_s dg_s \right| = \left| \int_0^t (D_{0+}^\alpha f)(s) \left( D_{t-}^{1-\alpha} g_{t-} \right)(s) \, ds \right|$$

$$\leq \sup_{0 \leq s \leq t \leq T} \left| \left( D_{t-}^{1-\alpha} g_{t-} \right)(s) \right| \int_0^t \left| (D_{0+}^\alpha f)(s) \right| ds$$

$$\leq \Lambda_\alpha(g) \, \|f\|_{\alpha,1} \, .$$

The following proposition is the main estimate for the generalized Stieltjes integral.

**Proposition 5.3.1** *Fix* $0 < \alpha < \frac{1}{2}$. *Given two functions* $g \in W_T^{1-\alpha,\infty}(0,T)$ *and* $f \in W_0^{\alpha,1}(0,T)$ *we set*

$$h_t = \int_0^t f_s dg_s.$$

*Then for all* $s < t \leq T$ *we have*

$$|h_t| + \int_0^t \frac{|h_t - h_s|}{(t-s)^{\alpha+1}} \, ds \;\; \leq \;\; \Lambda_\alpha(g) \, c_{\alpha,T}^{(1)} \int_0^t \left( (t-r)^{-2\alpha} + r^{-\alpha} \right)$$

$$\times \left( |f_r| + \int_0^r \frac{|f_r - f_y|}{(r-y)^{\alpha+1}} dy \right) dr, \quad (5.65)$$

*where* $c_{\alpha,T}^{(1)}$ *is a constant depending on* $\alpha$ *and* $T$.

*Proof:* Using the definition and additivity property of the indefinite integral we obtain

$$|h_t - h_s| = \left| \int_s^t f dg \right| = \left| \int_s^t D_{s+}^\alpha (f)(r) \left( D_{t-}^{1-\alpha} g_{t-} \right)(r) \, dr \right|$$

$$\leq \Lambda_\alpha(g) \left( \int_s^t \frac{|f_r|}{(r-s)^\alpha} dr \right.$$

$$\left. + \alpha \int_s^t \int_s^r \frac{|f_r - f_y|}{(r-y)^{\alpha+1}} dy dr \right). \quad (5.66)$$

Taking $s = 0$ we obtain the desired estimate for $|h_t|$. Multiplying (5.66) by $(t-s)^{-\alpha-1}$ and integrating in $s$ yields

$$\int_0^t \frac{|h_t - h_s|}{(t-s)^{\alpha+1}} \, ds \leq \Lambda_\alpha(g) \int_0^t (t-s)^{-\alpha-1} \quad (5.67)$$

$$\times \left( \int_s^t \frac{|f_r|}{(r-s)^\alpha} dr + \alpha \int_s^t \int_s^r \frac{|f_r - f_y|}{(r-y)^{\alpha+1}} dy dr \right) ds.$$

By the substitution $s = r - (t - r)y$ we have

$$\int_0^r (t - s)^{-\alpha-1}(r - s)^{-\alpha}ds \le (t - r)^{-2\alpha}\int_0^\infty (1 + y)^{-\alpha-1}y^{-\alpha}dy \quad (5.68)$$

and, on the other hand,

$$\int_0^y (t - s)^{-\alpha-1}ds = \alpha^{-1}\left[(t - y)^{-\alpha} - t^{-\alpha}\right] \le \alpha^{-1}(t - y)^{-\alpha}. \quad (5.69)$$

Substituting (5.68) and (5.69) into (5.67) yields

$$\int_0^t \frac{|h_t - h_s|}{(t - s)^{\alpha+1}}\,ds \le \Lambda_\alpha(g)\left[c_\alpha^{(1)}\int_0^t \frac{|f_r|}{(t - r)^{2\alpha}}dr \right.$$
$$\left. + \int_0^t\int_0^r \frac{|f(r) - f(y)|}{(r - y)^{\alpha+1}}(t - y)^{-\alpha}dydr\right],$$

where

$$c_\alpha^{(1)} = \int_0^\infty (1 + y)^{-\alpha-1}y^{-\alpha}dy = B(2\alpha, 1 - \alpha).$$

$\square$

As a consequence of this estimate, if $f \in W_T^{1-\alpha,\infty}(0, T)$ we have

$$\left|\int_s^t f_s dg_s\right| \le \Lambda_\alpha(g)\,c_{\alpha,T}^{(2)}\,(t - s)^{1-\alpha}\,\|f\|_{\alpha,\infty}, \quad (5.70)$$

and

$$\left\|\int_0^\cdot f_s dg_s\right\|_{\alpha,\infty} \le \Lambda_\alpha(g)\,c_{\alpha,T}^{(3)}\,\|f\|_{\alpha,\infty}. \quad (5.71)$$

### 5.3.2 Deterministic differential equations

Let $0 < \alpha < \frac{1}{2}$ be fixed. Let $g \in W_T^{1-d,\infty}(0, T; \mathbb{R}^d)$. Consider the deterministic differential equation on $\mathbb{R}^d$

$$x_t = x_0 + \int_0^t b(s, x_s)ds + \sum_{j=1}^d \sigma_j(s, x_s)\,dg_s^j, \quad t \in [0, T], \quad (5.72)$$

where $x_0 \in \mathbb{R}^m$.

Let us introduce the following assumptions on the coefficients:

**H1** $\sigma(t, x)$ is differentiable in $x$, and there exist some constants $0 < \beta, \delta \le 1$ and for every $N \ge 0$ there exists $M_N > 0$ such that the following properties hold:

$$|\sigma(t, x) - \sigma(t, y)| \le M_0|x - y|, \quad \forall x \in \mathbb{R}^m, \forall t \in [0, T],$$
$$|\partial_{x_i}\sigma(t, x) - \partial_{x_i}\sigma(t, y)| \le M_N|x - y|^\delta, \forall |x|, |y| \le N, \forall t \in [0, T],$$
$$|\sigma(t, x) - \sigma(s, x)| + |\partial_{x_i}\sigma(t, x) - \partial_{x_i}\sigma(s, x)| \le M_0|t - s|^\beta,$$
$$\forall x \in \mathbb{R}^m, \forall t, s \in [0, T].$$

for each $i = 1, \ldots, m$.

**H2** The coefficient $b(t, x)$ satisfies for every $N \geq 0$

$$
\begin{aligned}
|b(t, x) - b(t, y)| &\leq L_N |x - y|, \ \forall |x|, |y| \leq N, \ \forall t \in [0, T], \\
|b(t, x)| &\leq L_0 |x| + b_0(t), \ \forall x \in \mathbb{R}^m, \ \forall t \in [0, T],
\end{aligned}
$$

where $b_0 \in L^p(0, T; \mathbb{R}^m)$, with $\rho \geq 2$ and for some constant $L_N > 0$.

**Theorem 5.3.1** *Suppose that the coefficients $\sigma$ and $b$ satisfy the assumptions H1 and H2 with $\rho = \frac{1}{\alpha}$, $0 < \beta, \delta \leq 1$ and $0 < \alpha < \alpha_0 = \min\left(\frac{1}{2}, \beta, \frac{\delta}{\delta+1}\right)$. Then Equation (5.72) has a unique continuous solution such that $x^i \in W_0^{\alpha, \infty}(0, T)$ for all $i = 1, \ldots, m$.*

*Sketch of the proof:* Suppose $d = m = 1$. Fix $\lambda > 1$ and define the seminorm in $W_0^{\alpha, \infty}(0, T)$ by

$$
\|f\|_{\alpha, \lambda} = \sup_{t \in [0, T]} e^{-\lambda t} \left( |f_t| + \int_0^t \frac{|f_t - f_s|}{(t - s)^{\alpha+1}} \, ds \right).
$$

Consider the operator $\mathcal{L}$ defined by

$$
(\mathcal{L}f)_t = x_0 + \int_0^t b(s, f_s) ds + \int_0^t \sigma(s, f_s) \, dg_s.
$$

There exists $\lambda_0$ such that for $\lambda \geq \lambda_0$ we have

$$
\|\mathcal{L}f\|_{\alpha, \lambda} \leq |x_0| + 1 + \frac{1}{2} \|f\|_{\alpha, \lambda}.
$$

Hence, the operator $\mathcal{L}$ leaves invariant the ball $B_0$ of radius $2(|x_0| + 1)$ in the norm $\|\cdot\|_{\alpha, \lambda_0}$ of the space $W_0^{\alpha, \infty}(0, T)$. Moreover, there exists a constant $C$ depending on $g$ such that for any $\lambda \geq 1$ and $u, v \in B_0$

$$
\|\mathcal{L}(u) - \mathcal{L}(v)\|_{\alpha, \lambda} \leq \frac{C}{\lambda^{1-2\alpha}} (1 + \Delta(u) + \Delta(v)) \|u - v\|_{\alpha, \lambda}, \qquad (5.73)
$$

where

$$
\Delta(u) = \sup_{r \in [0, T]} \int_0^r \frac{|u_r - u_s|^\delta}{(r - s)^{\alpha+1}} \, ds.
$$

A basic ingredient in the proof of this inequality is the estimate

$$
\begin{aligned}
&|\sigma(r, f_r) - \sigma(s, f_s) - \sigma(r, h_r) + \sigma(s, h_s)| \\
&\leq M_0 |f_r - f_s - h_r + h_s| + M_0 |f_r - h_r| (r - s)^\beta \\
&\quad + M_N |f_r - h_r| \left( |f_r - f_s|^\delta + |h_r - h_s|^\delta \right),
\end{aligned}
$$

which is an immediate consequence of the properties of the function $\sigma$. The seminorm $\Delta$ is bounded on $\mathcal{L}(B_0)$, and, as a consequence, (5.74) implies that $\mathcal{L}$ is a contraction operator in $\mathcal{L}(B_0)$ with respect to a different norm $\|\cdot\|_{\alpha,\lambda_2}$ for a suitable value of $\lambda_2 > 1$. Finally, the the existence of a solution follows from a suitable fixed point argument (see Lemma 5.3.2). The uniqueness is proved again using the main estimate (5.65).

The following lemma has been used in the proof of Theorem 4.1.1 (for its proof see [254]).

**Lemma 5.3.2** *Let* $(X, \rho)$ *be a complete metric space and* $\rho_0, \rho_1, \rho_2$ *some metrics on* $X$ *equivalent to* $\rho$. *If* $\mathcal{L} : X \to X$ *satisfies:*
*i)    there exists* $r_0 > 0$, $x_0 \in X$ *such that if* $B_0 = \{x \in X : \rho_0(x_0, x) \le r_0\}$ *then*

$$\mathcal{L}(B_0) \subset B_0,$$

*ii)    there exists* $\varphi : (X, \rho) \to [0, +\infty]$ *lower semicontinuous function and some positive constants* $C_0, K_0$ *such that denoting* $N_\varphi(a) = \{x \in X : \varphi(x) \le a\}$

    *a)*    $\mathcal{L}(B_0) \subset N_\varphi(C_0)$,
    *b)*    $\rho_1(\mathcal{L}(x), \mathcal{L}(y)) \le K_0 \rho_1(x, y)$, $\forall x, y \in N_\varphi(C_0) \cap B_0$,

*iii)    there exists* $a \in (0, 1)$ *such that*

$$\rho_2(\mathcal{L}(x), \mathcal{L}(y)) \le a\, \rho_2(x, y), \quad \forall x, y \in \mathcal{L}(B_0),$$

*then there exists* $x^* \in \mathcal{L}(B_0) \subset X$ *such that*

$$x^* = \mathcal{L}(x^*).$$

*Estimates of the solution*

Suppose that the coefficient $\sigma$ satisfies the assumptions of the Theorem 4.1.1 and

$$|\sigma(t, x)| \le K_0 (1 + |x|^\gamma),\tag{5.74}$$

where $0 \le \gamma \le 1$. Then, the solution $f$ of Equation (5.72) satisfies

$$\|f\|_{\alpha,\infty} \le C_1 \exp(C_2 \Lambda_\alpha(g)^\kappa),\tag{5.75}$$

where

$$\kappa = \begin{cases} \frac{1}{1-2\alpha} & \text{if} & \gamma = 1 \\ > \frac{\gamma}{1-2\alpha} & \text{if} & \frac{1-2\alpha}{1-\alpha} \le \gamma < 1 \\ \frac{1}{1-\alpha} & \text{if} & 0 \le \gamma < \frac{1-2\alpha}{1-\alpha} \end{cases}$$

and the constants $C_1$ and $C_2$ depend on $T$, $\alpha$, and the constants that appear in conditions H1, H2 and (5.74).

The proof of (5.75) is based on the following version of Gronwall lemma:

**Lemma 5.3.3** *Fix* $0 \leq \alpha < 1$, $a, b \geq 0$. *Let* $x : [0, \infty) \rightarrow [0, \infty)$ *be a continuous function such that for each $t$*

$$x_t \leq a + bt^\alpha \int_0^t (t - s)^{-\alpha} s^{-\alpha} x_s ds. \qquad (5.76)$$

*Then*

$$x_t \leq a + a \sum_{n=1}^{\infty} b^n \frac{\Gamma(1-\alpha)^{n+1} t^{n(1-\alpha)}}{\Gamma[(n+1)(1-\alpha)]}.$$

$$\leq a d_\alpha \exp\left[c_\alpha t b^{1/(1-\alpha)}\right], \qquad (5.77)$$

*where $c_\alpha$ and $d_\alpha$ are positive constants depending only on $\alpha$ (as an example, one can set $c_\alpha = 2\left(\Gamma(1-\alpha)\right)^{1/(1-\alpha)}$ and $d_\alpha = 4e^2 \frac{\Gamma(1-\alpha)}{1-\alpha}$ ).*

This implies that there exists a constants $c_\alpha, d_\alpha > 0$ such that

$$x_t \leq a d_\alpha \exp\left[c_\alpha t b^{1/(1-\alpha)}\right].$$

## 5.3.3   Stochastic differential equations with respect to fBm

Let $B = \{B_t, t \geq 0\}$ be a $d$-dimensional fractional Brownian motion of Hurst parameter $H \in \left(\frac{1}{2}, 1\right)$. This means that the components of $B$ are independent fBm with the same Hurst parameter $H$. Consider the equation on $\mathbb{R}^m$

$$X_t = X_0 + \sum_{j=1}^{d} \int_0^t \sigma_j(s, X_s) \circ dB_s^j + \int_0^t b(s, X_s) ds, \quad t \in [0, T], \qquad (5.78)$$

where $X_0$ is an $m$-dimensional random variable. The integral with respect to $B$ is a path-wise Riemann-Stieltjes integral, and we know that this integral exists provided that the process $\sigma_j(s, X_s)$ has Hölder continuous trajectories of order larger that $1 - H$.

Choose $\alpha$ such that $1 - H < \alpha < \frac{1}{2}$. By Fernique's theorem, for any $0 < \delta < 2$ we have

$$E\left(\exp\left(\Lambda_\alpha(B)^\delta\right)\right) < \infty.$$

As a consequence, if $u = \{u_t, \ t \in [0, T]\}$ is a stochastic process whose trajectories belong to the space $W_T^{\alpha,1}(0, T)$, almost surely, the path-wise generalized Stieltjes integral integral $\int_0^T u_s \circ dB_s$ exists and we have the estimate

$$\left|\int_0^T u_s \circ dB_s\right| \leq G \|u\|_{\alpha,1}.$$

Moreover, if the trajectories of the process $u$ belong to the space $W_0^{\alpha,\infty}(0, T)$, then the indefinite integral $U_t = \int_0^t u_s \circ dB_s$ is Hölder continuous of order $1 - \alpha$, and its trajectories also belong to the space $W_0^{\alpha,\infty}(0, T)$.

Consider the stochastic differential equation (5.78) on $\mathbb{R}^m$ where the process $B$ is a $d$-dimensional fBm with Hurst parameter $H \in \left(\frac{1}{2}, 1\right)$ and $X_0$ is an $m$-dimensional random variable. Suppose that the coefficients $\sigma^i_j, b^i : \Omega \times [0, T] \times \mathbb{R}^m \to \mathbb{R}$ are measurable functions satisfying conditions H1 and H2, where the constants $M_N$ and $L_N$ may depend on $\omega \in \Omega$, and $\beta > 1 - H$, $\delta > \frac{1}{H} - 1$. Fix $\alpha$ such that

$$1 - H < \alpha < \alpha_0 = \min\left(\frac{1}{2}, \beta, \frac{\delta}{\delta + 1}\right)$$

and $\alpha \leq \frac{1}{\rho}$. Then the stochastic equation (5.65) has a unique continuous solution such that $X^i \in W_0^{\alpha,\infty}(0, T)$ for all $i = 1, \ldots, m$. Moreover the solution is Hölder continuous of order $1 - \alpha$.

Assume that $X_0$ is bounded and the constants do not depend on $\omega$. Suppose that

$$|\sigma(t, x)| \leq K_0 \left(1 + |x|^\gamma\right),$$

where $0 \leq \gamma \leq 1$. Then,

$$\|X\|_{\alpha,\infty} \leq C_1 \exp\left(C_2 \Lambda_\alpha(B)^\kappa\right).$$

Hence, for all $p \geq 1$

$$E\left(\|X\|^p_{\alpha,\infty}\right) \leq C_1^p E\left(\exp\left(pC_2 \Lambda_\alpha(B)^\kappa\right)\right) < \infty$$

provided $\kappa < 2$, that is,

$$\frac{\gamma}{4} + \frac{1}{2} \leq H$$

and

$$1 - H < \alpha < \frac{1}{2} - \frac{\gamma}{4}.$$

- If $\gamma = 1$ this means $\alpha < \frac{1}{4}$ and $H > \frac{3}{4}$.
- If $\gamma < 2 - \frac{1}{H}$ we can take any $\alpha$ such that $1 - H < \alpha < \frac{1}{2}$.

## Exercises

**5.3.1** Show the estimates (5.70) and (5.71).

**5.3.2** Show Lemma 5.3.1.

## 5.4   Vortex filaments based on fBm

The observations of three-dimensional turbulent fluids indicate that the vorticity field of the fluid is concentrated along thin structures called vortex filaments. In his book Chorin [67] suggests probabilistic descriptions of vortex filaments by trajectories of self-avoiding walks on a lattice. Flandoli [102] introduced a model of vortex filaments based on a three-dimensional Brownian motion. A basic problem in these models is the computation of the kynetic energy of a given configuration.

Denote by $u(x)$ the velocity field of the fluid at point $x \in \mathbb{R}^3$, and let $\xi = \text{curl} u$ be the associated vorticity field. The kynetic energy of the field will be

$$\mathbb{H} = \frac{1}{2} \int_{\mathbb{R}^3} |u(x)|^2 dx = \frac{1}{8\pi} \int_{\mathbb{R}^3} \int_{\mathbb{R}^3} \frac{\xi(x) \cdot \xi(y)}{|x - y|} dx dy. \qquad (5.79)$$

We will assume that the vorticity field is concentrated along a thin tube centered in a curve $\gamma = \{\gamma_t, 0 \le t \le T\}$. Moreover, we will choose a random model and consider this curve as the trajectory of a three-dimensional fractional Brownian motion $B = \{B_t, 0 \le t \le T\}$. This can be formally expressed as

$$\xi(x) = \Gamma \int_{\mathbb{R}^3} \left( \int_0^T \delta(x - y - B_s) \dot{B}_s ds \right) \rho(dy), \qquad (5.80)$$

where $\Gamma$ is a parameter called the circuitation, and $\rho$ is a probability measure on $\mathbb{R}^3$ with compact support.

Substituting (5.80) into (5.79) we derive the following formal expression for the kynetic energy:

$$\mathbb{H} = \int_{\mathbb{R}^3} \int_{\mathbb{R}^3} \mathbb{H}_{xy} \rho(dx) \rho(dy), \qquad (5.81)$$

where the so-called interaction energy $\mathbb{H}_{xy}$ is given by the double integral

$$\mathbb{H}_{xy} = \frac{\Gamma^2}{8\pi} \sum_{i=1}^{3} \int_0^T \int_0^T \frac{1}{|x + B_t - y - B_s|} \circ dB_s^i \circ dB_t^i. \qquad (5.82)$$

We are interested in the following problems: Is $\mathbb{H}$ a well defined random variable? Does it have moments of all orders and even exponential moments?

In order to give a rigorous meaning to the double integral (5.82) let us introduce the regularization of the function $|\cdot|^{-1}$:

$$\sigma_n = |\cdot|^{-1} * p_{1/n}, \qquad (5.83)$$

where $p_{1/n}$ is the Gaussian kernel with variance $\frac{1}{n}$. Then, the smoothed interaction energy

$$\mathbb{H}_{xy}^n = \frac{\Gamma^2}{8\pi} \sum_{i=1}^{3} \int_0^T \left( \int_0^T \sigma_n(x + B_t - y - B_s) \circ dB_s^i \right) \circ dB_t^i, \qquad (5.84)$$

is well defined, where the integrals are path-wise Riemann-Stieltjes integrals. Set

$$\mathbb{H}^n = \int_{\mathbb{R}^3} \int_{\mathbb{R}^3} \mathbb{H}^n_{xy} \rho(dx)\rho(dy). \tag{5.85}$$

The following result has been proved in [255]:

**Theorem 5.4.1** *Suppose that the measure $\rho$ satisfies*

$$\int_{\mathbb{R}^3} \int_{\mathbb{R}^3} |x - y|^{1 - \frac{1}{H}} \rho(dx)\rho(dy) < \infty. \tag{5.86}$$

*Let $\mathbb{H}^n_{xy}$ be the smoothed interaction energy defined by (5.84). Then $\mathbb{H}^n$ defined in (5.85) converges, for all $k \geq 1$, in $L^k(\Omega)$ to a random variable $\mathbb{H} \geq 0$ that we call the energy associated with the vorticity field (5.80).*

If $H = \frac{1}{2}$, fBm $B$ is a classical three-dimensional Brownian motion. In this case condition (5.86) would be $\int_{\mathbb{R}^3} \int_{\mathbb{R}^3} |x - y|^{-1} \rho(dx)\rho(dy) < \infty$, which is the assumption made by Flandoli [102] and Flandoli and Gubinelli [103]. In this last paper, using Fourier approach and Itô's stochastic calculus, the authors show that $Ee^{-\beta \mathbb{H}} < \infty$ for sufficiently small negative $\beta$.

The proof of Theorem 5.4.1 is based on the stochastic calculus of variations with respect to fBm and the application of Fourier transform.

*Sketch of the proof:* The proof will be done in two steps:

*Step 1* (Fourier transform) Using

$$\frac{1}{|z|} = \int_{\mathbb{R}^3} (2\pi)^3 \frac{e^{-i\langle \xi, z \rangle}}{|\xi|^2} d\xi$$

we get

$$\sigma_n(x) = \int_{\mathbb{R}^3} |\xi|^{-2} e^{i\langle \xi, x \rangle - |\xi|^2/2n} \, d\xi.$$

Substituting this expression in (5.84), we obtain the following formula for the smoothed interaction energy

$$\mathbb{H}^n_{xy} = \frac{\Gamma^2}{8\pi} \sum_{j=1}^{3} \int_0^T \int_0^T \left( \int_{\mathbb{R}^3} e^{i\langle \xi, x + B_t - y - B_s \rangle} \frac{e^{-|\xi|^2/2n}}{|\xi|^2} \right) \circ dB_s^j \circ dB_t^j$$

$$= \frac{\Gamma^2}{8\pi} \int_{\mathbb{R}^3} |\xi|^{-2} e^{i\langle \xi, x - y \rangle - |\xi|^2/2n} \|Y_\xi\|_{\mathbb{C}}^2 d\xi, \tag{5.87}$$

where

$$Y_\xi = \int_0^T e^{i\langle \xi, B_t \rangle} \circ dB_t$$

and $\|Y_\xi\|_{\mathbb{C}}^2 = \sum_{i=1}^{3} Y_\xi^i \overline{Y_\xi}^i$. Integrating with respect to $\rho$ yields

$$\mathbb{H}^n = \frac{\Gamma^2}{8\pi} \int_{\mathbb{R}^3} \|Y_\xi\|_{\mathbb{C}}^2 |\xi|^{-2} |\hat{\rho}(\xi)|^2 e^{-|\xi|^2/2n} d\xi \geq 0. \tag{5.88}$$

From Fourier analysis and condition (5.86) we know that

$$\int_{\mathbb{R}^3} \int_{\mathbb{R}^3} |x-y|^{1-\frac{1}{H}} \rho(dx)\rho(dy) = C_H \int_{\mathbb{R}^3} |\hat{\rho}(\xi)|^2 |\xi|^{\frac{1}{H}-4} d\xi < \infty. \tag{5.89}$$

Then, taking into account (5.89) and (5.88), in order to show the convergence in $L^k(\Omega)$ of $\mathbb{H}^n$ to a random variable $\mathbb{H} \geq 0$ it suffices to check that

$$E\left(\|Y_\xi\|_{\mathbb{C}}^{2k}\right) \leq C_k \left(1 \wedge |\xi|^{k\left(\frac{1}{H}-2\right)}\right). \tag{5.90}$$

*Step 2* (Stochastic calculus) We will present the main arguments for the proof of the estimate (5.90) for $k = 1$. Relation (5.37) applied to the process $u_t = e^{i\langle\xi,B_t\rangle}$ allows us to decompose the path-wise integral $Y_\xi = \int_0^T e^{i\langle\xi,B_t\rangle} \circ dB_t$ into the sum of a divergence plus a trace term:

$$Y_\xi = \int_0^T e^{i\langle\xi,B_t\rangle} dB_t + H \int_0^T i\xi e^{i\langle\xi,B_t\rangle} t^{2H-1} dt. \tag{5.91}$$

On the other hand, applying the three dimensional version of Itô's formula (5.44) we obtain

$$e^{i\langle\xi,B_T\rangle} = 1 + \sum_{j=1}^3 \int_0^T i\xi_j e^{i\langle\xi,B_t\rangle} \delta B_t^j - H \int_0^T t^{2H-1} |\xi|^2 e^{i\langle\xi,B_t\rangle} dt. \tag{5.92}$$

Multiplying both members of (5.92) by $i\xi|\xi|^{-2}$ and adding the result to (5.91) yields

$$Y_\xi = p_\xi \left(\int_0^T e^{i\langle\xi,B_t\rangle} dB_t\right) - \frac{i\xi}{|\xi|^2}\left(e^{i\langle\xi,B_T\rangle} - 1\right) := Y_\xi^{(1)} + Y_\xi^{(2)},$$

where $p_\xi(v) = v - \frac{\xi}{|\xi|^2}\langle\xi, v\rangle$ is the orthogonal projection of $v$ on $\langle\xi\rangle^\perp$. It suffices to derive the estimate (5.90) for the term $Y_\xi^{(1)}$. Using the duality relationship (1.42) for each $j = 1, 2, 3$ we can write

$$E\left(Y_\xi^{(1),j}\overline{Y}_\xi^{(1),j}\right) = E\left(\left\langle e^{i\langle\xi,B.\rangle}, p_\xi^j D. \left(\overline{p_\xi^j} \int_0^T e^{-i\langle\xi,B_t\rangle} dB_t\right)\right\rangle_{\mathcal{H}}\right). \tag{5.93}$$

The commutation relation $\langle D(\delta(u)), h\rangle_{\mathcal{H}} = \langle u, h\rangle_{\mathcal{H}} + \delta(\langle Du, h\rangle_{\mathcal{H}})$ implies

$$D_r^k \left( \int_0^T e^{-i\langle \xi, B_t\rangle} dB_t^j \right) = e^{-i\langle \xi, B_r^k\rangle} \delta_{k,j} + (-i\xi^k) \int_0^T \mathbf{1}_{[0,t]}(r) e^{-i\langle \xi, B_t\rangle} dB_t^j.$$

Applying the projection operators yields

$$p_\xi^j D_r \left( p_\xi^j \int_0^T e^{-i\langle \xi, B_t\rangle} dB_t \right) = e^{-i\langle \xi, B_r\rangle} \left( I - \frac{\xi^*\xi}{|\xi|^2} \right)_{j,j}$$

$$= e^{-i\langle \xi, B_r\rangle} \left( 1 - \frac{(\xi^j)^2}{|\xi|^2} \right).$$

Notice that the term involving derivatives in the expectation (5.93) vanishes. This cancellation is similar to what happens in the computation of the variance of the divergence of an adapted process, in the case of the Brownian motion. Hence,

$$\sum_{j=1}^3 E\left( Y_\xi^{(1),j} \overline{Y}_\xi^{(1),j} \right) = 2\, E\left( \left\langle e^{-i\langle \xi, B.\rangle}, e^{-i\langle \xi, B.\rangle} \right\rangle_{\mathcal{H}} \right)$$

$$= 2\alpha_H \int_0^T \int_0^T E\left( e^{i\langle \xi, B_s - B_r\rangle} \right) |s - r|^{2H-2}\, ds dr$$

$$= 2\alpha_H \int_0^T \int_0^T e^{-\frac{|s-r|^{2H}}{2}|\xi|^2} |s - r|^{2H-2}\, ds dr,$$

which behaves as $|\xi|^{\frac{1}{H}-2}$ as $|\xi|$ tends to infinity. This completes the proof of the desired estimate for $k = 1$.

In the general case $k \geq 2$ the proof makes use of the *local nondeterminism property* of fBm:

$$\mathrm{Var}\left( \sum_i (B_{t_i} - B_{s_i}) \right) \geq k_H \sum_i (t_i - s_i)^{2H}.$$

$\square$

*Decomposition of the interaction energy*

Assume $\frac{1}{2} < H < \frac{2}{3}$. For any $x \neq y$, set

$$\widehat{\mathbb{H}}_{xy} = \sum_{i=1}^3 \int_0^T \left( \int_0^t \frac{1}{|x + B_t - y - B_s|} \circ dB_s^i \right) \circ dB_t^i. \qquad (5.94)$$

Then $\widehat{\mathbb{H}}_{xy}$ exists as the limit in $L^2(\Omega)$ of the sequence $\widehat{\mathbb{H}}_{xy}^n$ defined using the approximation $\sigma_n(x)$ of $|x|^{-1}$ introduced in (5.83) and the following decomposition holds

$$\widehat{\mathbb{H}_{xy}} = \sum_{i=1}^{3} \int_0^T \int_0^t \frac{1}{|x - y + B_t - B_r|} dB_r^i dB_t^i.$$

$$-H^2 \int_0^T \int_0^t \delta_0(x - y + B_t - B_r)(t - r)^{2(2H-1)} dr\, dt.$$

$$+H(2H-1) \int_0^T \left( \int_0^t \frac{1}{|x - y + B_t - B_r|}(t-r)^{2H-2} dr \right) dt$$

$$+H \int_0^T \left( \frac{1}{|x - y + B_T - B_r|}(T - r)^{2H-2} + \frac{1}{|x - y + B_r|} r^{2H-1} \right) dr.$$

Notice that in comparison with $\mathbb{H}_{xy}$, in the definition of $\widehat{\mathbb{H}_{xy}}$ we chose to deal with the half integral over the domain

$$\{0 \le s \le t \le T\},$$

and to simplify the notation we have omitted the constant $\frac{\Gamma^2}{8\pi}$. Nevertheless, it holds that $\mathbb{H}_{xy} = \frac{\Gamma^2}{8\pi} \left( \widehat{\mathbb{H}_{xy}} + \widehat{\mathbb{H}_{yx}} \right)$, and we have proved using Fourier analysis that $\mathbb{H}_{xy}$ has moments of any order.

The following results have been proved in [255]:

1. All the terms in the above decomposition of $\widehat{\mathbb{H}_{xy}}$ exists in $L^2(\Omega)$ for $x \neq y$.

2. If $|x - y| \to 0$, then the terms behave as $|x - y|^{\frac{1}{H} - 1}$, so they can be integrated with respect to $\rho(dx)\rho(dy)$.

3. The bound $H < \frac{2}{3}$ is sharp: For $H = \frac{2}{3}$ the weighted self-intersection local time diverges.

**Notes and comments**

**[5.1]**    The fractional Brownian motion was first introduced by Kolmogorov [171] and studied by Mandelbrot and Van Ness in [217], where a stochastic integral representation in terms of a standard Brownian motion was established.

Hurst developed in [141] a statistical analysis of the yearly water run-offs of Nile river. Suppose that $x_1, \ldots, x_n$ are the values of $n$ successive yearly water run-offs. Denote by $X_n = \sum_{k=1}^{n} x_k$ the cumulative values. Then, $X_k - \frac{k}{n} X_n$ is the deviation of the cumulative value $X_k$ corresponding to $k$ successive years from the empirical means as calculated using data for $n$ years. Consider the range of the amplitude of this deviation:

$$R_n = \max_{1 \le k \le n} \left( X_k - \frac{k}{n} X_n \right) - \min_{1 \le k \le n} \left( X_k - \frac{k}{n} X_n \right)$$

and the empirical mean deviation

$$\mathcal{S}_n = \sqrt{\frac{1}{n}\sum_{k=1}^{n}\left(x_k - \frac{X_n}{n}\right)^2}.$$

Based on the records of observations of Nile flows in 622-1469, Hurst discovered that $\frac{\mathcal{R}_n}{\mathcal{S}_n}$ behaves as $cn^H$, where $H = 0.7$. On the other hand, the partial sums $x_1 + \cdots + x_n$ have approximately the same distribution as $n^H x_1$, where again $H$ is a parameter larger than $\frac{1}{2}$.

These facts lead to the conclusion that one cannot assume that $x_1, \ldots, x_n$ are values of a sequence of independent and identically distributed random variables. Some alternative models are required in order to explain the empirical facts. One possibility is to assume that $x_1, \ldots, x_n$ are values of the increments of a fractional Brownian motion. Motivated by these empirical observations, Mandelbrot has given the name of Hurst parameter to the parameter $H$ of fBm.

The fact that for $H > \frac{1}{2}$ fBm is not a semimartingale has been first proved by [198] (see also Example 4.9.2 in Liptser and Shiryaev [201]). Rogers in [296] has established this result for any $H \neq \frac{1}{2}$.

[5.2]   Different approaches have been used in the literature in order to define stochastic integrals with respect to fBm. Lin [198] and Dai and Heyde [73] have defined a stochastic integral $\int_0^T \phi_s dB_s$ as limit in $L^2$ of Riemann sums in the case $H > \frac{1}{2}$. This integral does not satisfy the property $E(\int_0^T \phi_s dB_s) = 0$ and it gives rise to change of variable formulae of Stratonovich type. A new type of integral with zero mean defined by means of Wick products was introduced by Duncan, Hu and Pasik-Duncan in [86], assuming $H > \frac{1}{2}$. This integral turns out to coincide with the divergence operator (see also Hu and Øksendal [140]).

A construction of stochastic integrals with respect to fBm with parameter $H \in (0,1)$ by a regularization technique was developed by Carmona and Coutin in [58]. The integral is defined as the limit of approximating integrals with respect to semimartingales obtained by smoothing the singularity of the kernel $K_H(t,s)$. The techniques of Malliavin Calculus are used in order to establish the existence of the integrals. The ideas of Carmona and Coutin were further developed by Alòs, Mazet and Nualart in the case $0 < H < \frac{1}{2}$ in [8].

The interpretation of the divergence operator as a stochastic integral was introduced by Decreusefont and Üstünel in [78]. A stochastic calculus for the divergence process has been developed by Alòs, Mazet and Nualart in [9], among others.

A basic reference for the stochastic calculus with respect to the fBM is the recent monograph by Hu [139]. An Itô's formula for $H \in (0,1)$ in the framework of white noise analysis has been established by Bender in [22].

We refer to [124] and [123] for the stochastic calculus with respect to fBM based on symmetric integrals.

The results on the stochastic calculus with respect to the fBm are based on the papers [11] (case $H > \frac{1}{2}$), [7] and [66] (case $H < \frac{1}{2}$).

**[5.3]**    In [202], Lyons considered deterministic integral equations of the form

$$x_t = x_0 + \int_0^t \sigma(x_s) dg_s,$$

$0 \le t \le T$, where the $g : [0, T] \to \mathbb{R}^d$ is a continuous functions with bounded $p$-variation for some $p \in [1, 2)$. This equation has a unique solution in the space of continuous functions of bounded $p$-variation if each component of $g$ has a Hölder continuous derivative of order $\alpha > p - 1$. Taking into account that fBm of Hurst parameter $H$ has locally bounded $p$-variation paths for $p > 1/H$, the result proved in [202] can be applied to Equation (5.78) in the case $\sigma(s, x) = \sigma(x)$, and $b(s, x) = 0$, provided the coefficient $\sigma$ has a Hölder continuous derivative of order $\alpha > \frac{1}{H} - 1$.

The *rough path analysis* developed by Lyons in the [203] is a powerful technique based on the notion of $p$-variation which permits to handle differential equations driven by irregular functions (see also the monograph [204] by Lyons and Qian). In [70] Coutin and Qian have established the existence of strong solutions and a Wong-Zakai type approximation limit for stochastic differential equations driven by a fractional Brownian motion with parameter $H > \frac{1}{4}$ using the approach of rough path analysis.

In [299] Ruzmaikina establishes an existence and uniqueness theorem for ordinary differential equations driven by a Hölder continuous function using Hölder norms.

The generalized Stieltjes integral defined in (5.64), based on the fractional integration by parts formula, was introduced by Zähle in [355]. In this paper, the author develops an approach to stochastic calculus based on the fractional calculus. As an application, in [356] the existence and uniqueness of solutions is proved for differential equations driven by a fractional Brownian motion with parameter $H > \frac{1}{2}$, in a small random interval, provided the diffusion coefficient is a contraction in the space $W_{2,\infty}^{\beta}$, where $\frac{1}{2} < \beta < H$. Here $W_{2,\infty}^{\beta}$ denotes the Besov-type space of bounded measurable functions $f : [0, T] \to \mathbb{R}$ such that

$$\int_0^T \int_0^T \frac{(f(t) - f(s))^2}{|t - s|^{2\beta + 1}} ds dt < \infty.$$

In [254] Nualart and Rascanu have established the existence and uniqueness of solution for Equation (5.78) using an a priori estimate based on the fractional integration by parts formula, following the approach of Zähle.

**[5.4]**    The results of this section have been proved by Nualart, Rovira and Tindel in [255].

# 6
# Malliavin Calculus in finance

In this chapter we review some applications of Malliavin Calculus to mathematical finance. First we discuss a probabilistic method for numerical computations of price sensitivities (Greeks) based on the integration by parts formula. Then, we discuss the use of Clark-Ocone formula to find hedging portfolios in the Black-Scholes model. Finally, the last section deals with the computation of additional expected utility for insider traders.

## 6.1   Black-Scholes model

Consider a market consisting of one stock (risky asset) and one bond (riskless asset). The price process of the risky asset is assumed to be of the form $S_t = S_0 e^{H_t}$, $t \in [0, T]$, with

$$H_t = \int_0^t (\mu_s - \frac{\sigma_s^2}{2}) ds + \int_0^t \sigma_s dW_s, \qquad (6.1)$$

where $W = \{W_t, t \in [0, T]\}$ is a Brownian motion defined in a complete probability space $(\Omega, \mathcal{F}, P)$. We will denote by $\{\mathcal{F}_t, t \in [0, T]\}$ the filtration generated by the Brownian motion and completed by the $P$-null sets. The *mean rate of return* $\mu_t$ and the *volatility* process $\sigma_t$ are supposed to be measurable and adapted processes satisfying the following integrability conditions

$$\int_0^T |\mu_t| dt < \infty, \quad \int_0^T \sigma_t^2 dt < \infty$$

almost surely.

By Itô's formula we obtain that $S_t$ satisfies a linear stochastic differential equation:

$$dS_t = \mu_t S_t dt + \sigma_t S_t dW_t. \tag{6.2}$$

The price of the bond $B_t$, $t \in [0, T]$, evolves according to the differential equation

$$dB_t = r_t B_t dt, \quad B_0 = 1,$$

where the *interest rate* process is a nonnegative measurable and adapted process satisfying the integrability condition

$$\int_0^T r_t dt < \infty,$$

almost surely. That is,

$$B_t = \exp\left( \int_0^t r_s ds \right).$$

Imagine an investor who starts with some initial endowment $x \geq 0$ and invests in the assets described above. Let $\alpha_t$ be the number of non-risky assets and $\beta_t$ the number of stocks owned by the investor at time $t$. The couple $\phi_t = (\alpha_t, \beta_t)$, $t \in [0, T]$, is called a *portfolio* or *trading strategy*, and we assume that $\alpha_t$ and $\beta_t$ are measurable and adapted processes such that

$$\int_0^T |\beta_t \mu_t| dt < \infty, \quad \int_0^T \beta_t^2 \sigma_t^2 dt < \infty, \quad \int_0^T |\alpha_t| r_t dt < \infty \tag{6.3}$$

almost surely. Then $x = \alpha_0 + \beta_0 S_0$, and the investor's wealth at time $t$ (also called the *value* of the portfolio) is

$$V_t(\phi) = \alpha_t B_t + \beta_t S_t.$$

The gain $G_t(\phi)$ made by the investor via the portfolio $\phi$ up to time $t$ is given by

$$G_t(\phi) = \int_0^t \alpha_s dB_s + \int_0^t \beta_s dS_s.$$

Notice that both integrals are well defined thanks to condition (6.3).

We say that the portfolio $\phi$ is *self-financing* if there is no fresh investment and there is no consumption. This means that the value equals to the intial investment plus the gain:

$$V_t(\phi) = x + \int_0^t \alpha_s dB_s + \int_0^t \beta_s dS_s. \tag{6.4}$$

From now on we will consider only self-financing portfolios.

We introduce the discounted prices defined by

$$\widetilde{S}_t = B_t^{-1} S_t = S_0 \exp\left( \int_0^t \left( \mu_s - r_s - \frac{\sigma_s^2}{2} \right) ds + \int_0^t \sigma_s dW_s \right).$$

Then, the discounted value of a portfolio will be

$$\widetilde{V}_t(\phi) = B_t^{-1} V_t(\phi) = \alpha_t + \beta_t \widetilde{S}_t. \qquad (6.5)$$

Notice that

$$\begin{aligned}
d\widetilde{V}_t(\phi) &= -r_t B_t^{-1} V_t(\phi) dt + B_t^{-1} dV_t(\phi) \\
&= -r_t \beta_t \widetilde{S}_t dt + B_t^{-1} \beta_t dS_t \\
&= \beta_t d\widetilde{S}_t,
\end{aligned}$$

that is

$$\begin{aligned}
\widetilde{V}_t(\phi) &= x + \int_0^t \beta_s d\widetilde{S}_s \\
&= x + \int_0^t (\mu_s - r_s) \beta_s \widetilde{S}_s ds + \int_0^t \sigma_s \beta_s \widetilde{S}_s dW_s. \qquad (6.6)
\end{aligned}$$

Equations (6.5) and (6.6) imply that the composition $\alpha_t$ on non-risky assets in a self-financing portfolio is determined by the initial value $x$ and $\beta_t$:

$$\begin{aligned}
\alpha_t &= \widetilde{V}_t(\phi) - \beta_t \widetilde{S}_t \\
&= x + \int_0^t \beta_s d\widetilde{S}_s - \beta_t \widetilde{S}_t.
\end{aligned}$$

On the other hand, (6.6) implies that if $\mu_t = r_t$ for $t \in [0, T]$, then the value process $V_t(\phi)$ of any self-financing portfolio is a local martingale.

## 6.1.1 Arbitrage opportunities and martingale measures

**Definition 6.1.1** *An arbitrage is a self-financing strategy $\phi$ such that $V_0(\phi) = 0$, $V_T(\phi) \geq 0$ and $P(V_T(\phi) > 0) > 0$.*

In general, we are interested in having models for the stock price process without opportunities of arbitrage. In the case of discrete time models, the absence of opportunities of arbitrage is equivalent to the existence of martingale measures:

**Definition 6.1.2** *A probability measure $Q$ on the $\sigma$-field $\mathcal{F}_T$, which is equivalent to $P$, is called a martingale measure (or a non-risky probability measure) if the discounted price process $\{\widetilde{S}_t, 0 \leq t \leq T\}$ is a martingale in the probability space $(\Omega, \mathcal{F}_T, Q)$.*

In continuous time, the relation between the absence of opportunities of arbitrage and existence of martingale measures is more complex. Let us assume the following additional conditions:

$$\sigma_t > 0$$

for all $t \in [0, T]$ and

$$\int_0^T \theta_s^2 ds < \infty$$

almost surely, where $\theta_t = \frac{\mu_t - r_t}{\sigma_t}$. Then, we can define the process

$$Z_t = \exp\left(-\int_0^t \theta_s dW_s - \frac{1}{2}\int_0^t \theta_s^2 ds\right),$$

which is a positive local martingale.

If $E(Z_T) = 1$, then, by Girsanov theorem (see Proposition 4.1.2), the process $Z$ is a martingale and the measure $Q$ defined by $\frac{dQ}{dP} = Z_T$ is a probability measure, equivalent to $P$, such that under $Q$ the process

$$\widetilde{W}_t = W_t + \int_0^t \theta_s ds$$

is a Brownian motion. Notice that in terms of the process $\widetilde{W}_t$ the price process can be expressed as

$$S_t = S_0 \exp\left(\int_0^t (r_s - \frac{\sigma_s^2}{2})ds + \int_0^t \sigma_s d\widetilde{W}_s\right),$$

and the discounted prices form a local martingale:

$$\widetilde{S}_t = B_t^{-1} S_t = S_0 \exp\left(\int_0^t \sigma_s d\widetilde{W}_s - \frac{1}{2}\int_0^t \sigma_s^2 ds\right).$$

Moreover, the discounted value process of any self-financing strategy is also a local martingale, because from (6.6) we obtain

$$\widetilde{V}_t(\phi) = x + \int_0^t \beta_s d\widetilde{S}_s = x + \int_0^t \sigma_s \beta_s \widetilde{S}_s d\widetilde{W}_s. \tag{6.7}$$

Condition (6.7) implies that there are no arbitrage opportunities verifying $E_Q\left(\int_0^T \left(\sigma_s \beta_s \widetilde{S}_s\right)^2 ds\right) < \infty$. In fact, this condition implies that the discounted value process $\widetilde{V}_t(\phi)$ is a martingale under $Q$. Then, using the martingale property we obtain

$$E_Q\left(\widetilde{V}_T(\phi)\right) = V_0(\phi) = 0,$$

so $V_T(\phi) = 0$, $Q$-almost surely, which contradicts the fact that $P(\,V_T(\phi) > 0) > 0$.

A portfolio $\phi$ is said to be *admissible* for the intial endowment $x \geq 0$ if its value process satisfies $V_t(\phi) \geq 0$, $t \in [0,T]$, almost surely. Then, there are no arbitrage opportunities in the class of admissible portfolios. In fact, such an arbitrage will have a discounted value process $\widetilde{V}_t(\phi)$ which is a nonnegative local martingale. Thus, it is a supermartingale, and, hence,

$$E_Q\left(\widetilde{V}_T(\phi)\right) \leq V_0(\phi) = 0,$$

so $V_T(\phi) = 0$, $Q$-almost surely, which contradicts the fact that $P(\,V_T(\phi) > 0) > 0$.

Note that, if we assume that the process $\sigma_t$ is uniformly bounded, then the discounted price process $\{\widetilde{S}_t, 0 \leq t \leq T\}$ is a martingale under $Q$, and $Q$ is a martingale measure in the sense of Definition 6.1.2.

## 6.1.2 Completeness and hedging

A *derivative* is a contract on the risky asset that produces a payoff $H$ at maturity time $T$. The payoff is, in general, an $\mathcal{F}_T$-measurable nonnegative random variable $H$.

**Example 6.1.1** *European Call-Option with maturity $T$ and exercise price $K > 0$: The buyer of this contract has the option to buy, at time $T$, one share of the stock at the specified price $K$. If $S_T \leq K$ the contract is worthless to him and he does not exercise his option. If $S_T > K$, the seller is forced to sell one share of the stock at the price $K$, and thus the buyer can make a profit $S_T - K$ by selling then the share at its market price. As a consequence, this contract effectively obligates the seller to a payment of $H = (S_T - K)^+$ at time $T$.*

**Example 6.1.2** *European Put-Option with maturity $T$ and exercise price $K > 0$: The buyer of this contract has the option to sell, at time $T$, one share of the stock at the specified price $K$. If $S_T \geq K$ the contract is worthless to him and he does not exercise his option. If $S_T < K$, the seller is forced to buy one share of the stock at the price $K$, and thus the buyer can make a profit $S_T - K$ by buying first the share at its market price. As a consequence, this contract effectively obligates the seller to a payment of $H = (K - S_T)^+$ at time $T$.*

**Example 6.1.3** *Barrier Option: $H = (S_T - K)^+ \mathbf{1}_{\{T_a \leq T\}}$, for some $a > K > 0$, $a > S_0$, where $T_a = \inf\{t \geq 0 : S_t \geq a\}$. This contract is similar to the European call-option with exercise price $K$ and maturity $T$, except that now the stock price has to reach a certain barrier level $a > \max(K, S_0)$ for the option to become activated.*

**Example 6.1.4** *Assian Option:* $H = \left( \frac{1}{T} \int_0^T S_t dt - K \right)^+$. *This contract is similar to the European call-option with exercise price $K$ and maturity $T$, except that now the average stock price is used in place of the terminal stock price.*

We shall say that a nonnegative $\mathcal{F}_T$-measurable payoff $H$ can be replicated if there exists a self-financing portfolio $\phi$ such that $V_T(\phi) = H$. The following proposition asserts that any derivative satisfying $E(B_T^{-2} Z_T^2 H^2) < \infty$ is replicable, that is, the Black and Scholes model is *complete*. Its proof is a consequence of the integral representation theorem (see Theorem 1.1.3).

**Proposition 6.1.1** *Let $H$ be a nonnegative $\mathcal{F}_T$-measurable random variable such that $E(B_T^{-2} Z_T^2 H^2) < \infty$. Then, there exists a self-financing portfolio $\phi$ such that $V_T(\phi) = H$.*

*Proof:*     By the integral representation theorem (Theorem 1.1.3) there exists an adapted and measurable process $u = \{u_t, t \in [0,T]\}$ such that $E \left( \int_0^T u_s^2 ds \right) < \infty$ and

$$B_T^{-1} Z_T H = E \left( B_T^{-1} Z_T H \right) + \int_0^T u_s dW_s.$$

Set

$$M_t = E \left( B_T^{-1} Z_T H | \mathcal{F}_t \right) = E \left( B_T^{-1} Z_T H \right) + \int_0^t u_s dW_s. \tag{6.8}$$

Notice that $B_T Z_T^{-1} M_T = H$. We claim that there exists a self-financing portfolio $\phi$ such that $\widetilde{V}_t(\phi) = H$. Define the self-financing portfolio $\phi_t = (\alpha_t, \beta_t)$ by

$$\beta_t = \frac{Z_t^{-1} (u_t + M_t \theta_t)}{\sigma_t \widetilde{S}_t},$$

$$\alpha_t = M_t Z_t^{-1} - \beta_t \widetilde{S}_t.$$

The discounted value of this portfolio is

$$\widetilde{V}_t(\phi) = \alpha_t + \beta_t \widetilde{S}_t = Z_t^{-1} M_t, \tag{6.9}$$

so, its final value will be

$$V_T(\phi) = B_T \widetilde{V}_T(\phi) = B_T Z_T^{-1} M_T = H.$$

Let us show that this portfolio is self-financing. By Itô's formula we have

$$d(Z_t^{-1}) = Z_t^{-1}(\theta_t dW_t + \theta_t^2 dt)$$

and

$$
\begin{aligned}
d(Z_t^{-1}M_t) &= Z_t^{-1}dM_t + M_t d(Z_t^{-1}) + dM_t d(Z_t^{-1}) \\
&= Z_t^{-1}u_t dW_t + M_t Z_t^{-1}(\theta_t dW_t + \theta_t^2 dt) + Z_t^{-1}u_t\theta_t dt \\
&= Z_t^{-1}\left(u_t + M_t\theta_t\right) d\widetilde{W}_t \\
&= \sigma_t \widetilde{S}_t \beta_t d\widetilde{W}_t.
\end{aligned}
$$

Hence,

$$
\begin{aligned}
dV_t(\phi) &= d(B_t Z_t^{-1} M_t) = B_t \sigma_t \widetilde{S}_t \beta_t d\widetilde{W}_t + B_t Z_t^{-1} M_t r_t dt \\
&= B_t \sigma_t \widetilde{S}_t \beta_t dW_t + B_t \sigma_t \widetilde{S}_t \beta_t \theta_t dt + B_t(\alpha_t + \beta_t \widetilde{S}_t) r_t dt \\
&= S_t \sigma_t \beta_t dW_t + S_t \beta_t \left(\mu_t - r_t\right) dt + \beta_t S_t r_t dt + B_t \alpha_t r_t dt \\
&= S_t \sigma_t \beta_t dW_t + \beta_t S_t \mu_t dt + B_t \alpha_t r_t dt \\
&= \alpha_t dB_t + \beta_t dS_t.
\end{aligned}
$$

$\square$

The price of a derivative with payoff $H$ at time $t \leq T$ is given by the value at time $t$ of a portfolio which replicates $H$. Under the assumptions of Proposition 6.1.1, from (6.8) and (6.9) we deduce

$$
V_t(\phi) = B_t Z_t^{-1} E\left(B_T^{-1} Z_T H | \mathcal{F}_t\right) = Z_t^{-1} E(Z_T e^{-\int_t^T r_s ds} H | \mathcal{F}_t).
$$

Assume now $E(Z_T) = 1$ and let $Q$ be given by $\frac{dQ}{dP} = Z_T$. Then, using Exercise 4.2.11 we can write

$$
V_t(\phi) = E_Q(e^{-\int_t^T r_s ds} H | \mathcal{F}_t).
$$

In particular,

$$
V_0(\phi) = E_Q(e^{-\int_0^T r_s ds} H). \tag{6.10}
$$

### 6.1.3  Black-Scholes formula

Suppose that the parameters $\sigma_t = \sigma$, $\mu_t = \mu$ and $r_t = r$ are constant. In that case we obtain that the dynamics of the stock price is described by a *geometric Brownian motion:*

$$
S_t = S_0 \exp\left(\left(\mu - \frac{\sigma^2}{2}\right) t + \sigma W_t\right).
$$

Moreover, $\theta_t = \theta = \frac{\mu - r}{\sigma}$, and

$$
Z_t = \exp\left(-\theta W_t - \frac{\theta^2}{2} t\right).
$$

So, $E(Z_T) = 1$, and $\widetilde{W}_t = W_t + \theta t$ is a Brownian motion under $Q$, with $\frac{dQ}{dP} = Z_T$, on the time interval $[0, T]$.

This model is complete in the sense that any payoff $H \geq 0$ satisfying $E_Q(H^2) < \infty$ is replicable. In this case, we simply apply the integral representation theorem to the random variable $e^{-rT} H \in L^2(\Omega, \mathcal{F}_T, Q)$ with respect to the Wiener process $\widetilde{W}$. In this way we obtain

$$e^{-rT} H = E_Q \left( e^{-rT} H \right) + \int_0^T u_s d\widetilde{W}_s,$$

and the self-financing replicating portfolio is given by

$$\beta_t = \frac{u_t}{\sigma \widetilde{S}_t},$$

$$\alpha_t = M_t - \beta_t \widetilde{S}_t,$$

where

$$M_t = E_Q \left( e^{-rT} H | \mathcal{F}_t \right) = E_Q \left( e^{-rT} H \right) + \int_0^t u_s d\widetilde{W}_s.$$

Consider the particular case of an European option, that is, $H = \Phi(S_T)$, where $\Phi$ is a measurable function with linear growth. The value of this derivative at time $t$ will be

$$
\begin{aligned}
V_t(\phi) &= E_Q \left( e^{-r(T-t)} \Phi(S_T) | \mathcal{F}_t \right) \\
&= e^{-r(T-t)} E_Q \left( \Phi(S_t e^{r(T-t)} e^{\sigma(\widetilde{W}_T - \widetilde{W}_t) - \sigma^2/2(T-t)}) | \mathcal{F}_t \right).
\end{aligned}
$$

Hence,

$$V_t = F(t, S_t), \tag{6.11}$$

where

$$F(t, x) = e^{-r(T-t)} E_Q \left( \Phi(x e^{r(T-t)} e^{\sigma(\widetilde{W}_T - \widetilde{W}_t) - \sigma^2/2(T-t)}) \right). \tag{6.12}$$

Under general hypotheses on $\Phi$ (for instance, if $\Phi$ has linear growth, is continuous and piece-wise differentiable) which include the cases

$$
\begin{aligned}
\Phi(x) &= (x - K)^+, \\
\Phi(x) &= (K - x)^+,
\end{aligned}
$$

the function $F(t, x)$ is of class $C^{1,2}$. Then, applying Itô's formula to (6.11) we obtain

$$
\begin{aligned}
V_t(\phi) &= V_0(\phi) + \int_0^t \sigma \frac{\partial F}{\partial x}(u, S_u) S_u d\widetilde{W}_u + \int_0^t r \frac{\partial F}{\partial x}(u, S_u) S_u du \\
&\quad + \int_0^t \frac{\partial F}{\partial u}(u, S_u) du + \frac{1}{2} \int_0^t \frac{\partial^2 F}{\partial x^2}(u, S_u) \sigma^2 S_u^2 du.
\end{aligned}
$$

On the other hand, we know that $V_t(\phi)$ has the representation

$$V_t(\phi) = V_0(\phi) + \int_0^t \sigma \beta_u S_u d\widetilde{W}_u + \int_0^t r V_u du.$$

Comparing these expressions, and taking into account the uniqueness of the representation of an Itô process, we deduce the equations

$$\beta_t = \frac{\partial F}{\partial x}(t, S_t),$$

$$rF(t, S_t) = \frac{\partial F}{\partial t}(t, S_t) + \frac{1}{2}\sigma^2 S_t^2 \frac{\partial^2 F}{\partial x^2}(t, S_t)$$

$$+ rS_t \frac{\partial F}{\partial x}(t, S_t).$$

The support of the probability distribution of the random variable $S_t$ is $[0, \infty)$. Therefore, the above equalities lead to the following partial differential equation for the function $F(t, x)$, where $0 \le t \le T, x \ge 0$

$$\frac{\partial F}{\partial t}(t, x) + rx\frac{\partial F}{\partial x}(t, x) + \frac{1}{2}\frac{\partial^2 F}{\partial x^2}(t, x)\,\sigma^2\,x^2 = rF(t, x),$$

$$F(T, x) = \Phi(x).$$

The replicating portfolio is given by

$$\beta_t = \frac{\partial F}{\partial x}(t, S_t), \tag{6.13}$$

$$\alpha_t = e^{-rt}\left(F(t, S_t) - \beta_t S_t\right). \tag{6.14}$$

Formula (6.12) can be written as

$$F(t, x) = e^{-r(T-t)}E_Q\left(\Phi(xe^{r(T-t)}e^{\sigma(\widetilde{W}_T - \widetilde{W}_t) - \sigma^2/2(T-t)})\right)$$

$$= e^{-r\tau}\frac{1}{\sqrt{2\pi}}\int_{-\infty}^{\infty}\Phi(xe^{r\tau - \frac{\sigma^2}{2}\tau + \sigma\sqrt{\tau}y})e^{-y^2/2}dy,$$

where $\tau = T - t$ is the time to maturity.

In the particular case of an European call-option with exercise price $K$ and maturity $T$, $\Phi(x) = (x - K)^+$, and we get

$$F(t, x) = \frac{1}{\sqrt{2\pi}}\int_{-\infty}^{\infty}e^{-y^2/2}\left(xe^{-\frac{\sigma^2}{2}\tau + \sigma\sqrt{\tau}y} - Ke^{-r\theta}\right)^+ dy$$

$$= x\Phi(d_+) - Ke^{-r(T-t)}\Phi(d_-), \tag{6.15}$$

where

$$d_- = \frac{\log\frac{x}{K} + \left(r - \frac{\sigma^2}{2}\right)\tau}{\sigma\sqrt{\tau}},$$

$$d_+ = \frac{\log\frac{x}{K} + \left(r + \frac{\sigma^2}{2}\right)\tau}{\sigma\sqrt{\tau}}.$$

## Exercises

**6.1.1** Consider the Black-Scholes model with constant volatility, mean rate of return and interest rate. Compute the price at time $t_0$, $0 \le t_0 < T$ of a derivative whose payoff is:

(i) $H = \frac{1}{T} \int_0^T S_t dt$.

(ii) $H = S_T^{4/3}$.

**6.1.2** Using
$$(S_T - K)^+ - (S_T - K)^- = S_T - K \qquad (6.16)$$
deduce a relation between the price of an European call-option and that of an European put-option. In particular, in the case of the Black-Scholes model with constant volatility, mean rate of return and interest rate, obtain a formula for the price of an European put-option from (6.15).

**6.1.3** From (6.15) show that
$$\frac{\partial F}{\partial x} = \Phi(d_+)$$
$$\frac{\partial^2 F}{\partial x^2} = \frac{1}{x\sigma\sqrt{T}}\Phi'(d_+)$$
$$\frac{\partial F}{\partial \sigma} = x\sqrt{T}\Phi'(d_+).$$

Deduce that $F$ is a nondecreasing and convex function of $x$.

## 6.2 Integration by parts formulas and computation of Greeks

In this section we will present a general integration by parts formula and we will apply it to the computation of Greeks. We will assume that the price process follows a Black-Scholes model with constant coefficients $\sigma$, $\mu$, and $r$.

Let $W = \{W(h), h \in H\}$ denote an isonormal Gaussian process associated with the Hilbert space $H$. We assume that $W$ is defined on a complete probability space $(\Omega, \mathcal{F}, P)$, and that $\mathcal{F}$ is generated by $W$.

**Proposition 6.2.1** *Let $F$, $G$ be two random variables such that $F \in \mathbb{D}^{1,2}$. Consider an $H$-valued random variable $u$ such that $D^u F = \langle DF, u \rangle_H \ne 0$ a.s. and $Gu(D^u F)^{-1} \in \mathrm{Dom}\delta$. Then, for any  continuously differentiable function function $f$ with bounded derivative we have*

$$E(f'(F)G) = E(f(F)H(F,G)),$$

*where $H(F,G) = \delta(Gu(D^u F)^{-1})$.*

*Proof:*     By the chain rule (see Proposition 1.2.3) we have

$$D^u(f(F)) = f'(F)D^u F.$$

Hence, by the duality relationship (1.42) we get

$$
\begin{aligned}
E(f'(F)G) &= E\left(D^u(f(F))(D^u F)^{-1}G\right) \\
&= E\left(\langle D(f(F)), u(D^u F)^{-1}G\rangle_H\right) \\
&= E(f(F)\delta(Gu(D^u F)^{-1})).
\end{aligned}
$$

This completes the proof.                                                    □

**Remarks:**

**1.** If the law of $F$ is absolutely continuous, we can assume that the function $f$ is Lipschitz.

**2.** Suppose that $u$ is deterministic. Then, for $Gu(D^u F)^{-1} \in \mathrm{Dom}\delta$ it suffices that $G(D^u F)^{-1} \in \mathbb{D}^{1,2}$. Sufficient conditions for this are given in Exercise 6.2.1.

**3.** Suppose we take $u = DF$. In this case

$$H(F,G) = \delta\left(\frac{GDF}{\|DF\|_H^2}\right),$$

and Proposition 6.2.1 yields

$$E(f'(F)G) = E\left(f(F)\delta\left(\frac{GDF}{\|DF\|_H^2}\right)\right). \tag{6.17}$$

A Greek is a derivative of a financial quantity, usually an option price, with respect to any of the parameters of the model. This derivative is useful to measure the stability of this quantity under variations of the parameter. Consider an option with payoff $H$ such that $E_Q(H^2) < \infty$. From (6.10) its price  at time $t = 0$ is given by

$$V_0 = E_Q(e^{-rT}H).$$

We are interested in computing the derivative of this expectation with respect to a parameter $\alpha$, $\alpha$ being one of the parameters of the problem, that is, $S_0$, $\sigma$, or $r$. Suppose that we can write $H = f(F_\alpha)$. Then

$$\frac{\partial V_0}{\partial \alpha} = e^{-rT}E_Q\left(f'(F_\alpha)\frac{dF_\alpha}{d\alpha}\right). \tag{6.18}$$

Using Proposition 6.2.1 we obtain

$$\frac{\partial V_0}{\partial \alpha} = e^{-rT}E_Q\left(f(F_\alpha)H\left(F_\alpha, \frac{dF_\alpha}{d\alpha}\right)\right). \tag{6.19}$$

In some cases the function $f$ is not smooth and formula (6.19) provides better result in combination with Montecarlo simultation that (6.18). We are going to discuss several examples of this type.

### 6.2.1 Computation of Greeks for European options

The most important Greek is the Delta, denoted by $\Delta$, which by definition is the derivative of $V_0$ with respect to the initial price of the stock $S_0$.

Suppose that the payoff $H$ only depends on the price of the stock at the maturity time $T$. That is, $H = \Phi(S_T)$. We call these derivative European options. From (6.13) it follows that $\Delta$ coincides with the composition in risky assets of the replicating portfolio.

If $\Phi$ is a Lipschitz function we can write

$$\Delta = \frac{\partial V_0}{\partial S_0} = E_Q(e^{-rT}\Phi'(S_T)\frac{\partial S_T}{\partial S_0}) = \frac{e^{-rT}}{S_0}E_Q(\Phi'(S_T)S_T).$$

Now we will apply Proposition 6.2.1 with $u = 1$, $F = S_T$ and $G = S_T$. We have

$$D^u S_T = \int_0^T D_t S_T dt = \sigma T\, S_T.$$

Hence, all the conditions appearing in Remark 2 above are satisfies in this case and we we have

$$\delta\left(S_T\left(\int_0^T D_t S_T dt\right)^{-1}\right) = \delta\left(\frac{1}{\sigma T}\right) = \frac{W_T}{\sigma T}.$$

As a consequence,

$$\Delta = \frac{e^{-rT}}{S_0\sigma T}E_Q(\Phi(S_T)W_T). \tag{6.20}$$

The Gamma, denoted by $\Gamma$, is the second derivative of the option price with respect to $S_0$. As before we obtain

$$\Gamma = \frac{\partial^2 V_0}{\partial S_0^2} = E_Q\left(e^{-rT}\Phi''(S_T)\left(\frac{\partial S_T}{\partial S_0}\right)^2\right) = \frac{e^{-rT}}{S_0^2}E_Q(\Phi''(S_T)S_T^2).$$

Assuming that $\Phi'$ is Lipschitz we obtain, taking $G = S_T^2$, $F = S_T$ and $u = 1$ and applying Proposition 6.2.1

$$\delta\left(S_T^2\left(\int_0^T D_t S_T dt\right)^{-1}\right) = \delta\left(\frac{S_T}{\sigma T}\right) = S_T\left(\frac{W_T}{\sigma T} - 1\right)$$

and, as  a consequence,

$$E_Q(\Phi''(S_T)S_T^2) = E_Q\left(\Phi'(S_T)S_T\left(\frac{W_T}{\sigma T} - 1\right)\right).$$

Finally, applying again Proposition 6.2.1 with $G = S_T \left( \frac{W_T}{\sigma T} - 1 \right)$, $F = S_T$ and $u = 1$ yields

$$\delta \left( S_T \left( \frac{W_T}{\sigma T} - 1 \right) \left( \int_0^T D_t S_T dt \right)^{-1} \right) = \delta \left( \frac{W_T}{\sigma^2 T^2} - \frac{1}{\sigma T} \right)$$
$$= \left( \frac{W_T^2}{\sigma^2 T^2} - \frac{1}{\sigma^2 T} - \frac{W_T}{\sigma T} \right)$$

and, as a consequence,

$$E_Q \left( \Phi'(S_T) S_T \left( \frac{W_T}{\sigma T} - 1 \right) \right) = E_Q \left( \Phi(S_T) \left( \frac{W_T^2}{\sigma^2 T^2} - \frac{1}{\sigma^2 T} - \frac{W_T}{\sigma T} \right) \right).$$

Therefore, we obtain

$$\Gamma = \frac{e^{-rT}}{S_0^2 \sigma T} E_Q \left( \Phi(S_T) \left( \frac{W_T^2}{\sigma T} - \frac{1}{\sigma} - W_T \right) \right). \tag{6.21}$$

The derivative with respect to the volatility is called Vega, and denoted by $\vartheta$:

$$\vartheta = \frac{\partial V_0}{\partial \sigma} = E_Q(e^{-rT} \Phi'(S_T) \frac{\partial S_T}{\partial \sigma}) = e^{-rT} E_Q(\Phi'(S_T) S_T (W_T - \sigma T)).$$

Applying Proposition 6.2.1 with $G = S_T W_T$, $F = S_T$ and $u = 1$ yields

$$\delta \left( S_T (W_T - \sigma T) \left( \int_0^T D_t S_T dt \right)^{-1} \right) = \delta \left( \frac{W_T}{\sigma T} - 1 \right)$$
$$= \left( \frac{W_T^2}{\sigma T} - \frac{1}{\sigma} - W_T \right).$$

As a consequence,

$$\vartheta = e^{-rT} E_Q \left( \Phi(S_T) \left( \frac{W_T^2}{\sigma T} - \frac{1}{\sigma} - W_T \right) \right). \tag{6.22}$$

By means of an approximation procedure these formulas still hold although the function $\Phi$ and its derivative are not Lipschitz. We just need $\Phi$ to be piecewise continuous with jump discontinuities and with linear growth. In particular, we can apply these formulas to the case of and European call-option ($\Phi(x) = (x - K)^+$), and European put-option ($\Phi(x) = (K - x)^+$), or a digital option ($\Phi(x) = \mathbf{1}_{\{x > K\}}$).

We can compute the values of the previous derivatives with a Monte Carlo numerical procedure. We refer to [110] and [169] for a discussion of the numerical simulations.

## 6.2.2  Computation of Greeks for exotic options

Consider options whose payoff is a function of the average of the stock price $\frac{1}{T}\int_0^T S_t dt$, that is

$$H = \Phi\left(\frac{1}{T}\int_0^T S_t dt\right).$$

For instance, an Asiatic call-option with exercise price $K$, is a derivative of this type, where $H = \left(\frac{1}{T}\int_0^T S_t dt - K\right)^+$. In this case there is no closed formula for the density of the random variable $\frac{1}{T}\int_0^T S_t dt$. From (6.10) the price of this option at time $t = 0$ is given by

$$V_0 = e^{-rT}E_Q\left(\Phi\left(\frac{1}{T}\int_0^T S_t dt\right)\right).$$

Let us compute the Delta for this type of options. Set $\overline{S}_T = \frac{1}{T}\int_0^T S_t dt$. We have

$$\Delta = \frac{\partial V_0}{\partial S_0} = E_Q(e^{-rT}\Phi'(\overline{S}_T)\frac{\partial \overline{S}_T}{\partial S_0}) = \frac{e^{-rT}}{S_0}E_Q(\Phi'(\overline{S}_T)\overline{S}_T).$$

We are going to apply Proposition 6.2.1 with $G = \overline{S}_T$, $F = \overline{S}_T$ and $u_t = S_t$. Let us compute

$$D_t F = \frac{1}{T}\int_0^T D_t S_r dr = \frac{\sigma}{T}\int_t^T S_r dr,$$

and

$$\delta\left(\frac{GS.}{\int_0^T S_t D_t F dt}\right) = \frac{2}{\sigma}\delta\left(\frac{S.}{\int_0^T S_t dt}\right)$$

$$= \frac{2}{\sigma}\left(\frac{\int_0^T S_t dW_t}{\int_0^T S_t dt} + \frac{\int_0^T S_t\left(\int_t^T \sigma S_r dr\right)dt}{\left(\int_0^T S_t dt\right)^2}\right)$$

$$= \frac{2}{\sigma}\frac{\int_0^T S_t dW_t}{\int_0^T S_t dt} + 1.$$

Notice that

$$\int_0^T S_t dW_t = \frac{1}{\sigma}\left(S_T - S_0 - r\int_0^T S_t dt\right).$$

Thus,

$$\delta\left(\frac{GS.}{\int_0^T S_t D_t F dt}\right) = \frac{2(S_T - S_0)}{\sigma^2\int_0^T S_t dt} + 1 - \frac{2r}{\sigma^2} = \frac{2}{\sigma^2}\left(\frac{S_T - S_0}{\int_0^T S_t dt} - m\right),$$

where $m = r - \frac{\sigma^2}{2}$. Finally, we obtain the following expression for the Delta:

$$\Delta = \frac{2e^{-rT}}{S_0\sigma^2} E_Q \left( \Phi\left(\overline{S}_T\right) \left( \frac{S_T - S_0}{T\overline{S}_T} - m \right) \right).$$

For other type of path dependent options it is not convenient to take $u = 1$ in the integration by parts formula. Consider the general case of an option depending on the prices at a finite number of times, that is,

$$H = \Phi(S_{t_1}, \ldots, S_{t_m}),$$

where $\Phi : \mathbb{R}^m \to \mathbb{R}$ is a continuously differentiable function with bounded partial derivatives and $0 < t_1 < t_2 < \cdots < t_m = T$. We introduce the set $\Gamma_m$ defined by

$$\Gamma_m = \left\{ a \in L^2([0,T]) : \int_0^{t_j} a_t dt = 1 \; \forall j = 1, \ldots, m \right\}.$$

Then we have

$$\Delta = E_Q \left( H \int_0^T a_t dW_t \right).$$

In fact, we have

$$\begin{aligned}
D^a H &= \sum_{j=1}^{m} \partial_j \Phi(S_{t_1}, \ldots, S_{t_m}) D^a S_{t_j} \\
&= \sigma \sum_{j=1}^{m} \partial_j \Phi(S_{t_1}, \ldots, S_{t_m}) S_{t_j}.
\end{aligned}$$

As a consequence,

$$\frac{\partial H}{\partial S_0} = \frac{1}{S_0} \sum_{j=1}^{m} \partial_j \Phi(S_{t_1}, \ldots, S_{t_m}) S_{t_j} = \frac{D^a H}{\sigma S_0}$$

and we obtain

$$\Delta = e^{-rT} E_Q \left( \frac{\partial H}{\partial S_0} \right) = \frac{e^{-rT}}{\sigma S_0} E_Q \left( D^a H \right) = \frac{e^{-rT}}{\sigma S_0} E_Q \left( H \int_0^T a_t dW_t \right).$$

We can take for instance $a = \frac{1}{t_1} \mathbf{1}_{[0,t_1]}$ and we get

$$\Delta = \frac{e^{-rT}}{\sigma S_0} E_Q \left( H \frac{W_{t_1}}{t_1} \right). \tag{6.23}$$

Formula (6.23) is not very useful for simulation due to the inestability of $\frac{W_{t_1}}{t_1}$ if $t_1$ is small. For this reason, specific alternative methods should be used to handle every case. For example, in [169] the authors consider an up in and down call option with payoff

$$H = \mathbf{1}_{\{\inf_{i=1,\ldots,m} S_{t_i} \leq D\}} \mathbf{1}_{\{\sup_{i=1,\ldots,m} S_{t_i} \geq U\}} \mathbf{1}_{\{S_T < K\}},$$

and apply an integration by parts formula using a dominating process defined as

$$Y_t = \sqrt{m \sum_{\substack{1 \leq i \leq m \\ t_i \leq t}} (S_{t_i} - S_{t_{i-1}})^2}.$$

It is proved in [169] that if $\Psi : [0, \infty) \to [0, 1]$ is a function in $C_b^\infty$ such that $\Psi(x) = 1$ if $x \leq a/2$ and $\Psi(x) = 0$ if $x > a$, where $U > S_0 + \frac{a}{2} > S_0 - \frac{a}{2} > D$, then,

$$\Delta = \frac{S_0 e^{-rT}}{\sigma} E_Q \left( H \, \delta \left( \frac{\Psi(Y_.)}{\int_0^T \Psi(Y_t) dt} \right) \right).$$

## Exercises

**6.2.1** Suppose that $G \in \mathbb{D}^{1,4}$, $F \in \mathbb{D}^{2,2}$ and $u$ is an $H$-valued random variable such that: $E(G^6) < \infty$, $E((D^u F)^{-12}) < \infty$, and $E(\|DD^u F\|_H^6) < \infty$. Show that $G(D^u F)^{-1} \in \mathbb{D}^{1,2}$.

**6.2.2** Using formulas (6.20), (6.21), and (6.22) compute the values of $\Delta$, $\Gamma$ and $\vartheta$ for an European call option with exercise price $K$ and compare the results with those obtained in Exercise 6.1.3.

**6.2.3** Compute $\Delta$, $\Gamma$ and $\vartheta$ for a digital option using formulas (6.20), (6.21), and (6.22).

## 6.3   Application of the Clark-Ocone formula in hedging

In this section we discuss the application of Clark-Ocone formula to find explicit formulas for a replicating portfolio in the Black-Scholes model.

### 6.3.1   A generalized Clark-Ocone formula

Suppose that

$$\widetilde{W}_t = W_t + \int_0^t \theta_s ds,$$

where $\theta = \{\theta_t, t \in [0, T]\}$ is an adapted and measurable process such that $\int_0^T \theta_t^2 dt < \infty$ almost surely. Suppose that $E(Z_T) = 1$, where the process

$Z_t$ is given by

$$Z_t = \exp\left(-\int_0^t \theta_s dW_s - \frac{1}{2}\int_0^t \theta_s^2 ds\right).$$

Then, by Girsanov Theorem (see Proposition 4.1.2), the process $\widetilde{W} = \{\widetilde{W}_t, t \in [0,T]\}$ is a Brownian motion under the probability $Q$ on $\mathcal{F}_T$ given by $\frac{dQ}{dP} = Z_T$.

The Clark-Ocone formula established in Proposition 1.3.14 can be generalized in oder to represent an $\mathcal{F}_T$-measurable random variable $F$ as a stochastic integral with respect to the process $\widetilde{W}$. Notice that, in general, we have $\mathcal{F}_T^{\widetilde{W}} \subset \mathcal{F}_T$ (where $\{\mathcal{F}_t^{\widetilde{W}}, 0 \le t \le T\}$ denotes the family of $\sigma$-fields generated by $\widetilde{W}$) and usually $\mathcal{F}_T^{\widetilde{W}} \ne \mathcal{F}_T$. Thus, an $\mathcal{F}_T$-measurable random variable $F$ may not be $\mathcal{F}_T^{\widetilde{W}}$-measurable and we cannot obtain a representation of $F$ as an integral with respect to $\widetilde{W}$ simply by applying the Clark-Ocone formula to the Brownian motion $\widetilde{W}$ on the probaiblity space $(\Omega, \mathcal{F}_T^{\widetilde{W}}, Q)$.

In order to establish a generalized Clark-Ocone formula we need the following technical lemma. Its proof is left as an exercise (Exercise 6.3.1).

**Lemma 6.3.1** *Let $F$ be an $\mathcal{F}_T$-measurable random variable such that $F \in \mathbb{D}^{1,2}$ and let $\theta \in \mathbb{L}^{1,2}$. Assume*

*(i)* $E(Z_T^2 F^2) + E(Z_T^2 \int_0^T (D_t F)^2 dt) < \infty,$

*(ii)* $E\left(Z_T^2 F^2 \int_0^T \left(\theta_t + \int_t^T D_t\theta_s dW_s + \int_t^T \theta_s D_t\theta_s ds\right)^2 dt\right) < \infty.$

*Then $Z_T F \in \mathbb{D}^{1,2}$ and*

$$D_t(Z_T F) = Z_T D_t F - Z_T F\left(\theta_t + \int_t^T D_t\theta_s dW_s + \int_t^T \theta_s D_t\theta_s ds\right).$$

**Theorem 6.3.1** *Let $F$ be an $\mathcal{F}_T$-measurable random variable such that $F \in \mathbb{D}^{1,2}$ and let $\theta \in \mathbb{L}^{1,2}$. Suppose that conditions (i) and (ii) of Lemma 6.3.1 hold. Then*

$$F = E_Q(F) + \int_0^T E_Q\left(D_t F - F\int_t^T D_t\theta_s d\widetilde{W}_s \Big| \mathcal{F}_t\right) d\widetilde{W}_t.$$

*Proof:*    Put $Y_t = E_Q(F|\mathcal{F}_t)$. Using Exercise 4.2.11 we can write

$$Y_t = Z_t^{-1}E(Z_T F|\mathcal{F}_t),$$

where

$$Z_t^{-1} = \exp\left(\int_0^t \theta_s dW_s + \frac{1}{2}\int_0^t \theta_s^2 ds\right).$$

Then, by Clark-Ocone formula, we have

$$E(Z_T F | \mathcal{F}_t) = E(Z_T F) + \int_0^t E(D_s(Z_T F)|\mathcal{F}_s) dW_s.$$

Hence,

$$Y_t = Z_t^{-1} E_Q(F) + Z_t^{-1} \int_0^t E(D_s(Z_T F)|\mathcal{F}_s) dW_s. \qquad (6.24)$$

From Lemma 6.3.1 we obtain

$$
\begin{aligned}
E(D_t(Z_T F)|\mathcal{F}_t) &= E\left( Z_T \left( D_t F - F \left( \theta_t + \int_t^T D_t \theta_s d\widetilde{W}_s \right) \right) | \mathcal{F}_t \right) \\
&= Z_t E_Q \left( D_t F - F \left( \theta_t + \int_t^T D_t \theta_s d\widetilde{W}_s \right) | \mathcal{F}_t \right) \\
&= Z_t \Psi_t - Z_t Y_t \theta_t, \qquad (6.25)
\end{aligned}
$$

where

$$\Psi_t = E_Q(D_t F - F \int_t^T D_t \theta_s d\widetilde{W}_s | \mathcal{F}_t).$$

Substituting (6.25) into (6.24) yields

$$Y_t = Z_t^{-1} E_Q(F) + Z_t^{-1} \int_0^t Z_s \Psi_s dW_s - Z_t^{-1} \int_0^t Z_s Y_s \theta_s dW_s.$$

Applying Itô's formula and using

$$d(Z_t^{-1}) = Z_t^{-1}(\theta_t dW_t + \theta_t^2 dt)$$

we get

$$
\begin{aligned}
dY_t &= Y_t(\theta_t dW_t + \theta_t^2 dt) + \Psi_t dW_t - Y_t \theta_t dW_t + \theta_t \Psi_t dt - Y_t \theta_t^2 dt \\
&= \Psi_t d\widetilde{W}_t.
\end{aligned}
$$

This completes the proof.     □

## 6.3.2 Application to finance

Let $H$ be the payoff of a derivative in the Black-Scholes model (6.2). Suppose that $E(B_T^{-2} Z_T^2 H^2) < \infty$. We have seen that $H$ is replicable, as a consequence of the integral representation theorem. Furthermore, if $\phi_t = (\alpha_t, \beta_t)$ is a replicating portfolio, we have seen that

$$\widetilde{V}_t(\phi) = E_Q(B_T^{-1} H) + \int_0^t \sigma_s \widetilde{S}_s \beta_s d\widetilde{W}_s.$$

Suppose now that $B_T^{-1}H \in \mathbb{D}^{1,2}$, $\theta \in \mathbb{L}^{1,2}$ and the conditions (i) and (ii) of Lemma 6.3.1 are satisfied by $F = B_T^{-1}H$ and $\theta$. Then, we conclude that

$$\sigma_t \widetilde{S}_t \beta_t = E_Q\left(D_t\left(B_T^{-1}H\right) - B_T^{-1}H\int_t^T D_t\theta_s d\widetilde{W}_s | \mathcal{F}_t\right).$$

Hence,

$$\beta_t = \frac{B_t}{\sigma_t \widetilde{S}_t}E_Q\left(D_t\left(B_T^{-1}H\right) - B_T^{-1}H\int_t^T D_t\theta_s d\widetilde{W}_s | \mathcal{F}_t\right).$$

If the interest rate $r_t = r$ is constant this formula reduces to

$$\beta_t = \frac{e^{-r(T-t)}}{\sigma_t S_t}E_Q\left(D_t H - H\int_t^T D_t\theta_s d\widetilde{W}_s | \mathcal{F}_t\right).$$

In the particular case $\mu_t = \mu$, $\sigma_t = \sigma$, we obtain $D_t\theta_s = 0$, and

$$\beta_t = \frac{e^{-r(T-t)}}{\sigma S_t}E_Q\left(D_t H | \mathcal{F}_t\right).$$

In that case, the only hypothesis required is $H \in \mathbb{D}^{1,2,\widetilde{W}}$. In fact, it suffices to apply the ordinary Clark-Ocone formula for the Brownian motion $\widetilde{W}$ and use that $\beta_t = \frac{e^{rt}u_t}{\sigma S_t}$, where

$$e^{-rT}H = E_Q\left(e^{-rT}H\right) + \int_0^T u_t d\widetilde{W}_t.$$

Consider the particular case of an European option with payoff $H = \Phi(S_T)$. Then

$$\begin{aligned}
\beta_t &= \frac{e^{-r(T-t)}}{\sigma S_t}E_Q\left(\Phi'(S_T)\sigma S_T | \mathcal{F}_t\right) \\
&= e^{-r(T-t)}E_Q\left(\Phi'(\frac{S_T}{S_t}S_t)\frac{S_T}{S_t} | \mathcal{F}_t\right) \\
&= e^{-r(T-t)}E_Q\left(\Phi'(xS_{T-t})S_{T-t}\right)|_{x=S_t}.
\end{aligned}$$

In this way we recover the fact that $\beta_t$ coincides with $\frac{\partial F}{\partial x}(t, S_t)$, where $F(t,x)$ is the price function.

Consider now an option whose payoff is a function of the average of the stock price $\overline{S}_T = \frac{1}{T}\int_0^T S_t dt$, that is $H = \Phi\left(\overline{S}_T\right)$. In this case we obtain

$$\beta_t = \frac{e^{T-t}}{S_t}E_Q\left(\Phi'(\overline{S}_T)\frac{1}{T}\int_t^T S_r dr | \mathcal{F}_t\right).$$

We can write

$$\overline{S}_T = \frac{t}{T}\overline{S}_t + \frac{1}{T}\int_t^T S_r dr,$$

where $\overline{S}_t = \frac{1}{t}\int_0^t S_r dr$. As a consequence we obtain

$$\beta_t = \frac{e^{-r(T-t)}}{S_t}E_Q\left(\Phi'\left(\frac{tx}{T} + \frac{y(T-t)}{T}\overline{S}_{T-t}\right)\left(\frac{y(T-t)}{T}\overline{S}_{T-t}\right)\right)\Big|_{x=\overline{S}_t, y=S_t}.$$

## Exercises

**6.3.1** Prove Lemma 6.3.1.

**6.3.2** Let $\widetilde{W}_t = W_t + \int_0^t \theta_s ds$, where $\theta = \{\theta_t, t \in [0, T]\}$ is an adapted and measurable process. Use the generalized Clark-Ocone formula to find the integral representation in terms of the Wiener process $\widetilde{W}$ of the following random variables:

(i) $F = W_T^2$, and $\theta$ is bounded and belongs to $\mathbb{D}^{1,p}$ for some $p > 2$,
(ii) $F = \exp(aW_T)$, and $\theta$ is bounded and belongs to $\mathbb{D}^{1,p}$ for some $p > 2$,
(iii) $F = \exp(aW_T)$ and $\theta_t = W_t$.

**6.3.3** Consider the Black-Scholes model with constant volatility, mean rate of return and interest rate. Consider the particular case of an European option with payoff $H = \exp(aW_T)$. Find a replicating portfolio using the Clark-Ocone formula.

## 6.4 Insider trading

Suppose that in a financial market there are two types of agents: a *natural agent* whose information coincides with the natural filtration of the price process, and an *insider* who possesses some extra information from the beginning of the trading interval $[0, T]$. The simpler modelization of this additional information consists in assume the knowledge of a given random variable $L$. Two important questions in this context are:

i) How can one calculate the additional utility of the insider?

ii) Does the insider have arbitrage opportunities?

Consider the Black-Scholes model for the price process $S_t$ with a measurable and adapted mean rate of return $\mu_t$ satisfying $E\left(\int_0^T |\mu_t| dt\right) < \infty$ and a measurable and adapted volatility process satisfying $E\left(\int_0^T \sigma_t^2 dt < \infty\right)$ and $\sigma_t > 0$ a.s. The price process is then given by

$$dS_t = S_t(\mu_t dt + \sigma_t dW_t).$$

As usual, we denote, $\{\mathcal{F}_t, t \geq 0\}$ the filtration generated by the Wiener process $W$ and the $P$-null sets. The price of the bond is given by

$$B_t = \exp\left(\int_0^t r_s ds\right),$$

where $r_t$ is a non negative measurable and adapted process satisfying

$$E\left(\int_0^T r_t dt\right) < \infty.$$

A natural insider will use a self-financing portfolio $\phi_t = (\alpha_t, \beta_t)$ where the processes $\alpha_t$ and $\beta_t$ are measurable and $\mathcal{F}_t$-adapted, satisfying

$$\int_0^T |\beta_t \mu_t| dt < \infty, \quad \int_0^T |\alpha_t| r_t dt < \infty, \quad \int_0^T \beta_t^2 \sigma_t^2 dt < \infty \qquad (6.26)$$

almost surely. That is, the value process

$$V_t(\phi) = \alpha_t B_t + \beta_t S_t$$

satisfies the self-financing condition

$$dV_t(\phi) = r_t \alpha_t B_t dt + \beta_t dS_t. \qquad (6.27)$$

We restrict ourselves to strictly admissible portfolios, that is to portfolios $\phi$ satisfying $V_t(\phi) > 0$ for all $t \in [0, T]$. The quantity $\pi_t = \frac{\beta_t S_t}{V_t(\phi)}$ is the proportion of the wealth invested in stocks, and it determines the portfolio. In terms of $\pi_t$ the value process (denoted by $V_t$) satisfies the following linear stochastic differential equation

$$dV_t = (\rho_t + (\mu_t - r_t)\pi_t)V_t dt + \pi_t \sigma_t V_t dW_t.$$

The solution of this equation is

$$V_t = V_0 B_t \exp\left(\int_0^t \left[(\mu_s - r_s)\pi_s - \frac{(\pi_s \sigma_s)^2}{2}\right] ds + \int_0^t \pi_s \sigma_s dW_s\right). \qquad (6.28)$$

One of the possible objectives of an investor is to maximize the expected utility from terminal wealth, by means of choosing an appropriate portfolio. We will consider the logarithmic utility function $u(x) = \log x$ to measure the utility that the trader draws from his wealth at the maturity time $T$. Then, the problem is to find a portfolio $\pi_t$ which maximizes the expected utility:

$$\Phi_{\mathcal{F}} = \max_{\pi} E\left(\log V_T\right), \qquad (6.29)$$

where

$$E\left(\log V_T\right) = \log V_0 + E\left(\int_0^T r_s ds\right)$$

$$+ E\left(\int_0^T \left[(\mu_s - r_s)\pi_s - \frac{(\pi_s \sigma_s)^2}{2}\right] ds + \int_0^T \pi_s \sigma_s dW_s\right).$$

Due to the local martingale property of the stochastic integral $\int_0^t \pi_s \sigma_s dW_s$, the utility maximization probllem reduces to find the maximum of

$$E\left(\int_0^T \left[(\mu_s - r_s)\pi_s - \frac{(\pi_s \sigma_s)^2}{2}\right] ds\right).$$

We can write this expression as

$$-\frac{1}{2} E\left(\int_0^T \left(\pi_s \sigma_s - \frac{\mu_s - r_s}{\sigma_s}\right)^2 ds\right) + E\left(\int_0^T \frac{(\mu_s - r_s)^2}{2\sigma_s^2} ds\right),$$

and the solution to the maximization problem (6.29) is (Merton's formula)

$$\pi_s = \frac{\mu_s - r_s}{\sigma_s^2}, \tag{6.30}$$

and

$$\Phi_{\mathcal{F}} = E\left(\int_0^T \frac{(\mu_s - r_s)^2}{2\sigma_s^2} ds\right).$$

Consider now the problem of an insider trader. Assume that the investor is allowed to use a portfolio $\phi_t = (\alpha_t, \beta_t)$ which is measurable and $\mathcal{G}_t$-adapted, where $\{\mathcal{G}_t, t \in [0,T]\}$ is a filtration which is modelling the information of the investor. In principle no assumption will be made on this filtration. In the particular case where the additional information is given by a random variable $L$ which is $\mathcal{F}_T$-measurable (or more generally, $\mathcal{F}_{T+\varepsilon}$-measurable for some $\varepsilon > 0$) we have

$$\mathcal{G}_t = \mathcal{F}_t \vee \sigma(L).$$

We also assume condition (6.26). Now the self-financing condition (6.27), by definition, will be

$$V_t(\phi) = V_0 + \int_0^t r_s \alpha_s B_s ds + \int_0^t \beta_s \mu_s S_s ds + \int_0^t \beta_s \sigma_s S_s d^- W_s, \tag{6.31}$$

where $\int_0^t \beta_s S_s \sigma_s d^- W_s$ denotes the *forward stochastic integral* introduced in Definition 3.2.1, and the process $\beta_t \sigma_t S_t$ satisfies the assumptions of

Proposition 3.2.3. That is, the mapping $t \to \beta_t \sigma_t S_t$ is continuous from $[0, T]$ into $\mathbb{D}^{1,2}$, and $\beta \sigma S \in \mathbb{L}^{1,2}_{1-}$.

Assume $V_t(\phi) > 0$ for all $t \in [0, T]$ and set $\pi_t = \frac{\beta_t S_t}{V_t(\phi)}$ as before. Then, from (6.31) we obtain

$$dV_t = r_t V_t + (\mu_t - r_t) \pi_t V_t dt + \pi_t \sigma_t V_t d^- W_t. \tag{6.32}$$

Under some technical conditions on the processes $\pi_t$, $\sigma_t$ and $\mu_t$ (see Exercise 6.4.1), the process $V_t$ defined by

$$V_t = V_0 e^{\int_0^T r_s ds} \exp\left( \int_0^t \left[ (\mu_s - r_s) \pi_s - \frac{(\pi_s \sigma_s)^2}{2} \right] ds + \int_0^t \pi_s \sigma_s d^- W_s \right) \tag{6.33}$$

satisfies the linear stochastic differential equation (6.32).

Let us denote by $\mathcal{A}_{\mathcal{G}}$ the class of portfolios $\pi = \{\pi_t, t \in [0, T]\}$ satisfying the following conditions:

(i) $\pi$ is $\mathcal{G}_t$-adapted.

(ii) The processes $\pi$ and $\sigma$ are continuous in the norm of $\mathbb{D}^{1,4}$ and $\pi \in \mathbb{L}^{1,2}_{2-}$.

(iii) $E\left( \int_0^T |(\mu_s - \rho_s) \pi_s| ds \right) < \infty$.

We need the following technical lemma.

**Lemma 6.4.1** *Let $\pi$ and $\sigma$ be measurable process which are continuous in the norm of $\mathbb{D}^{1,4}$ and $\pi \in \mathbb{L}^{1,2}_{2-}$. Suppose that $\sigma$ is $\mathcal{F}_t$-adapted. Then the forward integral $\int_0^T \pi_s \sigma_s d^- W_s$ exists and*

$$\int_0^T \pi_s \sigma_s d^- W_s = \int_0^T \pi_s \sigma_s \, dW_s - \int_0^T \left( D^- \pi \right)_s \sigma_s \, ds.$$

*Moreover, $\int_0^T \pi_s \sigma_s d^- W_s$ equals to the following limit in $L^1(\Omega)$*

$$\int_0^T \pi_s \sigma_s d^- W_s = \lim_{|\pi| \downarrow 0} \sum_{i=0}^{n-1} \pi_{t_i} \left( \int_{t_i}^{t_{i+1}} \sigma_s dW_s \right).$$

*Proof:* Notice first that $\pi \sigma \in \mathbb{L}^{1,2}_{2-}$ and $(D^-(\pi\sigma))_t = (D^- \pi)_t \sigma_t$. This follows from the fact that for any $s > t$ we have

$$D_s(\pi_t \sigma_t) = \sigma_t D_s \pi_t.$$

The proof of the lemma can be done using the same ideas as in the proof of Proposition 3.2.3 (Exercise 6.4.2). $\qquad \square$

We aim to solve the optimization problem

$$\Phi_{\mathcal{G}} = \max_{\pi \in \mathcal{A}_{\mathcal{G}}} E\left(\log V_T^{\pi}\right). \tag{6.34}$$

The following theorem provides a characterization of optimal portfolios.

**Theorem 6.4.1** *The following conditions are equivalent for a portfolio* $\pi^* \in \mathcal{A}_{\mathcal{G}}$:

(i) *The portfolio* $\pi^*$ *is optimal for problem (6.34).*

(ii) *The function* $s \to E\left(\int_0^s \sigma_r dW_r | \mathcal{G}_t\right)$, *is absolutely continuous in* $[t, T]$ *for any* $t \in [0, T)$ *and there exists*

$$E\left(\mu_s - \rho_s - \sigma_s^2 \pi_s^* | \mathcal{G}_t\right) = -\frac{d}{ds} E\left(\int_0^s \sigma_r dW_r | \mathcal{G}_t\right) \tag{6.35}$$

*for almost all* $s \in [t, T]$.

*Proof:*    Set

$$
\begin{aligned}
J(\pi) &= E\left(\log V_T^{\pi}\right) - V_0 - E\left(\int_0^T r_s ds\right) \\
&= E\left(\int_0^T \left[(\mu_s - r_s)\pi_s - \frac{(\pi_s \sigma_s)^2}{2}\right] ds\right) \\
&\quad + E\left(\int_0^T \pi_s \sigma_s d^- W_s\right). \tag{6.36}
\end{aligned}
$$

Assume first that $\pi^* \in \mathcal{A}_{\mathcal{G}}$ is optimal. We have $J(\pi^*) \geq J(\pi^* + \varepsilon\beta)$ for any $\beta \in \mathcal{A}_{\mathcal{G}}$ and $\varepsilon \in \mathbb{R}$. Therefore

$$\frac{d}{d\varepsilon} J(\pi^* + \varepsilon\beta)|_{\varepsilon=0} = 0.$$

Hence,

$$E\left(\int_0^T \left(\mu_s - r_s - \sigma_s^2 \pi_s^*\right)\beta_s ds + \int_0^T \beta_s \sigma_s d^- W_s\right) = 0, \tag{6.37}$$

for all $\beta \in \mathcal{A}_{\mathcal{G}}$. In particular, applying this to $\beta_u = G\mathbf{1}_{(s,t]}(u)$, where $0 \leq r < t \leq T$ and $G$ is $\mathcal{G}_t$-measurable and bounded, we obtain

$$E\left(\int_r^t \left(\mu_s - \rho_s - \sigma_s^2 \pi_s^*\right) ds + \int_r^t \sigma_s dW_s | \mathcal{G}_r\right) = 0, \tag{6.38}$$

which implies (ii). Conversely, integrating (6.35) we obtain (6.38). For any $\beta \in \mathcal{A}_{\mathcal{G}}$ we have

$$
\begin{aligned}
E\left(\int_0^T \beta_s d^- X_s\right) &= \lim_{|\pi| \downarrow 0} \sum_{i=0}^{n-1} E\left(\beta_{t_i} \left(\int_{t_i}^{t_{i+1}} \sigma_s dW_s\right)\right) \\
&= \lim_{|\pi| \downarrow 0} \sum_{i=0}^{n-1} E\left(\beta_{t_i} E\left(\int_{t_i}^{t_{i+1}} \sigma_s dW_s | \mathcal{G}_{t_i}\right)\right) \\
&= \lim_{|\pi| \downarrow 0} \sum_{i=0}^{n-1} E\left(\beta_{t_i} E\left(\int_{t_i}^{t_{i+1}} \left(\mu_s - r_s - \sigma_s^2 \pi_s^*\right) ds | \mathcal{G}_{t_i}\right)\right) \\
&= \lim_{|\pi| \downarrow 0} \sum_{i=0}^{n-1} E\left(\beta_{t_i} \int_{t_i}^{t_{i+1}} \left(\mu_s - r_s - \sigma_s^2 \pi_s^*\right) ds\right) \\
&= E\left(\int_0^T \left(\mu_s - r_s - \sigma_s^2 \pi_s^*\right) \beta_s ds\right)
\end{aligned}
$$

and (6.37) holds. This implies that the directional derivative of $J$ at $\pi^*$ with respect to the direction $\beta$ denoted by $D_\beta J(\pi^*)$ is zero. Note that $J : \mathcal{A}_{\mathcal{G}} \to \mathbb{R}$ is concave. Therefore, for all $\alpha, \beta \in \mathcal{A}_{\mathcal{G}}$ and $\varepsilon \in (0, 1)$, we have

$$
\begin{aligned}
J(\alpha + \varepsilon\beta) - J(\alpha) &= J\left((1 - \varepsilon)\frac{\alpha}{1 - \varepsilon} + \varepsilon\beta\right) - J(\alpha) \\
&\geq (1 - \varepsilon)J\left(\frac{\alpha}{1 - \varepsilon}\right) + \varepsilon J(\beta) - J(\alpha) \\
&= J\left(\frac{\alpha}{1 - \varepsilon}\right) - J(\alpha) + \varepsilon\left(J(\beta) - J\left(\frac{\alpha}{1 - \varepsilon}\right)\right).
\end{aligned}
$$

Now, with $\frac{1}{1-\varepsilon} = 1 + \eta$ we have

$$
\lim_{\varepsilon \to 0} \frac{1}{\varepsilon}\left(J\left(\frac{\alpha}{1 - \varepsilon}\right) - J(\alpha)\right) = \lim_{\eta \to 0} \frac{1 + \eta}{\eta}\left(J(\alpha + \eta\alpha) - J(\alpha)\right) = D_\alpha J(\alpha),
$$

and we obtain

$$
D_\beta J(\alpha) = \lim_{\varepsilon \to 0} \frac{1}{\varepsilon}\left(J(\alpha + \varepsilon\beta) - J(\alpha)\right) \geq D_\alpha J(\alpha) + J(\beta) - J(\alpha).
$$

In particular, applying this to $\alpha = \pi^*$ and using that $D_\beta J(\pi^*) = 0$ we get

$$
J(\beta) - J(\pi^*) \leq 0
$$

for all $\beta \in \mathcal{A}_{\mathcal{G}}$ and this proves that $\pi^*$ is optimal.     □

The characterization theorem provides a closed formula for the optimal portfolio $\pi^*$. In fact, we have

$$
\pi_t^* E(\sigma_t^2 | \mathcal{G}_t) = E(\mu_t - r_t | \mathcal{G}_t) + a(t), \tag{6.39}
$$

where
$$a(t) = \lim_{h \downarrow 0} \frac{1}{h} E\left(\int_t^{t+h} \sigma_r dW_r | \mathcal{G}_t\right).$$

Note that the optimal portfolio has a similar form as the solution of the Merton problem (6.30).

We compute now the value of the optimal portfolio when it exists. From (6.39) we have
$$\pi_t^* = \frac{E(\nu_t | \mathcal{G}_t) + a(t)}{E(\sigma_t^2 | \mathcal{G}_t)}, \tag{6.40}$$

where $\nu_t = \mu_t - r_t$. Substituting (6.40) into (6.36) yields

$$
\begin{aligned}
J(\pi^*) \;=\; & E\left(\int_0^T \left[\nu_s \frac{E(\nu_s | \mathcal{G}_s) + a(s)}{E(\sigma_s^2 | \mathcal{G}_s)} - \frac{\sigma_s^2}{2} \left(\frac{E(\nu_s | \mathcal{G}_s) + a(s)}{E(\sigma_s^2 | \mathcal{G}_s)}\right)^2\right] ds\right) \\
& + E\left(\int_0^T \frac{E(\nu_s | \mathcal{G}_s) + a(s)}{E(\sigma_s^2 | \mathcal{G}_s)} \sigma_s d^- W_s\right).
\end{aligned}
$$

Now, using the properties of the conditional expectation and applying Proposition 3.2.3 we get

$$
\begin{aligned}
J(\pi^*) \;=\; & \frac{1}{2} E\left(\int_0^T \left[\frac{(E(\mu_s - r_s | \mathcal{G}_s))^2}{E(\sigma_s^2 | \mathcal{G}_s)} - \frac{a(s)^2}{E(\sigma_s^2 | \mathcal{G}_s)}\right] ds\right) \\
& + E\left(\int_0^T \left(D^- \frac{E(\mu_{\cdot} - r_{\cdot} | \mathcal{G}_{\cdot}) + a(\cdot)}{E(\sigma_{\cdot}^2 | \mathcal{G}_{\cdot})} \sigma_{\cdot}\right)_s ds\right). \tag{6.41}
\end{aligned}
$$

**Example 1** *(Partial observation case).* Assume $\mathcal{G}_t \subset \mathcal{F}_t$. Then,

$$\frac{d}{ds} E\left(\int_0^s \sigma_r dW_r | \mathcal{G}_t\right) = 0$$

for all $s > t$. That is $a(t) = 0$, and the optimal portfolio is given by

$$\pi_t^* = \frac{E(\mu_t - r_t | \mathcal{G}_t)}{E(\sigma_t^2 | \mathcal{G}_t)}$$

if the right-hand side is well defined as an element of $\mathcal{A}_\mathcal{G}$. In this case, the optimal utility is

$$J(\pi^*) = \frac{1}{2} E\left(\int_0^T \left[\frac{(E(\mu_s - r_s | \mathcal{G}_s))^2}{E(\sigma_s^2 | \mathcal{G}_s)}\right] ds\right).$$

**Example 2** *(Insider strategy).* Suppose $\mathcal{F}_t \subset \mathcal{G}_t$. Then

$$\pi_t^* = \frac{\mu_t - r_t}{\sigma_t^2} + \frac{a(t)}{\sigma_t^2}$$

and

$$J(\pi^*) = \frac{1}{2} E \left( \int_0^T \left[ \frac{(\mu_s - r_s)^2}{\sigma_s^2} - \frac{a(s)^2}{\sigma_s^2} \right] ds \right) + E \left( \int_0^T \frac{(D^- a)_s}{\sigma_s} ds \right).$$

(6.42)

Consider the particular case where the $\sigma$-field $\mathcal{G}_t$ is generated by an $\mathcal{F}_T$-measurable random variable $L$. We can apply the approach of the enlargement of filtrations and deduce that $W$ is a semimartingale with respect to the filtration $\mathcal{G}_t$. Suppose that there exists a $\mathcal{G}_t$-progresively measurable process $\mu_t^{\mathcal{G}}$ such that $\int_0^T |\mu_t^{\mathcal{G}}| dt < \infty$ almost surely and

$$\widetilde{W}_t = W_t - \int_0^t \mu_s^{\mathcal{G}} ds$$

is a $\mathcal{G}_t$-Brownian motion. Then, for any $\mathcal{G}_t$-progresively measurable portfolio $\pi$ such that $E \left( \int_0^T \pi_s^2 \sigma_s^2 ds \right) < \infty$ we can write

$$E \left( \int_0^T \pi_s \sigma_s d^- W_s \right) = E \left( \int_0^T \pi_s \sigma_s \mu_s^{\mathcal{G}} ds \right)$$

and, as a consequence,

$$J(\pi) = E \left( \int_0^T \left[ \left( \mu_s - r_s + \sigma_s \mu_s^{\mathcal{G}} \right) \pi_s - \frac{(\pi_s \sigma_s)^2}{2} \right] ds \right)$$

$$= -\frac{1}{2} \int_0^T \left( \frac{\mu_s - r_s}{\sigma_s} + \mu_s^{\mathcal{G}} - \sigma_s \pi_s \right)^2 ds + \frac{1}{2} \int_0^T \left( \frac{\mu_s - r_s}{\sigma_s} + \mu_s^{\mathcal{G}} \right)^2 ds.$$

Thus, the optimal portfolio is given by

$$\pi_t^* = \frac{\mu_s - r_s}{\sigma_s^2} + \frac{\mu_s^{\mathcal{G}}}{\sigma_s}.$$

On the other hand, the additional expected logarithmic utility will be

$$\frac{1}{2} E \left( \int_0^T \left( \frac{\mu_s - r_s}{\sigma_s} + \mu_s^{\mathcal{G}} \right)^2 ds \right) - \frac{1}{2} E \left( \int_0^T \left( \frac{\mu_s - r_s}{\sigma_s} \right)^2 ds \right)$$

$$= \frac{1}{2} E \left( \int_0^T \left( \mu_s^{\mathcal{G}} \right)^2 ds \right).$$

because

$$E \left( \int_0^T \left( \frac{\mu_s - r_s}{\sigma_s} \right) \mu_s^{\mathcal{G}} ds \right) = E \left( \int_0^T \left( \frac{\mu_s - r_s}{\sigma_s} \right) \left( dW_s - d\widetilde{W}_s \right) \right) = 0,$$

provided we assume $E\left(\int_0^T \left(\frac{\mu_s - r_s}{\sigma_s}\right)^2 ds\right) < \infty$. Notice that

$$
\begin{aligned}
a(t) &= \lim_{h\downarrow 0} \frac{1}{h} E\left(\int_t^{t+h} \sigma_s d\widetilde{W}_s | \mathcal{G}_t\right) + \lim_{h\downarrow 0} \frac{1}{h} E\left(\int_t^{t+h} \sigma_s \mu_s^{\mathcal{G}} ds | \mathcal{G}_t\right) \\
&= \sigma_t \mu_t^{\mathcal{G}}.
\end{aligned}
$$

**Example 6.4.1** *Suppose that* $L = W_S$, *where* $S \geq T$. *Then*

$$
\mu_t^{\mathcal{G}} = \frac{W_S - W_t}{S - t},
$$

*and the additional expected utillity is infinite because*

$$
E\left(\int_0^T \left(\frac{W_S - W_t}{S - t}\right)^2 dt\right) = \int_0^T \frac{dt}{S - t} = \log \frac{S}{S - T},
$$

*which is finite if and only if* $S > T$.

## Exercises

**6.4.1** Suppose that $\pi\sigma \in \mathbb{L}^{2,4}_{-loc}$, $V_0 \in \mathbb{D}^{1,2}_{loc}$, $(\mu_t - \rho_t)\pi_t - \frac{(\pi_t \sigma_t)^2}{2} \in \mathbb{L}^{1,2}_{loc}$ and the value process $V_t$ has continuous trajectories. Then, appy the Itô's formula (3.36) in order to deduce (6.32).

**6.4.2** Complete the proof of Lemma 6.4.1.

**6.4.3** Compute the additional expected logarithmic utility when $L = \mathbf{1}_{\{S_{t_0} \geq K\}}$, where $t_0 \geq T$.

### Notes and comments

**[6.1]**    The Black-Sholes formula for option pricing was derived by Black and Scholes in a paper published in 1973 [44]. Their results were influenced by the research developed by Samuelson and Merton. There are many reference books on mathematical finance where the techniques of stochastic calculus are applied to derive the basic formulas for pricing and hedging of derivatives (see Karatzas [160], Karatzas and Shreve [165], Lamberton and Lapeyre [189], and Shreve [312]).

The recent monograph by Malliavin and Thalmaier [216] discusses a variety of applications of Malliavin calculus in mathematical finance.

**[6.2]**    We refer to [169] as a basic expository paper on the applications of the integration by parts formula of Malliavin calculus to Monte Carlo simulations of greeks. In [110] and [111] the Malliavin calculus is applied to derive formulas for the derivative of the expectation of a diffusion process with respect to the parameters of the equation. The results on the case of an option depending on the price at a finite number have been taken from [122] and [31].

**[6.3]**    The generalized Clark-Ocone formula and its applications in hedging in the multidimensional case has been studied by Karatzas and Ocone in [161].

**[6.4]**    A pioneering paper in the study of the additional utility for insider traders is the work by Karatzas and Pikovsky [163]. They assume that the extra information is hidden in a random variable $L$ from the beginning of the trading interval, and they make use of the technique of enlargement of filtrations. For another work on this topic using the technique of enlargement of filtrations we refer to [12]. The existence of opportunities of arbitrage for the insider has been investigated in [127] and [151]. For a review of the role of Malliavin calculus in the problem of insider trading we refer to the work by Imkeller [149]. The approach of anticipating stochastic calculus in the problem of insider trading has been developed in [196].

# A

# Appendix

## A.1  A Gaussian formula

Let $\{a_n, n \geq 1\}$ be a sequence of real numbers such that $\sum_{n=1}^{\infty} a_n^2 < \infty$. Suppose that $\{\xi_n, n \geq 1\}$ is a sequence of independent random variables defined in the probability space $(\Omega, \mathcal{F}, P)$ with distribution $N(0, 1)$. Then, for each $p > 0$

$$E\left(\left|\sum_{n=1}^{\infty} a_n \xi_n(t)\right|^p\right) = A_p \left(\sum_{n=1}^{\infty} a_n^2\right)^{\frac{p}{2}}, \tag{A.1}$$

where

$$A_p = \int_{\mathbb{R}} \frac{|x|^p}{\sqrt{2\pi}} e^{-\frac{x^2}{2}} dx.$$

## A.2  Martingale inequalities

Let $\{M_t, t \in [0, T]\}$ be a continuous local martingale with respect to an increasing family of $\sigma$-fields $\{\mathcal{F}_t, t \geq 0\}$. The following inequalities play a fundamental role in the stochastic calculus:

*Doob's maximal inequality*

For any $p > 1$ we have

$$E \left( \sup_{0 \le t \le T} |M_t|^p \right) \le \left( \frac{p}{p-1} \right)^p E(|M_T|^p). \tag{A.2}$$

Actually (A.2) holds if we replace $|M_t|$ by any continuous nonnegative submartingale.

*Burkholder-Davis-Gundy inequality*

For any $p > 0$ there exist constants $c_1(p)$ and $c_2(p)$ such that

$$c_1(p)E \left( \langle M \rangle_T^{\frac{p}{2}} \right) \le E \left( \sup_{0 \le t \le T} |M_t|^p \right) \le c_2(p)E \left( \langle M \rangle_T^{\frac{p}{2}} \right). \tag{A.3}$$

This inequality still holds if we replace $T$ by a stopping time $S : \Omega \to [0, T]$.

On the other hand, applying the Gaussian formula (A.1) and (A.3) one can show Burkholder's inequality for Hilbert-valued martingales. That is, if $\{M_t, t \in [0, T]\}$ is a continuous local martingale with values in a separable Hilbert space $K$, then for any $p > 0$ one has

$$E \left( \|M_T\|^p \right) \le c_p E \left( \langle M \rangle_T^{\frac{p}{2}} \right), \tag{A.4}$$

where

$$\langle M \rangle_T = \sum_{i=1}^{\infty} \langle \langle M, e_i \rangle_K \rangle_T,$$

$\{e_i, i \ge 1\}$ being a complete orthonormal system in $K$.

*Exponential inequality*

For any $\delta > 0$ and $\rho > 0$ we have

$$P \left\{ \langle M \rangle_T < \rho, \ \sup_{0 \le t \le T} |M_t| \ge \delta \right\} \le 2 \exp \left( -\frac{\delta^2}{2\rho} \right). \tag{A.5}$$

*Two-parameter martingale inequalities*

Let $W = \{W(z), z \in \mathbb{R}_+^2\}$ be a two-parameter Wiener process. Consider the $\sigma$-fields

$$
\begin{aligned}
\mathcal{F}_z &= \sigma\{W(z'), z' \le z\}, \\
\mathcal{F}_s^1 &= \sigma\{W(s', t'), 0 \le s' \le s, t' \ge 0\}, \\
\mathcal{F}_t^2 &= \sigma\{W(s', t'), 0 \le t' \le t, s' \ge 0\}.
\end{aligned}
$$

We will say that a random field $\xi = \{\xi(z), z \in \mathbb{R}_+^2\}$ is adapted if $\xi(z)$ is $\mathcal{F}_z$-measurable for all $z$. Also, we will say that $\xi$ is 1-adapted (resp. 2-adapted) if $\xi(s,t)$ is $\mathcal{F}_s^1$-measurable (resp. $\mathcal{F}_t^2$-measurable). Let $\xi = \{\xi(z), z \in \mathbb{R}_+^2\}$ be a measurable and adapted process such that

$$E \left( \int_0^S \int_0^T \xi^2(z) dz \right) < \infty$$

for all $(S,T)$. Define, for $s \leq S$, $t \leq T$,

$$M(s,t) = \int_0^s \int_0^t \xi(z) W(dz). \tag{A.6}$$

Applying Doob's inequality (A.2) (see [55]) twice, we obtain for any $p > 1$

$$E \left( \sup_{\substack{0 \leq s \leq S \\ 0 \leq t \leq T}} |M(s,t)|^p \right) \leq \left( \frac{p}{p-1} \right)^{2p} E \left( \left| \int_0^S \int_0^T \xi(z) W(dz) \right|^p \right). \tag{A.7}$$

Moreover, if $\xi$ is 1-adapted (resp. 2-adapted), the stochastic integral (A.6) can also be defined, and it is a continuous martingale with respect to the coordinate $s$ (resp. $t$), whose quadratic variation is given by

$$\langle M \rangle(s,t) = \int_0^s \int_0^t \xi(z)^2 dz.$$

As a consequence, (A.3) and (A.7) yield, for any $p > 1$,

$$c_1(p) E \left( \left| \int_0^s \int_0^t \xi(z)^2 dz \right|^{\frac{p}{2}} \right) \leq E(| \int_0^s \int_0^t \xi(z) W(dz)|^p)$$

$$\leq c_2(p) E \left( \left| \int_0^s \int_0^t \xi(z)^2 dz \right|^{\frac{p}{2}} \right). \tag{A.8}$$

## A.3   Continuity criteria

The real analysis lemma due to Garsia, Rodemich, and Rumsey's [114] is a powerful tool to deduce results on the modulus of continuity of stochastic processes from estimates of the moments of their increments. The following version of this lemma has been taken from Stroock and Varadhan [322, p. 60].

**Lemma A.3.1** *Let $p, \Psi : \mathbb{R}_+ \to \mathbb{R}_+$ be continuous and strictly increasing functions vanishing at zero and such that $\lim_{t \uparrow \infty} \Psi(t) = \infty$. Suppose that*

$\phi : \mathbb{R}^d \to E$ *is a continuous function with values on a separable Banach space* $(E, \| \cdot \|)$. *Denote by* $B$ *the open ball in* $\mathbb{R}^d$ *centered at* $x_0$ *with radius* $r$. *Then, provided*

$$\Gamma = \int_B \int_B \Psi \left( \frac{\|\phi(t) - \phi(s)\|}{p(|t-s|)} \right) ds\, dt < \infty,$$

*it holds, for all* $s, t \in B$,

$$\|\phi(t) - \phi(s)\| \leq 8 \int_0^{2|t-s|} \Psi^{-1} \left( \frac{4^{d+1}\Gamma}{\lambda_d u^{2d}} \right) p(du),$$

*where* $\lambda_d$ *is a universal constant depending only on* $d$.

Now suppose that $X = \{X(t), t \in \mathbb{R}^d\}$ is a continuous stochastic process with values on a separable Banach space $(E, \| \cdot \|)$ such that the following estimate holds:

$$E(\|X(t) - X(s)\|^\gamma) \leq H|t-s|^\alpha \tag{A.9}$$

for some $H > 0$, $\gamma > 0$, $\alpha > d$, and for all $s, t \in B$.

Then taking $\Psi(x) = x^\gamma$ and $p(x) = x^{\frac{m+2d}{\gamma}}$, with $0 < m < \alpha - d$, one obtains

$$\|X(t) - X(s)\|^\gamma \leq C_{d,\gamma,m}|t-s|^m \Gamma, \tag{A.10}$$

for all $s, t \in B$, where $\Gamma = \int_B \int_B \frac{\|X(t)-X(s)\|^\gamma}{|t-s|^{m+2d}} ds\, d$. Moreover, if $E(\|X(t_0)\|^\gamma) < \infty$ for some $t_0 \in E$, then we can conclude that

$$E(\sup_{|t| \leq a} \|X(t)\|^\gamma) < \infty \tag{A.11}$$

for any $a > 0$.

If $X$ is not supposed to be continuous, one can show by an approximation argument that (A.9) implies the existence of a continuous version of $X$ satisfying (A.10) (see [315]). This proves the classical Kolmogorov continuity criterion. Actually, (A.10) follows from $E(\Gamma) < \infty$ if we assume $\gamma \geq 1$, as it can be proved by an approximation of the identify argument.

## A.4   Carleman-Fredholm determinant

Let $(T, \mathcal{B}, \mu)$ be a measure space, and let $K \in L^2(T \times T)$. Assume that the Hilbert space $H = L^2(T)$ is separable and let $\{e_i, i \geq 1\}$ be a complete orthonormal system in $H$. In the particular case $K = \sum_{i,j=1}^{n} a_{ij} e_i \otimes e_j$ the Carleman-Fredholm determinant of $I + K$ is defined as

$$\det_2(I + K) = \det(I + A)\exp(-\mathrm{T}\, A),$$

where $A = (a_{ij})_{1 \le i,j \le n}$. It can be proved that the mapping $K \to \det_2(I + K)$ is continuous in $L^2(T \times T)$. As a consequence it can be extended to the whole space $L^2(T \times T)$. If the operator on $L^2(T)$ associated with the kernel $K$ (denoted again by $K$) is nuclear, then one can write

$$\det_2(I + K) = \det(I + K) \exp(-\mathrm{T}\,K).$$

A useful formula to compute the Carleman-Fredholm determinant $\det_2(I + K)$, where $K \in L^2(T \times T)$, is the following:

$$\det_2(I + K) = 1 + \sum_{n=2}^{\infty} \frac{\gamma_n}{n!}, \tag{A.12}$$

where

$$\gamma_n = \int_{T^n} \det(\hat{K}(t_i, t_j)) \mu(dt_1) \dots \mu(dt_n).$$

Here $\hat{K}(t_i, t_j) = K(t_i, t_j)$ if $i \ne j$ and $\hat{K}(t_i, t_i) = 0$.

We refer to [316] and [314] for a more detailed analysis of the properties of this determinant.

## A.5   Fractional integrals and derivatives

We recall the basic definitions and properties of the fractional calculus. For a detailed presentation of these notions we refer to [300].

Let $a, b \in \mathbb{R}$, $a < b$. Let $f \in L^1(a, b)$ and $\alpha > 0$. The left and right-sided *fractional integrals* of $f$ of order $\alpha$ are defined for almost all $x \in (a, b)$ by

$$I_{a+}^{\alpha} f(x) = \frac{1}{\Gamma(\alpha)} \int_a^x (x - y)^{\alpha - 1} f(y)\, dy \tag{A.13}$$

and

$$I_{b-}^{\alpha} f(x) = \frac{1}{\Gamma(\alpha)} \int_x^b (y - x)^{\alpha - 1} f(y)\, dy, \tag{A.14}$$

respectively. Let $I_{a+}^{\alpha}(L^p)$ (resp. $I_{b-}^{\alpha}(L^p)$) the image of $L^p(a, b)$ by the operator $I_{a+}^{\alpha}$ (resp. $I_{b-}^{\alpha}$).

If $f \in I_{a+}^{\alpha}(L^p)$ (resp. $f \in I_{b-}^{\alpha}(L^p)$) and $0 < \alpha < 1$ then the left and right-sided *fractional derivatives* are defined by

$$D_{a+}^{\alpha} f(x) = \frac{1}{\Gamma(1 - \alpha)} \left( \frac{f(x)}{(x - a)^{\alpha}} + \alpha \int_a^x \frac{f(x) - f(y)}{(x - y)^{\alpha + 1}} dy \right), \tag{A.15}$$

and

$$D_{b-}^{\alpha} f(x) = \frac{1}{\Gamma(1 - \alpha)} \left( \frac{f(x)}{(b - x)^{\alpha}} + \alpha \int_x^b \frac{f(x) - f(y)}{(y - x)^{\alpha + 1}} dy \right) \tag{A.16}$$

for almost all $x \in (a, b)$ (the convergence of the integrals at the singularity $y = x$ holds point-wise for almost all $x \in (a, b)$ if $p = 1$ and moreover in $L^p$-sense if $1 < p < \infty$).

Recall the following properties of these operators:

- If $\alpha < \dfrac{1}{p}$ and $q = \dfrac{p}{1 - \alpha p}$ then

$$I_{a+}^{\alpha} (L^p) = I_{b-}^{\alpha} (L^p) \subset L^q (a, b) .$$

- If $\alpha > \dfrac{1}{p}$ then

$$I_{a+}^{\alpha} (L^p) \cup I_{b-}^{\alpha} (L^p) \subset C^{\alpha - \frac{1}{p}} (a, b) ,$$

where $C^{\alpha - \frac{1}{p}} (a, b)$ denotes the space of $\left( \alpha - \dfrac{1}{p} \right)$-Hölder continuous functions of order $\alpha - \dfrac{1}{p}$ in the interval $[a, b]$.

The following inversion formulas hold:

$$I_{a+}^{\alpha} \left( D_{a+}^{\alpha} f \right) = f$$

for all $f \in I_{a+}^{\alpha} (L^p)$, and

$$D_{a+}^{\alpha} \left( I_{a+}^{\alpha} f \right) = f$$

for all $f \in L^1 (a, b)$. Similar inversion formulas hold for the operators $I_{b-}^{\alpha}$ and $D_{b-}^{\alpha}$.

The following *integration by parts formula* holds:

$$\int_a^b \left( D_{a+}^{\alpha} f \right) (s) g(s) ds = \int_a^b f(s) \left( D_{b-}^{\alpha} g \right) (s) ds, \tag{A.17}$$

for any $f \in I_{a+}^{\alpha} (L^p)$, $g \in I_{b-}^{\alpha} (L^q)$, $\dfrac{1}{p} + \dfrac{1}{q} = 1$.

# References

[1] R. A. Adams: *Sobolev Spaces*, Academic Press, 1975.

[2] H. Airault: Differential calculus on finite codimensional submanifolds of the Wiener space. *J. Functional Anal.* **100** (1991) 291–316.

[3] H. Airault and P. Malliavin: Intégration géométrique sur l'espace de Wiener. *Bull. Sciences Math.* **112** (1988) 3–52.

[4] H. Airault and Van Biesen: Le processus d'Ornstein Uhlenbeck sur une sous–variété de l'espace de Wiener. *Bull. Sciences Math.* **115** (1991) 185–210.

[5] A. Alabert, M. Ferrante, and D. Nualart: Markov field property of stochastic differential equations. *Ann. Probab.* **23** (1995) 1262–1288.

[6] A. Alabert and D. Nualart: Some remarks on the conditional independence and the Markov property. In: *Stochastic Analysis and Related Topics*, eds.: H. Korezlioglu and A. S. Üstünel, Birkhäuser, 1992, 343–364.

[7] E. Alòs, J. A. León, and D. Nualart: Stratonovich stochastic calculus with respect to fractional Brownian motion with Hurst parameter less than 1/2. *Taiwanesse Journal of Mathematics* **5** (2001), 609–632.

[8] E. Alòs, O. Mazet, and D. Nualart: Stochastic calculus with respect to fractional Brownian motion with Hurst parameter lesser than $\frac{1}{2}$. *Stoch. Proc. Appl.* **86** (1999) *121–139*.

[9] E. Alòs, O. Mazet, and D. Nualart: Stochastic calculus with respect to Gaussian processes. *Annals of Probability* **29** (2001) 766–801.

[10] E. Alòs and D. Nualart: An extension of Itô's formula for anticipating processes. *J. Theoret. Probab.* **11** (1998) 493–514.

[11] E. Alòs and D. Nualart: Stochastic integration with respect to the fractional Brownian motion. *Stoch. Stoch. Rep.* **75** (2003) 129–152.

[12] J. Amendinger, P. Imkeller, and M. Schweizer: Additional logarithmic utility of an insider. *Stochastic Process. Appl.* **75** (1998) 263–286.

[13] N. Aronszajn: Theory of reproducing kernels. *Trans. Amer. Math. Soc.* **68** (1950) 337–404.

[14] J. Asch and J. Potthoff: Itô's lemma without nonanticipatory conditions. *Probab. Theory Rel. Fields* **88** (1991) 17–46.

[15] D. Bakry: L'hypercontractivité et son utilisation en théorie des semi-groupes. In: *Ecole d'Eté de Probabilités de Saint Flour XXII-1992*, Lecture Notes in Math. **1581** (1994) 1–114.

[16] V. V. Balkan: Integration of random functions with respect to a Wiener random measure. *Theory Probab. Math. Statist.* **29** (1984) 13–17.

[17] V. Bally: On the connection between the Malliavin covariance matrix and Hörmander's condition. *J. Functional Anal.* **96** (1991) 219–255.

[18] V. Bally, I. Gyöngy, and E. Pardoux: White noise driven parabolic SPDEs with measurable drift. *J. Functional Anal.* **120** (1994) 484–510.

[19] V. Bally and E. Pardoux: Malliavin calculus for white noise driven parabolic SPDEs. *Potential Anal.* **9** (1998) 27–64.

[20] V. Bally and B. Saussereau: A relative compactness criterion in Wiener-Sobolev spaces and application to semi-linear stochastic PDEs. *J. Functional Anal.* **210** (2004), 465–515.

[21] R. F. Bass and M. Cranston: The Malliavin calculus for pure jump processes and applications to local time. *Ann. Probab.* **14** (1986) 490–532.

[22] D. R. Bell: *The Malliavin Calculus*, Pitman Monographs and Surveys in Pure and Applied Math. 34, Longman and Wiley, 1987.

[23] D. R. Bell and S. E. A. Mohammed: The Malliavin calculus and stochastic delay equations. *J. Functional Anal.* **99** (1991) 75–99.

[24] D. R. Bell and S. E. A. Mohammed: Hypoelliptic parabolic operators with exponential degeneracies. *C. R. Acad. Sci. Paris* **317** (1993) 1059–1064.

[25] G. Ben Arous and R. Léandre: Annulation plate du noyau de la chaleur. *C. R. Acad. Sci. Paris* **312** (1991) 463–464.

[26] G. Ben Arous and R. Léandre: Décroissance exponentielle du noyau de la chaleur sur la diagonale I. *Probab. Theory Rel. Fields* **90** (1991) 175–202.

[27] G. Ben Arous and R. Léandre: Décroissance exponentielle du noyau de la chaleur sur la diagonale II. *Probab. Theory Rel. Fields* **90** (1991) 377–402.

[28] C. Bender: An Itô formula for generalized functionals of a fractional Brownian motion with arbitrary Hurst parameter. *Stochastic Processes Appl.* **104** (2003) 81–106.

[29] M. A. Berger: A Malliavin–type anticipative stochastic calculus. *Ann. Probab.* **16** (1988) 231–245.

[30] M. A. Berger and V. J. Mizel: An extension of the stochastic integral. *Ann. Probab.* **10** (1982) 435–450.

[31] H. P. Bermin, A. Kohatsu-Higa and M. Montero: Local vega index and variance reduction methods. *Math. Finance* **13** (2003) 85–97.

[32] F. Biagini, B. Øksendal, A. Sulem, and N. Wallner: An introduction to white-noise theory and Malliavin calculus for fractional Brownian motion. *Proc. R. Soc. Lond. Ser. A Math. Phys. Eng. Sci.* **460** (2004) 347–372.

[33] K. Bichteler and D. Fonken: A simple version of the Malliavin calculus in dimension one. In: *Martingale Theory on Harmonic Analysis and Banach Spaces*, Lecture Notes in Math. **939** (1981) 6–12.

[34] K. Bichteler and D. Fonken: A simple version of the Malliavin calculus in dimension $N$. In: *Seminar on Stochastic Processes* 1982, eds.: Cinlar et al., Birkhäuser, 1983, 97–110.

[35] K. Bichteler, J. B. Gravereaux, and J. Jacod: *Malliavin Calculus for Processes with Jumps*, Stochastic Monographs Vol. 2, Gordon and Breach Publ., 1987.

[36] K. Bichteler and J. Jacod: Calcul de Malliavin pour les diffusions avec sauts; existence d'une densité dans le cas unidimensionnel. In: *Seminaire de Probabilités XVII*, Lecture Notes in Math. **986** (1983) 132–157.

[37] J. M. Bismut: *Mécanique aléatoire*, Lecture Notes in Math. **866**, 1981.

[38] J. M. Bismut: Martingales, the Malliavin Calculus and hypoellipticity under general Hörmander's condition. *Z. für Wahrscheinlichkeitstheorie verw. Gebiete* **56** (1981) 469–505.

[39] J. M. Bismut: Calcul des variations stochastiques et processus de sauts. *Z. für Wahrscheinlichkeitstheorie verw. Gebiete* **63** (1983) 147–235.

[40] J. M. Bismut: The calculus of boundary processes. *Ann. Sci. Ecole Norm. Sup.* **17** (1984) 507–622.

[41] J. M. Bismut: *Large deviations and the Malliavin calculus*, Progress in Math. 45, Birkhäuser, 1984.

[42] J. M. Bismut: The Atiyah-Singer theorems: A probabilistic approach. I. The index theorem, II. The Lefschetz fixed point formulas. *J. Functional Anal.* **57** (1984) 56–99; 329–348.

[43] J. M. Bismut and D. Michel: Diffusions conditionnelles. I. Hypoellipticité partielle, II. Générateur conditionnel. Application au filtrage. *J. Functional Anal.* Part I **44** (1981) 174–211, Part II **45** (1982) 274–292.

[44] F. Black and M. Scholes: The Pricing of Options and Corporate Liabilities. *Journal of Political Economy* **81** (1973) 637–654.

[45] V. I. Bogachev and O. G. Smolyanov: Analytic properties of infinite-dimensional distributions. *Russian Math. Surveys* **45** (1990) 1–104.

[46] N. Bouleau and F. Hirsch: Propriétés d'absolue continuité dans les espaces de Dirichlet et applications aux équations différentielles stochastiques. In: *Seminaire de Probabilités XX*, Lecture Notes in Math. **1204** (1986) 131–161.

[47] N. Bouleau and F. Hirsch: *Dirichlet Forms and Analysis on Wiener Space*, de Gruyter Studies in Math. 14, Walter de Gruyter, 1991.

[48] R. Buckdahn: Anticipative Girsanov transformations. *Probab. Theory Rel. Fields* **89** (1991) 211–238.

[49] R. Buckdahn: Linear Skorohod stochastic differential equations. *Probab. Theory Rel. Fields* **90** (1991) 223–240.

[50] R. Buckdahn: Anticipative Girsanov transformations and Skorohod stochastic differential equations. Seminarbericht Nr. 92–2 (1992).

[51] R. Buckdahn: Skorohod stochastic differential equations of diffusion type. *Probab. Theory Rel. Fields* **92** (1993) 297–324.

[52] R. Buckdahn and H. Föllmer: A conditional approach to the anticipating Girsanov transformation. *Probab. Theory Rel. Fields* **95** (1993) 31–330.

[53] R. Buckdahn and D. Nualart: Linear stochastic differential equations and Wick products. *Probab. Theory Rel. Fields.* **99** (1994) 501–526.

[54] R. Buckdahn and E. Pardoux: Monotonicity methods for white noise driven SPDEs. In: *Diffusion Processes and Related Problems in Analysis*, Vol. I, ed. : M. Pinsky, Birkhäuser, 1990, 219–233.

[55] R. Cairoli and J. B. Walsh: Stochastic integrals in the plane. *Acta Mathematica* **134** (1975) 111–183.

[56] R. H. Cameron and W. T. Martin: Transformation of Wiener integrals under a general class of linear transformations. *Trans. Amer. Math. Soc.* **58** (1945) 148–219.

[57] R. H. Cameron and W. T. Martin: Transformation of Wiener integrals by nonlinear transformations. *Trans. Amer. Math. Soc.* **66** (1949) 253–283.

[58] Ph. Carmona, L. Coutin, and G. Montseny: Stochastic integration with respect to fractional Brownian motion. *Ann. Inst. H. Poincaré* **39** (2003) 27–68.

[59] R. Carmona and D. Nualart: Random nonlinear wave equations: Smoothness of the solution. *Probab. Theory Rel. Fields* **79** (1988) 469—-580.

[60] P. Cattiaux: Hypoellipticité et hypoellipticité partielle pour les diffusions avec une condition frontière. *Ann. Inst. Henri Poincaré* **22** (1986) 67–112.

[61] M. Chaleyat–Maurel and D. Michel: Hypoellipticity theorems and conditional laws. *Z. für Wahrscheinlichkeitstheorie verw. Gebiete* **65** (1984) 573–597.

[62] M. Chaleyat–Maurel and D. Nualart: The Onsager–Machlup functional for a class of anticipating processes. *Probab. Theory Rel. Fields* **94** (1992) 247–270.

[63] M. Chaleyat-Maurel and D. Nualart: Points of positive density for smooth functionals. *Electron. J. Probab.* **3** (1998) 1–8.

[64] M. Chaleyat-Maurel and M. Sanz-Solé: Positivity of the density for the stochastic wave equation in two spatial dimensions. *ESAIM Probab. Stat.* **7** (2003) 89–114.

[65] P. Cheridito: Mixed fractional Brownian motion. *Bernoulli* **7** (2001) 913–934.

[66] P. Cheridito and D. Nualart: Stochastic integral of divergence type with respect to fractional Brownian motion with Hurst parameter $H \in (0, 1/2)$. *Ann. Inst. Henri Poincaré* **41** (2005) 1049–1081.

[67] A. Chorin: *Vorticity and Turbulence*, Springer-Verlag, 1994.

[68] J. M. C. Clark: The representation of functionals of Brownian motion by stochastic integrals. *Ann. Math. Statist.* **41** (1970) 1282–1295; **42** (1971) 1778.

[69] L. Coutin, D. Nualart, and C. A. Tudor: Tanaka formula for the fractional Brownian motion. *Stochastic Processes Appl.* **94** (2001) 301–315.

[70] L. Coutin and Z. Qian: Stochastic analysis, rough paths analysis and fractional Brownian motions. *Probab. Theory Rel. Fields* **122** (2002) 108–140.

[71] A. B. Cruzeiro: Equations différentielles sur l'espace de Wiener et formules de Cameron–Martin nonlinéaires. *J. Functional Anal.* **54** (1983) 206–227.

[72] A. B. Cruzeiro: Unicité de solutions d'équations différentielles sur l'espace de Wiener. *J. Functional Anal.* **58** (1984) 335–347.

[73] W. Dai and C. C. Heyde: Itô's formula with respect to fractional Brownian motion and its application. *Journal of Appl. Math. and Stoch. An.* **9** (1996) 439–448.

[74] R. C. Dalang: Extending the martingale measure stochastic integral with applications to spatially homogeneous s.p.d.e.'s. *Electron. J. Probab.* **4** (1999) 1–29.

[75] R. C. Dalang and N. E. Frangos: The stochastic wave equation in two spatial dimensions. *Ann. Probab.* **26** (1998) 187–212.

[76] R. C. Dalang and E. Nualart: Potential theory for hyperbolic SPDEs. *Ann. Probab.* **32** (2004) 2099–2148.

[77] Yu. A. Davydov: A remark on the absolute continuity of distributions of Gaussian functionals. *Theory Probab. Applications* **33** (1988) 158–161.

[78] L. Decreusefond and A. S. Üstünel: Stochastic analysis of the fractional Brownian motion. *Potential Analysis* **10** (1998) 177–214.

[79] C. Dellacherie and P. A. Meyer: *Probabilités et Potentiel. Théorie des Martingales*, Hermann, Paris, 1980.

[80] C. Donati-Martin: Equations différentielles stochastiques dans ℝ avec conditions au bord. *Stochastics and Stochastics Reports* **35** (1991) 143–173.

[81] C. Donati-Martin: Quasi-linear elliptic stochastic partial differential equation. Markov property. *Stochastics and Stochastics Reports* **41** (1992) 219–240.

[82] C. Donati-Martin and D. Nualart: Markov property for elliptic stochastic partial differential equations. *Stochastics and Stochastics Reports* **46** (1994) 107–115.

[83] C. Donati-Martin and E. Pardoux: White noise driven SPDEs with reflection. *Probab. Theory Rel. Fields* **95** (1993) 1–24.

[84] H. Doss: Liens entre équations différentielles stochastiques et ordinaires. *Ann. Inst. Henri Poincaré* **13** (1977) 99–125.

[85] B. Driver: A Cameron–Martin type quasi–invariance theorem for Brownian motion on a compact Riemannian manifold. *Journal Functional Analysis* **110** (1992) 272–376.

[86] T. E. Duncan, Y. Hu, and B. Pasik-Duncan: Stochastic calculus for fractional Brownian motion I. Theory. *SIAM J. Control Optim.* **38** (2000) 582–612.

[87] N. Dunford, J. T. Schwartz: *Linear Operators, Part II*, Interscience Publishers, 1963.

[88] M. Eddahbi, R. Lacayo, J. L. Solé, C. A. Tudor, and J. Vives: Regularity of the local time for the *d*-dimensional fractional Brownian motion with *N*-parameters. *Stoch. Anal. Appl.* **23** (2005) 383–400.

[89] K. D. Elworthy: *Stochastic Differential Equations on Manifolds*, Cambridge Univ. Press, 1982.

[90] K. D. Elworthy: Stochastic flows in Riemannian manifolds. In: *Diffusion Problems and Related Problems in Analysis*, vol. II, eds.: M. A. Pinsky and V. Vihstutz, Birkhäuser, 1992, 37–72.

[91] O. Enchev: Nonlinear transformations on the Wiener space. *Ann. Probab.* **21** (1993) 2169–2188.

[92] O. Enchev and D. W. Stroock: Rademacher's theorem for Wiener functionals. *Ann. Probab.* **21** (1993) 25–33.

[93] O. Enchev and D. W. Stroock: Anticipative diffusions and related change of measures. *J. Functional Anal.* **116** (1993) 449–477.

[94] C. O. Ewald and A. Zhang: A new technique for calibrating stochastic volatility models: The Malliarin gradient method. Preprint

[95] S. Fang: Une inégalité isopérimétrique sur l'espace de Wiener. *Bull. Sciences Math.* **112** (1988) 345–355.

[96] H. Federer: *Geometric Measure Theory*, Springer-Verlag, 1969.

[97] M. Ferrante: Triangular stochastic differential equations with boundary conditions. *Rend. Sem. Mat. Univ. Padova* **90** (1993) 159–188.

[98] M. Ferrante and D. Nualart: Markov field property for stochastic differential equations with boundary conditions. *Stochastics and Stochastics Reports* **55** (1995) 55–69.

[99] M. Ferrante, C. Rovira, and M. Sanz-Solé: Stochastic delay equations with hereditary drift: estimates of the density. *J. Funct. Anal.* **177** (2000) 138–177.

[100] D. Feyel and A. de La Pradelle: Espaces de Sobolev Gaussiens. *Ann. Institut Fourier* **39** (1989) 875–908.

[101] D. Feyel and A. de La Pradelle: Capacités Gaussiennes. *Ann. Institut Fourier* **41** (1991) 49–76.

[102] F. Flandoli: On a probabilistic description of small scale structures in 3D fluids. *Ann. Inst. Henri Poincaré* **38** (2002) 207–228.

[103] F. Flandoli and M. Gubinelli: The Gibbs ensemble of a vortex filament. *Probab. Theory Relat. Fields* **122** (2001) 317–340.

[104] P. Florchinger: Malliavin calculus with time-dependent coefficients and application to nonlinear filtering. *Probab. Theory Rel. Fields* **86** (1990) 203–223.

[105] P. Florchinger and R. Léandre: Décroissance non exponentielle du noyau de la chaleur. *Probab. Theory Rel. Fields* **95** (1993) 237–262.

[106] P. Florchinger and R. Léandre: Estimation de la densité d'une diffusion trés dégénérée. Etude d'un exemple. *J. Math. Kyoto Univ.* **33–1** (1993) 115–142.

[107] C. Florit and D. Nualart: A local criterion for smoothness of densities and application to the supremum of the Brownian sheet. *Statistics and Probability Letters* **22** (1995) 25–31.

[108] H. Föllmer: Calcul d'Itô sans probabilités. *Lecture Notes in Math.* **850** (1981) 143–150.

[109] H. Föllmer: Time reversal on Wiener space. In: *Stochastic Processes-Mathematics and Physics, Proc. Bielefeld, 1984*, Lecture Notes in Math. **1158** (1986) 119–129.

[110] E. Fournié, J. M. Lasry, J. Lebuchoux, P. L. Lions, and N. Touzi: Applications of Malliavin calculus to Monte Carlo methods in finance. *Finance Stoch.* **3** (1999) 391–412.

[111] E. Fournié, J. M. Lasry, J. Lebuchoux, and P. L. Lions: Applications of Malliavin calculus to Monte-Carlo methods in finance. II. *Finance Stoch.* **5** (2001) 201–236.

[112] A. Friedman: *Stochastic Differential Equations and Applications*, Vol. 1, Academic Press, 1975.

[113] M. Fukushima: *Dirichlet Forms and Markov Processes*, North-Holland, 1980.

[114] A. Garsia, E. Rodemich, and H. Rumsey: A real variable lemma and the continuity of paths of some Gaussian processes. *Indiana Univ. Math. Journal* **20** (1970/71) 565–578.

[115] B. Gaveau and J. M. Moulinier: Intégrales oscillantes stochastiques: Estimation asymptotique de fonctionnelles caractéristiques. *J. Functional Anal.* **54** (1983) 161–176.

[116] B. Gaveau and P. Trauber: L'intégrale stochastique comme opérateur de divergence dans l'espace fonctionnel. *J. Functional Anal.* **46** (1982) 230–238.

[117] D. Geman and J. Horowitz: Occupation densities. *Annals of Probability* **8** (1980) *1–67*.

[118] E. Getzler: Degree theory for Wiener maps. *J. Functional Anal.* **68** (1986) 388–403.

[119] I. I. Gihman and A. V. Skorohod: *Stochastic Differential Equations*, Springer-Verlag, 1972.

[120] D. Gilbarg and N. S. Trudinger: *Elliptic Partial Differential Equations of Second Order*, Springer-Verlag, 1977.

[121] I. V. Girsanov: On transformations of one class of random processes with the help of absolutely continuous substitution of the measure. *Theory Probab. Appl.* **3** (1960) 314–330.

[122] E. Gobet and A. Kohatsu-Higa: Computation of Greeks for barrier and look-back optionsusing Malliavin calculus. *Electron. Comm. Probab.* **8** (2003) 51–62.

[123] M. Gradinaru, I. Nourdin, F. Russo, and P. Vallois: m-order integrals and generalized Itô's formula: the case of a fractional Brownian motion with any Hurst index. *Ann. Inst. Henri Poincaré* **41** (2005) 781–806.

[124] M. Gradinaru, F. Russo, and P. Vallois: Generalized covariations, local time and Stratonovich Itô's formula for fractional Brownian motion with Hurst index $H \geq 1/4$. *Ann. Probab* **31** (2003) 1772–1820.

[125] A. Grorud, D. Nualart, and M. Sanz: Hilbert-valued anticipating stochastic differential equations. *Ann. Inst. Henri Poincaré* **30** (1994) 133–161.

[126] A. Grorud and E. Pardoux: Intégrales hilbertiennes anticipantes par rapport à un processus de Wiener cylindrique et calcul stochastique associé. *Applied Math. Optimization* **25** (1992) 31–49.

[127] A. Grorud and M. Pontier: Asymmetrical information and incomplete markets. Information modeling in finance. *Int. J. Theor. Appl. Finance* **4** (2001) 285–302.

[128] L. Gross: Abstract Wiener spaces. In: *Proc. Fifth Berkeley Symp. Math. Stat. Prob. II*, Part 1, Univ. California Press, Berkeley, 1965, 31–41.

[129] R. F. Gundy: *Some topics in probability and analysis*, American Math. Soc., CBMS **70**, Providence, 1989.

[130] I. Gyöngy: On non-degenerate quasi-linear stochastic partial differential equations. *Potential Analysis* **4** (1995) 157–171.

[131] I. Gyöngy and E. Pardoux: On quasi-linear stochastic partial differential equations. *Probab. Theory Rel. Fields* **94** (1993) 413–425.

[132] M. Hairer, J. Mattingly, and E. Pardoux: Malliavin calculus for highly degenerate 2D stochastic Navier-Stokes equations. *C.R. Math. Acad. Sci. Paris* **339** (2004) 793–796.

[133] U. Haussmann: On the integral representation of functionals of Itô processes. *Stochastics* **3** (1979) 17–28.

[134] F. Hirsch: Propriété d'absolue continuité pour les équations différentielles stochastiques dépendant du passé. *J. Functional Anal.* **76** (1988) 193–216.

[135] M. Hitsuda: Formula for Brownian partial derivatives. In: Second Japan-USSR *Symp. Probab. Th.* **2** (1972) 111–114.

[136] M. Hitsuda: Formula for Brownian partial derivatives. *Publ. Fac. of Integrated Arts and Sciences Hiroshima Univ.* **3** (1979) 1–15.

[137] R. Holley and D. W. Stroock: Diffusions on an infinite dimensional torus. *J. Functional Anal.* **42** (1981) 29–63.

[138] L. Hörmander: Hypoelliptic second order differential equations. *Acta Math.* **119** (1967) 147–171.

[139] Y. Hu: *Integral transformations and anticipative calculus for fractional Brownian motion*, American Mathematical Society, 2005.

[140] Y. Hu and B. Øksendal: Fractional white noise calculus and applications to finance. *Infin. Dimens. Anal. Quantum Probab. Relat. Top.* **6** (2003) 1–32.

[141] H. E. Hurst: Long-term storage capacity in reservoirs. *Trans. Amer. Soc. Civil Eng.* **116** (1951) 400–410.

[142] N. Ikeda: Probabilistic methods in the study of asymptotics. In: *Ecole d'Eté de Probabilités de Saint Flour XVIII*, Lecture Notes in Math. **1427** (1990) 197–325.

[143] N. Ikeda and I. Shigekawa: The Malliavin calculus and long time asymptotics of certain Wiener integrals. *Proc. Center for Math. Anal. Australian Nat. Univ.* **9**, Canberra (1985).

[144] N. Ikeda and S. Watanabe: An introduction to Malliavin's calculus. In: *Stochastic Analysis, Proc. Taniguchi Inter. Symp. on Stoch. Analysis*, Katata and Kyoto 1982, ed. : K. Itô, Kinokuniya/North-Holland, Tokyo, 1984, 1–52.

[145] N. Ikeda and S. Watanabe: Malliavin calculus of Wiener functionals and its applications. In: *From Local Times to Global Geometry, Control, and Physics*, ed.: K. D. Elworthy, Pitman Research, Notes in Math. Ser. 150, Longman Scientific and Technical, Horlow, 1987, 132–178.

[146] N. Ikeda and S. Watanabe: *Stochastic Differential Equations and Diffusion Processes*, second edition, North-Holland, 1989.

[147] P. Imkeller: Occupation densities for stochastic integral processes in the second chaos. *Probab. Theory Rel. Fields* **91** (1992) 1–24.

[148] P. Imkeller: Regularity of Skorohod integral processes based on integrands in a finite Wiener chaos. *Probab. Theory Rel. Fields* **98** (1994) 137–142.

[149] P. Imkeller: Malliavin's calculus in insider models: additional utility and free lunches. *Math. Finance* **13** (2003) 153–169.

[150] P. Imkeller and D. Nualart: Integration by parts on Wiener space and the existence of occupation densities. *Ann. Probab.* **22** (1994) 469–493.

[151] P. Imkeller, M. Pontier, and F. Weisz: Free lunch and arbitrage possibilities in a financial market model with an insider. *Stochastic Process. Appl.* **92** (2001) 103–130.

[152] K. Itô: Stochastic integral. *Proc. Imp. Acad. Tokyo* **20** (1944) 519–524.

[153] K. Itô: Multiple Wiener integral. *J. Math. Soc. Japan* **3** (1951) 157–169.

[154] K. Itô: Malliavin calculus on a Segal space. In: *Stochastic Analysis, Proc. Paris 1987*, Lecture Notes in Math. **1322** (1988) 50–72.

[155] K. Itô and H. P. McKean, Jr. : *Diffusion Processes and their Sample Paths*, Springer-Verlag, 1974.

[156] J. Jacod: *Calcul stochastique et problèmes de martingales*, Lecture Notes in Math. **714**, Springer–Verlag, 1979.

[157] T. Jeulin: *Semimartingales et grossissement d'une filtration*, Lecture Notes in Math. **833**, Springer–Verlag, 1980.

[158] M. Jolis and M. Sanz: On generalized multiple stochastic integrals and multiparameter anticipative calculus. In: *Stochastic Analysis and Related Topics II*, eds.: H. Korezlioglu and A. S. Üstünel, Lecture Notes in Math. **1444** (1990) 141–182.

[159] Yu. M. Kabanov and A. V. Skorohod: Extended stochastic integrals (in Russian). Proc. of the School–seminar on the theory of random processes, Druskininkai 1974, pp. 123–167, Vilnius 1975.

[160] I. Karatzas: *Lectures on the mathematics of finance*, CRM Monograph Series, 8. American Mathematical Society, 1997.

[161] I. Karatzas and D. Ocone: A generalized Clark representation formula, with application to optimal portfolios. *Stochastics and Stochastics Reports* **34** (1991) 187–220.

[162] I. Karatzas, D. Ocone, and Jinju Li: An extension of Clark's formula. *Stochastics and Stochastics Reports* **37** (1991) 127–131.

[163] I. Karatzas and I. Pikovski: Anticipative portfolio optimization. *Adv. Appl. Prob.* **28** (1996) 1095–1122.

[164] I. Karatzas and S. E. Shreve: *Brownian Motion and Stochastic Calculus*, Springer-Verlag, 1988.

[165] I. Karatzas and S. E. Shreve: *Methods of mathematical finance*, Springer-Verlag, 1998.

[166] A. Kohatsu-Higa: Lower bounds for densities of uniformly elliptic random variables on Wiener space. *Probab. Theory Related Fields* **126** (2003) 421–457.

[167]  A. Kohatsu-Higa, D. Márquez-Carreras, and M. Sanz-Solé: Asymptotic behavior of the density in a parabolic SPDE. *J. Theoret. Probab.* **14** (2001) 427–462.

[168]  A. Kohatsu-Higa, D. Márquez-Carreras, and M. Sanz-Solé: Logarithmic estimates for the density of hypoelliptic two-parameter diffusions. *J. Funct. Anal.* **190** (2002) 481–506.

[169]  A. Kohatsu-Higa and M. Montero: Malliavin calculus in finance. In: *Handbook of computational and numerical methods in finance,* Birkhäuser Boston, Boston, 2004, 111–174.

[170]  J. J. Kohn: Pseudo-differential operators and hypoellipticity. *Proc. Symp. Pure Math.* **23**, A. M. S. (1973) 61–69.

[171]  A. N. Kolmogorov: Wienersche Spiralen und einige andere interessante Kurven im Hilbertschen Raum. *C. R. (Doklady) Acad. URSS (N.S.)* **26** (1940) 115–118.

[172]  M. Krée and P. Krée: Continuité de la divergence dans les espaces de Sobolev relatifs à l'espace de Wiener. *C. R. Acad. Sci. Paris* **296** (1983) 833–836.

[173]  H. Kunita: Stochastic differential equations and stochastic flow of diffeomorphisms. In: *Ecole d'Eté de Probabilités de Saint Flour XII,* 1982, Lecture Notes in Math. **1097** (1984) 144–305.

[174]  H. Kunita: *Stochastic Flows and Stochastic Differential Equations,* Cambridge Univ. Press, 1988.

[175]  H. H. Kuo: *Gaussian Measures in Banach Spaces,* Lecture Notes in Math. **463**, Springer-Verlag, 1975.

[176]  H. H. Kuo and A. Russek: White noise approach to stochastic integration. *Journal Multivariate Analysis* **24** (1988) 218–236.

[177]  H. Künsch: Gaussian Markov random fields. *Journal Fac. Sci. Univ. Tokyo,* I. A. Math. **7** (1982) 567–597.

[178]  S. Kusuoka: The non–linear transformation of Gaussian measure on Banach space and its absolute continuity (I). *J. Fac. Sci. Univ. Tokyo IA* **29** (1982) 567–597.

[179]  S. Kusuoka: On the absolute continuity of the law of a system of multiple Wiener integrals. *J. Fac. Sci. Univ. Tokyo Sec. IA Math.* **30** (1983) 191–197.

[180]  S. Kusuoka: The generalized Malliavin calculus based on Brownian sheet and Bismut's expansion for large deviations. In: *Stochastic Processes–Mathematics and Physics, Proc. Bielefeld, 1984,* Lecture Notes in Math. **1158** (1984) 141–157.

[181]  S. Kusuoka: On the fundations of Wiener Riemannian manifolds. In: *Stochastic Analysis,* Pitman, **200**, 1989, 130–164.

[182]  S. Kusuoka: Analysis on Wiener spaces. I. Nonlinear maps. *J. Functional Anal.* **98** (1991) 122–168.

[183]  S. Kusuoka: Analysis on Wiener spaces. II. Differential forms. *J. Functional Anal.* **103** (1992) 229–274.

[184]  S. Kusuoka and D. W. Stroock: Application of the Malliavin calculus I. In: *Stochastic Analysis, Proc. Taniguchi Inter. Symp. on Stochastic Analysis,*

*Katata and Kyoto* 1982, ed.: K. Itô, Kinokuniya/North-Holland, Tokyo, 1984, 271–306.

[185]  S. Kusuoka and D. W. Stroock: The partial Malliavin calculus and its application to nonlinear filtering. *Stochastics* **12** (1984) 83–142.

[186]  S. Kusuoka and D. W. Stroock: Application of the Malliavin calculus II. *J. Fac. Sci. Univ. Tokyo Sect IA Math.* **32** (1985) 1–76.

[187]  S. Kusuoka and D. W. Stroock: Application of the Malliavin calculus III. *J. Fac. Sci. Univ. Tokyo Sect IA Math.* **34** (1987) 391–442.

[188]  S. Kusuoka and D. W. Stroock: Precise asymptotic of certain Wiener functionals. *J. Functional Anal.* **99** (1991) 1–74.

[189]  D. Lamberton and B. Lapeyre: *Introduction to stochastic calculus applied to finance*, Chapman & Hall, 1996.

[190]  N. Lanjri Zadi and D. Nualart: Smoothness of the law of the supremum of the fractional Brownian motion. *Electron. Comm. Probab.* **8** (2003) 102–111.

[191]  B. Lascar: Propriétés locales d'espaces du type Sobolev en dimension infinie. *Comm. Partial Differential Equations* **1** (1976) 561–584.

[192]  R. Léandre: Régularité des processus de sauts dégénérés. *Ann. Institut Henri Poincaré* **21** (1985) 125–146.

[193]  R. Léandre: Majoration en temps petit de la densité d'une diffusion dégénérée. *Probab. Theory Rel. Fields* **74** (1987) 289–294.

[194]  R. Léandre and F. Russo: Estimation de Varadhan pour les diffusions à deux paramètres. *Probab. Theory Rel. Fields* **84** (1990) 429–451.

[195]  M. Ledoux and M. Talagrand: *Probability on Banach Spaces*, Springer-Verlag, 1991.

[196]  J. A. León, R. Navarro, and D. Nualart: An anticipating calculus approach to the utility maximization of an insider. *Math. Finance* **13** (2003) 171–185.

[197]  P. Lescot: Un théorème de désintégration en analyse quasi–sžre. In: *Séminaire de Probabilités XXVII*, Lecture Notes in Math. **1557** (1993) 256–275.

[198]  S. J. Lin: Stochastic analysis of fractional Brownian motions. *Stochastics Stochastics Reports* **55** (1995) 121–140.

[199]  J. L. Lions: *Quelques méthodes de résolution des problèmes aux limites non–linéaires*, Dunod, 1969.

[200]  R. S. Liptser and A. N. Shiryaev: *Statistics of Random Processes I. General Theory*. Springer–Verlag, 1977.

[201]  R. S. Lipster and A. N. Shiryaev: *Theory of Martingales*, Kluwer Acad. Publ., Dordrecht, 1989.

[202]  T. Lyons: Differential equations driven by rough signals (I): An extension of an inequality of L. C. Young. *Mathematical Research Letters* **1** (1994) 451–464.

[203]  T. Lyons: Differential equations driven by rough signals. *Rev. Mat. Iberoamericana* **14** (1998) 215–310.

[204]  T. Lyons and Z. Qian: *System control and rough paths*, Oxford University Press, 2002.

[205]  Z. M. Ma and M. Röckner: *Introduction to the Theory of (Non–Symmetric) Dirichlet Forms.* Springer–Verlag, 1991.

[206]  M. P. Malliavin and P. Malliavin: Integration on loop groups I, quasi–invariant measures. *J. Functional Anal.* **93** (1990) 207–237.

[207]  P. Malliavin: Stochastic calculus of variations and hypoelliptic operators. In: *Proc. Inter. Symp. on Stoch. Diff. Equations, Kyoto* 1976, Wiley 1978, 195–263.

[208]  P. Malliavin: Implicit functions of finite corank on the Wiener space. *Stochastic Analysis, Proc. Taniguchi Inter. Symp. on Stoch. Analysis*, Katata and Kyoto 1982, ed.: K. Itô, Kinokuniya/North-Holland, Tokyo, 1984, 369–386.

[209]  P. Malliavin: Sur certaines intégrales stochastiques oscillantes. *C. R. Acad. Sci. Paris* **295** (1982) 295–300.

[210]  P. Malliavin: Estimation du rayon d'analyticité transverse et calcul des variations. *C. R. Acad. Sci. Paris* **302** (1986) 359–362.

[211]  P. Malliavin: Minoration de l'état fondamental de l'équation de Schrödinger du magnetisme et calcul des variations stochastiques. *C. R. Acad. Sci. Paris* **302** (1986) 481–486.

[212]  P. Malliavin: *Stochastic Analysis.* Grundlehren der Mathematischen Wissenschaften, 313. Springer-Verlag, Berlin, 1997.

[213]  P. Malliavin and D. Nualart: Quasi–sure analysis and Stratonovich anticipative stochastic differential equations. *Probab. Theory Rel. Fields* **96** (1993) 45–55.

[214]  P. Malliavin and D. Nualart: Quasi–sure analysis of stochastic flows and Banach space valued smooth functionals on the Wiener space. *J. Functional Anal.* **112** (1993) 287–317.

[215]  P. Malliavin and D. W. Stroock: Short time behavior of the heat kernel and its logarithmic derivatives. *J. Differential Geom.* **44** (1996) 550–570.

[216]  P. Malliavin and A. Thalmaier: *Stochastic calculus of variations in mathematical finance*, Springer, 2004.

[217]  B. B. Mandelbrot and J. W. Van Ness: Fractional Brownian motions, fractional noises and applications. *SIAM Review* **10** (1968) 422–437.

[218]  T. Masuda: Absolute continuity of distributions of solutions of anticipating stochastic differential equations. *J. Functional Anal.* **95** (1991) 414–432.

[219]  J. C. Mattingly and E. Pardoux: Malliavin calculus for the stochastic 2D Navier-Stokes equation. To appear in *Comm. in Pure and Appl. Math.*

[220]  G. Mazziotto and A. Millet: Absolute continuity of the law of an infinite dimensional Wiener functional with respect to the Wiener probability. *Probab. Theory Rel. Fields* **85** (1990) 403–411.

[221]  H. P. McKean, Jr: *Stochastic Integrals*, Academic Press, 1969.

[222]  J. Memin, Y. Mishura, and E. Valkeila: Inequalities for the moments of Wiener integrals with respecto to fractional Brownian motions. *Statist. Prob. Letters* **55** (2001) 421–430.

[223]  R. C. Merton: Theory of rational option pricing. *Bell Journal of Economics and Management Science* **4** (1973) 141–183.

[224]  P. A. Meyer: *Probabilités et Potentiel*, Hermann, Paris, 1966.

[225] P. A. Meyer: Transformations de Riesz pour les lois gaussiennes. In: *Seminaire de Probabilités XVIII*, Lecture Notes in Math. **1059** (1984) 179–193.

[226] D. Michel: Régularité des lois conditionnelles en théorie du filtrage non–linéaire et calcul des variations stochastiques. *J. Functional Anal.* **41** (1981) 1–36.

[227] D. Michel: Conditional laws and Hörmander's condition. *Stochastic Analysis, Proc. Taniguchi Inter. Symp. on Stoch. Analysis*, Katata and Kyoto 1982, ed.: K. Itô, Kinokuniya/North-Holland, Tokyo, 1984, 387–408.

[228] A. Millet and D. Nualart: Support theorems for a class of anticipating stochastic differential equations. *Stochastics and Stochastics Reports* **39** (1992) 1–24.

[229] A. Millet, D. Nualart, and M. Sanz: Integration by parts and time reversal for diffusion processes. *Ann. Probab.* **17** (1989) 208–238.

[230] A. Millet, D. Nualart, and M. Sanz: Time reversal for infinite dimensional diffusions. *Probab. Theory Rel. Fields* **82** (1989) 315–347.

[231] A. Millet and M. Sanz-Solé: A stochastic wave equation in two space dimension: smoothness of the law. *Ann. Probab.* **27** (1999) 803–844.

[232] J. M. Moulinier: Absolue continuité de probabilités de transtition par rapport à une mesure gaussienne dans un espace de Hilbert. *J. Functional Anal.* **64** (1985) 275–295.

[233] J. M. Moulinier: Fonctionnelles oscillantes stochastiques et hypoellipticité. *Bull. Sciences Math.* **109** (1985) 37–60.

[234] C. Mueller: On the support of solutions to the heat equation with noise. *Stochastics and Stochastics Reports* **37** (1991) 225–245.

[235] E. Nelson: The free Markov field. *J. Functional Anal.* **12** (1973) 217–227.

[236] J. Neveu: Sur l'espérance conditionnelle par rapport à un mouvement Brownien. *Ann. Inst. Henri Poincaré* **12** (1976) 105–109.

[237] Nguyen Minh Duc and D. Nualart: Stochastic processes possessing a Skorohod integral representation. *Stochastics* **30** (1990) 47–60.

[238] Nguyen Minh Duc, D. Nualart, and M. Sanz: Application of the Malliavin calculus to a class of stochastic differential equations. *Probab. Theory Rel. Fields* **84** (1990) 549–571.

[239] J. Norris: Simplified Malliavin calculus. In: *Seminaire de Probabilités XX*, Lecture Notes in Math. **1204** (1986) 101–130.

[240] I. Norros, E. Valkeila, and J. Virtamo: An elementary approach to a Girsanov formula and other analytical resuls on fractional Brownian motion. *Bernoulli* **5** (1999) 571–587.

[241] D. Nualart: Application du calcul de Malliavin aux équations différentielles stochastiques sur le plan. In: *Seminaire de Probabilités XX*, Lecture Notes in Math. **1204** (1986) 379–395.

[242] D. Nualart: Malliavin calculus and stochastic integrals. *Lecture Notes in Math.* **1221** (1986) 182–192.

[243] D. Nualart: Some remarks on a linear stochastic differential equation. *Statistics and Probability Letters* **5** (1987) 231–234.

[244] D. Nualart: Noncausal stochastic integrals and calculus. In: *Stochastic Analysis and Related Topics*, eds.: H. Korezlioglu and A. S. Üstünel, Lecture Notes in Math. **1316** (1988) 80–129.

[245] D. Nualart: *Cálculo de Malliavin*. Cuadernos de Probabilidad y Estadística Matemática. Fondo Editorial Acta Científica Venezolana y Sociedad Bernouilli, Venezuela, 1990.

[246] D. Nualart: Nonlinear transformations of the Wiener measure and applications. In: *Stochastic Analysis, Liber Amicorum for Moshe Zakai*, Academic Press, 1991, 397–432.

[247] D. Nualart: Markov fields and transformations of the Wiener measure. In: *Stochastic Analysis and Related Topics*, eds.: T. Lindstrøm et al., Gordon and Breach, 1993, 45–88.

[248] D. Nualart: Analysis on Wiener space and anticipating stochastic calculus. In: *Lectures on probability theory and statistics* (Saint-Flour, 1995), Lecture Notes in Math. **1690** (1998) 123–227.

[249] D. Nualart and E. Pardoux: Stochastic calculus with anticipating integrands. *Probab. Theory Rel. Fields* **78** (1988) 535–581.

[250] D. Nualart and E. Pardoux: Boundary value problems for stochastic differential equations. *Ann. Probab.* **19** (1991) 1118–1144 .

[251] D. Nualart and E. Pardoux: Second order stochastic differential equations with Dirichlet boundary conditions. *Stochastic Processes and Their Applications* **39** (1991) 1–24 .

[252] D. Nualart and E. Pardoux: Stochastic differential equations with bondary conditions. In: *Stochastic Analysis and Applications*, ed.: A. B. Cruzeiro and J. C. Zambrini, Birkhäuser, Progress in Prob. Series, 26, 1991, 155–175.

[253] D. Nualart and E. Pardoux: Markov field properties of solutions of white noise driven quasi-linear parabolic PDEs. *Stochastics and Stochastics Reports* **48** (1994) 17–44.

[254] D. Nualart and A. Rascanu: Differential equations driven by fractional Brownian motion. *Collectanea Mathematica* **53** (2002) 55–81.

[255] D. Nualart, C. Rovira, and S. Tindel: Probabilistic models for vortex filaments based on fractional Brownian motion. *Ann. Probab.* **31** (2003) 1862–1899.

[256] D. Nualart and M. Sanz: Malliavin calculus for two-parameter Wiener functionals. *Z. für Wahrscheinlichkeitstheorie verw. Gebiete* **70** (1985) 573–590.

[257] D. Nualart and M. Sanz: Stochastic differential equations on the plane: Smoothness of the solution. *Journal Multivariate Analysis* **31** (1989) 1–29.

[258] D. Nualart and A. S. Üstünel: Geometric analysis of conditional independence on Wiener space. *Probab. Theory Rel. Fields* **89** (1991) 407–422.

[259] D. Nualart, A. S. Üstünel, and M. Zakai: Some relations among classes of $\sigma$–fields on Wiener space. *Probab. Theory Rel. Fields* **85** (1990) 119–129.

[260] D. Nualart, A. S. Üstünel, and M. Zakai: Some remarks on independence and conditioning on Wiener space. In: *Stochastic Analysis and Related Topics II*, eds.: H. Korezlioglu and A. S. Üstünel, Lecture Notes in Math. **1444** (1990) 122–127.

[261] D. Nualart and J. Vives: Continuité absolue de la loi du maximum d'un processus continu. *C. R. Acad. Sci. Paris* **307** (1988) 349–354.

[262] D. Nualart and M. Wschebor: Intégration par parties dans l'espace de Wiener et approximation du temps local. *Probab. Theory Rel. Fields* **90** (1991) 83–109.

[263] D. Nualart and M. Zakai: Generalized stochastic integrals and the Malliavin calculus. *Probab. Theory Rel. Fields* **73** (1986) 255–280.

[264] D. Nualart and M. Zakai: Generalized multiple stochastic integrals and the representation of Wiener functionals. *Stochastics* **23** (1988) 311–330.

[265] D. Nualart and M. Zakai: A summary of some identities of the Malliavin calculus. Lecture Notes in Math. **1390** (1989) 192–196.

[266] D. Nualart and M. Zakai: The partial Malliavin calculus. In: *Seminaire de Probabilités XXIII*, Lecture Notes in Math. **1372** (1989) 362–381.

[267] D. Nualart and M. Zakai: On the relation between the Stratonovich and Ogawa integrals. *Ann. Probab.* **17** (1989) 1536–1540.

[268] D. Nualart and M. Zakai: Multiple Wiener-Itô integrals possessing a continuous extension. *Probab Theory Rel. Fields* **18** (1990) 131–145.

[269] D. Ocone: Malliavin calculus and stochastic integral representation of diffusion processes. *Stochastics* **12** (1984) 161–185.

[270] D. Ocone: A guide to the stochastic calculus of variations. In: *Stochastic Analysis and Related Topics*, eds.: H. Korezlioglu and A. S. Üstünel, Lecture Notes in Math. **1316** (1987) 1–79.

[271] D. Ocone: Stochastic calculus of variations for stochastic partial differential equations. *J. Functional Anal.* **79** (1988) 231–288.

[272] D. Ocone and E. Pardoux: A generalized Itô–Ventzell formula. Application to a class of anticipating stochastic differential equations. *Ann. Inst. Henri Poincaré* **25** (1989) 39–71.

[273] D. Ocone and E. Pardoux: Linear stochastic differential equations with boundary conditions. *Probab. Theory Rel. Fields* **82** (1989) 489–526.

[274] S. Ogawa: Quelques propriétés de l'intégrale stochastique de type noncausal. *Japan J. Appl. Math.* **1** (1984) 405–416.

[275] S. Ogawa: The stochastic integral of noncausal type as an extension of the symmetric integrals. *Japan J. Appl. Math.* **2** (1984) 229–240.

[276] S. Ogawa: Une remarque sur l'approximation de l'intégrale stochastique du type noncausal par une suite des intégrales de Stieltjes. *Tohoku Math. J.* **36** (1984) 41–48.

[277] O. A. Oleĭnik and E. V. Radkevič: *Second order equations with nonnegative characteristic form*, A. M. S. and Plenum Press, 1973.

[278] E. Pardoux: Applications of anticipating stochastic calculus to stochastic differential equations. In: *Stochastic Analysis and Related Topics II*, eds.: H. Korezlioglu and A. S. Üstünel, Lecture Notes in Math. **1444** (1990) 63–105.

[279] E. Pardoux and S. Peng: Adapted solution of a backward stochastic differential equation. *Systems and Control Letters* **14** (1990) 55–61.

[280] E. Pardoux and S. Peng: Backward doubly stochastic differential equations and systems of quasilinear SPDEs. *Probab. Theory Rel. Fields* **98** (1994) 209–228.

[281] E. Pardoux and P. Protter: Two-sided stochastic integrals and calculus. *Probab. Theory Rel. Fields* **76** (1987) 15–50.

[282] E. Pardoux and T. S. Zhang: Absolute continuity of the law of the solution of a parabolic SPDE. *J. Functional Anal.* **112** (1993) 447–458.

[283] V. Pipiras and M. S. Taqqu: Integration questions related to fractional Brownian motion. *Probab. Theory Rel. Fields* **118** (2000) 121–291.

[284] V. Pipiras and M. S. Taqqu: Are classes of deterministic integrands for fractional Brownian motion on a interval complete? *Bernoulli* **7** (2001) 873–897

[285] G. Pisier: Riesz transforms. A simple analytic proof of P. A. Meyer's inequality. In: *Seminaire de Probabilités XXIII*, Lecture Notes in Math. **1321** (1988) 485–501.

[286] L. D. Pitt: A Markov property for Gaussian processes with a multidimensional parameter. *Arch. Rational. Mech. Anal.* **43** (1971) 367–391.

[287] J. Potthoff: White noise approach to Malliavin's calculus. *J. Functional Anal.* **71** (1987) 207–217.

[288] Ll. Quer-Sardanyons and M. Sanz-Solé: Absolute continuity of the law of the solution to the 3-dimensional stochastic wave equation. *J. Funct. Anal.* **206** (2004) 1–32.

[289] Ll. Quer-Sardanyons and M. Sanz-Solé: A stochastic wave equation in dimension 3: smoothness of the law. *Bernoulli* **10** (2004) 165–186.

[290] R. Ramer: On nonlinear transformations of Gaussian measures. *J. Functional Anal.* **15** (1974) 166–187.

[291] J. Ren: Analyse quasi–sûre des équations différentielles stochastiques. *Bull. Sciences Math.* **114** (1990) 187–214.

[292] D. Revuz and M. Yor: *Continuous Martingales and Brownian Motion*, Springer–Verlag, 1991.

[293] J. Rosinski: On stochastic integration by series of Wiener integrals. *Appl. Math. Optimization* **19** (1989) 137–155.

[294] L. C. G. Rogers: Arbitrage with fractional Brownian motion. *Math. Finance* **7** (1997) 95–105.

[295] C. Rovira and M. Sanz-Solé: Stochastic Volterra equations in the plane: smoothness of the law. *Stochastic Anal. Appl.* **19** (2001) 983–1004.

[296] Yu. Rozanov: *Markov Random Fields*, Springer–Verlag, 1982.

[297] A. Russek: On series expansions of stochastic integrals. In: *Probability Theory on Vector Spaces IV*, Lecture Notes in Math. **1391** (1989) 337–352.

[298] F. Russo and P. Vallois: Forward, backward and symmetric stochastic integration. *Probab. Theory Rel. Fields* **97** (1993) 403–421.

[299] A. A. Ruzmaikina: Stieltjes integrals of Hölder continuous functions with applications to fractional Brownian motion. *J. Statist. Phys.* **100** (2000) 1049–1069.

[300] S. G. Samko, A. A. Kilbas O. L. and Marichev: *Fractional Integrals and Derivatives. Theory and Applications*, Gordon and Breach, 1993.

[301] G. Samorodnitsky and M. S. Taqqu: *Stable non-Gaussian random processes*, Chapman and Hall, 1994.

[302] M. Sanz-Solé: Applications of Malliavin calculus to SPDE's. In: *Stochastic partial differential equations and applications* (Trento, 2002), Lecture Notes in Pure and Appl. Math. **227** (2002) 429–442.

[303] I. E. Segal: Abstract probability spaces and a theorem of Kolmogorov. *Amer. J. Math.* **76** (1954) 721–732.

[304] I. E. Segal: Tensor algebras on Hilbert spaces I. *Trans. Amer. Soc.* **81** (1956) 106–134.

[305] T. Sekiguchi and Y. Shiota: $L^2$-theory of noncausal stochastic integrals. *Math. Rep. Toyama Univ.* **8** (1985) 119–195.

[306] A. Ju. Sevljakov: The Itô formula for the extended stochastic integral. *Theory Prob. Math. Statist.* **22** (1981) 163–174.

[307] I. Shigekawa: Derivatives of Wiener functionals and absolute continuity of induced measures. *J. Math. Kyoto Univ.* **20–2** (1980) 263–289.

[308] I. Shigekawa: The de Rahm–Hodge Kodaira decomposition on an abstract Wiener space. *J. Math. Kyoto Univ.* **26** (1986) 191–202.

[309] I. Shigekawa: Existence of invariant measures of diffusions on an abstract Wiener space. *Osaka J. Math.* **24** (1987) 37–59.

[310] I. Shigekawa: *Stochastic analysis*, American Mathematical Society, 2004.

[311] Y. Shiota: A linear stochastic integral equation containing the extended Itô integral. *Math. Rep. Toyama Univ.* **9** (1986) 43–65.

[312] S. E. Shreve: *Stochastic calculus for finance. II. Continuous-time models*, Springer-Verlag, 2004.

[313] B. Simon: *The P[φ]-2 Euclidean Quantum Field Theory*, Princeton Univ. Press, 1974.

[314] B. Simon: *Trace Ideals and Their Applications*, Cambridge Univ. Press, 1979.

[315] A. V. Skorohod: On a generalization of a stochastic integral. *Theory Probab. Appl.* **20** (1975) 219–233.

[316] F. Smithies: *Integral Equations*, Cambridge Univ. Press, 1958.

[317] E. M. Stein: *Singular Integrals and Differentiability of Functions*, Princeton Univ. Press, 1970.

[318] D. W. Stroock: The Malliavin calculus, a functional analytic approach. *J. Functional Anal.* **44** (1981) 212–257.

[319] D. W. Stroock: The Malliavin calculus and its applications to second order parabolic differential equations. *Math. Systems Theory*, Part I, **14** (1981) 25–65; Part II, **14** (1981) 141–171.

[320] D. W. Stroock: Some applications of stochastic calculus to partial differential equations. In: *Ecole d'Eté de Probabilités de Saint Flour*, Lecture Notes in Math. **976** (1983) 267–382.

[321] D. W. Stroock: Homogeneous chaos revisited. In: *Seminaire de Probabilités XXI*, Lecture Notes in Math. **1247** (1987) 1–8.

[322] D. W. Stroock and S. R. S. Varadhan: *Multidimensional diffusion processes*, Springer–Verlag, 1979.

[323] H. Sugita: Sobolev spaces of Wiener functionals and Malliavin's calculus. *J. Math. Kyoto Univ.* **25–1** (1985) 31–48.

[324] H. Sugita: On a characterization of the Sobolev spaces over an abstract Wiener space. *J. Math. Kyoto Univ.* **25–4** (1985) 717–725.

[325] H. Sugita: Positive generalized Wiener functionals and potential theory over abstract Wiener spaces. *Osaka J. Math.* **25** (1988) 665–696.

[326] S. Taniguchi: Malliavin's stochastic calculus of variations for manifold valued Wiener functionals. *Z. für Wahrscheinlichkeitstheorie verw. Gebiete* **65** (1983) 269–290.

[327] S. Taniguchi: Applications of Malliavin's calculus to time–dependent systems of heat equations. *Osaka J. Math.* **22** (1985) 307320.

[328] M. Thieullen: Calcul stochastique nonadapté pour des processus à deux paramètres. *Probab. Theory Rel. Fields* **89** (1991) 457–485.

[329] A. S. Üstünel: Representation of the distributions on Wiener space and stochastic calculus of variations. *J. Functional Anal.* **70** (1987) 126–139.

[330] A. S. Üstünel: The Itô formula for anticipative processes with nonmonotonous time via the Malliavin calculus. *Probab. Theory Rel. Fields* **79** (1988) 249–269.

[331] A. S. Üstünel: Some comments on the filtering of diffusions and the Malliavin calculus. In: *Stochastic Analysis and Related Topics*, eds.: H. Korezlioglu and A. S. Üstünel, Lecture Notes in Math **1316** (1988) 247–266.

[332] A. S. Üstünel: Intégrabilité exponentielle de fonctionnelles de Wiener. *C. R. Acad. Sci. Paris* **315** (1992) 997–1000.

[333] A. S. Üstünel and M. Zakai: On independence and conditioning on the Wiener space. *Ann. Probab.* **17** (1989) 1441–1453.

[334] A. S. Üstünel and M. Zakai: On the structure of independence on the Wiener space. *J. Functional Anal.* **90** (1990) 113–137.

[335] A. S. Üstünel and M. Zakai: Transformation of Wiener measure under anticipative flows. *Probab. Theory Rel. Fields* **93** (1992) 91–136.

[336] A. S. Üstünel and M. Zakai: Applications of the degree theorem to absolute continuity on Wiener space. *Probab. Theory Rel. Fields* **95** (1993) 509–520.

[337] A. S. Üstünel and M. Zakai: The composition of Wiener functionals with non absolutely continuous shifts. *Probab. Theory Rel. Fields* **98** (1994) 163–184.

[338] A. S. Üstünel and M. Zakai: Transformations of the Wiener measure under non–invertible shifts. *Probab. Theory Rel. Fields* **99** (1994) 485–500.

[339] A. S. Üstünel and M. Zakai: *Transformation of measure on Wiener space*, Springer-Verlag, 2000.

[340] J. Van Biesen: The divergence on submanifolds of the Wiener space. *J. Functional Anal.* **113** (1993) 426–461.

[341] A. Yu. Veretennikov: A probabilistic approach to hypoellipticity. *Russian Math. Surveys*, **38**:3 (1983) 127–140.

[342] J. B. Walsh: An introduction to stochastic partial differential equations. In: *Ecole d'Eté de Probabilités de Saint Flour XIV*, Lecture Notes in Math. **1180** (1986) 265–438.

[343] S. Watanabe: *Lectures on Stochastic Differential Equations and Malliavin Calculus*, Tata Institute of Fundamental Research, Springer-Verlag, 1984.

[344] S. Watanabe: Analysis of Wiener functionals (Malliavin Calculus) and its applications to heat kernels. *Ann. Probab.* **15** (1987) 1–39.

[345] S. Watanabe: Short time asymptotic problems in Wiener functional integration. Application to heat kernel and index theorem. In: *Stochastic Analysis*

*and Related Topics II*, eds.: H. Korezlioglu and A. S. Üstünel, Lecture Notes in Math. **1444** (1990) 1–62.

[346] S. Watanabe: Donsker's δ-functions in the Malliavin calculus. In: *Stochastic Analysis, Liber Amicorum for Moshe Zakai*, eds.: E. Mayer-Wolf, E. Merzbach, and A. Shwartz, Academic Press, 1991, 495–502.

[347] S. Watanabe: Fractional order Sobolev spaces on Wiener space. *Probab. Theory Rel. Fields* **95** (1993) 175–198.

[348] N. Wiener: Differential space. *J. Math. Phys.* **2** (1923) 131–174.

[349] N. Wiener: The homogeneous chaos. *Amer. J. Math.* **60** (1938) 879–936.

[350] D. Williams: To begin at the beginning. In *Stochastic Integrals*, Lecture Notes in Math. **851** (1981) 1–55.

[351] M. Wschebor: *Surfaces Aléatoires*, Lecture Notes in Math. **1147**. Springer–Verlag, 1985.

[352] D. Ylvisaker: The expected number of zeros of a stationary Gaussian process. *Annals Math. Stat.* **36** (1965) 1043–1046.

[353] D. Ylvisaker: A note on the absence of tangencies in Gaussian sample paths. *Annals Math. Stat.* **39** (1968) 261–262.

[354] L. C. Young: An inequality of the Hölder type connected with Stieltjes integration. *Acta Math.* **67** (1936), 251–282.

[355] M. Zähle: Integration with respect to fractal functions and stochastic calculus. I, *Prob. Theory Relat. Fields* **111** (1998) 333-374.

[356] M. Zähle: On the link between fractional and stochastic calculus. In: *Stochastic Dynamics*, Bremen 1997, ed.: H. Crauel and M. Gundlach, pp. 305-325, Springer 1999.

[357] M. Zakai: The Malliavin calculus. *Acta Appl. Math.* **3** (1985) 175–207.

[358] M. Zakai: Malliavin derivatives and derivatives of functionals of the Wiener process with respect to a scale parameter. *Ann. Probab.* **13** (1985) 609–615.

# Index